Recent Advances in Electromagnetic Theory

Contents

Charles Herach Papas

Introduction

Charles Herach Papas was born on March 29, 1918, in Troy, New York. Following his family, which was involved in the import/export trade, he had the unusual opportunity to spend his early childhood in Tien-Ching (Tientsin), mainland China. After returning to the United States and completing high school, he matriculated at the Massachusetts Institute of Technology where he received a B.S. in electrical engineering in 1941. Harvard University awarded him his M.S. in communications engineering in 1946, and a Ph.D. in electrodynamics in 1948.

During World War II, he served in the U.S. Navy. His tour of duty took him to the Naval Ordnance Laboratory, where he investigated the degaussing problem, and to the Bureau of Ships, where he designed radio and radar antennas.

In 1948 Papas returned to civilian life at Harvard as a Research Fellow. He collaborated with such renowned scientists and scholars as R.W.P. King and Professor L. Brillouin on projects involving electromagnetic boundary-value problems and antenna theory. In the early 1950s Professor Papas moved westward, seeking intellectual challenge and stimulation at the University of California. He served jointly as an assistant professor at Berkeley and as a staff member at Los Alamos. The dual responsibilities introduced him to Professors Louis Alvarez and Enrico Fermi.

In 1952 Papas accepted a faculty position at the California Institute of Technology where he continued his teaching and research and collaborated with Professor W.R. Smythe. After remaining there for 36 years, he retired in 1989; he is currently an Emeritus Professor of Electrical Engineering.

Throughout a long and productive career, Papas has devoted his research activities to the study of electromagnetic theory, extending its application to include analysis of antennas, propagation, gravitational electromagnetics, astrophysics, guided waves, and remote sensing. As a result of his pioneering work in radiophysics and the electrodynamics of flare stars, Papas was honored with election to the Academy of Sciences in Yerevan, Armenia, and in Bologna, Italy.

Professor Papas has published numerous technical papers on electromagnetics and optics. He coauthored, with Fritz Borgnis, a book *Randwertprobleme der Mikrowellenphysik* (Springer-Verlag), and a book-length article, *Waveguides and Cavity Resonators*, in *Handbuch der Physik* (Springer-Verlag). His *Theory of Electromagnetic Wave Propagation* (originally McGraw-Hill, 1965) is a recognized classic, which has been translated into Russian and reprinted by Dover.

Professor Papas's most important and far-reaching contribution is that he has been an inspiration to pupils and peers for over three decades. The contributors to this volume are especially appreciative of their association with him. His outstanding teaching style has encouraged beginners to delve into the fundamentals of physics and mathematics; his unique style of seeking simplicity and elegance in the physical world, and his deep sense of human values in dealing with his students and colleagues has been a lasting source of inspiration for all of those who have had the unique privilege to be associated with him.

Philadelphia, Pennsylvania

HARALAMBOS N. KRITIKOS
DWIGHT L. JAGGARD

1
Electromagnetic Waves in Chiral Media

Sassan Bassiri*

Any man who,
upon looking at his bare feet,
doesn't laugh,
has either no sense of symmetry
or no sense of humor.

Descartes

1.1. Introduction

The phenomenon of natural optical activity has played a major role in the study and development of such diverse areas of science as the physics of light, the structure of molecules, and the nature of life itself. Organic optical activity has been looked upon as a unique characteristic of life. The present living matter on Earth is asymmetric, as it contains mainly L-amino acids and D-sugars. An optically active molecule exists in two distinct mirror image forms, it either has the left-handed L- or the right-handed D-type structure. On a macroscopic scale, chirality occurs in nature and in man-made articles. Screws, gloves, golf clubs, and springs are some examples of manufactured chiral objects; whereas flowers, winding vegetations, and snails are a few examples of natural chiral objects.

Optical activity is a manifestation of spatial dispersion; that is, it occurs because the polarization of a medium at a given point depends on the field, not only at that point but also in its vicinity. Such an effect occurs in chiral media. An isotropic chiral medium is a macroscopically continuous medium composed of equivalent chiral objects that are uniformly distributed and randomly oriented. A chiral object is a three-dimensional body that cannot be brought into congruence with its mirror image by translation and rotation. An object of this sort has the property of handedness and must be either left-handed or right-handed. An object that is not chiral is said to be achiral, and thus all objects are either chiral or achiral. Some chiral objects occur naturally in two versions related to each other as a chiral object and its mirror image. Objects so related are said to be enantiomorphs of each other. If a chiral object is left-handed, its enantiomorph is right-handed, and vice versa.

* Jet Propulsion Laboratory, California Institute of Technology, 4800 Oak Grove Drive, Pasadena, CA 91109, USA.

Simple examples of chiral objects include the helix, the Möbius strip, and the irregular tetrahedron.

When a linearly polarized wave is incident normally upon a slab of chiral medium, two waves are generated in the medium. One is a left-circularly polarized wave and the other is a right-circularly polarized wave of a different phase velocity. Beyond the slab, the two waves combine to yield a linearly polarized wave whose plane of polarization is rotated with respect to the plane of polarization of the incident wave. The amount of rotation depends on the distance traveled in the medium, implying that the optical activity occurs not only at the surfaces of the slab but throughout the medium. Optical activity in a chiral medium differs from the phenomenon of Faraday rotation in, say, a magnetically biased plasma, in that the former is independent of the direction of propagation whereas the latter is not. The optical activity is invariant under time reversal and the Faraday rotation is invariant under spatial inversion.

A new era in physics began when Arago [1] discovered the phenomenon of optical activity. He observed that a crystal of quartz rotates the plane of polarization of linearly polarized light which has passed along the crystal's optic axis. Arago did not appreciate the nature of the phenomenon and it was the subsequent experiments by Biot [2–4] on plates of quartz which established:

(i) the dependence of optical activity on the thickness of the plate;
(ii) the unequal rotation of the plane of polarization of light of different wavelengths; and
(iii) the absence of any optical activity when two plates of quartz of the same thickness but opposite handedness are used.

Biot [5] also discovered that optical activity is not restricted to crystalline solids but appears as well in organic liquids such as oils of turpentine and laurel, alcoholic solutions of camphor, and aqueous solutions of sugar and tartaric acid.

It was Fresnel [6] who showed that a ray of light traveling along the axis of a quartz crystal is resolved into two circularly polarized rays of opposite handednesses that travel with unequal phase velocities. He argued that the difference in the two wave velocities is the cause of optical activity. Fresnel [7] also offered an explanation of why the phase velocities might be different for the two circularly polarized rays:

> This may result from a particular constitution of the refracting medium or of its integral molecules which establishes a difference between the sense of right to left and that of left to right; such would be, for example a helical arrangement of molecules of the medium which would present opposite properties according as these helices are right-handed or left-handed.

Pasteur [8] postulated that molecules are three-dimensional objects and that the optical activity of a medium is caused by the chirality of its molecules. Through his extensive experiments, he established the following fundamental connection between molecular chirality and life, which can be stated as: *living organisms discriminate between enantiomorphs.* Based on his researches, Pasteur [9] drew the conclusion:

> Life as manifested to us is a function of the asymmetry of the universe and of the consequences of this fact. The universe is asymmetrical. Life is dominated by asymmetrical actions. I can even imagine that all living species are primordially in their structure, in their external forms a function of cosmic asymmetry.

In time Pasteur's conjectures were proven, as it was shown a century later that matter is really asymmetric. Lindman [10], [11] introduced a new approach to the study of chirality when he devised macroscopic models of chiral media by using wire spirals instead of chiral molecules, and demonstrated the phenomenon of optical activity using microwaves instead of light. Many other experiments were performed, and a very thorough account of them is contained in the book by Lowry [12].

By the end of the nineteenth century, experimental evidence and empirical facts on optical activity were well established, and physicists had started to develop theories in order to explain the interaction of electromagnetic waves with chiral media. Born [13], Oseen [14], and Gray [15] put forward, independently and almost simultaneously, explanations of optical activity. In their work, an optically active molecule is modeled as a spatial distribution of coupled oscillators. Kuhn [16] also contributed greatly to the problem by considering the most simple case of the coupled oscillator model to demonstrate optical activity. Condon et al. [17] showed that it is possible to explain optical activity by considering a single oscillator moving in a dissymmetric field. A detailed account of these microscopic theories is contained in Condon's paper [18].

More recently, the paper by Bokut and Federov [19] studied light reflection from chiral surfaces, two papers by Bohren [20], [21] examined the reflection of electromagnetic waves from chiral spheres and cylinders, and the book by Kong [22], and numerous references therein, discussed general bianisotropic media. Shortly thereafter, a macroscopic treatment of the interaction of electromagnetic waves with chiral structures, which is the theoretical counterpart of Lindman's experiments, was given by Jaggard et al. [23].

Some of the most recent theoretical studies of optical activity include the work on transition radiation at a chiral–achiral interface by Engheta and Michelson [24], and that on the reflection of waves from a chiral–achiral interface by Silverman [25]. In addition to the extensive literature on chiral media, the dyadic Green's function and dipole radiation for an unbounded chiral medium by Bassiri et al. [26], the treatment of canonical sources in chiral media by Jaggard et al. [27], and the reflection and transmission of waves at a chiral–achiral interface and through a chiral slab by Bassiri et al. [28] are of direct relevance to this work.

1.2. Constitutive Relations

A formal derivation of the constitutive relations for chiral media is given by Post [29]. A more physical, though heuristic, approach is to derive the constitutive relations for a chiral medium composed of randomly oriented equivalent, short, perfectly conducting wire helices. A typical short helix of this type consists of a circular loop of wire whose two ends are straight wires extended perpendicular to the plane of the loop in opposite directions.

According to Jaggard et al. [23], when an incident electromagnetic wave falls on the helix it induces both electric and magnetic dipole moments. These dipole moments are directed parallel to the axis of the helix. The incident electric field induces currents in the straight portion of the helix, and by continuity these currents must also flow in the circular portion of the helix. The current in the straight portion contributes to the electric dipole moment of the object and the current in the circular portion contributes to its magnetic dipole moment. In a complementary manner, the incident magnetic field induces currents in the circular portion and by continuity in the straight portion. Thus, the magnetic field also contributes to the electric and magnetic dipole moments of the object.

When these electric and magnetic dipole moments are averaged over all possible orientation angles of the randomly oriented helices, the macroscopic polarization and magnetization of the chiral medium, when $e^{-i\omega t}$ time-dependence is assumed for the fields, can be written as

$$\mathbf{P} = \chi_e \varepsilon_0 \mathbf{E} + \gamma_e \mathbf{B}, \tag{1.1}$$

$$\mathbf{M} = -\gamma_m \mathbf{E} + \chi_m \frac{1}{\mu_0} \mathbf{B}, \tag{1.2}$$

where **P** and **M** are the polarization and magnetization of the medium, and χ_e, χ_m are the electric and magnetic self-susceptibilities, and γ_e, γ_m are the cross-susceptibilities. The permittivity and permeability of free space are denoted by ε_0 and μ_0, respectively. The polarization vectors are defined (Stratton [30]) by the equations

$$\mathbf{P} = \mathbf{D} - \varepsilon_0 \mathbf{E}, \tag{1.3}$$

$$\mathbf{M} = \frac{1}{\mu_0} \mathbf{B} - \mathbf{H}. \tag{1.4}$$

When **P** and **M** are substituted in (1.3) and (1.4) from (1.1) and (1.2), it can be shown that

$$\mathbf{D} = \varepsilon \mathbf{E} + \gamma_e \mathbf{B}, \tag{1.5}$$

$$\mathbf{H} = \gamma_m \mathbf{E} + \frac{1}{\mu} \mathbf{B}, \tag{1.6}$$

where $\varepsilon = \varepsilon_0(1 + \chi_e)$ and $\mu = \mu_0/(1 - \chi_m)$, and ε, μ, γ_e, γ_m can be complex quantities. When $\gamma_e = \gamma_m = i\gamma$ (γ real) the chiral media is reciprocal, and when $\gamma_e = -\gamma_m^*$ the chiral media is lossless; the asterisk denotes complex conjugate.

Therefore, in the case of a chiral medium composed of lossless, reciprocal, short wire helices that are all of the same handedness, the constitutive relations for time-harmonic fields have the form

$$\mathbf{D} = \varepsilon\mathbf{E} + i\gamma\mathbf{B}, \tag{1.7}$$

$$\mathbf{H} = i\gamma\mathbf{E} + \frac{1}{\mu}\mathbf{B}, \tag{1.8}$$

where ε, μ, and γ are real quantities. Moreover, it was conjectured (Post [29]) that (1.7) and (1.8) apply not only to chiral media composed of helices but also to lossless, reciprocal, isotropic, chiral media composed of chiral objects of arbitrary shape.

Since **D** and **E** are polar vectors and **B** and **H** are axial vectors, it follows that ε and μ are true scalars and γ is a pseudo-scalar. This means that when the axes of a right-handed Cartesian coordinate system are reversed to form a left-handed Cartesian coordinate system, γ changes in sign whereas ε and μ remain unchanged. Thus the handedness of the medium is manifested by the quantity γ. When $\gamma > 0$, the medium is right-handed and the sense of polarization is right-handed; when $\gamma < 0$, the medium is left-handed and the sense of polarization is left-handed; and when $\gamma = 0$, the medium is simple and there is no optical activity.

When $(1/i\omega)\nabla \times \mathbf{E}$ is substituted for **B** in (1.7), it is evident that the value of **D** at any given point in space depends not only on the value of **E** at that particular point, but also on the derivatives of **E**; that is to say, **D** depends also on the behavior of **E** in the vicinity of this point (Sommerfeld [31]). This nonlocal spatial relationship between **D** and **E** is called spatial dispersion. Therefore the medium described by the constitutive relations (1.7) and (1.8) is a spatially dispersive, isotropic, lossless, reciprocal, chiral medium.

1.2.1. *A Lower Bound on Chirality*

A lower bound on the magnitude of the chirality, γ, of a medium composed of randomly oriented equivalent short wire helices can be found. It has been shown that the parameters χ_e, χ_m, γ_e, and γ_m are given by (Jaggard et al. [23])

$$\chi_e = \frac{Nl^2C}{\varepsilon_0}, \tag{1.9}$$

$$\chi_m = \frac{-N\mu_0(\pi a^2/2)^2}{L}, \tag{1.10}$$

$$\gamma_e = \gamma_m = iN\frac{l}{2}(\pi a^2)\omega C, \tag{1.11}$$

where N is the number of short wire helices per unit volume, C and L are, respectively, the capacitance and inductance of the helix, and $2l$ and $2a$ represent the length and the width of the short helix. From physical considerations and eqs. (1.9)–(1.11), it follows that the constraint $LC = \omega^{-2}$ is placed upon the inductance and capacitance of the helix.

From the definitions of ε and μ and eqs. (1.9)–(1.11), it can be deduced that

$$\varepsilon = \varepsilon_0 + Nl^2C, \tag{1.12}$$

$$\frac{1}{\mu} = \frac{1}{\mu_0} + \frac{N(\pi a^2/2)^2}{L}, \tag{1.13}$$

$$\gamma = N\frac{l}{2}(\pi a^2)\omega C. \tag{1.14}$$

A lower bound on the capacitance C is given by (Jaggard et al. [23])

$$C \geq \varepsilon_0(4\pi)^{2/3}(3V_h)^{1/3}, \tag{1.15}$$

where V_h $(= 2\pi b^2[l + \pi a])$ is the volume occupied by the short wire helix and b is the wire radius. When C from eq. (1.15) is substituted into eq. (1.14), the following lower bound on the magnitude of γ is obtained

$$|\gamma| \geq 3^{1/3}4^{1/6}\pi^{5/3}\sqrt{\frac{\varepsilon_0}{\mu_0}}\frac{N}{\lambda}(2\pi la^2)V_h^{1/3}. \tag{1.16}$$

This bound is proportional to the product of the third root of the volume V_h of the wire helix and the cylindrical volume $(= 2l\pi a^2)$ containing the helix and is inversely proportional to the wavelength.

1.3. Plane Waves in Chiral Media

Plane waves, the simplest solutions of Maxwell's equations, are not only good approximations to waves at large distances from their sources, but they can also be used to represent complicated waves by utilizing Fourier analysis. Plane wave solution reveals fundamental features of wave propagation in chiral media, such as the propagation constants, the wave impedance, the double-mode propagation, and the polarization properties of the principle modes of propagation. The source-free plane wave solutions of Maxwell's equations in a homogeneous unbounded chiral medium, in a semi-infinite chiral medium, and in an infinite chiral slab are now examined.

1.3.1. *Unbounded Chiral Media*

Maxwell's equations, for fields with $e^{-i\omega t}$ time-dependence, in source-free regions are

$$\nabla \times \mathbf{E} = i\omega\mathbf{B}, \tag{1.17}$$

$$\nabla \times \mathbf{H} = -i\omega\mathbf{D}. \tag{1.18}$$

When **H** and **D** from the constitutive relations (1.7) and (1.8) are substituted in eq. (1.18) and the resulting equation is used with the curl of eq. (1.17), the following differential equation for the electric field is obtained:

$$\nabla \times \nabla \times \mathbf{E} - k^2\mathbf{E} - 2\omega\mu\gamma\nabla \times \mathbf{E} = 0, \tag{1.19}$$

where $k^2 = \omega^2\mu\varepsilon$, and ε and μ are the permittivity and permeability of the chiral medium, respectively. Using the vector identity $\nabla \times \nabla \times \mathbf{E} = \nabla\nabla\cdot\mathbf{E} - \nabla^2\mathbf{E}$ and noting that $\nabla\cdot\mathbf{E} = 0$, eq. (1.19) is reduced to

$$\nabla^2\mathbf{E} + k^2\mathbf{E} + 2\omega\mu\gamma\nabla \times \mathbf{E} = 0. \tag{1.20}$$

Since the chiral medium is isotropic, there is no preferred direction of propagation. With no loss of generality, it is assumed that a monochromatic plane wave propagates along the positive z-axis of a Cartesian coordinate system (x, y, z). The unit vectors of this coordinate system are denoted by $\mathbf{e_x}$, $\mathbf{e_y}$, and $\mathbf{e_z}$. By definition, the electric field vector of such a monochromatic plane wave has the form

$$\mathbf{E} = \mathbf{E}_0 e^{ihz} = (E_{0x}\mathbf{e_x} + E_{0y}\mathbf{e_y} + E_{0z}\mathbf{e_z})e^{ihz}, \tag{1.21}$$

where $\mathbf{E}_0$ is a complex-constant amplitude vector, h is the propagation constant, and z is measured along the z-axis. Using eq. (1.21) and the fact that $\nabla\cdot\mathbf{E} = 0$, it can be shown that $E_{0z} = 0$. When eq. (1.21) is substituted into (1.20) and it is noted that $E_{0z} = 0$, the following two homogeneous equations, in the two unknowns E_{0x} and E_{0y}, are obtained

$$(k^2 - h^2)E_{0x} - \alpha h E_{0y} = 0, \tag{1.22}$$

$$\alpha h E_{0x} + (k^2 - h^2)E_{0y} = 0, \tag{1.23}$$

where $\alpha = 2i\omega\mu\gamma$. This system of equations has a nontrivial solution if and only if

$$(k^2 - h^2)^2 + \alpha^2 h^2 = 0. \tag{1.24}$$

The values of h that satisfy (1.24) are the possible propagation constants. They are

$$h = \pm\omega\mu\gamma \pm \sqrt{\omega^2\mu^2\gamma^2 + k^2}. \tag{1.25}$$

Since the plane waves are assumed to propagate along the positive z-axis, the permissible values of h are

$$h_1 = \omega\mu\gamma + \sqrt{\omega^2\mu^2\gamma^2 + k^2}, \tag{1.26}$$

$$h_2 = -\omega\mu\gamma + \sqrt{\omega^2\mu^2\gamma^2 + k^2}. \tag{1.27}$$

Therefore, there are two modes of propagation, one being a wave of phase velocity ω/h_1, and the other a wave of phase velocity ω/h_2. For $\gamma > 0$, the former is the slower mode whereas for $\gamma < 0$ the latter is the slower mode. The electric fields corresponding to these modes are

$$\mathbf{E}_1 = E_{01}(\mathbf{e_x} + i\mathbf{e_y})e^{ih_1z}, \tag{1.28}$$

which is a right-circularly polarized wave, and

$$\mathbf{E}_2 = E_{02}(\mathbf{e_x} - i\mathbf{e_y})e^{ih_2z}, \tag{1.29}$$

which is a left-circularly polarized wave. The total electric field **E**, which is given by

$$\mathbf{E} = \mathbf{E}_1 + \mathbf{E}_2 = [E_{01}(\mathbf{e_x} + i\mathbf{e_y})e^{ih_1z} + E_{02}(\mathbf{e_x} - i\mathbf{e_y})e^{ih_2z}], \quad (1.30)$$

is an elliptically polarized wave. The complex-constant amplitudes E_{01}, E_{02} can be written as $E_{01} = \rho_1 e^{i\varphi_1}$ and $E_{02} = \rho_2 e^{i\varphi_2}$, where their moduli and phases ρ_j and φ_j ($j = 1, 2$) are real numbers. The length of the semimajor axis of the polarization ellipse is $(\rho_1 + \rho_2)$, and that of the semiminor axis is $|\rho_1 - \rho_2|$. The orientation angle ψ of the ellipse is given by

$$\psi_{\mathrm{E}} = \frac{h_2 - h_1}{2} z + \frac{\varphi_2 - \varphi_1}{2} = -\omega\mu\gamma z + \frac{\varphi_2 - \varphi_1}{2}. \quad (1.31)$$

The magnetic field is obtained when **E** from (1.30) is substituted into

$$\mathbf{H} = i\gamma\mathbf{E} + \frac{1}{\mu}\mathbf{B} = i\gamma\mathbf{E} + \frac{1}{i\omega\mu}\nabla \times \mathbf{E}, \quad (1.32)$$

and it is found to be

$$\mathbf{H} = -i\sqrt{\gamma^2 + \frac{\varepsilon}{\mu}}[E_{01}(\mathbf{e_x} + i\mathbf{e_y})e^{ih_1z} - E_{02}(\mathbf{e_x} - i\mathbf{e_y})e^{ih_2z}]. \quad (1.33)$$

The orientation angle ψ for the magnetic field is given by

$$\psi_{\mathrm{H}} = \frac{h_2 - h_1}{2} z + \frac{\varphi_2 - \varphi_1}{2} - \frac{\pi}{2} = -\omega\mu\gamma z + \frac{\varphi_2 - \varphi_1}{2} - \frac{\pi}{2}. \quad (1.34)$$

The polarization ellipse of **H** is similar and oriented perpendicularly to the polarization ellipse of **E**. That is, the polarization ellipses of **E** and **H** are perpendicular to each other and have the same sense of rotation. The sense of rotation is the same as that of the circularly polarized component with the larger modulus.

When the notation $\mathbf{E}(t) = \mathrm{Re}(\mathbf{E}e^{-i\omega t})$ and $\mathbf{H}(t) = \mathrm{Re}(\mathbf{H}e^{-i\omega t})$ is used, it is seen from (1.30) and (1.33) that at any point (x, y, z) and at any time t the vectors $\mathbf{E}(t)$ and $\mathbf{H}(t)$ are perpendicular to each other and to the direction of propagation. Hence, the monochromatic plane waves in a chiral medium are transverse electromagnetic waves. Specifically,

$$\mathbf{e}_z \times \mathbf{E}(t) = \frac{1}{\sqrt{\gamma^2 + (\varepsilon/\mu)}}\mathbf{H}(t). \quad (1.35)$$

It follows from this relation that the wave impedance Z of the chiral medium is given by (Bassiri [32])

$$Z = \sqrt{\frac{\mu}{\varepsilon}} \frac{1}{\sqrt{1 + (\mu/\varepsilon)\gamma^2}}. \quad (1.36)$$

Therefore, the wave impedance decreases as the medium becomes more chiral.

The time-averaged power density **P** carried by the plane waves, is given by (Papas [33])

$$\mathbf{P} = \mathrm{Re}\{\mathbf{S}\} = \tfrac{1}{2}\,\mathrm{Re}\{\mathbf{E} \times \mathbf{H}^*\}. \tag{1.37}$$

When **E** and **H** from (1.30) and (1.33) are substituted into (1.37), it is found that

$$\mathbf{P} = \sqrt{\gamma^2 + \frac{\varepsilon}{\mu}}\,[|E_{01}|^2 + |E_{02}|^2]\mathbf{e}_z. \tag{1.38}$$

This shows that for fixed E_{01} and E_{02}, as the medium becomes more chiral, the time-averaged power density increases.

1.3.2. *Semi-Infinite Chiral Media*

When a plane wave is incident upon a boundary between a dielectric and a chiral medium it splits into two transmitted waves proceeding into the chiral medium, and a reflected wave propagating back into the dielectric. To formulate mathematically the problem of reflection from and transmission through a semi-infinite chiral medium, a Cartesian coordinate system (x, y, z) is introduced. As shown in Figure 1.1, the xy-plane is the plane of interface of a sim-

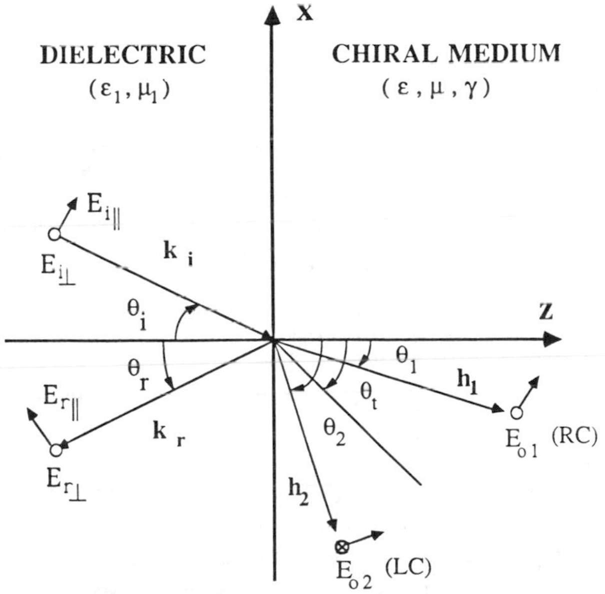

FIGURE 1.1. Orientation of the wave vectors of the incident, the reflected, and the transmitted waves at an oblique incidence on a semi-infinite chiral medium. In the chiral medium the h_1- and the h_2-waves are right-circularly and left-circularly polarized waves, respectively. The two angles of refraction are denoted by θ_1 and θ_2. For $\gamma = 0$, these two angles approach θ_t.

ple homogeneous dielectric with permittivity ε_1 and permeability μ_1, and a homogeneous chiral medium described by the constitutive relations (1.7) and (1.8).

It is assumed that a monochromatic plane wave is obliquely incident upon the interface. The complex-constant amplitude vectors of the incident, reflected, and transmitted plane waves always lie on planes perpendicular to the directions of their propagation. Therefore, it is always possible to decompose any one of these amplitudes into a component normal to the plane of incidence, and a second component lying in the plane of incidence (the xz-plane), as shown in Figure 1.1. The plane of incidence is the plane containing the normal to the interface and the wave vector of the incident wave.

From the boundary conditions, i.e., the continuity of the tangential electric field and the tangential magnetic field at the interface, it can be shown that

$$\mathbf{k}_{\mathrm{i}} \times \mathbf{e}_z = \mathbf{k}_{\mathrm{r}} \times \mathbf{e}_z = \mathbf{h}_1 \times \mathbf{e}_z = \mathbf{h}_2 \times \mathbf{e}_z, \tag{1.39}$$

where $\mathbf{k}_{\mathrm{i}}$, $\mathbf{k}_{\mathrm{r}}$, $\mathbf{h}_1$, and $\mathbf{h}_2$ are the wave vectors of the incident wave, the reflected wave, and the two transmitted waves, respectively (Born and Wolf [34]).

When the magnitude of the vector equation (1.39) is taken, it is found that

$$k_{\mathrm{i}} \sin \theta_{\mathrm{i}} = k_{\mathrm{r}} \sin \theta_{\mathrm{r}} = h_1 \sin \theta_1 = h_2 \sin \theta_2, \tag{1.40}$$

and since $k_{\mathrm{i}} = k_{\mathrm{r}}$, then $\theta_{\mathrm{i}} = \theta_{\mathrm{r}}$. From eq. (1.40), the angles θ_1 and θ_2 corresponding to the two transmitted waves can be expressed as (see Figure 1.1)

$$\theta_1 = \arcsin\left(\frac{k_{\mathrm{i}} \sin \theta_{\mathrm{i}}}{h_1}\right), \tag{1.41}$$

$$\theta_2 = \arcsin\left(\frac{k_{\mathrm{i}} \sin \theta_{\mathrm{i}}}{h_2}\right), \tag{1.42}$$

where θ_{i} is the angle of incidence, $k_{\mathrm{i}} = \omega\sqrt{\mu_1 \varepsilon_1}$, and h_1 and h_2 are given by (1.26) and (1.27). When $\gamma = 0$, the angle of the transmitted wave is given by

$$\theta_{\mathrm{t}} = \arcsin\left(\frac{k_{\mathrm{i}} \sin \theta_{\mathrm{i}}}{k}\right), \tag{1.43}$$

where θ_{t} is the refraction angle of a dielectric–dielectric interface.

From eq. (1.43) it can be seen that as $\gamma \to 0$, angles θ_1 and θ_2 approach θ_{t}. As $\gamma \to \infty$, angles θ_1 and θ_2 approach 0 and $\pi/2$, respectively. That is, the h_2-wave will become evanescent and only the h_1-wave will propagate, and its direction of propagation is along the positive z-axis. As $\gamma \to -\infty$, angles θ_1 and θ_2 approach $\pi/2$ and 0, respectively. In this case, the h_1-wave will become evanescent and the h_2-wave will propagate along the positive z-axis.

1.3.2.1. Total Internal Reflection

In general, two transmitted waves propagate inside the chiral medium, namely the h_1-wave and the h_2-wave. When neither of these waves propagate inside the chiral medium, the phenomenon of total internal reflection occurs. By letting θ_1 and θ_2 be equal to $\pi/2$ in eqs. (1.41) and (1.42), the critical angles of incidence θ_{c1} and θ_{c2} can be found. They are

$$\theta_{c1} = \arcsin\left(\frac{\mu\gamma + \sqrt{\mu^2\gamma^2 + \mu\varepsilon}}{\sqrt{\mu_1\varepsilon_1}}\right), \tag{1.44}$$

$$\theta_{c2} = \arcsin\left(\frac{-\mu\gamma + \sqrt{\mu^2\gamma^2 + \mu\varepsilon}}{\sqrt{\mu_1\varepsilon_1}}\right). \tag{1.45}$$

When $\gamma > 0$ and $h_2 < h_1 < k_i$, then θ_{c1} is always greater than θ_{c2}. There are three possibilities:

(i) If $\theta_i < \theta_{c2} < \theta_{c1}$, then both the h_1- and h_2-waves propagate. Their direction of propagation is given by (1.41) and (1.42).
(ii) If $\theta_{c2} \leq \theta_i < \theta_{c1}$, then only the h_1-wave propagates and the h_2-wave becomes evanescent. The direction of propagation for the h_1-wave is given by eq. (1.41).
(iii) If $\theta_{c2} < \theta_{c1} \leq \theta_i$, then neither of the waves propagates and total internal reflection of the two waves into the dielectric occurs.

When $\gamma < 0$ and $h_1 < h_2 < k_i$, then θ_{c1} is always less than θ_{c2}. There are three possibilities:

(i) If $\theta_i < \theta_{c1} < \theta_{c2}$, then both the h_1- and h_2-waves propagate. The direction of propagation of these waves is given by (1.41) and (1.42).
(ii) If $\theta_{c1} \leq \theta_i < \theta_{c2}$, then only the h_2-wave propagates and the h_1-wave becomes evanescent. The direction of propagation for the h_2-wave is given by eq. (1.42).
(iii) If $\theta_{c1} < \theta_{c2} \leq \theta_i$, then none of the waves propagates inside the chiral medium and the phenomenon of total internal reflection occurs.

Obviously, there are other possibilities. For example, when $\gamma > 0$ and $h_2 < k_i < h_1$, or $\gamma < 0$ and $h_1 < k_i < h_2$, there is only one real solution for the critical angle of incidence.

Depending on the incidence angle and the relative values of h_1, h_2, and k_i, one, both, or none of the transmitted waves propagate inside the chiral medium.

1.3.2.2. Fresnel Equations

In order to study the power carried by the reflected and transmitted waves and also the polarization properties of these waves, it is necessary to determine the complex-constant amplitude vectors associated with these waves. This is done by matching the fields at the interface using the boundary conditions (Papas [33])

$$(\mathbf{E}_{0i} + \mathbf{E}_{0r}) \times \mathbf{e}_z = (\mathbf{E}_{01} + \mathbf{E}_{02}) \times \mathbf{e}_z, \tag{1.46}$$

$$(\mathbf{H}_{0i} + \mathbf{H}_{0r}) \times \mathbf{e}_z = (\mathbf{H}_{01} + \mathbf{H}_{02}) \times \mathbf{e}_z, \tag{1.47}$$

where $\mathbf{E}_{0i}$, $\mathbf{E}_{0r}$ are the complex-constant amplitudes of the incident and the reflected electric fields, respectively. Similarly, $\mathbf{E}_{01}$ and $\mathbf{E}_{02}$ are the amplitudes of the electric fields associated with the right-circularly and the left-circularly polarized transmitted waves, respectively.

The incident electric and magnetic fields can be written as (see Figure 1.1)

$$\mathbf{E}_i = \mathbf{E}_{0i} \exp[ik_i(z \cos \theta_i - x \sin \theta_i)], \tag{1.48}$$

$$\mathbf{H}_i = \mathbf{H}_{0i} \exp[ik_i(z \cos \theta_i - x \sin \theta_i)], \tag{1.49}$$

where

$$\mathbf{E}_{0i} = E_{i\perp}\mathbf{e_y} + E_{i\parallel}(\cos \theta_i \mathbf{e_x} + \sin \theta_i \mathbf{e_z}), \tag{1.50}$$

$$\mathbf{H}_{0i} = \eta_1^{-1}[E_{i\parallel}\mathbf{e_y} - E_{i\perp}(\cos \theta_i \mathbf{e_x} + \sin \theta_i \mathbf{e_z})], \tag{1.51}$$

and $\eta_1 = \sqrt{\mu_1/\varepsilon_1}$. The reflected fields may be written as (see Figure 1.1)

$$\mathbf{E}_r = \mathbf{E}_{0r} \exp[-ik_i(z \cos \theta_i + x \sin \theta_i)], \tag{1.52}$$

$$\mathbf{H}_r = \mathbf{H}_{0r} \exp[-ik_i(z \cos \theta_i + x \sin \theta_i)], \tag{1.53}$$

where

$$\mathbf{E}_{0r} = E_{r\perp}\mathbf{e_y} + E_{r\parallel}(\cos \theta_i \mathbf{e_x} - \sin \theta_i \mathbf{e_z}), \tag{1.54}$$

$$\mathbf{H}_{0r} = \eta_1^{-1}[-E_{r\parallel}\mathbf{e_y} + E_{r\perp}(\cos \theta_i \mathbf{e_x} - \sin \theta_i \mathbf{e_z})]. \tag{1.55}$$

The subscript $\perp$ refers to the amplitude of the field component perpendicular to the plane of incidence, and the subscript $\parallel$ refers to the amplitude of the field component that lies in the plane of incidence. Since the two transmitted waves are circularly polarized, they can be written as

$$\mathbf{E}_t = \mathbf{E}_{01} \exp[ih_1(z \cos \theta_1 - x \sin \theta_1)] + \mathbf{E}_{02} \exp[ih_2(z \cos \theta_2 - x \sin \theta_2)], \tag{1.56}$$

$$\mathbf{H}_t = \mathbf{H}_{01} \exp[ih_1(z \cos \theta_1 - x \sin \theta_1)] + \mathbf{H}_{02} \exp[ih_2(z \cos \theta_2 - x \sin \theta_2)], \tag{1.57}$$

where

$$\mathbf{E}_{01} = E_{01}(\cos \theta_1 \mathbf{e_x} + \sin \theta_1 \mathbf{e_z} + i\mathbf{e_y}), \tag{1.58}$$

$$\mathbf{H}_{01} = -iZ^{-1}E_{01}(\cos \theta_1 \mathbf{e_x} + \sin \theta_1 \mathbf{e_z} + i\mathbf{e_y}), \tag{1.59}$$

and

$$\mathbf{E}_{02} = E_{02}(\cos \theta_2 \mathbf{e_x} + \sin \theta_2 \mathbf{e_z} - i\mathbf{e_y}), \tag{1.60}$$

$$\mathbf{H}_{02} = iZ^{-1}E_{02}(\cos \theta_2 \mathbf{e_x} + \sin \theta_2 \mathbf{e_z} - i\mathbf{e_y}), \tag{1.61}$$

and the wave impedance of the medium Z, is defined by eq. (1.36).

It is assumed that the amplitude, the polarization, the direction of propagation, and the frequency of the incident field are known. To find the complex-constant amplitude vectors of the reflected and transmitted waves, the boundary conditions at the interface must be applied to the x and y components of the electric and magnetic fields. A system of four nonhomogeneous equations in the four unknowns $E_{r\perp}$, $E_{r\parallel}$, E_{01}, and E_{02} is obtained.

The expressions for $E_{r\perp}$, $E_{r\parallel}$, E_{01} and E_{02} in terms of the components of the incident wave can be written as

$$\begin{pmatrix} E_{r\perp} \\ E_{r\parallel} \end{pmatrix} = \begin{pmatrix} R_{11} & R_{12} \\ R_{21} & R_{22} \end{pmatrix} \begin{pmatrix} E_{i\perp} \\ E_{i\parallel} \end{pmatrix}, \tag{1.62}$$

and

$$\begin{pmatrix} E_{01} \\ E_{02} \end{pmatrix} = \begin{pmatrix} T_{11} & T_{12} \\ T_{21} & T_{22} \end{pmatrix} \begin{pmatrix} E_{i\perp} \\ E_{i\parallel} \end{pmatrix}. \tag{1.63}$$

The 2×2 matrix in eq. (1.62) is the reflection coefficient matrix, and its entries are

$$R_{11} = \frac{\cos\theta_i(1 - g^2)(\cos\theta_1 + \cos\theta_2) + 2g(\cos^2\theta_i - \cos\theta_1\cos\theta_2)}{\cos\theta_i(1 + g^2)(\cos\theta_1 + \cos\theta_2) + 2g(\cos^2\theta_i + \cos\theta_1\cos\theta_2)}, \tag{1.64}$$

$$R_{12} = \frac{-2ig\cos\theta_i(\cos\theta_1 - \cos\theta_2)}{\cos\theta_i(1 + g^2)(\cos\theta_1 + \cos\theta_2) + 2g(\cos^2\theta_i + \cos\theta_1\cos\theta_2)}, \tag{1.65}$$

$$R_{21} = \frac{-2ig\cos\theta_i(\cos\theta_1 - \cos\theta_2)}{\cos\theta_i(1 + g^2)(\cos\theta_1 + \cos\theta_2) + 2g(\cos^2\theta_i + \cos\theta_1\cos\theta_2)}, \tag{1.66}$$

$$R_{22} = \frac{\cos\theta_i(1 - g^2)(\cos\theta_1 + \cos\theta_2) - 2g(\cos^2\theta_i - \cos\theta_1\cos\theta_2)}{\cos\theta_i(1 + g^2)(\cos\theta_1 + \cos\theta_2) + 2g(\cos^2\theta_i + \cos\theta_1\cos\theta_2)}, \tag{1.67}$$

where $g = \sqrt{(\mu_1/\varepsilon_1)\gamma^2 + (\mu_1\varepsilon/\varepsilon_1\mu)}$.

The 2×2 matrix in (1.63) is the transmission coefficient matrix, and its entries are

$$T_{11} = \frac{-2i\cos\theta_i(g\cos\theta_i + \cos\theta_2)}{\cos\theta_i(1 + g^2)(\cos\theta_1 + \cos\theta_2) + 2g(\cos^2\theta_i + \cos\theta_1\cos\theta_2)}, \tag{1.68}$$

$$T_{12} = \frac{2\cos\theta_i(\cos\theta_i + g\cos\theta_2)}{\cos\theta_i(1 + g^2)(\cos\theta_1 + \cos\theta_2) + 2g(\cos^2\theta_i + \cos\theta_1\cos\theta_2)}, \tag{1.69}$$

$$T_{21} = \frac{2i\cos\theta_i(g\cos\theta_i + \cos\theta_1)}{\cos\theta_i(1 + g^2)(\cos\theta_1 + \cos\theta_2) + 2g(\cos^2\theta_i + \cos\theta_1\cos\theta_2)}, \tag{1.70}$$

$$T_{22} = \frac{2\cos\theta_i(\cos\theta_i + g\cos\theta_1)}{\cos\theta_i(1 + g^2)(\cos\theta_1 + \cos\theta_2) + 2g(\cos^2\theta_i + \cos\theta_1\cos\theta_2)}. \tag{1.71}$$

When the incident wave falls normally on the interface, i.e., $\theta_i = 0$, expressions (1.64)–(1.71) reduce to

$$R_{11} = R_{22} = \frac{1 - g}{1 + g}, \tag{1.72}$$

$$R_{12} = R_{21} = 0, \tag{1.73}$$

and

$$T_{11} = -iT_{22} = \frac{-i}{1+g}, \tag{1.74}$$

$$T_{12} = -iT_{21} = \frac{1}{1+g}. \tag{1.75}$$

A study of the power and the polarization of the reflected wave as a function of the incident angle for different values of γ is contained in the paper by Bassiri et al. [28].

1.3.2.3. Brewster Angle

Under a certain condition a monochromatic plane wave of arbitrary polarization, on reflection from a chiral medium, becomes a linearly polarized wave. The angle of incidence at which this phenomenon occurs is called the Brewster angle. The plane containing the electric field vector and the direction of propagation is called the plane of polarization. For a linearly polarized wave, the angle between the plane of polarization and the plane of incidence is called the azimuthal angle. This angle lies in the range from $-\pi/2$ to $\pi/2$. It is defined to be positive whenever the sense of rotation of the plane of polarization toward the plane of incidence and the direction of wave propagation form a right-handed screw. Let α_i and α_r be the azimuthal angles of the incident and reflected waves, respectively. From the above definitions it can be shown that

$$\tan \alpha_i = \frac{E_{i\perp}}{E_{i\parallel}}, \tag{1.76}$$

$$\tan \alpha_r = \frac{E_{r\perp}}{E_{r\parallel}}, \tag{1.77}$$

where α_i and α_r can be complex angles. The amplitudes of the perpendicular and the parallel components of the incident and the reflected waves are related by the matrix equation (1.62). By using (1.76), (1.77), and (1.62), it can be shown that

$$\tan \alpha_r = \frac{R_{12} + R_{11} \tan \alpha_i}{R_{22} + R_{21} \tan \alpha_i}. \tag{1.78}$$

If the incident wave is incident upon the interface at Brewster's angle θ_B, then the reflected wave must be linearly polarized. Therefore, α_r at this angle must be a real constant for all α_i (Chen [35]). When eq. (1.78) is differentiated with respect to α_i, the following condition is obtained:

$$R_{11}R_{22} - R_{12}R_{21} = 0. \tag{1.79}$$

Under this condition, eq. (1.78) becomes

$$\tan \alpha_r = \frac{R_{12}}{R_{22}} = \frac{R_{11}}{R_{21}}. \tag{1.80}$$

Substitutions of eqs. (1.64)–(1.67) into eq. (1.79) results in the following equation:

$$(1-g^2)^2 \cos^2\theta_i(\cos\theta_1+\cos\theta_2)^2 = 4g^2(\cos^2\theta_i-\cos^2\theta_1)(\cos^2\theta_i-\cos^2\theta_2), \quad (1.81)$$

where θ_1 and θ_2 can be written in terms of θ_i, using eqs. (1.41) and (1.42). The incidence angle that satisfies the transcendental equation (1.81) is the Brewster angle, and it can be solved for numerically.

1.3.3. *Infinite Chiral Slab*

In this section, plane wave propagation through an infinite chiral slab of thickness d is considered. The slab (ε, μ, γ) is confined between two infinitely extended planes, $z = 0$ and $z = d$, and lies between two dielectrics with the same constitutive parameters (ε_1, μ_1), as shown in Figure 1.2. Let a plane wave be incident at an angle θ_i on the chiral slab from the dielectric that borders the slab at $z = 0$. The purpose of the analysis that follows is to find the amplitudes of the reflected and transmitted waves outside the slab.

The incident electric and magnetic fields can be written as (see Figure 1.2)

$$\mathbf{E}_i = \mathbf{E}_{0i} \exp[ik_i(z\cos\theta_i - x\sin\theta_i)], \quad (1.82)$$

$$\mathbf{H}_i = \mathbf{H}_{0i} \exp[ik_i(z\cos\theta_i - x\sin\theta_i)], \quad (1.83)$$

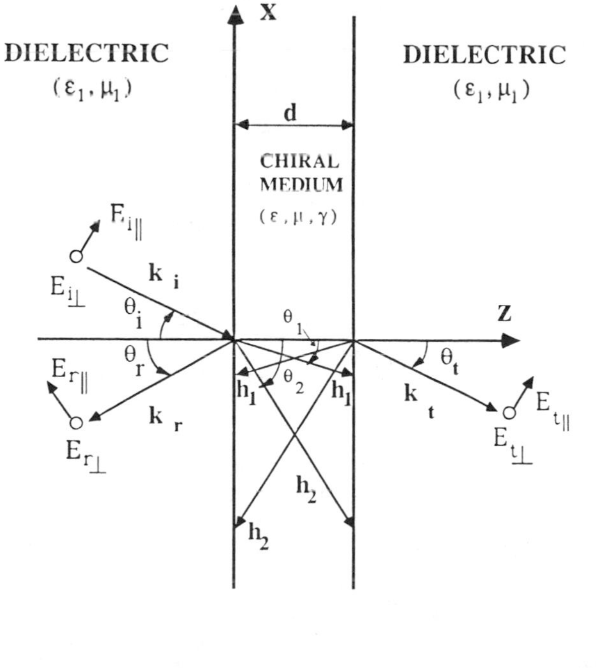

FIGURE 1.2. Oblique incidence on an infinite slab of chiral medium. The dielectrics occupying the regions $z < 0$ and $z > d$ have the same constitutive parameters. In the chiral slab, the h_1- and h_2-waves are right-circularly and left-circularly polarized waves, respectively.

where

$$\mathbf{E}_{0i} = E_{i\perp}\mathbf{e_y} + E_{i\parallel}(\cos\theta_i\mathbf{e_x} + \sin\theta_i\mathbf{e_z}), \tag{1.84}$$

$$\mathbf{H}_{0i} = \eta_1^{-1}[E_{i\parallel}\mathbf{e_y} - E_{i\perp}(\cos\theta_i\mathbf{e_x} + \sin\theta_i\mathbf{e_z})], \tag{1.85}$$

and $\eta_1 = \sqrt{\mu_1/\varepsilon_1}$. The reflected fields may be written as (see Figure 1.2)

$$\mathbf{E}_r = \mathbf{E}_{0r}\exp[-ik_i(z\cos\theta_i + x\sin\theta_i)], \tag{1.86}$$

$$\mathbf{H}_r = \mathbf{H}_{0r}\exp[-ik_i(z\cos\theta_i + x\sin\theta_i)], \tag{1.87}$$

where

$$\mathbf{E}_{0r} = E_{r\perp}\mathbf{e_y} + E_{r\parallel}(\cos\theta_i\mathbf{e_x} - \sin\theta_i\mathbf{e_z}), \tag{1.88}$$

$$\mathbf{H}_{0r} = \eta_1^{-1}[-E_{r\parallel}\mathbf{e_y} + E_{r\perp}(\cos\theta_i\mathbf{e_x} - \sin\theta_i\mathbf{e_z})]. \tag{1.89}$$

In the chiral slab, it is assumed that there are four total waves, two propagating toward the interface $z = d$, and the other two propagating toward the interface $z = 0$ (see Figure 1.2). The electric and magnetic fields of the two waves propagating inside the chiral medium toward the interface $z = d$ can be written as

$$\mathbf{E}_c^+ = \mathbf{E}_{01}^+\exp[ih_1(z\cos\theta_1 - x\sin\theta_1)] + \mathbf{E}_{02}^+\exp[ih_2(z\cos\theta_2 - x\sin\theta_2)], \tag{1.90}$$

$$\mathbf{H}_c^+ = \mathbf{H}_{01}^+\exp[ih_1(z\cos\theta_1 - x\sin\theta_1)] + \mathbf{H}_{02}^+\exp[ih_2(z\cos\theta_2 - x\sin\theta_2)], \tag{1.91}$$

where

$$\mathbf{E}_{01}^+ = E_{01}^+(\cos\theta_1\mathbf{e_x} + \sin\theta_1\mathbf{e_z} + i\mathbf{e_y}), \tag{1.92}$$

$$\mathbf{H}_{01}^+ = -iZ^{-1}E_{01}^+(\cos\theta_1\mathbf{e_x} + \sin\theta_1\mathbf{e_z} + i\mathbf{e_y}), \tag{1.93}$$

and

$$\mathbf{E}_{02}^+ = E_{02}^+(\cos\theta_2\mathbf{e_x} + \sin\theta_2\mathbf{e_z} - i\mathbf{e_y}), \tag{1.94}$$

$$\mathbf{H}_{02}^+ = iZ^{-1}E_{02}^+(\cos\theta_2\mathbf{e_x} + \sin\theta_2\mathbf{e_z} - i\mathbf{e_y}). \tag{1.95}$$

The electric and magnetic fields of the other two total waves propagating inside the chiral medium toward the interface $z = 0$ may be written as

$$\mathbf{E}_c^- = \mathbf{E}_{01}^-\exp[-ih_1(z\cos\theta_1 + x\sin\theta_1)] + \mathbf{E}_{02}^-\exp[-ih_2(z\cos\theta_2 + x\sin\theta_2)], \tag{1.96}$$

$$\mathbf{H}_c^- = \mathbf{H}_{01}^-\exp[-ih_1(z\cos\theta_1 + x\sin\theta_1)] + \mathbf{H}_{02}^-\exp[-ih_2(z\cos\theta_2 + x\sin\theta_2)], \tag{1.97}$$

where

$$\mathbf{E}_{01}^- = E_{01}^-(\sin\theta_1\mathbf{e_z} - \cos\theta_1\mathbf{e_x} + i\mathbf{e_y}), \tag{1.98}$$

$$\mathbf{H}_{01}^- = -iZ^{-1}E_{01}^-(\sin\theta_1\mathbf{e_z} - \cos\theta_1\mathbf{e_x} + i\mathbf{e_y}), \tag{1.99}$$

and

$$\mathbf{E}_{02}^{-} = E_{02}^{-}(\sin\theta_2\mathbf{e_z} - \cos\theta_2\mathbf{e_x} - i\mathbf{e_y}), \tag{1.100}$$

$$\mathbf{H}_{02}^{-} = iZ^{-1}E_{02}^{-}(\sin\theta_2\mathbf{e_z} - \cos\theta_2\mathbf{e_x} - i\mathbf{e_y}), \tag{1.101}$$

and the wave impedance Z, of the chiral medium, is defined by eq. (1.36). Outside the slab, in the dielectric which borders the slab at $z = d$, the total transmitted wave can be written as

$$\mathbf{E}_t = \mathbf{E}_{0t}\exp[ik_t(z\cos\theta_t - x\sin\theta_t)], \tag{1.102}$$

$$\mathbf{H}_t = \mathbf{H}_{0t}\exp[ik_t(z\cos\theta_t - x\sin\theta_t)], \tag{1.103}$$

where

$$\mathbf{E}_{0t} = E_{t\perp}\mathbf{e_y} + E_{t\parallel}(\cos\theta_t\mathbf{e_x} + \sin\theta_t\mathbf{e_z}), \tag{1.104}$$

$$\mathbf{H}_{0t} = \eta_1^{-1}[E_{t\parallel}\mathbf{e_y} - E_{t\perp}(\cos\theta_t\mathbf{e_x} + \sin\theta_t\mathbf{e_z})], \tag{1.105}$$

where $k_t = k_i$ and $\theta_t = \theta_i$.

To find the complex-constant amplitude vectors of the reflected and transmitted waves in the two dielectrics, and those of the waves inside the slab, the boundary conditions at the two interfaces $z = 0$ and $z = d$ must be applied to the x and y components of the electric and magnetic fields. When this is donc, a system of eight nonhomogeneous equations in the eight unknowns $E_{r\perp}$, $E_{r\parallel}$, E_{01}^{+}, E_{02}^{+}, E_{01}^{-}, E_{02}^{-}, $E_{t\perp}$, and $E_{t\parallel}$ is obtained. This system of equations can be written in the following matrix form:

$$\begin{pmatrix} E_{r\perp} \\ E_{r\parallel} \\ E_{01}^{+} \\ E_{02}^{+} \\ E_{01}^{-} \\ E_{02}^{-} \\ E_{t\perp} \\ E_{t\parallel} \end{pmatrix} = \mathbf{Q}^{-1}\cdot\begin{pmatrix} E_{i\parallel} \\ E_{i\perp} \\ E_{i\perp} \\ E_{i\parallel} \\ 0 \\ 0 \\ 0 \\ 0 \end{pmatrix}, \tag{1.106}$$

where $\mathbf{Q}$ is the following matrix:

$$\mathbf{Q} = \begin{pmatrix} 0 & -1 & R_1 & R_2 & -R_1 & -R_2 & 0 & 0 \\ -1 & 0 & i & -i & i & -i & 0 & 0 \\ 1 & 0 & igR_1 & -igR_2 & -igR_1 & igR_2 & 0 & 0 \\ 0 & 1 & g & g & g & g & 0 & 0 \\ 0 & 0 & R_1e^{i\delta_1} & R_2e^{i\delta_2} & -R_1e^{-i\delta_1} & -R_2e^{-i\delta_2} & 0 & -e^{i\delta_i} \\ 0 & 0 & ie^{i\delta_1} & -ie^{i\delta_2} & ie^{-i\delta_1} & -ie^{-i\delta_2} & -e^{i\delta_i} & 0 \\ 0 & 0 & -igR_1e^{i\delta_1} & igR_2e^{i\delta_2} & igR_1e^{-i\delta_1} & -igR_2e^{-i\delta_2} & e^{i\delta_i} & 0 \\ 0 & 0 & ge^{i\delta_1} & ge^{i\delta_2} & ge^{-i\delta_1} & ge^{-i\delta_2} & 0 & -e^{i\delta_i} \end{pmatrix},$$

and where $R_1 = \cos\theta_1/\cos\theta_i$, $R_2 = \cos\theta_2/\cos\theta_i$, $\delta_1 = h_1 d\cos\theta_1$, $\delta_2 = h_2 d\cos\theta_2$, and $\delta_i = k_i d\cos\theta_i$.

Since the analytical solution of this system of eight nonhomogeneous equations leads to involved expressions for the field amplitudes, it is therefore best to use numerical techniques to invert the matrix equation (1.106). However, it is interesting and important to obtain the analytical solution of this system for the case in which a linearly polarized wave is normally incident upon the interface. With no loss of generality, it is assumed that the incident electric field is directed along the positive x-axis. This field can be written as

$$\mathbf{E}_i = E_i \mathbf{e_x} e^{ik_i z}. \tag{1.107}$$

When θ_i and $E_{i\perp}$ are set to zero in the matrix equation (1.106), and the resulting matrix equation is solved, then the reflected and transmitted electric fields can be found (Stratton [30]). The reflected field can be written as

$$\mathbf{E}_r = E_r \mathbf{e_x} e^{-ik_i z}, \tag{1.108}$$

where

$$E_r = E_i\left(\frac{1+g}{1-g}\right)\frac{1-e^{i(\delta_1+\delta_2)}}{[(1+g)/(1-g)]^2 - e^{i(\delta_1+\delta_2)}}. \tag{1.109}$$

Therefore the polarization of the reflected wave is the same as that of the incident wave; that is, the chiral slab behaves as an ordinary dielectric as far as the polarization of the reflected wave is concerned. The transmitted wave can be written as

$$\mathbf{E}_t = E_t\left[\mathbf{e_x} + \tan\left(\frac{\delta_2-\delta_1}{2}\right)\mathbf{e_y}\right]e^{ik_i z}, \tag{1.110}$$

where

$$E_t = E_i\frac{2g}{(1-g)^2}\frac{e^{i(\delta_1-\delta_i)} + e^{i(\delta_2-\delta_i)}}{[(1+g)/(1-g)]^2 - e^{i(\delta_1+\delta_2)}}. \tag{1.111}$$

Since $\tan[(\delta_2-\delta_1)/2]$ is real, the transmitted wave is linearly polarized, and the ratio of its x and y components,

$$\frac{E_{ty}}{E_{tx}} = \tan\left(\frac{\delta_2-\delta_1}{2}\right) = \tan(-\omega\mu\gamma d), \tag{1.112}$$

shows that the plane of polarization of the transmitted wave is rotated by an angle of $-\omega\mu\gamma d$ with respect to the positive x-axis. If γ is positive, then the rotation is toward the negative y-axis; and if γ is negative, then the rotation is toward the positive y-axis.

1.4. Radiation in Chiral Media

In this section the determination of electromagnetic fields generated by a given distribution of sources in a chiral medium is considered. The method used is that of the dyadic Green's function, in which the electric field vector is written directly as a volume integral in terms of the source current vector. These two vectors are related via the dyadic Green's function. A detailed account of the derivation of this dyadic and its application to the determination of the field of a dipole antenna in a chiral medium is contained in this part of the chapter.

1.4.1. *Dyadic Green's Function*

The Maxwell equations in a region with an external source can be written as

$$\nabla \times \mathbf{E} = i\omega\mathbf{B}, \tag{1.113}$$

$$\nabla \times \mathbf{H} = \mathbf{J} - i\omega\mathbf{D}. \tag{1.114}$$

It is desired to calculate the radiation emitted by a monochromatic dipole antenna surrounded by an unbounded chiral medium (Bassiri et al. [26]). Proceeding from the constitutive relations (1.7) and (1.8) and the Maxwell equations, the differential equation for the emitted electric field $\mathbf{E}$ is found to be

$$\nabla \times \nabla \times \mathbf{E} - k^2\mathbf{E} - 2\omega\mu\gamma\nabla \times \mathbf{E} = i\omega\mu\mathbf{J}, \tag{1.115}$$

where $\mathbf{J}$ denotes the current density of the antenna and, as before, $k^2 = \omega^2\mu\varepsilon$. Since this equation is linear, the desired solution can be expressed as the volume integral

$$\mathbf{E}(\mathbf{r}) = i\omega\mu \int \mathbf{\Gamma}(\mathbf{r}, \mathbf{r}')\cdot\mathbf{J}(\mathbf{r}')\, dV', \tag{1.116}$$

where $\mathbf{\Gamma}(\mathbf{r}, \mathbf{r}')$ is the dyadic Green's function, $\mathbf{r} = x\mathbf{e}_x + y\mathbf{e}_y + z\mathbf{e}_z$ is the observation point position vector, and $\mathbf{r}' = x'\mathbf{e}_x + y'\mathbf{e}_y + z'\mathbf{e}_z$ is the source point position vector. The integration with respect to the primed coordinates extends throughout the volume V' occupied by $\mathbf{J}(\mathbf{r}')$. Thus, to carry out the calculations, $\mathbf{\Gamma}$ must first be determined and then integral (1.116) can be evaluated.

To find the Green's function a procedure that was used in the case of simple media and in the case of magnetically biased plasma (Papas [33]) will be followed. When (1.116) is substituted into (1.115), it is seen that $\mathbf{\Gamma}$ must satisfy the differential equation

$$(\nabla^2 + k^2)\mathbf{\Gamma}(\mathbf{r}, \mathbf{r}') + 2\omega\mu\gamma\nabla \times \mathbf{\Gamma}(\mathbf{r}, \mathbf{r}') = -\left(\mathbf{u} + \frac{1}{k^2}\nabla\nabla\right)\delta(\mathbf{r} - \mathbf{r}'), \tag{1.117}$$

where $\mathbf{u}$ is the unit dyadic and $\delta(\mathbf{r} - \mathbf{r}')$ is the Dirac delta function. To solve this equation, it is assumed that $\mathbf{\Gamma}(\mathbf{r}, \mathbf{r}')$ can be written as a Fourier integral, viz.

$$\mathbf{\Gamma}(\mathbf{r}, \mathbf{r}') = \frac{1}{8\pi^3}\int_{-\infty}^{\infty} \mathbf{\Lambda}(\mathbf{p})e^{i\mathbf{p}\cdot(\mathbf{r}-\mathbf{r}')}\, d\mathbf{p}, \tag{1.118}$$

where $\mathbf{p}$ is the position vector and $d\mathbf{p}$ the volume element in p-space, and where

$\mathbf{\Lambda}(\mathbf{p})$ is a dyadic function of $\mathbf{p}$. Also the Dirac delta function can be expressed as

$$\delta(\mathbf{r}-\mathbf{r}') = \frac{1}{8\pi^3}\int_{-\infty}^{\infty} e^{i\mathbf{p}\cdot(\mathbf{r}-\mathbf{r}')}\, d\mathbf{p}. \tag{1.119}$$

It follows from (1.117), (1.118), and (1.119) that the dyadic $\mathbf{\Lambda}(\mathbf{p})$ must satisfy the algebraic equation

$$(k^2-p^2)\mathbf{\Lambda} + 2i\omega\mu\gamma\mathbf{p}\times\mathbf{\Lambda} = -\left(\mathbf{u} - \frac{1}{k^2}\mathbf{pp}\right). \tag{1.120}$$

By dyadic algebra it is found from (1.120) that

$$\begin{aligned}\mathbf{\Lambda} &= \frac{k^2-p^2}{(k^2-p^2)^2+\alpha^2p^2}\left(\frac{1}{k^2}\mathbf{pp}-\mathbf{u}\right)\\ &\quad+\frac{1}{(k^2-p^2)^2+\alpha^2p^2}\left(\alpha\mathbf{p}\times\mathbf{u} - \frac{\alpha^2}{k^2}\mathbf{pp}\right),\end{aligned} \tag{1.121}$$

where $\alpha = 2i\omega\mu\gamma$ and $p^2 = \mathbf{p}\cdot\mathbf{p}$. When (1.121) is substituted into (1.118), the following expression is obtained for the dyadic Green's function in terms of three-dimensional integrals in p-space:

$$\begin{aligned}\mathbf{\Gamma}(\mathbf{r},\mathbf{r}') = &-\frac{1}{8\pi^3}\left[\left(\mathbf{u}+\frac{1}{k^2}\nabla\nabla\right)\int_{-\infty}^{\infty}\frac{k^2-p^2}{(k^2-p^2)^2+\alpha^2p^2}e^{i\mathbf{p}\cdot(\mathbf{r}-\mathbf{r}')}\,d\mathbf{p}\right]\\ &+\frac{1}{8\pi^3}\left[\frac{\alpha^2}{k^2}\nabla\nabla\int_{-\infty}^{\infty}\frac{1}{(k^2-p^2)^2+\alpha^2p^2}e^{i\mathbf{p}\cdot(\mathbf{r}-\mathbf{r}')}\,d\mathbf{p}\right]\\ &-\frac{1}{8\pi^3}\left[i\alpha\nabla\times\left(\mathbf{u}\int_{-\infty}^{\infty}\frac{1}{(k^2-p^2)^2+\alpha^2p^2}e^{i\mathbf{p}\cdot(\mathbf{r}-\mathbf{r}')}\,d\mathbf{p}\right)\right].\end{aligned} \tag{1.122}$$

To reduce the three-dimensional integrals to one-dimensional integrals the spherical coordinates (p, η, ξ), in p-space with polar axis along $\mathbf{R}$ $(= \mathbf{r}-\mathbf{r}')$, are introduced. In these coordinates each of the integrals in (1.122) has the form

$$I = \int_{-\infty}^{\infty} f(p)e^{i\mathbf{p}\cdot\mathbf{R}}\,d\mathbf{p}, \tag{1.123}$$

where $f(p)$ is an even function of p. That is,

$$I = \int_0^{\infty} p^2 f(p)\,dp\int_0^{\pi} e^{ipR\cos\eta}\sin\eta\,d\eta\int_0^{2\pi} d\xi, \tag{1.124}$$

where $R = \sqrt{\mathbf{R}\cdot\mathbf{R}}$. Clearly, this three-dimensional integral reduces to the one-dimensional integral

$$I = \frac{4\pi}{R}\int_0^{\infty} pf(p)\sin pR\,dp, \tag{1.125}$$

which, since $f(p)$ is an even function, can be written as

$$I = \frac{2\pi}{R}\int_{-\infty}^{\infty} pf(p)\sin pR\,dp$$

$$= \frac{2\pi}{iR}\int_{-\infty}^{\infty} pf(p)e^{ipR}\,dp. \tag{1.126}$$

From (1.123) and (1.126) it follows that

$$\int_{-\infty}^{\infty} f(p)e^{i\mathbf{p}\cdot\mathbf{R}}\,d\mathbf{p} = \frac{2\pi}{iR}\int_{-\infty}^{\infty} pf(p)e^{ipR}\,dp. \tag{1.127}$$

Using (1.127), eq. (1.122) can be written as

$$\begin{aligned}\boldsymbol{\Gamma}(\mathbf{r},\mathbf{r}') = &-\frac{1}{4\pi^2 iR}\left(\mathbf{u} + \frac{1}{k^2}\nabla\nabla\right)\int_{-\infty}^{\infty}\frac{p(k^2-p^2)}{(k^2-p^2)^2+\alpha^2p^2}e^{ipR}\,dp\\ &+\frac{1}{4\pi^2 iR}\frac{\alpha^2}{k^2}\nabla\nabla\int_{-\infty}^{\infty}\frac{p}{(k^2-p^2)^2+\alpha^2p^2}e^{ipR}\,dp\\ &-\frac{\alpha}{4\pi^2 R}\nabla\times\left(\mathbf{u}\int_{-\infty}^{\infty}\frac{p}{(k^2-p^2)^2+\alpha^2p^2}e^{ipR}\,dp\right).\end{aligned} \tag{1.128}$$

The integrals in this expression are one dimensional and can be evaluated by contour integration.

1.4.1.1. Radiation Condition

To evaluate the integrals in (1.128) by contour integration (theorem of residues) it is noted that $p = \pm h_1$ and $p = \pm h_2$ are the roots of

$$(k^2 - p^2)^2 + \alpha^2p^2 = 0. \tag{1.129}$$

Since $\alpha = 2i\omega\mu\gamma$, it is seen from (1.129) that

$$h_1 = \omega\mu\gamma + \sqrt{\omega^2\mu^2\gamma^2 + k^2}, \tag{1.130}$$

$$h_2 = -\omega\mu\gamma + \sqrt{\omega^2\mu^2\gamma^2 + k^2}. \tag{1.131}$$

Accordingly, the poles of the integrands in (1.128) lie along the real axis of the complex p-plane at $\pm h_1$ and $\pm h_2$. Expressions (1.130) and (1.131) show that for $\gamma > 0$, $h_1 > k > h_2 > 0$, and that for $\gamma < 0$, $h_2 > k > h_1 > 0$.

The path of integration for $\gamma > 0$ is shown in Figure 1.3. The contour integration consists of an integration along the real axis of the complex p-plane, with upward indentation at the poles $p = -h_2$ and $p = -h_1$ and downward indentation at the poles $p = h_2$ and $p = h_1$, and of an integration along the semicircle at infinity. The factor $\exp(ipR)$ makes the contribution of the integration along the semicircle disappear. The value of $\boldsymbol{\Gamma}$, as given by (1.128), is undetermined unless the manner of going around the poles is

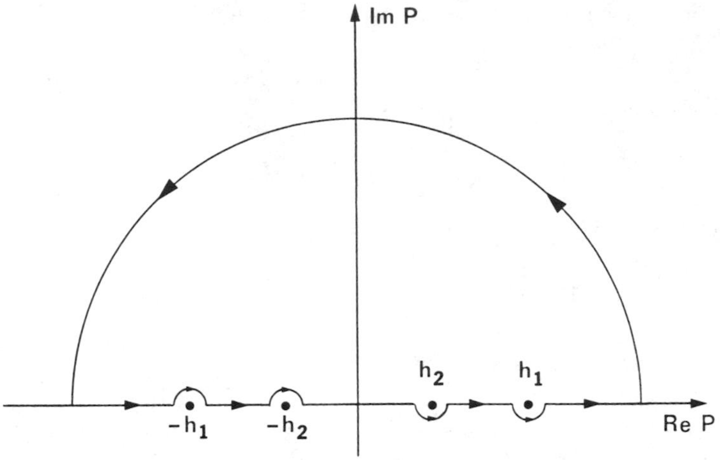

FIGURE 1.3. Path of integration in complex p-plane for a right-handed ($\gamma > 0$) chiral medium. The poles are located on the real axis and their values are given by eqs. (1.130) and (1.131).

specified. Here, the chosen path yields a description of $\mathbf{\Gamma}$ in terms of only outgoing waves and thus makes $\mathbf{\Gamma}$ satisfy the physically required radiation condition.

According to the theorem of residues, the integrals in (1.128) are equal to $2\pi i$ multiplied by the sum of residues at the poles $p = h_1$ and $p = h_2$. With the aid of this theorem the following equations are obtained:

$$\int_{-\infty}^{\infty} \frac{p(k^2 - p^2)}{(k^2 - p^2)^2 + \alpha^2 p^2} e^{ipR}\, dp = 2\pi i \left[\frac{(k^2 - h_1^2)e^{ih_1 R} - (k^2 - h_2^2)e^{ih_2 R}}{2(h_1^2 - h_2^2)} \right], \tag{1.132}$$

$$\int_{-\infty}^{\infty} \frac{p}{(k^2 - p^2)^2 + \alpha^2 p^2} e^{ipR}\, dp = 2\pi i \left[\frac{e^{ih_1 R} - e^{ih_2 R}}{2(h_1^2 - h_2^2)} \right]. \tag{1.133}$$

Finally, when (1.132) and (1.133) are substituted into (1.128) it is found that the desired three-dimensional dyadic Green's function is given by (Bassiri et al. [26])

$$\mathbf{\Gamma}(\mathbf{r}, \mathbf{r}') = a\left(\mathbf{u} + \frac{1}{h_1}\mathbf{u} \times \nabla + \frac{1}{h_1^2}\nabla\nabla\right)\frac{e^{ih_1 R}}{4\pi R} + b\left(\mathbf{u} - \frac{1}{h_2}\mathbf{u} \times \nabla + \frac{1}{h_2^2}\nabla\nabla\right)\frac{e^{ih_2 R}}{4\pi R}, \tag{1.134}$$

where

$$a = \frac{h_1^2 - k^2}{h_1^2 - h_2^2}, \qquad b = \frac{k^2 - h_2^2}{h_1^2 - h_2^2}. \tag{1.135}$$

This is the solution of (1.117) that satisfies the radiation condition and is valid for all γ (positive and negative). It involves two modes of propagation, one being an outgoing wave of phase velocity ω/h_1, and the other an outgoing wave of phase velocity ω/h_2. For $\gamma > 0$ the former is the slower mode, whereas for $\gamma < 0$ the latter is the slower mode.

Let **J** in eq. (1.116) be a function of two space variables. With no loss of generality, these two space variables can be taken to be x' and y'. Therefore, $\mathbf{J}(\mathbf{r}') = \mathbf{J}(\boldsymbol{\rho}') = \mathbf{J}(x', y')$. The two-dimensional dyadic Green's function can be written as

$$\boldsymbol{\Gamma}(\boldsymbol{\rho}, \boldsymbol{\rho}') = a\left(\mathbf{u} + \frac{1}{h_1}\mathbf{u} \times \nabla + \frac{1}{h_1^2}\nabla\nabla\right)\frac{i}{4}H_0^{(1)}(h_1|\boldsymbol{\rho} - \boldsymbol{\rho}'|)$$

$$+ b\left(\mathbf{u} - \frac{1}{h_2}\mathbf{u} \times \nabla + \frac{1}{h_2^2}\nabla\nabla\right)\frac{i}{4}H_0^{(1)}(h_2|\boldsymbol{\rho} - \boldsymbol{\rho}'|), \quad (1.136)$$

where $\boldsymbol{\rho} = x\mathbf{e_x} + y\mathbf{e_y}$, $\boldsymbol{\rho}' = x'\mathbf{e_x} + y'\mathbf{e_y}$, and $H_0^{(1)}(\cdot)$ is the zeroth-order Hankel function of the first kind.

When the electric current density **J** in eq. (1.116) is only a function of one space variable, say $\mathbf{J}(\mathbf{r}') = \mathbf{J}(x')$, the one-dimensional Green's function can be written as

$$\boldsymbol{\Gamma}(x, x') = a\left(\mathbf{u} + \frac{1}{h_1}\mathbf{u} \times \nabla + \frac{1}{h_1^2}\nabla\nabla\right)\frac{i}{2h_1}e^{ih_1|x-x'|}$$

$$+ b\left(\mathbf{u} - \frac{1}{h_2}\mathbf{u} \times \nabla + \frac{1}{h_2^2}\nabla\nabla\right)\frac{i}{2h_2}e^{ih_2|x-x'|}, \quad (1.137)$$

where a and b are given by eq. (1.135). The derivations of one- and two-dimensional dyadic Green's functions are similar to the derivation of the three-dimensional dyadic Green's function treated here (Engheta and Bassiri [36]).

The dyadic Green's functions (1.134), (1.136), and (1.137) satisfy the following reciprocity properties (Tai [37]):

$$\boldsymbol{\Gamma}(\mathbf{r}, \mathbf{r}') = [\boldsymbol{\Gamma}(\mathbf{r}', \mathbf{r})]^{\mathrm{T}}, \quad (1.138)$$

$$\nabla \times \boldsymbol{\Gamma}(\mathbf{r}, \mathbf{r}') = [\nabla' \times \boldsymbol{\Gamma}(\mathbf{r}', \mathbf{r})]^{\mathrm{T}}, \quad (1.139)$$

where the superscript T denotes the transpose of the dyadic. These reciprocity properties imply that the transmitting and receiving patterns of an antenna in the medium would be the same.

1.4.2. *Dipole Radiation*

Now consider an oscillating dipole antenna, of length l, located in the chiral medium. It is assumed that the antenna lies at the origin of a Cartesian coordinate system (x, y, z) and is directed parallel to the z-axis. It is convenient to introduce a spherical coordinate system (r, θ, φ) where $x = r \sin\theta \cos\varphi$,

$y = r \sin\theta \sin\varphi$, and $z = r\cos\theta$, because, as will be seen later, the field of the antenna is composed of spherical waves.

The current density of the antenna is assumed to be

$$\mathbf{J} = \mathbf{e_z} I_0 \delta(x')\delta(y')\delta(z') = \mathbf{e_z} Il\delta(\mathbf{r}'), \tag{1.140}$$

where $\mathbf{e_z}$ is the unit vector in the z direction and I is the antenna current. When this expression is substituted into (1.116) it is seen that the electric field emitted by the antenna can be obtained from

$$\mathbf{E}(\mathbf{r}) = i\omega\mu Il \boldsymbol{\Gamma}(\mathbf{r}, 0)\cdot \mathbf{e_z}, \tag{1.141}$$

where $\boldsymbol{\Gamma}(\mathbf{r}, 0)$ is given by (1.134) with R replaced by r.

From (1.134) and (1.141), the far-zone part or the radiation field of $\mathbf{E}$ which varies as $1/r$, the induction part of $\mathbf{E}$ which changes with distance as $(1/r^2)$, and the near-zone part of $\mathbf{E}$ which varies as $(1/r^3)$ can be found. The complete electric field can be written as (Bassiri [32])

$$\mathbf{E} = \mathbf{E}_{(1/r)} + \mathbf{E}_{(1/r^2)} + \mathbf{E}_{(1/r^3)}, \tag{1.142}$$

where

$$\mathbf{E}_{(1/r)} = \frac{-iZIl}{8\pi r} \sin\theta[h_1(\mathbf{e}_\theta + i\mathbf{e}_\varphi)e^{ih_1 r} + h_2(\mathbf{e}_\theta - i\mathbf{e}_\varphi)e^{ih_2 r}], \tag{1.143}$$

$$\mathbf{E}_{(1/r^2)} = \frac{ZIl}{8\pi r^2}[\sin\theta\{(\mathbf{e}_\theta + i\mathbf{e}_\varphi)e^{ih_1 r} + (\mathbf{e}_\theta - i\mathbf{e}_\varphi)e^{ih_2 r}\} + 2\cos\theta\{e^{ih_1 r} + e^{ih_2 r}\}\mathbf{e}_r], \tag{1.144}$$

and

$$\mathbf{E}_{(1/r^3)} = \frac{iZIl}{8\pi r^3}\left[(\sin\theta\mathbf{e}_\theta + 2\cos\theta\mathbf{e}_r)\left(\frac{e^{ih_1 r}}{h_1} + \frac{e^{ih_2 r}}{h_2}\right)\right]. \tag{1.145}$$

where $\mathbf{e}_\theta$ is the unit vector in the θ direction, $\mathbf{e}_\varphi$ is the unit vector in the φ direction, and $\mathbf{e}_r$ is the unit vector in the r direction.

1.4.2.1. Polarization and Wave Impedance

The electric field vector $\mathbf{E}$ of the dipole antenna's radiation field, eq. (1.143), can be written as

$$\mathbf{E} = \frac{-i\omega\mu I_0 \sin\theta}{4\pi r}[a(\mathbf{e}_\theta + i\mathbf{e}_\varphi)e^{ih_1 r} + b(\mathbf{e}_\theta - i\mathbf{e}_\varphi)e^{ih_2 r}]. \tag{1.146}$$

Taking $(-i\omega\mu I_0 \sin\theta/4\pi r)$ as a normalization factor, the quantity in the square brackets is considered to be descriptive of the state of polarization (the polarizaton ellipse) of the electric field. The first term in the square brackets describes a right-handed circularly polarized wave of amplitude a and phase velocity ω/h_1; and the second term describes a left-handed circularly polarized wave of amplitude b and phase velocity ω/h_2. From (1.130), (1.131), and (1.135)

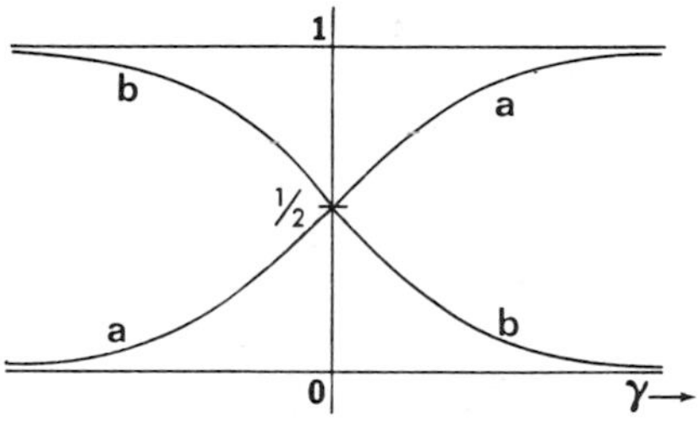

FIGURE 1.4. Sketch of amplitudes a and b versus γ, where a is the amplitude of a right-handed circularly polarized wave and b is the amplitude of a left-handed circularly polarized wave, and where γ is the measure of chirality.

it is clear that $a = b = \frac{1}{2}$ when $\gamma = 0$; that $a = 0$, $b = 1$ when $\gamma \to -\infty$; and that $a = 1$, $b = 0$ when $\gamma \to \infty$.

From Figure 1.4, which shows a sketch of a and b versus γ, it is seen that $a > b$ when $\gamma > 0$ and that $b > a$ when $\gamma < 0$. This means that the polarization ellipse of **E** has a right-handed sense of rotation when $\gamma > 0$ and a left-handed sense of rotation when $\gamma < 0$, and that the twist of the major axis of the ellipse is to the left when $\gamma > 0$ and to the right when $\gamma < 0$. The lengths of the semimajor and semiminor axes of the ellipse are given by $(a + b)$ and $|a - b|$, respectively. When (1.130) and (1.131) are recalled, it is seen from (1.135) that

$$\text{length of semimajor axis} = a + b = 1, \tag{1.147}$$

$$\text{length of semiminor axis} = |a - b| = \frac{\omega\mu|\gamma|}{\sqrt{\omega^2\mu^2\gamma^2 + k^2}}. \tag{1.148}$$

As $\gamma \to 0$, the ellipse gets thin and the polarization becomes linear; as $|\gamma| \to \infty$, the ellipse gets thick and the polarization becomes circular. The orientation angle ψ of the ellipse is given by

$$\psi_E = \frac{h_2 - h_1}{2} r = -\omega\mu\gamma r. \tag{1.149}$$

The magnetic vector **H** of the dipole's radiation field can be obtained by substituting (1.146) into

$$\mathbf{H} = i\gamma\mathbf{E} + \frac{1}{\mu}\mathbf{B} = i\gamma\mathbf{E} + \frac{1}{i\omega\mu}\nabla \times \mathbf{E}, \tag{1.150}$$

and retaining only terms in $1/r$. Thus, this field can be written as

$$\mathbf{H} = -\frac{i\omega\mu I_0 \sin\theta}{4\pi r}\sqrt{\gamma^2 + \frac{\varepsilon}{\mu}} \cdot [a(\mathbf{e}_\theta + i\mathbf{e}_\varphi)e^{ih_1 r}e^{i3\pi/2} + b(\mathbf{e}_\theta - i\mathbf{e}_\varphi)e^{ih_2 r}e^{i\pi/2}]. \tag{1.151}$$

The orientation angle ψ for the magnetic field is given by

$$\psi_H = \frac{h_2 - h_1}{2} r - \frac{\pi}{2} = -\omega\mu\gamma r - \frac{\pi}{2}. \tag{1.152}$$

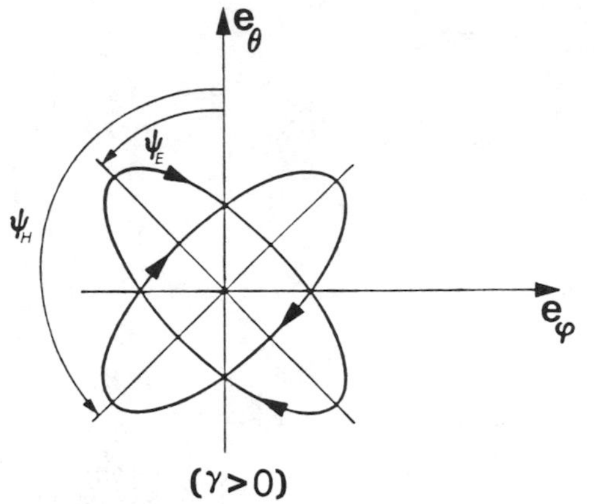

FIGURE 1.5. Polarization ellipses for **E** and **H** fields. These polarization ellipses are perpendicular to each other and have the same sense of rotation.

The polarization ellipse of **H** is similar and perpendicularly oriented to the polarization ellipse of **E**. That is, the polarization ellipses of **E** and **H** are perpendicular to each other and have the same sense of rotation, as shown in Figure 1.5.

When the notation $\mathbf{E}(t) = \mathrm{Re}(\mathbf{E}e^{-i\omega t})$ and $\mathbf{H}(t) = \mathrm{Re}(\mathbf{H}e^{-i\omega t})$ is used, it is seen from (1.146) and (1.151) that at any point (r, θ, φ) and at any time t the vectors $\mathbf{E}(t)$ and $\mathbf{H}(t)$ are perpendicular to each other and to the direction of propagation. Specifically,

$$\mathbf{e}_r \times \mathbf{E}(t) = \frac{1}{\sqrt{\gamma^2 + (\varepsilon/\mu)}}\mathbf{H}(t). \tag{1.153}$$

It follows from this relation that the wave impedance Z of the medium is given by

$$Z = \sqrt{\frac{\mu}{\varepsilon}}\frac{1}{\sqrt{1 + (\mu/\varepsilon)\gamma^2}}, \tag{1.154}$$

which shows that as the medium becomes more chiral the wave impedance decreases.

1.4.2.2. RADIATED POWER

From Maxwell's equations and the divergence theorem, the Poynting theorem can be written as

$$\tfrac{1}{2}\,\mathrm{Re}\int \mathbf{n}\cdot(\mathbf{E}\times\mathbf{H}^*)\,dS = -\tfrac{1}{2}\,\mathrm{Re}\int \mathbf{J}^*\cdot\mathbf{E}\,dV + \tfrac{1}{2}\,\mathrm{Re}\left[i\omega\int(\mathbf{B}\cdot\mathbf{H}^* - \mathbf{D}^*\cdot\mathbf{E})\,dV\right]. \tag{1.155}$$

The volume integrations are taken throughout a region bounded by a far-zone sphere centered on the antenna. The surface integration is taken over the far-zone sphere whose outward normal vector is **n**. From the constitutive relations (1.7) and (1.8) it follows that the integrand $\mathbf{B}\cdot\mathbf{H}^* - \mathbf{D}^*\cdot\mathbf{E}$ is purely real and hence the second term on the right is zero. The first term on the right side, by definition, is the time-average radiated power input. Consequently, the time-average radiated power P is expressed by the quantity on the left side. That is,

$$P = \tfrac{1}{2}\,\mathrm{Re}\int \mathbf{n}\cdot(\mathbf{E}\times\mathbf{H}^*)\,dS, \tag{1.156}$$

When (1.146) and (1.151) are substituted into (1.156), it is found that the radiated power is given by

$$P = \frac{\omega^2\mu^2 I^2 l^2}{12\pi}\,\frac{2\gamma^2 + (\varepsilon/\mu)}{\sqrt{\gamma^2 + (\varepsilon/\mu)}}. \tag{1.157}$$

When the radiation resistance R is defined by $P = \frac{1}{2}I^2R$, it is seen from (1.157) that

$$R = \frac{\omega^2\mu^2 l^2}{6\pi}\,\frac{2\gamma^2 + (\varepsilon/\mu)}{\sqrt{\gamma^2 + (\varepsilon/\mu)}}. \tag{1.158}$$

This shows that the radiation resistance increases as γ, or as the medium, becomes more chiral.

1.5. Conclusions

In this chapter, after a discussion on the propagation of electromagnetic waves in an unbounded chiral medium, the problem of reflection from and transmission through a semi-infinite chiral medium are analyzed by obtaining the Fresnel equations. Also, the conditions for total internal reflection and for the Brewster angle are obtained. The problem of electromagnetic wave propagation through an infinite slab of chiral medium is formulated for oblique incidence, and is solved analytically for the case of normal incidence. It is shown that, in this case, the plane of polarization of the transmitted wave is rotated with respect to that of the incident wave, and the plane of polarization of the reflected wave is unchanged with respect to that of the incident wave.

The one-, two-, and three-dimensional dyadic Green's functions for an unbounded chiral medium described by the constitutive relations (1.7) and (1.8) are derived. It is shown that the three-dimensional Green's function involves two spherical waves $\exp[(ih_1R)]/4\pi R$ and $\exp[(ih_2R)]/4\pi R$, which place in evidence that the medium supports double-mode propagation when $\gamma \neq 0$ and single-mode propagation when $\gamma = 0$. For a right-handed ($\gamma > 0$) medium the former wave is "slow" and the latter wave is "fast"; and for a left-handed ($\gamma < 0$) medium the converse is true. These waves are considered slow and fast in comparison to the velocity they have when $\gamma = 0$; that is, when the medium is achiral and the double-mode propagation is reduced to single-mode propagation.

In the case of a dipole antenna, it is found that the chirality of the ambient medium decreases the wave impedance of the medium, increases the radiation resistance of the dipole, but has no effect on the directivity of the radiation. The dominant effect of the chirality is the change it produces in the state of polarization of the dipole's field. However, although the directivity of a dipole is not changed by chirality, it cannot be concluded that the directivity of an antenna (that is not small compared to the wavelengths $2\pi/h_1$ and $2\pi/h_2$) is not changed by chirality. For example, if a two-element array is placed in a chiral medium, the directivity of the array would change with the chirality of the medium.

Due to mode-multiplicity, it appears that chirality could play an important role in the design of sizable antennas in general and antenna arrays in particular.

Acknowledgments

The author wishes to thank Professors H.N. Kritikos and D.L. Jaggard of the University of Pennsylvania for their diligence in editing this book. Appreciation is also expressed to Dr. T.P. Yunck of the Jet Propulsion Laboratory for his kind assistance and advice.

References

[1] D.F. Arago (1811), Sur une modification remarquable qu'éprouvent les rayons lumineux dans leur passage à travers certains corps diaphanes, et sur quelques autres nouveaux phénomènes d'optique, *Mém. Inst.*, **1**, 93–134.

[2] J.B. Biot (1812), Mémoire sur un nouveau genre d'oscillations que les molécules de la lumière éprouvent, en traversant certains cristaux, *Mém. Inst.*, **1**, 1–372.

[3] J.B. Biot (1817), Sur les rotations que certaines substances impriment aux axes de polarisation des rayons lumineux, *Mém. Acad. Sci.*, **2**, 41–136.

[4] J.B. Biot (1835), Mémoire sur la polarisation circulaire et sur ses applications à la chimie organique, *Mém. Acad. Sci.*, **13**, 39–175.

[5] J.B. Biot (1815), Phénomènes de polarisation successive, observés dans des fluides homogènes, *Bull. Soc. Philomat.*, 190–192.

[6] A. Fresnel (1822), Mémoire sur la double réfraction que les rayons lumineux éprouvent en traversant les aiguilles de cristal de roche suivant des directions parallèles à l'axe, *Oeuvres*, **1**, 731–751.

[7] A. Fresnel (1866), *Oeuvres Complètes*, Vol. 2, Imprimerie Impériale, Paris.

[8] L. Pasteur (1848), Sur les relations qui peuvent exister entre la forme cristalline, la composition chimique et le sens de la polarisation rotatoire, *Ann. Chim. Phys.*, **24**, 442–459.

[9] L. Pasteur (1860), Researches on the molecular asymmetry of natural organic products, Alembic Club Reprint No. 14, Livingston, Edinburgh and London.

[10] K.F. Lindman (1920), Über eine durch ein isotropes System von Spiralförmigen Resonatoren erzeugte Rotationspolarisation der elektromagnetischen Wellen, *Ann. Physik*, **63**, 621–644.

[11] K.F. Lindman (1922), Über die durch ein aktives Raumgitter erzeugte Rotationspolarisation der elektromagnetischen Wellen, *Ann. Physik*, **69**, 270–284.

[12] T.M. Lowry (1964), *Optical Rotatory Power*, Dover, New York.

[13] M. Born (1915), Über die natürliche optische Aktivität von Flüssigkeiten und Gasen, *Phys. Z.*, **16**, 251–258.

[14] C.W. Oseen (1915), Über die Wechselwirkung zwischen zwei elektrischen Dipolen und über in Kristallen und Flüssigkeiten, *Ann. Physik*, **48**, 1–56.

[15] F. Gray (1916), The optical activity of liquids and gases, *Phys. Rev.*, **7**, 472–488.

[16] W. Kuhn (1929), Quantitative Verhältnisse und Beziehungen bei der natürlichen optischen Aktivität, *Z. Phys. Chemie*, **B4**, 14–36.

[17] E.U. Condon, W. Altar, and H. Eyring (1937), One-electron rotatory power, *J. Chem. Phys.*, **5**, 753–775.

[18] E.U. Condon (1937), Theories of optical rotatory power, *Rev. Mod. Phys.*, **9**, 432–457.

[19] B.V. Bokut and F.I. Federov (1960), Reflection and refraction of light in optically isotropic active media, *Opt. Spektrosk.*, **9**, 334–336.

[20] C.F. Bohren (1974), Light scattering by an optically active sphere, *Chem. Phys. Lett.*, **29**, 458–462.

[21] C.F. Bohren (1975), Scattering of electromagnetic waves by an optically active cylinder, *J. Colloid. Interface. Sci.*, **66**, 105–109.

[22] J.A. Kong (1975), *Theory of Electromagnetic Waves*, Wiley–Interscience, New York.

[23] D.L. Jaggard, A.R. Mickelson, and C.H. Papas (1979), On electromagnetic waves in chiral media, *Appl. Phys.*, **18**, 211–216.

[24] N. Engheta and A.R. Mickelson (1982), Transition radiation caused by a chiral plate, *IEEE Trans. Antennas and Propagation*, **AP-30**, 1213–1216.

[25] M.P. Silverman (1986), Reflection and refraction at the surface of a chiral medium comparison of gyrotropic constitutive relations invariant or non-invariant under a duality transformation, *J. Opt. Soc. Amer. A*, **3**, 830–837.

[26] S. Bassiri, N. Engheta, and C.H. Papas (1986), Dyadic Green's function and dipole radiation in chiral media, *Alta Frequenza*, **LV-No. 2**, 83–88. Also Caltech Antenna Laboratory Report, No. 119, 1985, California Institute of Technology, Pasadena, CA.

[27] D.L. Jaggard, X. Sun, and N. Engheta (1988), Canonical sources and duality in chiral media, *IEEE Trans. Antennas and Propagation*, **AP-36**, 1007–1013.

[28] S. Bassiri, C.H. Papas, and N. Engheta (1988), Electromagnetic wave propagation through a dielectric–chiral interface and through a chiral slab, *J. Opt. Soc. Amer. A*, **5**, 1450–1459.

[29] E.J. Post (1962), *Formal Structure of Electromagnetics*, North-Holland, Amsterdam.

[30] J.A. Stratton (1941), *Electromagnetic Theory*, McGraw-Hill, New York.

[31] A. Sommerfeld (1954), *Optics*, Academic Press, New York.

[32] S. Bassiri (1987), Electromagnetic wave propagation and radiation in chiral media, Ph.D. dissertation. Also Caltech Antenna Laboratory Report, No. 121, California Institute of Technology, Pasadena, CA.

[33] C.H. Papas (1965), *Theory of Electromagnetic Wave Propagation*, McGraw-Hill, New York.

[34] M. Born and E. Wolf (1980), *Principles of Optics*, Pergamon Press, New York.

[35] H.C. Chen (1983), *Theory of Electromagnetic Waves*, McGraw-Hill, New York.

[36] N. Engheta and S. Bassiri (1989), One- and two-dimensional dyadic Green's functions in chiral media, *IEEE Trans. Antennas and Propagation*, **AP-37**, 512–515.

[37] C-T. Tai (1971), *Dyadic Green's Functions in Electromagnetic Theory*, International Textbook, Scranton, NJ.

2
Norms of Time-Domain Functions and Convolution Operators

CARL E. BAUM*

2.1. Introduction

This chapter develops various norms of time-domain functions and convolution operators to obtain bounds for transient system response. Besides the usual p-norm we can define another norm, the residue norm (or r-norm), based on the singularities in the complex-frequency (or Laplace-transform) plane.

In describing the interaction of electromagnetic fields with complex systems, electromagnetic topology is used to organize the system into parts which can be analyzed separately and the results subsequently combined [1], [2], [3]. Due to the complexity of the problem it is often more practical to look for bounds on the electromagnetic response instead of "exact" answers. In this context, the use of norms has been developed to obtain rigorous bounds which are in some sense also "tight" bounds. For this purpose, several papers discuss the use of vector norms and associated matrix norms of wave variables and associated scattering matrices, primarily in frequency domain [1], [4], [5], [6], [2].

More recently, these norm considerations have been extended into time domain [7], [8]. In this form the transient waveforms are characterized by positive scalars related to appropriate system response parameters. This chapter considers the p-norms of transient waveforms and the associated operator norms for time-domain convolution operators (what transfer functions become in time domain). Another type of norm called the residue norm is defined based on the singularities in the complex-frequency (or Laplace-transform) plane, and is related to the p-norm.

* Air Force Weapons Laboratory (NTAAB), Kirtland AFB, NM 87117-6008, USA.

2.2. Norms of Time-Domain Functions and Operators

Norms of vectors have been defined by the properties [9]

$$\begin{aligned}
&\|(x_n)\| \begin{cases} = 0 & \text{iff } (x_n) = (O_n), \\ > 0 & \text{otherwise,} \end{cases} \\
&\|\alpha(x_n)\| = |\alpha| \, \|(x_n)\|, \qquad \alpha \equiv \text{a complex scalar,} \\
&\|(x_n) + (y_n)\| \le \|(x_n)\| + \|(y_n)\| \quad \text{(triangle inequality),} \\
&\|(x_n)\| \text{ depends continuously on } (x_n).
\end{aligned} \tag{2.1}$$

Associated matrix norms are defined by

$$\|(A_{n,m})\| \equiv \sup_{(x_n) \neq (O_n)} \frac{\|(A_{n,m}) \cdot (x_n)\|}{\|(x_n)\|}, \tag{2.2}$$

where the matrices are allowed to be rectangular as long as there is a compatibility of numbers of rows and columns to allow dot multiplication. These matrix norms have the properties

$$\begin{aligned}
&\|(A_{n,m})\| \begin{cases} = 0 & \text{iff } (A_{n,m}) = (O_{n,m}), \\ > 0 & \text{otherwise,} \end{cases} \\
&\|\alpha(A_{n,m})\| = |\alpha| \, \|(A_{n,m})\|, \\
&\|(A_{n,m}) + (B_{n,m})\| \le \|(A_{n,m})\| + \|(B_{n,m})\| \quad \text{(triangle inequality),} \\
&\|(A_{n,m}) \cdot (B_{n,m})\| \le \|(A_{n,m})\| \, \|(B_{n,m})\| \quad \text{(Schwarz inequality),} \\
&\|(A_{n,m})\| \text{ depends continuously on } (A_{n,m}).
\end{aligned} \tag{2.3}$$

Here $(A_{n,m})$ and $(B_{n,m})$ are general rectangular matrices as long as they are compatible for addition and/or dot multiplication as required.

The vector norms are generalized to functional norms in the sense that a vector of infinitely many components (an infinite-dimensioned vector space) can be considered as a function of a real variable (taken as t (or time) in this case, but could represent other kinds of parameters). These have the properties [7], [10]

$$\begin{aligned}
&\|f(t)\| \begin{cases} = 0 & \text{iff } f(t) \equiv 0 \text{ or has zero “measure” per the particular norm,} \\ > 0 & \text{otherwise,} \end{cases} \\
&\|\alpha f(t)\| = |\alpha| \, \|f(t)\|, \qquad \alpha \equiv \text{a complex scalar,} \\
&\|f(t) + g(t)\| \le \|f(t)\| + \|g(t)\|.
\end{aligned} \tag{2.4}$$

In a manner similar to associated matrix norms we can define associated operator norms via

$$\|\Lambda(\;)\| = \sup_{f(t) \neq 0} \frac{\|\Lambda(f(t))\|}{\|f(t)\|}, \tag{2.5}$$

where now Λ is an operator which operates on a function to produce another function, say, as

$$F(t) = \Lambda(f(t)). \tag{2.6}$$

Here Λ can include integration, differentiation, or any kind of *linear* operation that results in the following properties of an operator norm:

$$\|\Lambda(\ )\| \begin{cases} = 0 & \text{iff } \Lambda(\ ) \equiv 0 \text{ or has zero "measure" per the particular norm,} \\ > 0 & \text{otherwise,} \end{cases}$$

$$\|\alpha\Lambda(\ )\| = |\alpha|\,\|\Lambda(\ )\|, \tag{2.7}$$

$$\|\Lambda(\ ) + \|\Upsilon(\ )\| \le \|\Lambda(\ )\| + \|\Upsilon(\ )\|,$$

$$\|\Lambda(\Upsilon(\ ))\| \le \|\Lambda(\ )\|\,\|\Upsilon(\ )\|.$$

Note that Λ and Υ are required to operate over the same range of t on which $f(t)$ is defined.

We should be careful in considering the norms of functions and operators. There are considerations of continuity, continuous derivatives, integrability, etc., which may need to be considered, depending on the particular norm under consideration. Note here that while we are defining function and operator norms as a natural extension of vector and matrix norms, this concept can be further generalized to vectors of functions, matrices of operators, functions of several variables and associated operators, etc.

There are various kinds of operators of physical interest. A very important one is the convolution operator (with respect to time) that characterizes linear, time-invariant systems. We symbolize this special operator by $g(t) \circ$ where

$$\begin{aligned} F(t) &= g(t) \circ f(t) \\ &= \int_{-\infty}^{\infty} g(t - t')f(t')\,dt' \\ &= \int_{-\infty}^{\infty} g(t')f(t - t')\,dt'. \end{aligned} \tag{2.8}$$

If (as we normally do) we assume that $g(t) \circ$ is causal (no response before an excitation), then

$$g(t) = 0 \qquad \text{for} \quad t < 0 \tag{2.9}$$

and

$$\begin{aligned} F(t) &= \int_{-\infty}^{t} g(t - t')f(t')\,dt' \\ &= \int_{0}^{\infty} g(t')f(t - t')\,dt'. \end{aligned} \tag{2.10}$$

The concept of convolution is closely related to the Laplace transform (two-sided) defined by

$$\tilde{f}(s) \equiv \int_{-\infty}^{\infty} f(t)e^{-st}\,dt,$$
$$f(t) = \frac{1}{2\pi j}\int_{\mathrm{Br}} \tilde{f}(s)e^{st}\,ds. \tag{2.11}$$

Br ≡ the Bromwich contour in the strip of convergence of the Laplace transform (parallel to the $j\omega$-axis) and $s = \Omega + j\omega$.

In terms of the Laplace transform, (2.8) (or (2.10)) becomes

$$\tilde{F}(s) = \tilde{g}(s)\tilde{f}(s), \tag{2.12}$$

so that convolution in time domain becomes multiplication in the complex-frequency (or Laplace or Fourier) domain

Still in time domain, convolution is an operation for which we consider a general convolution operator $g(t)\circ$ which is to be distinguished from the function $g(t)$. Following (2.5) we define the associated norm of a convolution operator as

$$\|g(t)\circ\| \equiv \sup_{f(t)\neq 0} \frac{\|g(t)\circ f(t)\|}{\|f(t)\|}, \tag{2.13}$$

which depends on the particular function norm chosen (and which requires that $f(t)$ be limited to functions for which such a norm exists). Note that, in general, the norm of the operator $g(t)\circ$ is not the same as the norm of the function $g(t)$.

2.3. p-Norm of Time-Domain Convolution Operators

The p-norm of time-domain waveforms is [7]

$$\|f(t)\|_p \equiv \left\{\int_{-\infty}^{\infty} |f(t)|^p\,dt\right\}^{1/p}, \qquad 1 \le p < \infty, \tag{2.14}$$

with a special case of the ∞-norm as

$$\|f(t)\|_\infty = \sup_{-\infty<t<\infty} |f(t)|, \tag{2.15}$$

where isolated values of $f(t)$ are excluded by considering limits from both sides of the values of $f(t)$ of concern.

Considering a convolution operator $g(t)\circ$, as in (2.2), we have from (2.13) and (2.10), for p-norms,

$$\|g(t)\circ\|_p = \sup_{f(t)\neq 0} \frac{\|g(t)\circ f(t)\|_p}{\|f(t)\|_p}$$
$$= \frac{\{\int_{-\infty}^{\infty} |\int_0^{\infty} g(t')f(t-t')\,dt'|^p\,dt\}^{1/p}}{\|f(t)\|_p}, \tag{2.16}$$

where $g(t)\circ$ has been assumed to be causal. Apply the Hölder inequality (Appendix A) with

$$\begin{aligned} f_1(t') &= |g(t')|^{(p-1)/p}, \\ f_2(t') &= |g(t')|^{1/p}|f(t-t')|, \\ p_1 &= \left[1-\frac{1}{p}\right]^{-1} = \frac{p}{p-1}, \\ p_2 &= p. \end{aligned} \tag{2.17}$$

Then

$$\left|\int_0^\infty g(t')f(t-t')\,dt'\right| \le \left\{\int_0^\infty |g(t')|\,dt'\right\}^{(p-1)/p} \left\{\int_0^\infty |g(t')|\,|f(t-t')|^p\,dt'\right\}^{1/p}. \tag{2.18}$$

Integrating over t of the pth power of the above (as in (2.16)) gives

$$\begin{aligned} &\int_{-\infty}^\infty \left|\int_0^\infty g(t')f(t-t')\,dt'\right|^p dt \\ &\le \int_{-\infty}^\infty \left\{\int_0^\infty |g(t')|\,dt'\right\}^{p-1} \left\{\int_0^\infty |g(t')|\,|f(t-t')|^p\,dt'\right\} dt \\ &= \int_{-\infty}^\infty \|g(t')\|_1^{p-1} \left\{\int_0^\infty |g(t')|\,|f(t-t')|^p\,dt'\right\} dt \\ &= \|g(t)\|_1^{p-1} \int_0^\infty |g(t')| \left\{\int_{-\infty}^\infty |f(t-t')|^p\,dt\right\} dt' \\ &= \|g(t)\|_1^{p-1} \int_0^\infty |g(t')|\,\|f(t)\|_p^p\,dt \\ &= \|g(t)\|_1^{p-1}\,\|f(t)\|_p^p \int_0^\infty |g(t')|\,dt' \\ &= \|g(t)\|_1^{p-1}\,\|f(t)\|_p^p\|g(t)\|_1 \\ &= g(t)\|_1^p\,\|f(t)\|_p^p. \end{aligned} \tag{2.19}$$

Substituting this result in (2.16) we have

$$\|g(t)\circ\|_p \le \|g(t)\|_1, \qquad 1 < p < \infty. \tag{2.20}$$

This is a remarkably compact result saying that the p-norm of $g(t)\circ$ (the convolution operator) is bounded by the 1-norm of $g(t)$ (the function).

This result is related to what in linear system theory [11] is called the bounded-input–bounded-output stability. Here the important point is that a convolution-operator bound is given by the 1-norm of the convolution function.

2.4. 1-Norm of Time-Domain Convolution Operators

In the case of the 1-norm the results of the previous section also apply, as can be seen from

$$\|g(t)\circ\|_1 \equiv \sup_{f(t)\neq 0} \frac{\int_{-\infty}^{\infty} |\int_0^{\infty} g(t')f(t-t')\,dt'|\,dt}{\|f(t)\|_1}, \tag{2.21}$$

$$\begin{aligned}\int_{-\infty}^{\infty}\left|\int_0^{\infty} g(t')f(t-t')\,dt'\right|dt &\le \int_{-\infty}^{\infty}\left[\int_0^{\infty} |g(t')||f(t-t')|\,dt'\right]dt\\ &= \int_0^{\infty} |g(t')|\left[\int_{-\infty}^{\infty} |f(t-t')|\,dt\right]dt'\\ &= \|f(t)\|_1 \int_0^{\infty} |g(t')|\,dt'\\ &= \|f(t)\|_1\|g(t)\|_1,\end{aligned}$$

giving

$$\|g(t)\circ\|_1 \le \|g(t)\|_1. \tag{2.22}$$

For a lower bound, consider the definition in (2.21) and substitute in a special $f(t)$

$$\begin{aligned}&f(t) = \delta(t) \quad \text{(delta function)},\\ &\|\delta(t)\|_1 = \int_{-\infty}^{\infty} |\delta(t)|\,dt = \int_{-\infty}^{\infty} \delta(t)\,dt = 1,\\ &\int_0^{\infty} g(t')\delta(t-t')\,dt' = g(t),\end{aligned} \tag{2.23}$$

giving

$$\|g(t)\circ\|_1 \ge \int_{-\infty}^{\infty} |g(t)|\,dt = \int_0^{\infty} |g(t)|\,dt = \|g(t)\|_1. \tag{2.24}$$

Note that it is important that the δ function have a 1-norm (in effect, be integrable) for this result.

Combining (2.22) and (2.24) we have

$$\|g(t)\circ\|_1 = \|g(t)\|_1. \tag{2.25}$$

2.5. 2-Norm of Time-Domain Convolution Operators

For the 2-norm it is convenient to use Laplace- (or Fourier-) transform concepts. The two-sided Laplace transform is

$$\tilde{f}(s) \equiv \int_{-\infty}^{\infty} f(t)e^{-st}\,dt,$$

$$s = \Omega + j\omega = \text{complex frequency}, \tag{2.26}$$

$$f(t) = \frac{1}{2\pi j}\int_{\mathrm{Br}} \tilde{f}(s)e^{st}\,dt.$$

Br $\equiv$ the Bromwich contour in the strip of convergence (parallel to the $j\omega$-axis).

In terms of the Laplace transform our convolution problem is just multiplication, i.e.,

$$\begin{aligned} F(t) &= g(t) \circ f(t), \\ \tilde{F}(s) &= \tilde{g}(s)\tilde{f}(s). \end{aligned} \tag{2.27}$$

As discussed in another paper [8], the 2-norm with respect to time is related to the 2-norm with respect to frequency ω. As

$$\begin{aligned} \|f(t)\|_2 &\equiv \left\{\int_{-\infty}^{\infty} |f(t)|^2\,dt\right\}^{1/2}, \\ \|\tilde{f}(j\omega)\|_2 &\equiv \left\{\int_{-\infty}^{\infty} |\tilde{f}(j\omega)|^2\,d\omega\right\}^{1/2}, \\ \|f(t)\|_2 &= \frac{1}{\sqrt{2\pi}}\|\tilde{f}(j\omega)\|_2, \end{aligned} \tag{2.28}$$

which is one way to state the Parseval theorem.

Now the 2-norm is expressible in both time and frequency domains as

$$\begin{aligned} \|g(t)\circ\|_2 &= \sup_{f(t)\neq 0} \frac{\|g(t)\circ f(t)\|_2}{\|f(t)\|_2} \\ &= \sup_{\tilde{f}(j\omega)\neq 0} \frac{\|\tilde{g}(j\omega)\tilde{f}(j\omega)\|_2}{\|\tilde{f}(j\omega)\|_2} \\ &= \sup_{\tilde{f}(j\omega)\neq 0} \frac{\{\int_{-\infty}^{\infty} |\tilde{g}(j\omega)\tilde{f}(j\omega)|^2\,d\omega\}^{1/2}}{\{\int_{-\infty}^{\infty} |\tilde{f}(j\omega)|^2\,d\omega\}^{1/2}}. \end{aligned} \tag{2.29}$$

We have the inequality

$$\begin{aligned} \int_{-\infty}^{\infty} |\tilde{g}(j\omega)\tilde{f}(j\omega)|^2\,d\omega &\leq \int_{-\infty}^{\infty} |\tilde{g}(j\omega)|_{\max}^2 |\tilde{f}(j\omega)|^2\,d\omega \\ &= |\tilde{g}(j\omega)|_{\max}^2 \int_{-\infty}^{\infty} |\tilde{f}(j\omega)|^2\,d\omega. \end{aligned} \tag{2.30}$$

Here the maximum of $|\tilde{g}(j\omega)|$ is over all real ω. This can be used to define an $\omega_{\max}$ via

$$|\tilde{g}(j\omega_{\max})| = |\tilde{g}(j\omega)|_{\max}. \tag{2.31}$$

Of course, there may be more than one $\omega_{\max}$ meeting this definition. This gives

$$\|g(t)\circ\|_2 \leq |\tilde{g}(j\omega)|_{\max} = |\tilde{g}(j\omega_{\max})|. \tag{2.32}$$

For a special case consider

$$|\tilde{f}(j\omega)|^2 = \delta(\omega - \omega_{\max}), \tag{2.33}$$

which gives

$$\|g(t)\circ\|_2 \geq |\tilde{g}(j\omega_{\max})|, \tag{2.34}$$

assuming that $|\tilde{g}(j\omega)|$ is sufficiently smooth near $\omega = \omega_{\max}$.
Combining (2.32) and (2.34) gives

$$\|g(t)\circ\|_2 = |\tilde{g}(j\omega)|_{\max} = |\tilde{g}(j\omega_{\max})|. \tag{2.35}$$

Combining this with (2.20) gives

$$|\tilde{g}(j\omega)|_{\max} \leq \|g(t)\|_1. \tag{2.36}$$

Interpreting (2.33) physically, this means that the spectrum of $f(t)$ is concentrated in some region near $\omega = \omega_{\max}$ leading to (2.34). This should be considered as some sort of limit process.

2.6. ∞-Norm of Time-Domain Convolution Operators

For the ∞-norm the results of Section 2.3 also apply, as can be seen from

$$\|g(t)\circ\|_\infty \equiv \sup_{f(t)\not\equiv 0} \frac{|\int_0^\infty g(t')f(t-t')\,dt'|_{\sup_t}}{|f(t)|_{\sup}}, \tag{2.37}$$

$$\begin{aligned}\left|\int_0^\infty g(t')f(t-t')\,dt'\right| &\leq \int_0^\infty |g(t')|\,|f(t-t')|\,dt' \\ &\leq |f(t)|_{\sup}\int_0^\infty |g(t')|\,dt' \\ &= |f(t)|_{\sup}\,\|g(t)\|_1,\end{aligned}$$

giving

$$\|g(t)\circ\|_\infty \leq \|g(t)\|_1. \tag{2.38}$$

For a lower bound, choose $f(t)$ in a special way. Think of fixing t and choosing

$$f(t-t') \equiv \begin{cases} +1 & \text{if } g(t') > 0, \\ 0 & \text{if } g(t') = 0, \\ -1 & \text{if } g(t') < 0. \end{cases} \tag{2.39}$$

This gives

$$|f(t)|_{\text{sup}} = 1,$$

$$\left|\int_0^\infty g(t')f(t-t')\,dt'\right| = \left|\int_0^\infty |g(t')|\,dt'\right|$$

$$= \|g(t)\|_1, \tag{2.40}$$

which, when substituted in (2.37), gives

$$\|g(t)\circ\|_\infty \geq \|g(t)\|_1. \tag{2.41}$$

Combining (2.38) and (2.41) we have

$$\|g(t)\circ\|_\infty = \|g(t)\|_1. \tag{2.42}$$

2.7. Residue Norm of Time-Domain Functions

Let us assume that our time-domain functions are of the form

$$f(t) = \sum_n R_n e^{s_n t} u(t) \tag{2.43}$$

with conjugate symmetry

$$\begin{aligned} s_{-n} &= s_n^*, \\ R_{-n} &= R_n^*, \end{aligned} \tag{2.44}$$

except that, for $n = 0$, another index is needed to allow more than one pole on the negative real axis, and so that in time domain the function is real.

In complex-frequency domain the above is

$$\tilde{f}(s) = \sum_n R_n [s - s_\alpha]^{-1}, \tag{2.45}$$

which is a sum of first-order poles. So that $f(t)$ may be bounded, let us require

$$\text{Re}[s_n] \leq 0 \qquad \text{for all } n. \tag{2.46}$$

Let us now define another norm as the residue norm or r-norm as

$$\|f(t)\|_r \equiv \sum_n |R_n| \tag{2.47}$$

with the restriction that this sum converges. Here the subscript r is purely symbolic and does not assume numerical values. We can verify that this is a norm by application of the required properties in (2.4). Let us assume that all the s_n are distinct so that no terms in (2.43) cancel (in particular, so that $f(t) \not\equiv 0$ unless all the R_n are zero). Then

$$\|f(t)\|_r \begin{cases} = 0 & \text{iff all } R_n = 0, \text{ or equivalently } f(t) \equiv 0, \\ > 0 & \text{otherwise.} \end{cases} \tag{2.48}$$

Also we have

$$\|\alpha f(t)\|_r = \sum_n |\alpha R_n| = |\alpha| \sum_n |R_n|$$
$$= |\alpha| \, \|f(t)\|_r. \tag{2.49}$$

The triangle inequality is verified by considering two separate functions, distinguished by superscripts, as

$$f^{(1)}(t) \equiv \sum_n R_n^{(1)} e^{s_n^{(1)}t} u(t),$$
$$f^{(2)}(t) \equiv \sum_n R_n^{(2)} e^{s_n^{(2)}t} u(t). \tag{2.50}$$

Then if the two sets of natural frequencies $\{s_n^{(1)}\}$ and $\{s_n^{(2)}\}$ are distinct (have no common elements)

$$\|f^{(1)}(t) + f^{(2)}(t)\|_r = \left\| \left[\sum_n R_n^{(1)} e^{s_n^{(1)}t} u(t) \right] + \left[\sum_n R_n^{(2)} e^{s_n^{(2)}t} u(t) \right] \right\|_r$$
$$= \sum_n |R_n^{(1)}| + \sum_n |R_n^{(2)}|$$
$$= \|f^{(1)}(t)\|_r + \|f_n^{(2)}(t)\|_r. \tag{2.51}$$

However, if some $s_n^{(2)} = s_{n'}^{(2)}$ then the associated residue is $R_n^{(1)} + R_{n'}^{(2)}$, and the term in the residue norm is

$$|R_n^{(1)} + R_{n'}^{(2)}| \leq |R_n^{(1)}| + |R_{n'}^{(2)}|. \tag{2.52}$$

This being true for all such pairs of $s_n^{(1)}$ and $s_{n'}^{(2)}$, then

$$\|f^{(1)}(t) + f^{(2)}(t)\|_r \leq \|f^{(1)}(t)\|_r + \|f^{(2)}(t)\|_r \tag{2.53}$$

and the r-norm has all the properties of a norm.

While (2.43) can describe many interesting waveforms, a general time-domain waveform can contain other types of terms as well. In the general theory of the singularity expansion method (SEM) there can be branch singularities which take the form [12]

$$f_{n'}(t) = \int_{C_{n'}} \tilde{R}_{n'}(s') e^{s't} u(t) \, ds',$$
$$\tilde{f}_{n'}(s) = \int_{C_{n'}} \tilde{R}_{n'}(s') [s - s']^{-1} \, ds', \tag{2.54}$$
$$C_{n'} \equiv n'\text{th contour in the left-half } s'\text{-plane.}$$

Near the branch point(s) we may need to be careful in defining the branch contribution. Here, for contours not on the negative real axis, we take the contours in pairs

$$C_{-n'} = C_{n'}^* \quad \text{(symbolic)},$$
$$\tilde{R}_{-n'}(s') = \tilde{R}_{n'}(s'^*) = R_{n'}^*(s'), \tag{2.55}$$

so that the resulting time-domain function is real valued.

Comparing (2.54) with (2.43) and (2.45) we note that the form is very similar. In particular, we can think of approximating an integral as in (2.54) by a sum as in (2.43) and (2.45). So let us define the r-norm of $f_{n'}(t)$ by

$$\|f_{n'}(t)\|_r \equiv \int_{C'_n} |\tilde{R}_{n'}(s')||ds'| \tag{2.56}$$

in agreement with (2.57), provided of course that this integral exists. Note that if C'_n is moved in the s-plane (even with fixed branch points), a different result may be obtained since the integrand is not analytic. Hence the definition of C'_n must in general be fixed for the problem at hand. The reader will note that a sum of such terms $f_{n'}(t)$ with those in (2.43), to give a more general $f(t)$ and the r-norm as defined by a sum of terms as in (2.47) and (2.56), is a legitimate norm satisfying (2.48), (2.49), and (2.53).

Another type of singularity (at ∞) is referred to as an entire function. As discussed in [12] this can be represented by a contour integral at ∞ of a form similar to that in (2.54). In this case, we need to be careful of the convergence of the integrals, particularly in the norm, as in (2.56).

The problem of higher-order poles can be addressed by noting that the r-norm of a single term in (2.43) is the same as the ∞-norm, i.e.,

$$\|R_n e^{s_n t} u(t)\|_\infty = |R_n| = \|R_n e^{s_n t} u(t)\|_r. \tag{2.57}$$

Then consider a multiple-order pole of the form

$$\begin{aligned} f_n^{(m)}(t) &= R_n^{(m)} \frac{t^{n-1}}{(m-1)!} e^{s_n t} u(t), \\ \tilde{f}_n^{(m)}(s) &= R_n^{(m)} [s - s_n]^{-m}, \qquad m = 1, 2, 3, \ldots. \end{aligned} \tag{2.58}$$

The peak magnitude is found from

$$|f_n^{(m)}(t)| = |R_n^{(m)}| \frac{t^{m-1}}{(m-1)!} e^{\mathrm{Re}[s_n]t} u(t), \tag{2.59}$$

where the time of the peak satisfies the equation

$$\begin{aligned} \frac{d}{dt} |f_n^{(m)}(t)| \bigg|_{t=t_p} &= 0 = \frac{|R_n^{(m)}|}{(m-1)!} [(m-1)t_p^{m-2} + t_p^{m-1} \mathrm{Re}[s_n]] e^{\mathrm{Re}[s_n]t_p} \\ t_p &= \frac{m-1}{-\mathrm{Re}[s_n]}, \end{aligned} \tag{2.60}$$

giving

$$\begin{aligned} \|f_n^{(m)}(t)\|_r &\equiv \sup_t |f_n^{(m)}(t)| \equiv \|f_n^{(m)}(t)\|_\infty \\ &= \frac{|R_n^{(m)}|}{(m-1)!} \left[\frac{m-1}{-\mathrm{Re}[s_n]}\right]^{m-1} e^{1-m}, \qquad m = 2, 3, 4, \ldots, \\ \|f_n^{(1)}(t)\|_r &= |R_n^{(1)}|. \end{aligned} \tag{2.61}$$

Note that for the higher-order poles we restrict

$$\mathrm{Re}[s_n] < 0 \qquad \text{for all } n. \tag{2.62}$$

Then our general form for the r-norm is

$$\begin{aligned} f(t) &= \sum_{n,m} f_n^{(m)}(t) + \sum_{n'} f_{n'}(t), \\ \|f(t)\|_r &= \sum_{n,m} \|f_n^{(m)}(t)\|_r + \sum_{n'} \|f_{n'}(t)\|_r, \end{aligned} \tag{2.63}$$

where the individual terms are defined in (2.56) and (2.61).

2.8. Relation Between the r-Norm and the p-Norm

Consider first the case of simple poles as in (2.43). If the p-norm of $f(t)$ is to exist, we require

$$\mathrm{Re}[s_n] \begin{Bmatrix} \le 0 & \text{for } p = \infty \\ < 0 & \text{for } 1 \le p < \infty \end{Bmatrix} \qquad \text{for all } n. \tag{2.64}$$

Then the p-norm of (2.43) can be bounded as

$$\begin{aligned} \|f(t)\|_p &= \left\| \sum_n R_n e^{s_n t} u(t) \right\|_p \\ &\le \sum_n \|R_n e^{s_n t} u(t)\|_p \\ &= \sum_n |R_n| \, \|e^{s_n t} u(t)\|_p \end{aligned} \tag{2.65}$$

using the fundamental properties of norms in (2.4). Considering the individual terms

$$\begin{aligned} \|e^{s_n t} u(t)\|_p &= \left\{ \int_0^\infty |e^{s_n t}|^p \, dt \right\}^{1/p} \\ &= \left\{ \int_0^\infty e^{p\,\mathrm{Re}[s_n] t} \, dt \right\}^{1/p} \\ &= \left[\frac{-1}{p\,\mathrm{Re}[s_n]} \right]^{1/p} = \left[\frac{1}{-\mathrm{Re}[s_n]} \right]^{1/p} p^{-1/p} \qquad \text{for} \quad 1 \le p < \infty. \end{aligned} \tag{2.66}$$

For $p = \infty$, we have

$$\|e^{s_n t} u(t)\|_\infty = |e^{s_n t} u(t)|_{\sup} = 1. \tag{2.67}$$

Then (2.65) becomes

$$\|f(t)\|_p \le \begin{cases} p^{1/p} \sum_n |R_n| \left[\dfrac{1}{-\mathrm{Re}[s_n]} \right]^{1/p} & \text{for} \quad 1 \le p < \infty, \\ \sum_n |R_n| = \|f(t)\|_r & \text{for} \quad p = \infty. \end{cases} \tag{2.68}$$

Note for these results to apply, not only must (2.64) apply, but also the series in (2.68) must converge. An interesting term in (2.68) is

$$p^{-1/p} = \begin{cases} 1 & \text{for} \quad p = 1, \\ 1/\sqrt{2} \simeq 0.707 & \text{for} \quad p = 2, \\ \lim\limits_{p\to\infty} p^{-1/p} = \lim\limits_{p\to\infty} e^{-(1/p)\ln(p)} = 1 & \text{for} \quad p = \infty, \end{cases} \tag{2.69}$$

$$\min_{1\le p\le\infty} p^{-1/p} = e^{-1/e} \simeq 0.692 \qquad \text{at} \quad p = e \simeq 2.718,$$

$$\max_{1\le p\le\infty} p^{-1/p} = 1 \qquad \text{at} \quad p = 1, \infty.$$

If we define

$$\Omega_{\text{max}} = \sup_n \operatorname{Re}[s_n] < 0, \tag{2.70}$$

then from (2.68) we have the looser (but simpler) bound

$$\|f(t)\|_p \le p^{-1/p}\left[\frac{1}{-\Omega_{\text{max}}}\right]^{1/p} \|f(t)\|_r. \tag{2.71}$$

Thus the p-norm can be bounded in terms of the r-norm.

A contour integral contribution as in (2.54) can be bounded provided the contour $C_{n'}$ has its location in the left-half s'-plane bounded to the left of the $j\omega'$-axis as

$$\sup_{s'\in C_{n'}} \operatorname{Re}[s'] = \Omega_{n'} < 0. \tag{2.72}$$

Then we have

$$\begin{aligned} \|f_{n'}(t)\|_p &= \left\{\int_0^\infty \left|\int_{C_{n'}} \tilde{R}_{n'}(s')e^{s't}u(t)\,ds'\right|^p dt\right\}^{1/p} \\ &\le \left\{\int_0^\infty \left[\int_{C_{n'}} |\tilde{R}_{n'}(s')|e^{\operatorname{Re}[s']t}u(t)|ds'|\right]^p dt\right\}^{1/p}. \end{aligned} \tag{2.73}$$

This can be bounded by regarding an integral as the limit of a summation, i.e., generalize (2.66) as

$$\begin{aligned} \|f_{n'}(t)\|_p &= \left\|\int_{C_{n'}} \tilde{R}_{n'}(s')e^{s't}u(t)ds'\right\|_p \\ &\le \int_{C_{n'}} \|\tilde{R}_{n'}(s')e^{s't}u(t)\|_p|ds'| \\ &= \int_{C_{n'}} |\tilde{R}_{n'}(s')|\,\|e^{s't}u(t)\|_p|ds'|. \end{aligned} \tag{2.74}$$

Then, applying (2.66) and (2.67), a result analogous to (2.68) is

$$\|f_{n'}(t)\|_p \le \begin{cases} p^{-1/p} \displaystyle\int_{C_{n'}} |\tilde{R}_{n'}(s')| \left[\frac{1}{-\mathrm{Re}[s']}\right]^{1/p} |ds'| & \text{for } 1 \le p < \infty, \\ \displaystyle\int_{C_{n'}} |\tilde{R}_{n'}(s')| \, |ds'| = \|f_{n'}(t)\|_r & \text{for } p = \infty. \end{cases} \tag{2.75}$$

Again, we need that $|\tilde{R}_n(s')$ be integrable on $C_{n'}$ with special attention paid to $\mathrm{Re}[s'] \to 0$ and $\mathrm{Re}[s'] \to -\infty$ (if such cases occur). For the bound on the general p-norm, it may be possible to allow $\mathrm{Re}[s'] \to 0$ and/or $-\infty$ provided the behavior of $|\tilde{R}_n(s')|$ is such as to allow integrability there (thereby loosening (2.72)).

A looser (but simpler) bound is found from (2.75), with the restriction of (2.72), as

$$\|f_{n'}(t)\|_p \le p^{-1/p} \left[\frac{1}{-Q_{n'}}\right]^{1/p} \|f_{n'}(t)\|_r. \tag{2.76}$$

Thus for a branch contribution as well, the p-norm can be bounded in terms of the r-norm.

We can also consider entire-function contributions which also have the form of a contour integral (at ∞) with bounds as above.

In the case of higher-order poles with the restriction (for $m > 1$) of

$$\mathrm{Re}[s_n] < 0 \qquad \text{for} \quad 1 \le p \le \infty, \tag{2.77}$$

we have the extension of (2.65) using (2.58)

$$\|f_n^{(m)}(t)\|_p = |R_n^{(m)}| \left\| \frac{t^{m-1}}{(m-1)!} e^{s_n t} u(t) \right\|_p. \tag{2.78}$$

The individual terms

$$\left\| \frac{t^{m-1}}{(m-1)!} e^{s_n t} u(t) \right\|_p \left\{ \int_0^\infty \left| \frac{t^{m-1}}{(m-1)!} e^{s_n t} \right|^p dt \right\}^{1/p}$$

$$= \frac{1}{(m-1)!} \left\{ \int_0^\infty t^{p(m-1)} e^{p\,\mathrm{Re}[s_n]t} \, dt \right\}^{1/p}, \tag{2.79}$$

are solved via a common integral as [13 (6.11)]

$$\Gamma(z) = k^2 \int_0^\infty t^{2-1} e^{-kt} \, dt \qquad \text{for} \quad \mathrm{Re}[2] > 0, \quad \mathrm{Re}[k] > 0, \tag{2.80}$$

$$k = -p\,\mathrm{Re}[s_n],$$

$$z = 1 + p(m-1),$$

$$\Gamma(z) \equiv \text{gamma function},$$

$$\Gamma(z) = (z-1)!.$$

This gives

$$\int_0^\infty t^{p(m-1)} e^{p\,\mathrm{Re}[s_n]t} \, dt = [-p\,\mathrm{Re}[s_n]]^{-1-p(m-1)} \Gamma(1 + p(m-1)) \tag{2.81}$$

and

$$\left\| \frac{t^{m-1}}{(m-1)!} e^{s_n t} u(t) \right\|_p = \frac{1}{(m-1)!} [-p \operatorname{Re}[s_n]]^{-(1/p)-m+1} \Gamma^{1/p}(1 + p(m-1))$$
$$= p^{-(1/p)-m+1} \left[\frac{1}{-\operatorname{Re}[s_n]} \right]^{(1/p)+m-1} \frac{\Gamma^{1/p}(1 + p(m-1))}{\Gamma(m)}, \tag{2.82}$$

which is a direct extension of (2.66). Then for higher-order poles, we have

$$\| f_n^{(m)}(t) \|_p = |R_n^{(m)}| p^{-(1/p)-m+1} \left[\frac{1}{-\operatorname{Re}[s_n]} \right]^{(1/p)+m-1} \frac{\Gamma^{1/p}(1 + p(m-1))}{\Gamma(m)}. \tag{2.83}$$

With the r-norm of a higher-order pole as in (2.61) (defined via the ∞-norm) we can write the p-norm as

$$\| f_n^{(m)}(t) \|_p = \| f_n^{(m)}(t) \|_r p^{-(1/p)-m+1} \left[\frac{1}{-\operatorname{Re}[s_n]} \right]^{1/p}$$
$$\times \left[\frac{e}{m-1} \right]^{m-1} \Gamma^{1/p}(1 + p(m-1))$$
$$= \| f_n^{(m)}(t) \|_r p^{-1/p} \left[\frac{1}{-\operatorname{Re}[s_n]} \right]^{1/p} A(p, m), \tag{2.84}$$
$$A(p, m) = \left[\frac{e}{p(m-1)} \right]^{m-1} \Gamma^{1/p}(1 + p(m-1)) \qquad \text{for} \quad m = 2, 3, \ldots,$$

where $p^{-(1/p)}$ has been considered in (2.69). The additional factor $A(p, m)$ can be considered for special cases. For $p = 1$ (the 1-norm), we have

$$A(1, m) = \left[\frac{e}{m-1} \right]^{m-1} = \exp\left\{ (m-1) \ln\left(\frac{e}{m-1} \right) \right\}$$
$$= \exp\{(m-1)[1 - \ln(m-1)]\}$$
$$= \begin{cases} 1 & \text{for } m = 1 \text{ (from (2.68))}, \\ e \simeq 2.718 & \text{for } m = 2, \\ \infty & \text{for } m = \infty. \end{cases} \tag{2.85}$$

For $p = 2$ (the 2-norm) we have

$$A(2, m) = \left[\frac{e}{2(m-1)} \right]^{m-1} [(2m-2)!]^{1/2}$$
$$= \exp\{(m-1)[1 - \ln[2(m-1)]]\} [(2m-2)!]^{1/2}$$
$$= \begin{cases} 1 & \text{for } m = 1 \text{ (from (2.68))}, \\ \dfrac{e}{\sqrt{2}} \simeq 1.922 & \text{for } m = 2, \\ \infty & \text{for } m = \infty. \end{cases} \tag{2.86}$$

For $p = \infty$ (the ∞-norm), use the Stirling approximation [13 (6.1.37)] as

$$\Gamma(z) = e^{-z} z^{z-(1/2)} \sqrt{2\pi}\,[1 + O(z^{-1})] \qquad \text{as} \quad z \to \infty,$$

$$\Gamma(1 + 1 + p(m-1)) = \sqrt{2\pi}\, \exp\{-[1 + p(m-1)] + [\tfrac{1}{2} + p(m-1)] \times \ln[1 + p(m-1)]\}\,[1 + O(p^{-1})] \qquad \text{as} \quad p \to \infty,$$

$$\begin{aligned}
\Gamma^{1/p}(1 + p(m-1)) &= (2\pi)^{1/2p} \exp\left\{-\left[\frac{1}{p} + (m-1)\right] + \left[\frac{1}{2p} + (m-1)\right] \times \ln[1 + p(m-1)]\right\}[1 + O(p^{-2})] \qquad \text{as} \quad p \to \infty, \\
&= \exp\{1 - m + (m-1)\ln[1 + p(m-1)]\}\,[1 + O(p^{-1})] \qquad \text{as} \quad p \to \infty, \\
&= [1 + p(m-1)]^{m-1} e^{1-m} [1 + O(p^{-1})] \qquad \text{as} \quad p \to \infty,
\end{aligned} \tag{2.87}$$

giving

$$\begin{aligned}
A(\infty, m) &= \lim_{p\to\infty} A(p, m) \\
&= \lim_{p\to\infty} \left[\frac{e}{p(m-1)}\right]^{m-1} [1 + p(m-1)]^{m-1} e^{1-m} \\
&= \lim_{p\to\infty} \left[\frac{1}{p(m-1)} + 1\right]^{m-1} \\
&= 1,
\end{aligned} \tag{2.88}$$

which agrees with our definition of the r-norm via the ∞-norm in (2.57).

Analogous to (2.68) and (2.75) we have the bound for a sum of higher-order poles

$$\|f(t)\|_p \le \sum_{n,m} \|f_n^{(m)}(t)\|_p$$

$$= \begin{cases} \displaystyle\sum_{n,m} |R_n^{(m)}|\, p^{-(1/p)-m+1} \left[\frac{1}{-\mathrm{Re}[s_n]}\right]^{(1/p)+m-1} \frac{\Gamma^{1/p}(1 + p(m-1))}{\Gamma(m)} \\ \quad \text{for } 1 \le p < \infty, \\ \displaystyle\sum_{n,m} |R_n^{(m)}| \left[\frac{m-1}{-\mathrm{Re}[s_n]}\right]^{m-1} \frac{e^{1-m}}{\Gamma(m)} = \|f(t)\|_r \\ \quad \text{for } p = \infty. \end{cases} \tag{2.89}$$

With the restriction of (2.70) this bound is loosened somewhat by replacing all the $\mathrm{Re}[s_n]$ by $\Omega_{\max}$.

Then our general form for the p-norm is

$$f(t) = \sum_{n,m} f_n^{(m)}(t) + \sum_{n'} f_{n'}(t), \tag{2.90}$$

$$\|f(t)\|_p \le \sum_{n,m} \|f_n^{(m)}(t)\|_p + \sum_{n'} \|f_{n'}(t)\|_p,$$

where the individual terms are defined in (2.68), (2.75), and (2.89).

2.9. Residue Norm of Time-Domain Convolution Operations

Now consider the r-norm of a time-domain convolution operator $g(t)\circ$ defined by

$$\|g(t)\circ\|_r \equiv \sup_{f(t)\neq 0} \frac{\|g(t)\circ f(t)\|_r}{\|f(t)\|_r}. \tag{2.91}$$

Section 2.7 has considered the r-norm of time-domain functions. For later use, let us define bounds on the real parts of the singularities of the functions, i.e.,

$$\begin{aligned} \tilde{f}(s) \text{ analytic} \quad &\text{for} \quad \mathrm{Re}[s] > \Omega_f, \\ \tilde{g}(s) \text{ analytic} \quad &\text{for} \quad \mathrm{Re}[s] > \Omega_g. \end{aligned} \tag{2.92}$$

Now for the r-norm of $f(t)$ to exist and for the response $g(t)\circ f(t)$ to be bounded we require

$$\Omega_f \leq 0, \qquad \Omega_g \leq 0. \tag{2.93}$$

Furthermore, it will be useful to bound one or both of these to the left of the imaginary axis for various applications.

2.9.1. *First-Order Poles*

Let us restrict the consideration at first to first-order poles as in (2.43)

$$f(t) = \sum_n R_n e^{s_n t} u(t) \tag{2.94}$$

with conjugate symmetry as in (2.44). Similarly, let $g(t)$ be represented by

$$g(t) = \sum_l G_l e^{s'_l t} u(t) \tag{2.95}$$

with conjugate symmetry as

$$\begin{aligned} s'_{-l} &= s_l'^{*}, \\ G_{-l} &= G_l^{*}, \end{aligned} \tag{2.96}$$

except that for $l = 0$ another index is needed to allow more than one pole on the negative real axis.

In complex-frequency domain, we have

$$\begin{aligned} \tilde{f}(s) &= \sum_n R_n[s - s_n]^{-1}, \\ \tilde{g}(s) &= \sum_l G_l[s - s'_l]^{-1}, \end{aligned} \tag{2.97}$$

which gives a product

$$\begin{aligned} \tilde{g}(s)\tilde{f}(s) &= \sum_{l,n} G_l R_n [s - s'_l]^{-1}[s - s_n]^{-1} \\ &= \sum_{l,n} G_l R_n \{[s - s'_l]^{-1}[s'_l - s_n]^{-1} + [s_n - s'_l]^{-1}[s - s_n]^{-1}\} \\ &\quad \text{for } s_n \neq s'_l \text{ for all } (n, l). \end{aligned} \tag{2.98}$$

In time domain, this is

$$g(t) \circ f(t) = \sum_{l,n} G_l R_n \{[s'_l - s_n]^{-1} e^{s'_l t} u(t) + [s_n - s'_l]^{-1} e^{s_n t} u(t)\}. \tag{2.99}$$

From this we can write the r-norm from (2.47) for first-order poles as

$$\begin{aligned} \|g(t) \circ f(t)\|_r &= 2 \sum_{l,n} |G_l||R_n||s'_l - s_n|^{-1} \\ &= \sum_n |R_n| \left\{ 2 \sum_l |G_l||s'_l - s_n|^{-1} \right\}. \end{aligned} \tag{2.100}$$

Defining

$$G^{(0)} \equiv \max_n 2 \sum_l |G_l||s'_l - s_n|^{-1} \tag{2.101}$$

(provided a sum exists), we have

$$\|g(t) \circ f(t)\|_r \leq G^{(0)} \|f(t)\|_r \tag{2.102}$$

and

$$\|g(t) \circ \|_r \leq G^{(0)} \tag{2.103}$$

as one way to consider the r-norm of a convolution operator.

However, note that $G^{(0)}$ is a function of the poles s_n of $\tilde{f}(s)$, not just of the G_l and s'_l. This problem can be alleviated if we can give a lower bound to the $|s'_l - s_n|$, say

$$\Delta = \inf_{l,n} |s'_l - s_n| > 0. \tag{2.104}$$

Then we have

$$G^{(0)} \leq \frac{2}{\Delta} \|g(t)\|_r \tag{2.105}$$

giving

$$\begin{aligned} \|g(t) \circ f(t)\|_r &\leq \frac{2}{\Delta} \|g(t)\|_r \|f(t)\|_r, \\ \|g(t) \circ \|_r &\leq \frac{2}{\Delta} \|g(t)\|_r, \end{aligned} \tag{2.106}$$

assuming the sum

$$\|g(t)\|_r = \sum_l |G_l| \tag{2.107}$$

converges. Note that the result in (106) is consistent with the symmetry of the convolution operation, i.e.,

$$g(t) \circ f(t) = f(t) \circ g(t) \tag{2.108}$$

so that either $g(t) \circ$ or $f(t) \circ$ can be considered as the convolution operator. However, the presence of Δ in (2.106) is still undesirable in that it depends on both $f(t)$ and $g(t)$ in the sense of the closest approach of corresponding poles.

2.9.2. *Second-Order Poles Appearing from Convolution*

Even though $\tilde{f}(s)$ and $\tilde{g}(s)$ have each been constrained to have only first-order poles, the product can, in principle, have two such first-order poles (or pole pairs) coincident, giving a second-order pole (or pair of second-order poles). Say, for some (n, l) pair,

$$\begin{aligned} s_n &= s_l' \\ s_n^* &= s_l'^* \quad (\text{or } s_{-n} = s_{-l}'). \end{aligned} \tag{2.109}$$

Then, considering one such case we have from (2.61)

$$\begin{aligned} \|[G_l e^{s_l' t} u(t)] \circ [R_n e^{s_n t} u(t)]\|_r &= \left\| \frac{1}{2\pi j} \int_{\text{Br}} G_l R_n [s - s_n]^{-2} e^{st} \, ds \right\|_r \\ &= \frac{1}{e} \frac{1}{-\text{Re}[s_n]} |G_l||R_n|. \end{aligned} \tag{2.110}$$

In this formula $2/\Delta$ in (2.106) has been replaced by $1/(-e \text{ Re}[s_n])$. So it is not the nearness of s_n and s_l' (coincidence in this case) which blows up the norm, but rather the nearness to the $j\omega$-axis. If we require Ω_f and Ω_g in (2.93) to be bounded to the left of the $j\omega$-axis, then such a coincidence causes no problem.

2.9.3. *Close Approach of Two Poles Appearing in a Convolution*

Well, if the coincidence of two poles in $\tilde{f}(s)$ and $\tilde{g}(s)$ does not cause the r-norm of the convolution to blow up, then the close approach of these two should not either, or rather the definition of the r-norm can be modified to take this into account. Let us say that s_n and s_l' are near to each other, and consider a term of the form

$$\begin{aligned} A(t) &\equiv [G_l e^{s_l' t} u(t)] \circ [R_n e^{s_n t} u(t)] \\ &= \frac{1}{2\pi j} \int_{\text{Br}} G_l R_n [s - s_l']^{-1} [s - s_n]^{-1} e^{st} \, ds. \end{aligned} \tag{2.111}$$

Now, back in Section 2.7, when considering the r-norm as a sum of norms of the s-plane singularity terms, the ∞-norm (or peak value) was used to define the norm of each term. Then, for the case of the close approach above, let us consider these two poles as a single term and find the ∞-norm and use this to define the r-norm for such a case.

Expanding the product of poles gives

$$\begin{aligned} A(t) &= G_l R_n \frac{1}{2\pi j} \int_{\text{Br}} \{[s - s_l']^{-1} [s_l' - s_n]^{-1} + [s_n - s_l']^{-1} [s - s_n]^{-1}\} e^{st} \, ds \\ &= G_l R_n [s_l' - s_n]^{-1} [e^{s_l' t} - e^{s_n t}] u(t). \end{aligned} \tag{2.112}$$

Defining

$$a \equiv \tfrac{1}{2}[s_l' + s_n],$$
$$b \equiv \tfrac{1}{2}[s_l' - s_n], \tag{2.113}$$

so that

$$s_l' = a + b,$$
$$s_n = a - b, \tag{2.114}$$

we have

$$\begin{aligned} A(t) &= G_l R_n \frac{1}{2b}[e^{(a+b)t} - e^{(a-b)t}]u(t) \\ &= G_l R_n \frac{e^{at}}{b} \sinh(bt) u(t). \end{aligned} \tag{2.115}$$

In magnitude this is [13 (4.5.49), (4.5.54)]

$$\begin{aligned} |A(t)| &= |G_l||R_n|\frac{|e^{at}|}{|b|}|\sinh(bt)|u(t) \\ &= |G_l||R_n|\frac{e^{\mathrm{Re}[a]t}}{|b|}\{\sinh^2(\mathrm{Re}[b]t)\cos^2[\mathrm{Im}[b]t) \\ &\qquad + \cosh^2(\mathrm{Re}[b]t)\sin^2(\mathrm{Im}[b]t)\}^{1/2} \\ &= |G_l||R_n|\frac{e^{\mathrm{Re}[a]t}}{|b|}\{\sinh^2(\mathrm{Re}[b]t) + \sin^2(\mathrm{Im}[b]t)\}^{1/2}. \end{aligned} \tag{2.116}$$

Let us now assume that

$$|b| \ll |a|, \tag{2.117}$$

since b represents the difference and a the sum of two complex frequencies that are assumed very close to each other. Noting that

$$\mathrm{Re}[a] \leq \Omega_f + \Omega_g, \tag{2.118}$$

assume that

$$\mathrm{Re}[a] < 0, \tag{2.119}$$

and take the limiting form for small $|b|$ in (2.116). This gives

$$|A(t) = |G_l||R_n|te^{\mathrm{Re}[a]t}u(t)[1 + O((bt)^2)] \qquad \text{as} \quad b \to 0. \tag{2.120}$$

Note now for, small b/a that $|A(t)|$ has the form of a second-order pole. From (2.61) we have

$$\|A(t)\|_\infty = \|A(t)\|_r = \frac{1}{e}\frac{1}{-\mathrm{Re}[a]}|G_l||R_n|, \tag{2.121}$$

which is the result for a second-order pole in (2.110), noting in (2.113) that

$$a \to s_n \qquad \text{for} \quad s'_l \to s_n. \tag{2.122}$$

Rewriting (2.121) for s'_l near s_n, we have a definition for the r-norm in such a case

$$\|A(t)\|_r \equiv \frac{1}{e}\frac{2}{-\mathrm{Re}[s'_l] - \mathrm{Re}[s_n]}|G_l||R_n|. \tag{2.123}$$

Thus as long as either (or both) s'_l and s'_n are bounded to the left of the $j\omega$-axis, then the $2/\Delta$ in (2.106) can be replaced by $2/(e[-\mathrm{Re}[s'_l] - \mathrm{Re}[s_n]])$ for the case of the closely approaching poles due to the product of $\tilde{g}(s)$ and $\tilde{f}(s)$. This allows even some $s'_l = s_n$ cases, without the r-norm blowing up. Note, however, that this does not allow $s'_l = s_n$ on the $j\omega$-axis.

Note that this case of the closely approaching poles due to a product is quite different from a case of, say, two s_n in a sum, such as represents $f(t)$ in (2.43) appearing close together. In such a case the residues of the two poles may be quite independently choosable. However, the product as in (2.112) inherently brings the s'_l and s_n into the effective compound residue or residues.

2.9.4. *Combination of Results*

Then in (2.112) let us exclude cases in which s'_l is near s_n and replace $2/|s'_l - s_n|$ by $2/(e[-\mathrm{Re}[s'_l] - \mathrm{Re}[s_n]])$. Note that nearness is defined by (2.117), i.e., $|s'_l - s_n| \ll |s'_l + s_n|$, which for both s'_l and s'_n in the second quadrant of the complex s-plane is approximately achieved. Note that for such a close approach in the second quadrant there is also another close approach of the conjugate poles in the third quadrant.

With this modification, then $G^{(0)}$ in (2.101) can be used to better bound the r-norm of a convolution, especially for the case of the closely approaching poles. In (2.106) we can use these results to replace $2/\Delta$ as above to remove cases of the closely approaching poles so that $G^{(0)}$ is bounded in (2.105).

2.10. Summary

This chapter has developed some of the norm properties of time-domain waveforms and convolution operators. This is done in the context of the usual p-norm and a new norm which we call the residue norm or r-norm.

The r-norm has been related to the p-norm, being defined basically as the ∞-norm on a termwise (singularity-by-singularity in the s-plane) basis. For use in bounding time-domain waveforms and operators, the r-norm has significant potential as it can be applied in the context of the singularity expansion method (SEM).

References

[1] C.E. Baum (1980), Electromagnetic topology: A formal approach to the analysis and design of complex electronic systems, Interaction Note 400, September 1980, and *Proc. EMC Symposium*, Zürich, March 1981, pp. 209–214.

[2] C.E. Baum (1985), On the use of electromagnetic topology for the decomposition of scattering matrices for complex physical structures, Interaction Note 454, July 1985.

[3] F.M. Tesche (1978), Topological concepts for internal EMP interaction, *IEEE Trans. Antennas and Propagation*, vol. 26, January 1978, pp. 60–64, and *IEEE Trans. EMC*, vol. 20, February 1978, pp. 60–64.

[4] A.K. Agrawal and C.E. Baum (1983), Bounding of signal levels at terminations of a multiconductor transmission-line network, Interaction Note 419, April 1983, and *Electromagnetics*, vol. 8, 1988, pp. 375–422.

[5] F.C. Yang and C.E. Baum (1983), Use of matrix norms of interaction supermatrix blocks for specifying electromagnetic performance of subshields, Interaction Note 427, April 1983, also as (same authors), Electromagnetic topology: Measurements and norms of scattering parameters of subshields, *Electromagnetics*, vol. 6, 1986, pp. 47–72.

[6] C.E. Baum, Bounds on norms of scattering matrices, Interaction Note 432, June 1983, and *Electromagnetics*, vol. 6, 1986, pp. 33–45.

[7] C.E. Baum (1983), Black box bounds, Interaction Note 429, May 1983, and *Proc. EMC Symposium*, Zürich, March 1985, pp. 381–386.

[8] C.E. Baum (1984), Some bounds concerning the response of linear systems with a nonlinear element, Interaction Note 438, June 1984.

[9] C.E. Baum (1979), Norms and Eigenvector Norms, Mathematics Note 63, November 1979.

[10] I. Stakgold (1979), *Green's Functions and Boundary Value Problems*, Wiley, New York.

[11] C.-T. Chen (1984), *Linear System Theory and Design*, Holt, Rinehart, and Winston, New York.

[12] C.E. Baum (1978), Toward an engineering theory of electromagnetic scattering: The singularity and eigenmode expansion methods, in *Electromagnetic Scattering*, P.L.E. Uslenghi, Academic Press, New York.

[13] M. Abramowitz and I.A. Stegun (1964), *Handbook of Mathematical Functions*, AMS 55, National Bureau of Standards, Washington, D.C.

[14] F. Riesz and B. Sz.-Nagy (1955), *Functional Analysis*, Frederick Ungar, New York.

[15] V.I. Krylov (1962), *Approximate Calculation of Integrals*, Macmillan, London.

Appendix A. The Hölder Inequality

For this chapter there is an important inequality known as the Hölder inequality. This is discussed in various texts such as [10], [11]. For vector p-norms we have

$$\begin{aligned} \|(x_n)\|_p &= \left\{ \sum_{n=1}^{N} |x_n|^p \right\}^{1/p} \qquad \text{for} \quad 1 \le p < \infty, \\ \|(x_n)\|_\infty &= \max_n |x_n|, \qquad 1 \le n \le N. \end{aligned} \tag{2.A.1}$$

The Hölder inequality is

$$|(x_n)\cdot(y_n)| \equiv \left| \sum_{n=1}^{N} x_n y_n \right| \le \left\{ \sum_{n=1}^{N} |x_n|^{p_1} \right\}^{1/p_1} \left\{ \sum_{n=1}^{N} |y_n|^{p_2} \right\}^{1/p_2}, \tag{2.A.2}$$

$$|(x_n)\cdot(y_n)| \le \|(x_n)\|_{p_1} \|(y_n)\|_{p_2},$$

$$1 = \frac{1}{p_1} + \frac{1}{p_2}, \qquad p_1 > 1, \quad p_2 > 1,$$

with equality if two conditions are met

$$\frac{|x_n|^{p_1}}{\|(x_n)\|_{p_1}^{p_1}} = \frac{|y_n|^{p_2}}{\|(y_n)\|_{p_2}^{p_2}},$$

$$x_n y_n \text{ has the same sign (all } + \text{ or all } -) \qquad \text{for all} \quad 1 \le n \le N. \tag{2.A.3}$$

A special case is that for the ∞-norm and 1-norm

$$\begin{aligned} |(x_n)\cdot(y_n)| &= \left| \sum_{n=1}^{\infty} x_n y_n \right| \\ &\le \sum_{n=1}^{\infty} |x_n|\,|y_n| \\ &\le \left\{ \max_n |x_n| \right\} \left\{ \sum_{n=1}^{\infty} |y_n| \right\} \\ &= \|(x_n)\|_\infty \|(y_n)\|_1, \\ |(x_n)\cdot(y_n)| &\le \|(x_n)\|_\infty \|(y_n)\|_1. \end{aligned} \tag{2.A.4}$$

Another case concerns the 2-norm which is also known as the Schwarz inequality

$$\begin{aligned} |(x_n)\cdot(y_n)| &\le \left\{ \sum_{n=1}^{N} |y_n|^2 \right\}^{1/2} \left\{ \sum_{n=1}^{N} |y_n|^2 \right\}^{1/2}, \\ |(x_n)\cdot(y_n)| &\le \|(x_n)\|_2 \|(y_n)\|_2. \end{aligned} \tag{2.A.5}$$

In terms of functions the vector p-norm is generalized (for real t) as

$$\|f(t)\|_p \equiv \left\{\int_a^b |f(t)|^p \, dt\right\}^{1/p} \quad \text{for} \quad 1 \leq p < \infty, \tag{2.A.6}$$
$$\|f(t)\|_\infty = \sup_{a \leq t \leq b} |f(t)|.$$

Here the supremum technically can exclude isolated values of $f(t)$ by considering limits from both sides of values of t of concern. The Hölder inequality is

$$\left|\int_a^b f_1(t)f_2(t)\,dt\right| \leq \left\{\int_a^b |f_1(t)|^{p_1}\,dt\right\}^{1/p_1} \left\{\int_a^b |f_2(t)|^{p_2}\,dt\right\}^{1/p_2},$$
$$\left|\int_a^b f_1(t)f_2(t)\,dt\right| \leq \|f_1(t)\|_{p_1} \|f_2(t)\|_{p_2} \tag{2.A.7}$$
$$1 = \frac{1}{p_1} + \frac{1}{p_2}, \qquad p_1 > 1, \quad p_2 > 1,$$

with equality if two conditions are met

$$\frac{|f_1(t)|^{p_1}}{\|f_1(t)\|_{p_1}^{p_1}} = \frac{|f_2(t)|^{p_2}}{\|f_2(t)\|_{p_2}^{p_2}}, \tag{2.A.8}$$

$f_1(t)f_2(t)$ has the same sign (all + or all −) "almost everywhere".

A special case is that for the ∞-norm and 1-norm

$$\left|\int_a^b f_1(t)f_2(t)\,dt\right| \leq \int_a^b |f_1(t)||f_2(t)|\,dt$$
$$\leq \left\{\sup_t |f_1(t)|\right\}\left\{\int_a^b |f_2(t)|\,dt\right\}$$
$$= \|f_1(t)\|_\infty \|f_2(t)\|_1, \tag{2.A.9}$$
$$\left|\int_a^b f_1(t)f_2(t)\,dt\right| \leq \|f_1(t)\|_\infty \|f_2(t)\|_1,$$
$$a < b.$$

The special case of the 2-norm is also known as the Schwarz inequality as

$$\left|\int_a^b f_1(t)f_2(t)\,dt\right| \leq \left\{\int_a^b |f_1(t)|^2\,dt\right\}^{1/2} \left\{\int_a^b |f_2(t)|^2\,dt\right\}^{1/2},$$
$$\left|\int_a^b f_1(t)f_2(t)\,dt\right| \leq \|f_1(t)\|_2 \|f_2(t)\|_2. \tag{2.A.10}$$

In this chapter, in dealing with time-domain waveforms, the case of interest has

$$\begin{aligned} a &= -\infty, \\ b &= \infty, \end{aligned} \tag{2.A.11}$$

so that we are dealing with integrals over all times (of interest). In general, such times, while of course finite, are much longer than times for which the waveforms of concern are significant so that $-\infty < t < \infty$ is a reasonable approximation.

3
An Overview of the Theory of the Near-Zone Doppler Effect

NADER ENGHETA*

3.1. Introduction

It is well known that when an observer in free space is in motion relative to a source of monochromatic electromagnetic radiation, the frequency of radiation as seen by the observer will be higher than that of the source (blue shift) as the source and observer approach each other and will be lower (red shift) as they get farther apart. This effect is known as the "Doppler effect" and was introduced by Christian Doppler in 1843 [1], [2].

However, these shifts in frequency do not always occur as expected. That is, a red shift does not necessarily mean that the source and observer are moving away from each other, and a blue shift does not always indicate that the source and observer are approaching one another. Indeed, Frank [3] and Lee [4], [5] have demonstrated that in some dispersive media, certain unusual effects, which resemble the inverse Doppler effects, can occur. In such media, a receding source produces a blue shift and an approaching source produces a red shift.

Basically, the Doppler effect is a result of the covariance of the Maxwell equations under Lorentz transformation. When the observer and source are in free space and the observer is in the far zone of the source, the exact relativistic formulation of the Doppler effect is known and well understood. But in the near zone of the source in free space or in the presence of material media, the Doppler effect is more complicated and may lead to results quite different from the normal Doppler effect. For the problem of calculating the Doppler effect in material media the reader is referred to the work of Papas [6] and Lee and Papas [4], [5], [7].

In this chapter the theory of the Doppler effect in the near zone of a monochromatic source of radiation in free space is reviewed, and as an illustrative example, the case where the source is an infinitesimal oscillating electric dipole is worked out in detail and some physical interpretations and

* The Moore School of Electrical Engineering, Department of Electrical Engineering, University of Pennsylvania, Philadelphia, PA 19104, USA.

insights are provided. Potential applications of the near-zone Doppler effect in navigation, direction finding, and target classification are also given, and the possibility of extension to the waves of other natures, e.g., acoustic waves, is discussed.

3.2. Formulation of the Problem of the Near-Zone Doppler Frequency

Fundamentally, the Doppler effect is a consequence of the covariance of Maxwell's equations under the Lorentz transformation. According to the theory of relativity, all the laws of physics must have the same form in all inertial frames of reference. Maxwell's equations are not exempted from this principle, i.e., they must be covariant under the Lorentz transformation. This covariance under the Lorentz transformation was originally introduced by Einstein, Lorentz, and Poincaré for the microscopic Maxwell–Lorentz equations, which is also known as the Maxwell equations of electron theory [8], [9], [10]. The theory was later generalized to the case of material media by Minkowski, from the postulate that the macroscopic Maxwell equations are also covariant under the Lotentz transformation [11], [12], [13]. In either case, the covariance of Maxwell's equations means that if we have two inertial frames, one K with coordinates x, y, z, and t, and the other K' with coordinates x', y', z', and t' moving at a uniform velocity $\mathbf{v}$ with respect to K (see Figure 3.1), and we write Maxwell's equations in the frame K and transform the coordinates x, y, z, t to the coordinates x', y', z', t' by a proper Lorentz transformation; the field vectors, the current density, and the charge density must transform in such a way that the transformed equations in the K' frame have the same forms as the original equations in the K frame.

Assuming that the spatial origins of these two reference frames be coincident at $t = t' = 0$, and that they have the same orientation, the Lorentz transformation law, which is a result of the postulate that the velocity of light in

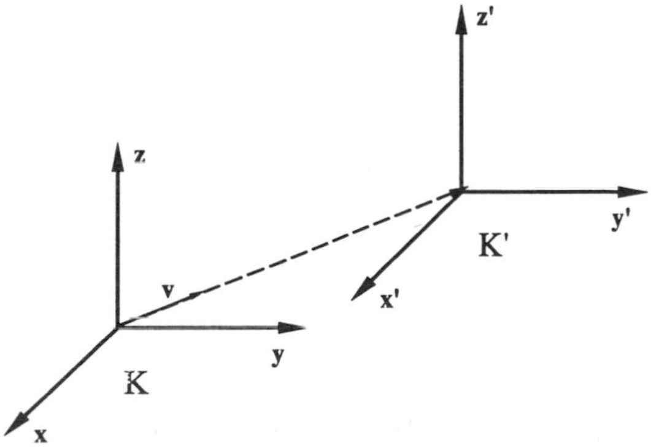

FIGURE 3.1. Reference frame K with coordinates x, y, z, t, and reference frame K' with coordinates x', y', z', t' moving at a uniform velocity v with respect to K.

vacuum has the same value c in all frames of reference, leads to the following relations for the coordinates x, y, z, and t of the K frame and the coordinates x', y', z', and t' of the K' frame [14]

$$\mathbf{r}' = \mathbf{r} - \gamma \mathbf{v} t + (\gamma - 1)\frac{\mathbf{r} \cdot \mathbf{v}}{v^2}\mathbf{v}, \tag{3.1}$$

$$t' = \gamma\left(t - \frac{\mathbf{r} \cdot \mathbf{v}}{c^2}\right), \tag{3.2}$$

where

$$\gamma = \frac{1}{\sqrt{1 - \beta^2}}, \tag{3.3}$$

$$\beta = \frac{v}{c}, \tag{3.4}$$

$$v = |\mathbf{v}|, \tag{3.5}$$

$$\mathbf{r} = x\mathbf{e}_x + y\mathbf{e}_y + z\mathbf{e}_z. \tag{3.6}$$

For the electric current density and electric charge density, the transformation law, in the rationalized MKS system of units, becomes

$$\mathbf{J}' = \mathbf{J} - \gamma \mathbf{v} \rho + (\gamma - 1)\frac{\mathbf{J} \cdot \mathbf{v}}{v^2}\mathbf{v}, \tag{3.7}$$

$$\rho' = \gamma\left(\rho - \frac{\mathbf{J} \cdot \mathbf{v}}{c^2}\right), \tag{3.8}$$

where $\mathbf{J}$ and ρ in the K frame are the electric current density and volume electric charge density, and $\mathbf{J}'$ and ρ' are their counterparts in the K' frame. Also, we find that the Lorentz transformation law leads to the following equations for the electromagnetic field quantities:

$$\mathbf{E}' = \gamma(\mathbf{E} + \mathbf{v} \times B) + (1 - \gamma)\frac{\mathbf{E} \cdot \mathbf{v}}{v^2}\mathbf{v}, \tag{3.9}$$

$$\mathbf{B}' = \gamma(\mathbf{B} - c^{-2}\mathbf{v} \times \mathbf{E}) + (1 - \gamma)\frac{\mathbf{B} \cdot \mathbf{v}}{v^2}\mathbf{v}, \tag{3.10}$$

$$\mathbf{D}' = \gamma(\mathbf{D} + c^{-2}\mathbf{v} \times \mathbf{H}) + (1 - \gamma)\frac{\mathbf{D} \cdot \mathbf{v}}{v^2}\mathbf{v}, \tag{3.11}$$

$$\mathbf{H}' = \gamma(\mathbf{H} - \mathbf{v} \times \mathbf{D}) + (1 - \gamma)\frac{\mathbf{H} \cdot \mathbf{v}}{v^2}\mathbf{v}. \tag{3.12}$$

Thus we see that when the coordinates, time, current density, charge density, and electromagnetic field quantities undergo the Lorentz transformations given in eqs. (3.1), (3.2) and (3.7)–(3.12), the Maxwell equations in the K frame (rationalized MKS system of units), viz.,

$$\nabla \times \mathbf{H} = \mathbf{J} + \frac{\partial \mathbf{D}}{\partial t}, \tag{3.13}$$

$$\nabla \times \mathbf{E} = -\frac{\partial \mathbf{B}}{\partial t}, \tag{3.14}$$

$$\nabla \cdot \mathbf{D} = \rho, \tag{3.15}$$

$$\nabla \cdot \mathbf{B} = 0 \tag{3.16}$$

transform into the Maxwell equations in the K' frame, viz.

$$\nabla' \times \mathbf{H}' = \mathbf{J}' + \frac{\partial \mathbf{D}'}{\partial t'}, \tag{3.17}$$

$$\nabla' \times \mathbf{E}' = -\frac{\partial \mathbf{B}'}{\partial t'}, \tag{3.18}$$

$$\nabla' \cdot \mathbf{D}' = \rho', \tag{3.19}$$

$$\nabla' \cdot \mathbf{B}' = 0. \tag{3.20}$$

We now consider a monochromatic source of electromagnetic radiation in the inertial frame K. This source can be described as electric current density $\mathbf{J}(\mathbf{r}, t) = \mathbf{J}(\mathbf{r}) \exp(-i\omega t)$, and the volume electric charge density $\rho(\mathbf{r}, t) = \rho(\mathbf{r}) \exp(-i\omega t)$ in a finite region of free space in the K frame. ω is the radian frequency. The two quantities $\mathbf{J}(\mathbf{r})$ and $\rho(\mathbf{r})$, which are phasors in the complex domain, are related via the equation of continuity, viz.,

$$\nabla \cdot \mathbf{J}(\mathbf{r}) = i\omega\rho(\mathbf{r}). \tag{3.21}$$

There are several methods of determining the electromagnetic fields radiated by given current and charge densities in an unbounded medium. One method is first to find the potentials of the source and then calculate the field from a knowledge of these potentials. The alternative method is that of the dyadic Green's function, which expresses the field directly in terms of the source current [15–19]. Using the latter method, the electric field generated by the source in the K frame is

$$\mathbf{E}(\mathbf{r}, t) = \left[i\omega\mu \int \mathbf{\Gamma}(\mathbf{r}, \mathbf{r}_s) \cdot \mathbf{J}(\mathbf{r}_s)\, dV_s \right] \exp(-i\omega t), \tag{3.22}$$

where $\mathbf{\Gamma}(\mathbf{r}, \mathbf{r}')$ is a dyadic Green's function of the coordinates of the observation point $\mathbf{r}$ and of the source point $\mathbf{r}_s$, and where the integration with respect to the source coordinates extends throughout the finite volume V_s occupied by the source. The quantity in the bracket in eq. (3.22) is a complex quantity and a vector function of the coordinate $\mathbf{r}$. Therefore, eq. (3.22) can be rewritten as the following:

$$\mathbf{E}(\mathbf{r}, t) = [E_x(\mathbf{r})\mathbf{e}_x + E_y(\mathbf{r})\mathbf{e}_y + E_z(\mathbf{r})\mathbf{e}_z] \exp(-i\omega t), \tag{3.23}$$

where E_j, $j = x, y, z$, are the scalar complex quantities, and $\mathbf{e}_j$, $j = x, y, z$, are the unit vectors in the inertial frame K. More explicitly, we have

$$\mathbf{E}(\mathbf{r}, t) = \{|E_x(\mathbf{r})| \exp[i\psi_{ex}(\mathbf{r})]\mathbf{e}_x + |E_y(\mathbf{r})| \exp[i\psi_{ey}(\mathbf{r})]\mathbf{e}_y + |E_z(\mathbf{r})| \exp[i\psi_{ez}(\mathbf{r})]\mathbf{e}_z\} \exp(-i\omega t). \tag{3.24}$$

In the above equation, the amplitudes $|E_j(\mathbf{r})|$ and the phases $\psi_{ej}(\mathbf{r})$, $j = x, y, z$, are real functions of $\mathbf{r}$. In the far zone of the source of radiation in the K frame, $|E_j(\mathbf{r})|$ and $\psi_{ej}(\mathbf{r})$ in the above expression have the following forms:

$$|E_j(\mathbf{r})| = \frac{|A_j(\mathbf{r}/r)|}{r}, \tag{3.25}$$

$$\psi_{ej}(\mathbf{r}) = \theta_{ej}(\mathbf{r}/r) + kr, \qquad j = x, y, z, \text{ in the far zone,} \tag{3.26}$$

where $k = \omega/c$, $r = \sqrt{\mathbf{r}\cdot\mathbf{r}}$; and $A_j(\ )$ and $\theta_{ej}(\ )$ are real functions of $\mathbf{r}/r$. Similar expressions can be written for other electromagnetic field vectors in K.

To find the expressions of the electromagnetic fields in the K' frame, we should use eq. (3.24) and the corresponding expressions for other fields in K, and substitute them in eqs. (3.9)–(3.12), and then transform the coordinates $\mathbf{r}$ and t of the K frame into the coordinates $\mathbf{r}'$ and t' of the K' frame. Following this procedure, the electric field in the K' frame can, in general, be written as follows:

$$\begin{aligned}\mathbf{E}'(\mathbf{r}', t') = \{&|E'_{x'}(\mathbf{r}', t')| \exp[i\psi'_{e'x'}(\mathbf{r}', t')]\mathbf{e}'_{x'} \\ &+ |E'_{y'}(\mathbf{r}', t')| \exp[i\psi'_{e'y'}(\mathbf{r}', t')]\mathbf{e}'_{y'} \\ &+ |E'_{z'}(\mathbf{r}', t')| \exp[i\psi'_{e'z'}(\mathbf{r}', t')]\mathbf{e}'_{z'}\} \exp(-i\gamma\omega t'),\end{aligned} \tag{3.27}$$

where $\mathbf{e}'_{j'}$, $j' = x', y', z'$, denote unit vectors in the K' frame, and $|E'_{j'}(\mathbf{r}', t')|$ and $\psi'_{e'j'}(\mathbf{r}', t')$ $(j' = x', y', z')$ are the amplitudes and phases of the components of the electric field in the K' frame. Note that $\psi'_{e'j'}(\mathbf{r}', t')$ are functions of both $\mathbf{r}'$ and t'.

In the K' frame, which is the rest frame of the observer, the phase of any field component is the important function in determing the observed frequency for that component. For instance, if the observer tends to measure the frequency for the z component of the electric field, the total phase of the z component of the electric field in the K' frame, i.e.,

$$\Phi'_{ez}(\mathbf{r}', t') = \psi'_{e'z'}(\mathbf{r}', t') - \gamma\omega t' \tag{3.28}$$

is the important function to evaluate. The frequency ω' can, in general, be defined as the time rate of change of the phase of a field component measured by the observer [20]. That is,

$$\omega' = -\frac{d\Phi'_{e'j'}(\mathbf{r}', t')}{dt'} = -\frac{d\psi'_{e'j'}(\mathbf{r}', t')}{dt'} + \gamma\omega \tag{3.29}$$

for the j' component of the electric field. In the K' frame, the observer is at rest at the origin. Therefore, $\mathbf{r}'$ is assumed to be zero for all time and we have

$$\omega' = -\frac{d\Phi'_{e'j'}(\mathbf{0}, t')}{dt'} = -\frac{d\psi'_{e'j'}(\mathbf{0}, t')}{dt'} + \gamma\omega. \tag{3.30}$$

The phase function $\psi'_{e'j'}$ can be expressed in terms of the coordinates $\mathbf{r}$ of the K frame. Thus we have

$$\psi'_{e'j'}(\mathbf{0}, t') = \varphi'_{e'j'}(\mathbf{r}(t')). \tag{3.31}$$

Equation (3.30) can be rewritten using eq. (3.31). That is,

$$\omega' = -\frac{d\Phi'_{e'j'}(\mathbf{0}, t')}{dt'} = -\frac{d\varphi'_{e'j'}(\mathbf{r}(t'))}{dt'} + \gamma\omega. \tag{3.32}$$

Expanding the above equation using chain rules [21], we obtain

$$\begin{aligned}\omega' &= -\frac{\partial\varphi'_{e'j'}(\mathbf{r}(t'))}{\partial t'} - \nabla\varphi'_{e'j'}(\mathbf{r}(t'))\cdot\frac{\partial\mathbf{r}}{\partial t'} + \gamma\omega \\ &= -\nabla\varphi'_{e'j'}(\mathbf{r}(t'))\cdot\frac{\partial\mathbf{r}}{\partial t'} + \gamma\omega.\end{aligned} \tag{3.33}$$

The coordinate $\mathbf{r}$ of the K frame is related to $\mathbf{r}'$ and t' of the K' frame through the following relation:

$$\mathbf{r} = \mathbf{r}' + \gamma\mathbf{v}t' + (\gamma - 1)\frac{\mathbf{r}'\cdot\mathbf{v}}{v^2}\mathbf{v}, \tag{3.34}$$

which is the transformation law in going from the K' frame to the K frame. Using eq. (3.34), we can write

$$\frac{\partial\mathbf{r}}{\partial t'} = \gamma\mathbf{v}, \tag{3.35}$$

and, finally, we obtain

$$\omega' = \gamma[\omega - \mathbf{v}\cdot\nabla\varphi'_{e'j'}], \qquad \text{measured for } E'_{j'}. \tag{3.36}$$

Introducing parameter η for a given field component, e.g., $E'_{j'}$, we can write eq. (3.36) as

$$\omega' = \gamma\omega\left[1 - \frac{\eta_{e'j'}v}{c}\right], \qquad \text{measured for } E_{j'}, \tag{3.37}$$

where

$$\eta_{e'j'} = \frac{c}{\omega}\,\frac{\mathbf{v}\cdot\nabla\varphi'_{e'j'}}{v}. \tag{3.38}$$

When $\eta_{e'j'} > 0$ we have $\omega' < \gamma\omega$, which is, for $\gamma \cong 1$, called a red shift, and when $\eta'_{e'j'} < 0$ we have $\omega' > \gamma\omega$, which is called a blue shift for $\gamma \cong 1$. Invoking the general definition of phase velocity measured in the direction of $\mathbf{v}$ [22], i.e.,

$$v_{\text{ph}} = \frac{\omega v}{\mathbf{v}\cdot\nabla\varphi}, \tag{3.39}$$

we can write eqs. (3.36) and (3.38) as follows:

$$\omega' = \gamma\omega\left[1 - \frac{v}{v_{\mathrm{phe'j'}}}\right], \qquad \text{measured for } E'_{j'}, \tag{3.40}$$

$$\eta_{e'j'} = \frac{c}{v_{\mathrm{phe'j'}}}, \qquad \text{measured for } E'_{j'}. \tag{3.41}$$

Equations (3.37) and (3.40) present the general expressions for the Doppler effects in the near zone as well as in the far zone of the source of radiation. In the far zone, the parameter η for all components approaches $\cos\theta$ and the so-called phase velocity for all components measured in the direction of $\mathbf{v}$ becomes $c/\cos\theta$, where θ is the angle between the direction of the wave propagation and the velocity of the observer. Hence, the measured frequency in the far zone reduces to

$$\omega' = \gamma\omega\left(1 - \frac{v\cos\theta}{c}\right) \tag{3.42}$$

which is the well-known relation for the conventional or normal Doppler effect. Also, in the far zone, the loci of constant phase are spherical surfaces which are approximated by plane surfaces. These surfaces are equidistant. In the near zone, however, the phase is distorted, the waves are no longer plane, i.e. the loci of constant phase are not, in general, spherical and quidistant. That is why the phase velocity is a function of position and field components. This variation is mainly responsible for the anomalous characteristics of the near-zone Doppler effect which will be discussed later.

Since the parameter η and the phase velocity defined in eqs. (3.38) and (3.39), in general depend on the field components and vary as functions of the position, the near-zone Doppler effect so defined depends on the position and the field components as well. In the far zone, however, η and v_{ph} are independent of the position and the field components. Therefore, the conventional, or so-called far-zone Doppler effect is neither a function of position nor a function of field components, and it only depends on the relative speed and direction of motion of the observer.

The above general analysis shows that as long as the observer is in the far-zone field of the source of radiation, the Doppler effect is quite normal. This means that the Doppler shift is the same for all field components, that it is independent of distance from the source, and that it is blue on approaching the source and red on receding from it. However, when the observer is in the source's near-zone field the Doppler effect is anomalous, i.e.,

(a) there, in general, exist several Doppler shifts, one for each field component;
(b) the Doppler shifts are functions of distance from the source; and
(c) the Doppler shifts are not necessarily red on receding from and blue on approaching the source.

To illustrate the near-zone Doppler effects, we present, in the next section, as an example, the case of an infinitesimal electric dipole as the source of radiation. For more details, the reader is refered to [23] and [24].

3.3. The Case of an Infinitesimal Electric Dipole

We consider an infinitesimal oscillating electric dipole at the origin of a Cartesian coordinate system (x, y, z) and lying along the z-axis. A spherical coordinate system (r, θ, φ) is also introduced which is related to the Cartesian system by $x = r \sin\theta \cos\varphi$, $y = r \sin\theta \sin\varphi$, and $z = r\cos\theta$ (see Figure 3.2). The electric and magnetic field vectors emitted by this dipole in the rest frame of the dipole are explicitly given by

$$E_r(r, t) = \frac{p}{4\pi\varepsilon_0 r}\left[-\frac{2ik}{r} + \frac{2}{r^2}\right]\cos\theta \exp[i(kr - \omega t)], \tag{3.43}$$

$$E_\theta(r, t) = -\frac{p}{4\pi\varepsilon_0 r}\left[\frac{ik}{r} - \frac{1}{r^2} + k^2\right]\sin\theta \exp[i(kr - \omega t)], \tag{3.44}$$

$$H_\varphi(r, t) = i\omega\frac{p}{4\pi\varepsilon_0 r}\left[ik - \frac{1}{r}\right]\sin\theta \exp[i(kr - \omega t)], \tag{3.45}$$

where p is the electric dipole moment, ε_0 is the dielectric constant of free space, $k = \omega/c$, and c is the vacuum speed of light [6]. For an observer at rest with respect to the dipole the frequency of the emitted field is ω.

As can be seen from (3.43)–(3.45), in the dipole's equatorial plane $(\theta = \pi/2)$, the electric vector has only a θ component E_θ, and the magnetic vector has only a φ component H_φ. These two field components in the equatorial plane can be expressed as

$$E_\theta(r, t) = -\frac{p}{4\pi\varepsilon_0 r}\left[\frac{ik}{r} - \frac{1}{r^2} + k^2\right]\exp[i(kr - \omega t)], \tag{3.46}$$

$$H_\varphi(r, t) = i\omega\frac{p}{4\pi\varepsilon_0 r}\left[ik - \frac{1}{r}\right]\exp[i(kr - \omega t)]. \tag{3.47}$$

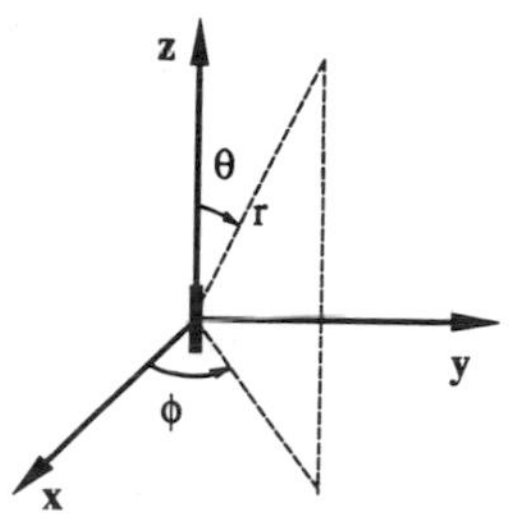

FIGURE 3.2. An infinitesimal oscillating electric dipole located at the origin of spherical and Cartesian coordinate systems.

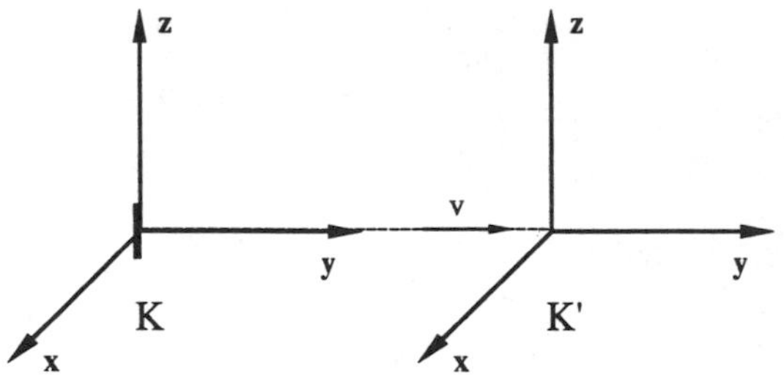

FIGURE 3.3. Coordinate system K for the rest frame of the dipole and coordinate system K' for the rest frame of observer moving at a uniform velocity $\mathbf{v}$ along the y-axis.

We now suppose that the observer is traveling at constant velocity $\mathbf{v}$ in the equatorial plane of the dipole. To simplify the problem further, we suppose that the observer is moving along the y-axis from $y = -\infty$ to $y = \infty$ with velocity $\mathbf{v} = v\mathbf{e}_y$, where $\mathbf{e}_y$ denotes a unit vector in the y direction. Moreover, we take the speed of the observer to be moderate, i.e., $\beta = v/c \ll 1$ (see Figure 3.3).

Substituting eqs. (3.46) and (3.47) into the Lorentz transformation laws given in eqs. (3.9) and (3.12), we obtain the electric and magnetic fields in the rest frame K' of the observer. To first order in β, they are given by

$$\mathbf{E}' = \mathbf{E} + \mu_0 \mathbf{v} \times \mathbf{H}, \tag{3.48}$$

$$\mathbf{H}' = \mathbf{H} - \varepsilon_0 \mathbf{v} \times \mathbf{E}, \tag{3.49}$$

where μ_0 and ε_0 are the permeability and dielectric constant of free space, respectively, and where $\mathbf{E}$ and $\mathbf{H}$ are the fields in the rest frame K of the dipole. Since $v/c \ll 1$, $\gamma \cong 1$, and the Lorentz transformation of the coordinates given in eqs. (3.1) and (3.2) reduces to

$$y = vt', \tag{3.50}$$

$$t = t'. \tag{3.51}$$

According to the description of the problem, when $t' < 0$ the observer is approaching the dipole, when $t' = 0$ the observer is at the dipole, and when $t' > 0$ the observer is receding from the dipole.

From a knowledge of E_θ and H_φ, as given in eqs. (3.46) and (3.47), we find from (3.48) and (3.49) that

$$E'_{\theta'} = -\frac{p}{4\pi\varepsilon_0 vt'}\left[\frac{ik}{vt'} - \frac{1}{(vt')^2} + k^2 + ik\beta\left(ik - \frac{1}{vt'}\right)\right]\exp[i(kv - \omega)t'], \tag{3.52}$$

$$H'_{\varphi'} = \frac{p}{4\pi vt'}\left[i\omega\left(ik - \frac{1}{vt'}\right) + v\left(\frac{ik}{vt'} - \frac{1}{(vt')^2} + k^2\right)\right]\exp[i(kv - \omega)t'], \tag{3.53}$$

for $t' > 0$, that is, for the observer moving away from the dipole. Similarly, we obtain

$$E'_{\theta'} = -\frac{p}{4\pi\varepsilon_0 vt'}\left[\frac{ik}{vt'} + \frac{1}{(vt')^2} - k^2 + ik\beta\left(ik + \frac{1}{vt'}\right)\right]\exp[-i(kv+\omega)t'], \tag{3.54}$$

$$H'_{\varphi'} = \frac{p}{4\pi vt'}\left[-i\omega\left(ik + \frac{1}{vt'}\right) - v\left(\frac{ik}{vt'} + \frac{1}{(vt')^2} - k^2\right)\right]\exp[-i(kv+\omega)t'], \tag{3.55}$$

for $t' < 0$, that is, for the observer moving toward the dipole. The measured fields can be expressed as

$$E'_{\theta'} = |E'_{\theta'}|\exp(i\psi'_{e'\theta'})\exp(-i\omega t'), \tag{3.56}$$

$$H'_{\varphi'} = |H'_{\varphi'}|\exp(i\psi'_{m'\varphi'})\exp(-i\omega t'), \tag{3.57}$$

where the amplitudes $|E'_{\theta'}|$ and $|H'_{\varphi'}|$ and the phases $\psi'_{e'\theta'}$ and $\psi'_{m'\varphi'}$ are real functions of t'. From eqs. (3.52)–(3.55), we obtain

$$\psi'_{e'\theta'} = \tan^{-1}\left[\frac{kvt'(1-\beta)}{(kvt')^2(1-\beta) - 1}\right] + \pi + kvt', \tag{3.58}$$

$$\psi'_{m'\varphi'} = \tan^{-1}\left[\frac{kvt'(1-\beta)}{(kvt')^2(1-\beta) + \beta}\right] + \pi + kvt', \tag{3.59}$$

for $t' > 0$; and

$$\psi'_{e'\theta'} = \tan^{-1}\left[\frac{-kvt'(1+\beta)}{(kvt')^2(1+\beta) - 1}\right] + \pi - kvt', \tag{3.60}$$

$$\psi'_{m'\varphi'} = \tan^{-1}\left[\frac{-kvt'(1+\beta)}{(kvt')^2(1+\beta) + \beta}\right] + \pi - kvt', \tag{3.61}$$

for $t' < 0$. Clearly, we can see that

$$\psi'_{e'\theta'} \to \pi + kvt', \tag{3.62}$$

$$\psi'_{m'\varphi'} \to \pi + kvt', \tag{3.63}$$

as $t' \to \infty$; and

$$\psi'_{e'\theta'} \to \pi - kvt', \tag{3.64}$$

$$\psi'_{m'\varphi'} \to \pi - kvt', \tag{3.65}$$

as $t' \to -\infty$.

In the rest frame K of the dipole, the phases of the electric field E_θ and magnetic field H_φ in the equatorial plane are given by

$$\psi_{e\theta} = \tan^{-1}\left(\frac{ky}{(ky)^2 - 1}\right) + \pi, \tag{3.66}$$

$$\psi_{m\varphi} = \tan^{-1}\left(\frac{1}{ky}\right) + \pi, \tag{3.67}$$

for $y > 0$; and

$$\psi_{e\theta} = \tan^{-1}\left(\frac{-ky}{(ky)^2 - 1}\right) + \pi, \tag{3.68}$$

$$\psi_{m\varphi} = \tan^{-1}\left(\frac{-1}{ky}\right) + \pi, \tag{3.69}$$

for $y < 0$. In the far-zone field of the dipole ($|ky| \gg 1$), we have

$$\psi_{e\theta} \to \pi, \tag{3.70}$$

$$\psi_{m\varphi} \to \pi. \tag{3.71}$$

Clearly, from (3.62)–(3.71) we see that, with respect to the transformation from K to K', the phases (excluding the shift term kvt') are invariant in the far zone of the dipole but not in the near zone, i.e., for $|kvt'| \ll 1$ and $|ky| \ll 1$. Thus for the near-zone field there is no phase invariance.

Having obtained the phases in the rest frame K' of the observer, we obtain the Doppler frequencies for the θ component of the electric field, the φ component of the magnetic field in the equatorial plane, and additionally for the r component of the electric field along the z-axis.

3.3.1. *The Doppler Frequency for the θ Component of the Dipole's Electric Field*

From eqs. (3.29), (3.31), (3.37), (3.58), and (3.60) we find that

$$\omega' = \omega\left[1 - \frac{\eta_{e'\theta'} v}{c}\right], \tag{3.72}$$

where

$$\eta_{e'\theta'} = \frac{1}{kv}\frac{d\psi'_{e'\theta'}}{dt'}, \tag{3.73}$$

and $\gamma \cong 1$. $\psi'_{e'\theta'}$ is given for $t' > 0$ and $t' < 0$ in eqs. (3.58) and (3.60), respectively. When $\eta_{e'\theta'} > 0$ we have a red shift, and when $\eta_{e'\theta'} < 0$ we have a blue shift.

Substituting (3.58) and (3.60) into (3.73), we obtain the following expressions for $\eta_{e'\theta'}$

$$\eta_{e'\theta'} = \frac{[(kvt')^2(1-\beta) - 1]^2 - (1-\beta)}{[(kvt')^2(1-\beta) - 1]^2 + [kvt'(1-\beta)]^2} \tag{3.74}$$

as the observer is receding from the dipole ($t' > 0$); and

$$\eta_{e'\theta'} = -\frac{[(kvt')^2(1+\beta) - 1]^2 - (1+\beta)}{[(kvt')^2(1+\beta) - 1]^2 + [kvt'(1+\beta)]^2} \tag{3.75}$$

as the observer is approaching the dipole ($t' < 0$). Figure 3.4 shows a plot of $\eta_{e'\theta'}$ as a function of kvt'. As can be seen from this figure, $\eta_{e'\theta'}$ has three zeros.

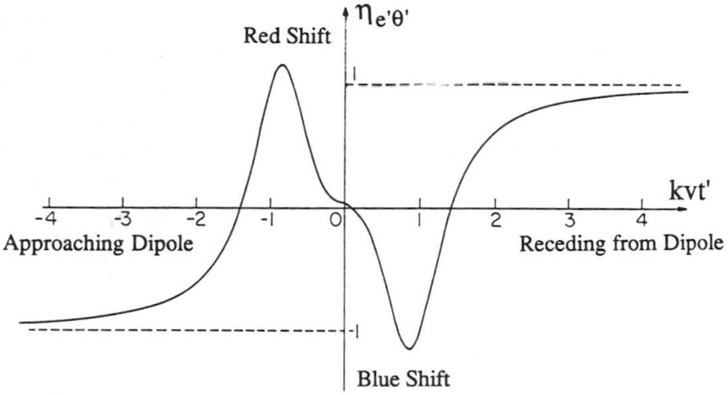

FIGURE 3.4. Function $\eta_{e'\theta'}$ for the θ component of the dipole's electric field in the equatorial plane as a function of kvt'. kvt' is positive when the observer is receding from the dipole, and is negative when the observer is approaching the dipole. Positive values of $\eta_{e'\theta'}$ correspond to red shifts, and negative values of $\eta_{e'\theta'}$ correspond to blue shifts. Here $\beta = v/c$ is taken to be 0.01 (after [23]).

For $v/c \ll 1$, the three zeros are

$$(kvt')_1 = \sqrt{\frac{1 + \sqrt{1 + \beta}}{1 + \beta}} \cong -\sqrt{2}(1 - \tfrac{3}{8}\beta), \tag{3.76}$$

$$(kvt')_2 = \sqrt{\frac{1 - \sqrt{1 - \beta}}{1 - \beta}} \cong \sqrt{\beta/2}, \tag{3.77}$$

$$(kvt')_3 = \sqrt{\frac{1 + \sqrt{1 - \beta}}{1 - \beta}} \cong \sqrt{2}(1 + \tfrac{3}{8}\beta). \tag{3.78}$$

From eq. (3.72) it follows that when $\eta_{e'\theta'}$ is zero, the frequency for the θ component of the electric field measured by the moving observer is the same as that of the source. Therefore, no Doppler shift is observed for those instances given in (3.76)–(3.78). From Figure 3.4, we also note that, as the observer travels from $y = -\infty$ to $y = \infty$, there is a blue shift for $-\infty \leq kvt' \leq (kvt')_1$. At $kvt' = (kvt')_1$, there is no shift. For $(kvt')_1 \leq kvt' \leq (kvt')_2$, although the observer is still approaching the source, an unexpected red shift is observed. Similarly, for $(kvt')_2 \leq kvt' \leq (kvt')_3$, there is a blue shift, although the observer is receding from the dipole. Finally, a red shift occurs for $(kvt')_3 \leq kvt' \leq \infty$. We note that for $|kvt'| \gg 1$, the function $\eta_{e'\theta'}$ approaches unity leading to the normal Doppler frequency shifts. Thus we see that, for the θ component of the electric field, there is an "inverse" Doppler effect in the vicinity of the dipole in free space. This effect is mainly due to anomalous behavior of the phase velocity in the near zone of the dipole. From eqs. (3.66) and (3.68), we see that the phase and, consequently, the phase velocity can be singular in the vicinity of the dipole, and its sign can change as one approaches the dipole. The phenomenon of the near-zone inverse Doppler effect is not observed in propagation of a uniform plane wave and in the far-zone field of a source. The peculiar behavior of the phase velocity, and its dependence on the field components in the near zone of a primary or of a secondary electromagnetic source, is responsible for the so-called near-zone inverse Doppler effect which was discovered by Engheta, Mickelson, and Papas in 1980 [23]. The peculiar inverse Doppler effect in an infinitesimal dipole in free space is within the zero crossings given in (3.76)–(3.78). The first and third of these zeros are approximately one-quarter of a wavelength λ away from the dipole, i.e.,

$$y_1 = (vt')_1 \cong -\frac{\sqrt{2}}{2\pi}\lambda, \tag{3.79}$$

$$y_3 = (vt')_3 \cong \frac{\sqrt{2}}{2\pi}\lambda. \tag{3.80}$$

and the second zero is very close to the dipole, i.e.,

$$y_2 = (vt')_2 \cong \sqrt{\frac{\beta}{2}}\frac{\lambda}{2\pi}. \tag{3.81}$$

3.3.2. *The Doppler Frequency for the φ Component of the Dipole's Magnetic Field*

Following the procedure used in the previous subsection, we find that the frequency for the φ component of the dipole's magnetic field, measured by the moving observer, is

$$\omega' = \omega\left[1 - \frac{\eta_{m'\varphi'}v}{c}\right], \tag{3.82}$$

where

$$\eta_{m'\varphi'} = \frac{1}{kv}\frac{d\psi'_{m'\varphi'}}{dt'} \tag{3.83}$$

is the parameter η for the φ component of the magnetic field in the K' frame, and $\gamma \cong 1$. $\psi'_{m'\varphi'}$ is given for $t' > 0$ and $t' < 0$ in eqs. (3.59) and (3.61), respectively.

Substituting (3.59) and (3.61) into (3.83), we obtain the following expressions for $\eta_{m'\varphi'}$:

$$\eta_{m'\varphi'} = \frac{[(kvt')^2(1-\beta)+\beta]^2 + \beta(1-\beta)}{[(kvt')^2(1-\beta)+\beta]^2 + [kvt'(1-\beta)]^2} \tag{3.84}$$

as the observer is receding from the dipole ($t' > 0$); and

$$\eta_{m'\varphi'} = \frac{[(kvt')^2(1+\beta)-\beta]^2 - \beta(1+\beta)}{[(kvt')^2(1+\beta)-\beta]^2 + [kvt'(1+\beta)]^2} \tag{3.85}$$

as the observer is approaching the dipole ($t' < 0$). This $\eta_{m'\varphi'}$ function is plotted versus kvt' in Figure 3.5.

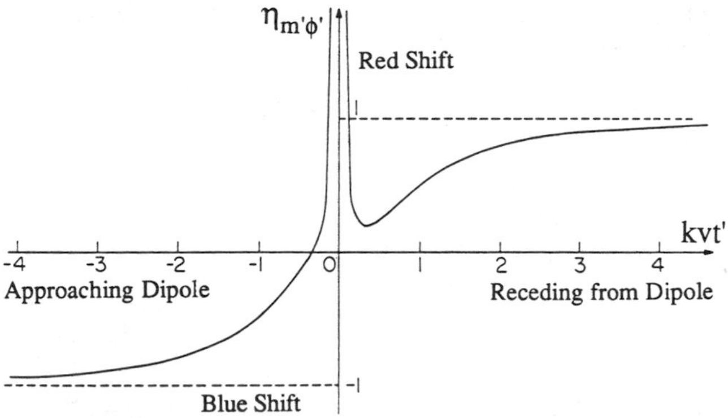

FIGURE 3.5. Function $\eta_{m'\varphi'}$ for the φ component of the dipole's magnetic field in the equatorial plane as a function of kvt'. kvt' is positive when the observer is receding from the dipole, and is negative when the observer is approaching the dipole. Positive values of $\eta_{m'\varphi'}$ correspond to red shifts, and negative values of $\eta_{m'\varphi'}$ correspond to blue shifts. Here $\beta = v/c$ is taken to be 0.01 (after [23]).

From this figure it is apparent that as the observer travels from $y = -\infty$ to $y = \infty$ and measures the frequency of the φ component of the magnetic field, first there is a blue shift for $-\infty \leq kvt' \leq (kvt')_0$, where

$$(kvt')_0 = -\sqrt{\frac{\beta + \sqrt{\beta(1+\beta)}}{1 + \beta}}, \tag{3.86}$$

and then a red shift for $(kvt')_0 \leq kvt' \leq \infty$. We note that the red shift appears when the observer is still approaching the dipole. Therefore, for the φ component of the dipole's magnetic field, there also exists a near-zone inverse Doppler effect in the vicinity of the dipole. However, for the φ component of the magnetic field, behavior of this inverse effect as a function of position of the observer (or equivalently the observer's time t') is different from that of the θ component of the electric field. The change from the blue to the red shift occurs at a distance y_0 from the dipole, where

$$y_0 = (vt')_0 \cong -\frac{\beta^{1/4}}{2\pi}\lambda \qquad \text{for} \quad \beta \ll 1. \tag{3.87}$$

3.3.3. *The Doppler Frequency for the r Component of the Dipole's Electric Field*

Since the radial component of the electric field is identically zero on the equatorial plane, to learn about the Doppler frequency for E_r we examine the Doppler frequency for an observer moving along the z-axis where E_r is most pronounced.

Following the procedure used in Subsections 3.3.1 and 3.3.2, we can show that the observed frequency ω' is given by

$$\omega' = \omega\left[1 - \frac{\eta'_{e'r'}v}{c}\right], \tag{3.88}$$

where

$$\eta_{e'r'} = \frac{1}{kv}\frac{d\psi'_{e'r'}}{dt'} \tag{3.89}$$

is the parameter η for the r component of the electric field in the K' frame, and $\gamma \cong 1$. $\psi'_{e'r'}$ is the corresponding phase of the r component of the electric field in the observer's rest frame K'. The function $\eta_{e'r'}$ is given by

$$\eta_{e'r'} = \frac{(kvt')^2}{(kvt')^2 + 1} \tag{3.90}$$

for $t' > 0$; and

$$\eta_{e'r'} = \frac{-(kvt')^2}{(kvt')^2 + 1} \tag{3.91}$$

for $t' < 0$. Thus as the observer travels from $z = -\infty$ to $z = \infty$, $\eta_{e'r'}$ goes smoothly from -1 to $+1$ and is zero at $z = 0$ ($t' = 0$) (see Figure 3.6). The near-zone effect is manifested through the smooth transition from a red shift to a blue shift or vice versa. However, there is no inverse Doppler effect for E_r.

As was shown in this section, the near-zone Doppler effect, in general, and the near-zone inverse Doppler effect, in particular, have been observed for the

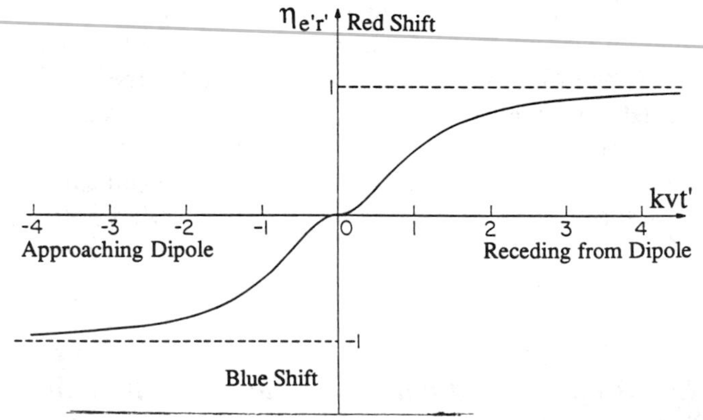

FIGURE 3.6. Function $\eta_{e'r'}$ for the r component of the dipole's electric field along the z-axis as a function of kvt'. kvt' is positive when the observer is receding from the dipole, and is negative when the observer is approaching the dipole. Positive values of $\eta_{e'r'}$ correspond to red shifts, and negative values of $\eta_{e'r'}$ correspond to blue shifts. Here $\beta = v/c$ is taken to be 0.01 (after [23]).

illustrative case of an infinitesimal electric dipole as a monochromatic source of electromagnetic energy. Prouty [24] extended the idea of the near-zone Doppler effect to other types of radiators and studied the interesting features of this effect for the two classes of antennas, viz., the prolate spheroidal antenna and the circular aperture antenna. The three attributes of the near-zone Doppler effect, namely: (a) the dependence of the Doppler effect on distance from the source; (b) several Doppler shifts, one for each field component; and (c) the possibility of an inverse Doppler shift, have been mathematically present, to some extent, for these two types of antennas.

3.4. Potential Applications

The anomalous characteristics of the near-zone Doppler effect can be potentially used in a variety of applications. Since the near-zone Doppler shift, so generally defined, is a function of distance from the source, and varies from one field component to the other, it is potentially more informative than the far-zone or normal Doppler effect. It can provide information about range and polarization, in addition to velocity. To illustrate this point more explicitly, consider a situation where an aircraft is flying toward a known primary source such as a transmitting antenna or toward a known secondary source such as a scatterer. The aircraft antenna will first sense a blue-shifted signal from which the velocity of the aircraft relative to the source can be determined. Then, as the aircraft enters the near zone of the antenna or scatterer, the aircraft's receiving antenna can, in principal, detect more that a single Doppler shift, one for each field component. From these shifts, the distance of the aircraft from the source can be determined.

The above example clearly illustrates potential applications of the near-zone Doppler effect in navigation and target detection and classification. If an unknown target is illuminated by a low-frequency signal with prescribed frequency and polarization, it will scatter the wave, and the scattered signal will be detected by the receiving antenna. The normal Doppler effect in the far zone, provides information on the velocity of the target, whereas the near-zone Doppler effect gives, in addition, range and polarization information. Polarization can, in general, reveal some information about the geometry of the targer. Thus, the velocity of the target, its range, and geometry can be obtained from the near-zone Doppler effect. This effect becomes most conceivable at low frequencies and/or for high-gain antennas where the near-zone field has a relatively large spatial extent [24].

Since the Doppler frequency has been defined in a rather general form in eqs. (3.37) and (3.40) in terms of the generalized phase velocity of the wave, the idea of an anomalous Doppler effect can be extended to the cases where the phase velocity in free space is known to be different from the speed of light in a vacuum. The near zone of a monochromatic source is a typical case which was discussed in this chapter. Another case is wave propagation in cylindrical waveguides. Work is under way to investigate the Doppler effect observed in such an environment.

The Doppler principle is not limited to the electromagnetic wave propagation. It is known that acoustic waves undergo a Doppler shift whenever an acoustic sensor and/or a source of acoustic wave are in motion. The idea of the near-zone Doppler effect described in this chapter can, in principal, be extended to the acoustic wave propagation.

Acknowledgments

I would like to dedicate this chapter to my mentor Professor Charles H. Papas of Caltech who has been an outstanding scholar and educator of science and humanity, and a dear friend. By his encouragement, continued guidance, and inspiration, he has taught me the fundamentals of electrodynamics and has shown me the elegance of the Maxwell equations.

I also wish to thank Professors H.N. Kritikos and D.L. Jaggard of the University of Pennsylvania for coming up with the idea of putting this book together and for their efforts as its editors.

References

[1] C. Doppler (1843), Uber das farbige Licht der Doppelsterne, *Abhandlungen der Koniglichen Bohmischen Gesellschaft der Wissenschaften.*

[2] E.N. Da and C. Andrade (1959), Doppler and the Doppler effect, *Endeaver*, **18**, 69.

[3] I.M. Frank (1943), Doppler effect in a refractive medium, *J. Phys. U.S.S.R.*, **2**, 49–67. See also, D.E.H. Rydbeck, Chalmers Research Report, No. 10, 1960.

[4] K.S.H. Lee (1968), Radiation from an oscillating source moving through a dispersive medium with particular reference to the complex Doppler effect, *Radio Science*, **3**, 1098–1104.

[5] K.S.H. Lee and C.H. Papas (1963), Doppler effects in inhomogeneous anisotropic ionized gases, *J. Math. Phys.*, **42**, 189–199.

[6] C.H. Papas (1965), *Theory of Electromagnetic Wave Propagation.* McGraw-Hill New York.

[7] K.S.H. Lee (1963), On the Doppler effect in a medium, *Caltech Antenna Laboratory Report*, N. 29, California Institute of Technology. See also, J.M. Jauch and K.M. Watson, Phenomenological quantum electrodynamics, *Phys. Rev.* **74**, 950, 1948.

[8] W. Pauli (1958), *Theory of Relativity*, Pergamon Press, New York.

[9] E. Whittaker (1953), *A History of the Theories of Aether and Electricity*, vol. 2, Harper & Row, New York.

[10] V. Fock (1952), *Theory of Space Time and Gravitation*, Pergamon Press, New York.

[11] A. Sommerfeld (1952), *Electrodynamics*, Academic Press, New York.

[12] C. Moller (1952), *The Theory of Relativity*, Oxford University Press, Fair Lawn, N.J.

[13] A. Einstein, H.A. Lorentz, H. Minkowski, and H. Weyl (1952), *The Principle of Relativity; A Collection of Original Memoirs*, Dover, New York.
[14] J.A. Kong (1986), *Electromagnetic Wave Theory*, Wiley, New York.
[15] C.T. Tai (1971), *Dyadic Green's Function in Electromagnetic Theory*, Intext, New York.
[16] C.H. Papas (1963), The role of dyadic Green's functions in the theory of electromagnetic wave propagation, *J. Geophys. Res.*, **68**, 1201.
[17] P.M. Morse and H. Feshbach (1953), *Methods of Theoretical Physics*, McGraw-Hill, New York.
[18] J. Van Bladel (1961), Some remarks on Green's dyadic for infinite space, *IRE Trans. Antennas and Propagation*, **AP-9**, 6, 563–566.
[19] H.C. Chen (1983), *Theory of Electromagnetic Waves*, McGraw-Hill, New York.
[20] A. Papoulis (1962), *The Fourier Integral and Its Applications*, McGraw-Hill, New York.
[21] F.B. Hildebrand (1976), *Advanced Calculus for Applications*, 2nd ed, Prentice-Hall, Englewood Cliffs, N.J.
[22] M. Born and E. Wolf (1975), *Principles of Optics*, Pergamon Press, Oxford.
[23] N. Engheta, A.R. Mickelson, and C.H. Papas (1980), On the near-zone inverse Doppler effect, *IEEE Trans. Antennas and Propagation*, **AP-28**, 519–522.
[24] D.A. Prouty (1982), Investigation of the near-zone Doppler effects, Ph.D. thesis, California Institute of Technology, Pasadena, CA. Also Caltech Antenna Laboratory Technical Report, No. 113, 1982.

4
Analysis of Channeled-Substrate-Planar Double-Heterostructure Lasers Using the Effective Index Technique

GARY A. EVANS* and JEROME K. BUTLER†

4.1. Introduction

Semiconductor lasers have applications as sources in optical communication systems, optical recording, consumer products, and interferometer systems. Because of the rapid advancement of the manufacture of optical fibers with low loss for the transmission of analog and digital signals, research in long-life and low-threshold single-mode laser sources has increased with high intensity over the last 25 years. The realization of high data-rate channels for optical fiber systems has been achieved using a combination of single-mode lasers, single-mode fibers, and high-speed detectors. Compact audio discs, along with other consumer applications such as laser printers, have driven the world production of laser diodes to over a million devices per month. The commercialization of optical memory for computers will provide a further surge in the production of semiconductor lasers. Semiconductor lasers are also being developed for space and satellite communication.

Laser sources fabricated with gaseous materials are realized by confining the optical field to a gain region whose boundaries are formed by a glass tube. The confinement mechanism occurs by properly forming the optical mirrors at each end of the optical tube. The electromagnetic fields in the cavity have no waveguiding mechanism so that the field is totally shaped by the end mirrors. (These fields are usually Gaussian.) On the other hand, the modes in a contemporary injection laser are shaped by the dielectric waveguide which is formed by the proper growth of semiconductor materials having different dielectric constants. Because the materials play a major role in the mode characterization of the laser, a majority of the research conducted in semiconductor laser design has been centered around the fabrication of single-mode devices. The most obvious way of achieving single-mode devices is to fabricate the waveguide with cross-sectional dimensions that are on the order

* David Sarnoff Research Center, Subsidiary of SRI International, Princeton, NJ 08543-5300, USA.

† Southern Methodist University, Dallas, TX 75275, USA.

of a wavelength in the material. However, such small dimensions would not only be difficult to fabricate, but such a small lasing volume would limit the output power. Consequently, it has been necessary to explore device structures that can achieve single-mode operation with practical dimensions of several microns. Typically, a gas laser has a circular cross section with a 1 mm or larger diameter and a length ranging from 30 cm to several meters. The waveguiding region in semiconductor lasers usually has a cross section of about 0.5 μm $\times$ 4 μm, and a length of 200–600 μm. The overal dimensions of a commercial semiconductor laser chip, before packaging, are comparable to a grain of salt, about 250-μm wide, 100-μm thick, and perhaps 300-μm long. Very recently, there has been research to extend the lateral and longitudinal dimensions of semiconductor lasers to several millimeters by forming edge-emitting [1] and surface-emitting [2] arrays.

Out of the several hundred types of semiconductor lasers that have been fabricated and studied, a laser structure that has obtained considerable attention is the channel-substrate-planar (CSP) laser fabricated from the AlGaAs material system. In addition to being important commercial products, CSP AlGaAs/GaAs lasers have been extensively studied both experimentally and theoretically [3–15]. They have single spatial mode output powers as high as any single element semiconductor laser [6], [7] and have demonstrated long life at very high power [8]. They have been used as the elements in linear [16], [17] and Y-guide [18], [19] arrays. Originally grown by liquid phase epitaxy (LPE) on *n*-type [3] and later *p*-type [4] GaAs substrates, functionally equivalent structures are also grown by metalorganic chemical vapor deposition (MOCVD) [20], [21] and molecular beam epitaxy (MBE) [22], [23]. Although GaAs substrates with GaAs buffer layers have been predominantly used for CSP lasers, AlGaAs buffer layers have also been used [17], [24].

In sections of this chapter we will:

(1) analyze the "cold cavity" modal characteristics of conventional CSP lasers which have GaAs and AlGaAs substrates; and
(2) present a physical explanation of the CSP guiding mechanism.

The waveguiding mechanisms of this extensively researched and highly developed laser structure (Figure 4.1(a)) have been qualitatively explained in physical terms [11], [15], and theoretical analyses [9–15] agree that lateral mode confinement results from the combination of a positive real index guide (Figure 4.1(b)) parallel to the junction, with high losses (related to a large imaginary component of the effective index) in the region outside the channel (Figure 4.1(c)). This resulting complex effective index profile provides mode confinement parallel to the junction.

Growth of CSP lasers on AlGaAs substrates or AlGaAs buffer layers is of interest to prevent meltback of the channel profile during the growth of both single devices and arrays [17], to prevent meltback of a grating incorporated in the structure that provides distributed feedback (DFB) [24], and to provide a transparent window for "junction-down" mounting of grating surface-

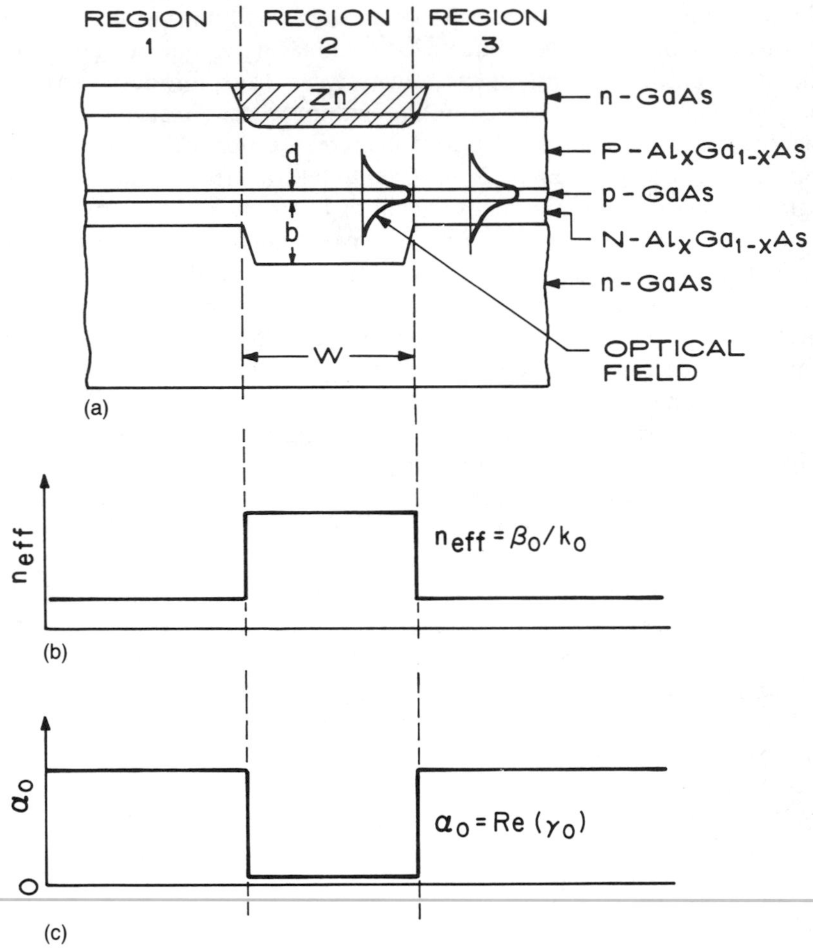

FIGURE 4.1. (a) Geometrical cross section of a CSP laser, (b) real part of the lateral effective index profile, and (c) imaginary part of the lateral effective index profile.

emitting lasers [2], [25]. In addition, AlGaAs buffer layers may be a more effective current-blocking layer than conventional GaAs-blocking layers [4], [20] and AlGaAs substrates reduce local heating.

The analysis of this chapter shows that CSP lasers emitting at wavelengths in the vicinity of 0.7–0.9 μm have a real positive index step even if the substrate has a mole fraction of AlAs approaching 0.3. Previously, a high absorption loss in the substrate was considered necessary for lateral guiding in CSP structures [11]. A common belief has been that CSP lasers would not be index guided at long wavelengths ($\geq$8600 Å) because of the reduced absorption of the lasing light by the GaAs substrate. (Depending on the dopant concentration of the substrate, the substrate absorption is reduced from about

5000 cm^{-1} at 8300 Å to about 100 cm^{-1} or less at 8800 Å [26].) The explanation of the guiding mechanism presented in this chapter is consistent with the experimental operation of index-guided CSP lasers at 8800 Å [27].

In the fabrication of any product, the manufacturing specifications allow for some tolerance. An ideal CSP laser would have a uniformly homogeneous composition throughout each layer, the channel shape would be symmetric, and the Zn diffusion region would be perfectly aligned. Experimentally this is seldom the case: within the channel of the *n*-clad layer (Figure 4.1(a)), significant nonuniformities in the direction perpendicular to the junction can exist in the Al/Ga ratio. Part of this chapter examines the consequences of these nonuniformities. We find that the AlAs nonuniformities in the channel can change a conventional CSP double heterostructure (DH) into either a CSP–LOC (large optical cavity) [28], [29] or a CSP structure with increased losses in the substrate which we have called an enhanced substrate loss (ESL) CSP. The resulting CSP–LOC laser generally has a wider perpendicular full-width half-power (FWHP) near-field distribution, and similar or larger perpendicular far-field beam divergence compared to a conventional CSP laser. The ESL–CSP laser often has a noticeable asymmetry in the perpendicular far field and can have either a larger or smaller FWHP perpendicular far field compared to a conventional CSP laser.

Noticeable asymmetries in the perpendicular far fields of conventional DH lasers are theoretically unexpected: The large real index steps perpendicular to the junction in AlGaAs DH lasers require that the near-field solution to the electromagnetic wave equation be almost real with only a negligible imaginary component due to active layer gain and material losses. Since the far-field pattern is related to the near-field distribution by a Fourier transform, the far-field pattern should be symmetric about an axis normal to the laser facet (the magnitude of the Fourier transform of a real function is symmetric). Although double-heterostructure lasers with a thin ($< 1.0\ \mu m$) cladding layer, as in the region outside the channel of CSP lasers, will have asymmetric, off-axis far-fields [30–32] due to radiation losses in the cap or substrate, we show in Section 4.2 that the contribution to the transverse far field from the regions outside the channel region is negligible.

Experimental measurements of some CSP lasers indicate that the Al concentration gradient may vary in the channel from high Al to low Al starting at the bottom of the channel, while in other channels on the same wafer, the gradient may be reversed. Both types of Al composition grading can occur in the channels of CSP lasers from portions of a wafer that otherwise produce conventional (uniform composition) CSP–DH lasers.

In this chapter we also analyze the influence of asymmetries in either the channel shape or the Zn diffusion on changes in the profile of the lateral gain and effective index distributions on the opto-electronic properties of CSP lasers. Alterations in gain profiles result from power-dependent interactions of the optical field with the electron-hole distribution. In an ideal symmetric structure, alterations of the gain profile produce only slight changes in the FWHP of the near- and far-field patterns. However, slight geometrical asymmetries can cause significant and undesirable shifts in the field patterns with increasing current. Furthermore, these slight asymmetries limit the ultimate emission power.

Several asymmetries can occur during the growth and fabrication of CSP–DH lasers. Asymmetry of the channel (shown in Figure 4.1(b)) can occur during etching of the channel into the substrate prior to growth or during the growth by meltback. Zinc diffused regions, used to confine current, may be misaligned with respect to the channel. Additionally, asymmetries introduced by nonuniform layer thicknesses and varying compositional properties generally produce variations in effective refractive index, and, if they exist in the active layer, strongly alter the local gain properties.

The first analysis of the static properties of stripe geometry lasers was given by Buus [33]. Analysis of different geometries and various laser materials have increased the understanding of the optical guiding mechanisms of semiconductor injection lasers [10], [11], [34–42]. In this chapter, we have modeled some of the optical characteristics of CSP–DH lasers with different device geometries associated with device fabrication. In one case we model the ideal device with symmetric current excitations as well as with symmetric device geometry. In the second case, the current is symmetric, but the channel profile is slightly asymmetric. In the third case, we consider asymmetries associated with misalignment of Zn diffusion relative to the channel. Our computations of the laser characteristics use a self-consistent calculation of the optical field and of the electron-hole pair distribution in the active layer. These detailed computations show how small asymmetries in the device geometry affect the optical and electrical properties of the CSP–DH laser. The parameters of the device model include the effects of the various layer thicknesses, refractive indices, and material absorption coefficients. Theoretical computations of the opto-electronic properties of symmetrical and asymmetrical configurations show good agreement with experimental measurements made on LPE grown CSP–DH lasers [7].

The lasers discussed in this chapter were grown by multibin LPE on a (100) GaAs substrate on which grooves had been chemically etched. Zinc stripe diffusion and final contacting were accomplished by using a chemical vapor-deposited SiO_2 diffusion mask and standard photolithographic techniques followed by e-beam evaporation of Ti–Pt–Au on the *p*-side and Ni–Au–Ge on the *n*-side. The diodes were mounted onto Au-plated Cu heat sinks using In or Sn solder.

4.2. The Effective Index Method

Early approaches of analyzing two-dimensional dielectric waveguides, other than those encountered in optical fibers, treated the two perpendicular waveguide modes independently [43]. More sophisticated approximate methods have been refined to what is generally now recognized as the effective index method [44], [45]. The effective index method yields field solutions which accurately approximate the actual fields when the transverse dielectric variations are large compared to the lateral ones. In semiconductor lasers, the lateral dielectric dependencies are due to the structure geometry, lateral gain variation, and temperature gradients. The lateral gain variations affect the imaginary part of the dielectric constant while temperature and carrier injection [46] affect the real part of the dielectric constant.

The relative dielectric constant κ in each layer of a semiconductor laser is generally assumed to be uniform, except in the active layer (the layer of thickness d_2 in Figures 4.1(a) and 4.2), which must reflect changes in the index of refraction due to carrier injection. The active layer dielectric constant can be written as

$$\kappa_2(y) = n_2^2 + 2n_2\delta n_2 + \frac{in_2 g(y)}{k_0}, \tag{4.1a}$$

$$= \kappa_{20} + \kappa_{2v}, \tag{4.1b}$$

where n_2 is the bulk refractive index of the active layer, δn_2 is the carrier-induced refractive index perturbation, $g(y)$ is the lateral gain distribution, k_0

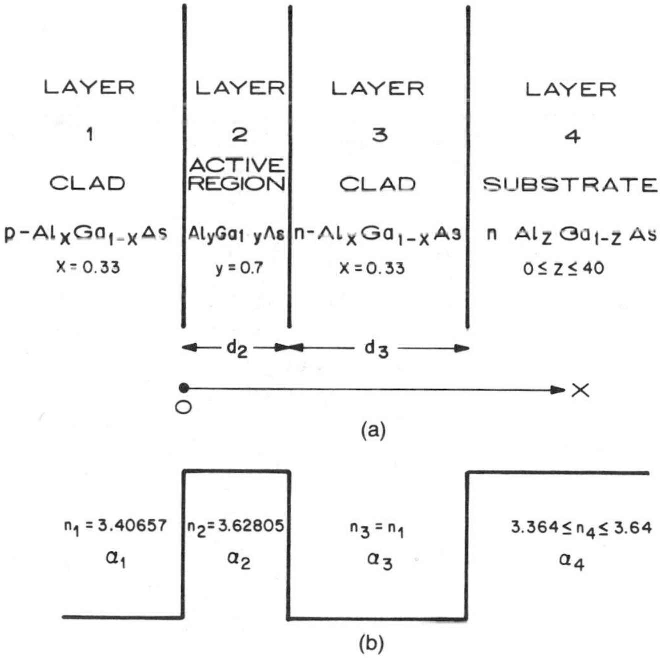

FIGURE 4.2. The four-layer waveguide structure with (a) the layer geometry and (b) the refractive index profile.

is the free-space propagation constant ($k_0 = 2\pi/\lambda$), and $\kappa_{20} = n_2^2$. κ_{2v} is the difference between the dielectric in the active layer with current injected ("hot cavity") and without current injected ("cold cavity"). Index changes due to temperature could be included in $\kappa_{2v}(y)$ as a real part. However, to make the analysis more general in nature, we write the dielectric constant of the ith layer as

$$\kappa_i(y) = \kappa_{i0} + \kappa_{iv}(y), \qquad i = 1, 2, 3, 4, \tag{4.2}$$

where κ_{iv} contains only "lateral" variations.

Most semiconductor lasers operate with fields polarized in the junction plane. Consequently, solutions to Maxwell's equations will be restricted to fields polarized along the junction plane and will be assumed to have the form

$$E_y(x, y) = E_0 \psi(x, y) \exp[i\omega t - \gamma z], \tag{4.3}$$

where $\gamma = \alpha + i\beta$ is the complex propagation constant, z is the longitudinal coordinate, ω is the radian frequency, E_0 is the amplitude of the electric field, and t is time.

The wave equation is

$$\nabla_t^2 \psi + [k_0^2 \kappa(x, y) + \gamma^2]\psi = 0, \tag{4.4}$$

where $\kappa(x, y)$ defines the complex dielectric constant of all space and ∇_t is the two-dimensional Laplacian operator. In the spirit of the effective index method, we write

$$\psi(x, y) = u(x, y)v(y), \tag{4.5}$$

where $u(x, y)$ describes the transverse fields and its y dependence appears there because of the layer thickness variations. The functional dependence of u is determined by the layer thickness $d_i(y)$ and the dielectric constant of each layer $\kappa_i(y)$. Substituting eq. (4.5) into eq. (4.4) gives

$$v\frac{\partial^2 u}{\partial x^2} + u\frac{\partial^2 v}{\partial y^2} + [k_0^2 \kappa(x, y) + \gamma^2]uv = 0, \tag{4.6}$$

where we have neglected derivatives of u with respect to y (the change in propagation constant due to layer thickness variations is very small compared to changes in the propagation constant due to dielectric variations in the x direction). Equation (4.6) can be rearranged in the following form:

$$v\left[\frac{\partial^2 u}{\partial x^2} + [\gamma_0^2 + k_0^2 \kappa_0(x, y)]u\right] + u\left[\frac{\partial^2 v}{\partial y^2} + [\gamma^2 - \gamma_0^2 + k_0^2 \kappa_v(x, y)]v\right] = 0 \tag{4.7}$$

which is satisfied if both

$$\frac{\partial^2 u}{\partial x^2} + [\gamma_0^2 + k_0^2 \kappa_0(x, y)]u = 0 \tag{4.8a}$$

and

$$\frac{\partial^2 v}{\partial y^2} + [\gamma^2 - \gamma_0^2 + k_0^2 \kappa_v(x, y)]v = 0 \tag{4.8b}$$

Equation (4.8a, b) defines the effective index concept and they have a simple physical interpretation. Equation (4.8a) is the wave equation for a planar dielectric waveguide at each point y. The effective index along the lateral direction is defined as $n_{e0}(y)$. It is obtained by repeated solutions of eq. (4.8a) for all values of y, yielding a lateral effective index profile which is now contained in eq. (4.8b). The x dependence of eq. (4.8b) can be eliminated by weighting with $|u(x, y)|^2$ and integrating over $(-\infty, \infty)$. In each layer we assume that κ_v has only y dependance. Therefore, upon integration, eq. (4.8b) becomes

$$\frac{\partial^2 v}{\partial y^2} + \left[\gamma^2 - \gamma_0^2 + k_0^2 \sum_i \Gamma_i(y)\kappa_{iv}(y)\right] v = 0 \tag{4.9a}$$

which is equivalent to the result derived by a more rigorous method [45].

The overlap parameter $\Gamma_i(y)$ satisfies

$$\Gamma_i(y) = \int_{i\text{th layer}} |u(x, y)|^2 \, dx. \tag{4.9b}$$

The transverse functions $u(x, y)$ are normalized to unity

$$1 = \int_{-\infty}^{\infty} u(x, y)u^*(x, y) \, dx \tag{4.9c}$$

along x so that $\sum_i \Gamma_i = 1$.

This above procedure characterizes the effective index method and reduces a two-dimensional dielectric waveguide problem (eq. (4.4)) into two one-dimensional problems (eqs. (4.8a) and (4.9a)).

For layered waveguides with no ohmic losses, the effective propagation constant $\gamma_0 = i\beta_0$ is imaginary for all proper modes.

For CSP structures, we can neglect temperature effects and consider that $\kappa_{iv}(y) = 0$ except for layer 2 (the active layer), and then (4.9a) assumes a form

$$\frac{\partial^2 v}{\partial y^2} + [\gamma^2 - \gamma_0^2 + k_0^2\Gamma_2(y)\kappa_{2v}(y)]v = 0, \tag{4.10}$$

where $\Gamma_2(y)$ is the active layer confinement factor (ratio of the mode power in the active layer and the total mode power). Defining $n_{e0} = -i\gamma_0/k_0$ as the complex effective index in the absence of injected current, the net lateral effective index is

$$n_{\text{eff}}(y)^2 = \left(\frac{i\gamma_0}{k_0}\right)^2 + \Gamma_2(y)\kappa_{2v}(y). \tag{4.11}$$

4.2.1. *The Four-Layer Slab Model*

The four-layer model can be applied to many contemporary laser structures. The four layers are associated with the epitaxial growth, the: n-type GaAs substrate, n-type AlGaAs cladding layer, an undoped AlGaAs active layer, and a p-type AlGaAs cladding layer, as shown in Figure 4.2. Although actual devices have a GaAs p-type "capping" layer for electrical contact, this final layer has negligible influence on the optical fields, assuming the top p-clad layer is greater than 1.0 μm in thickness.

The transverse field function $u(x, y)$ defined above is a solution of the wave equation (4.8a) and the secular equation for the modes in the four-layer waveguide is

$$[(r^2 - q^2) \tan rd_2 - 2qr](p + q) \exp(qd_2) + [(r^2 + q^2) \tan rd_2](p - q) \exp(-qd_2) = 0, \tag{4.12}$$

where

$$r^2 = [\gamma_0^2 + k_0^2 \kappa_2], \tag{4.13a}$$

$$q^2 = -[\gamma_0^2 + k_0^2 \kappa_i], \qquad i = 1, 3, \tag{4.13b}$$

$$p^2 = -[\gamma_0^2 + k_0^2 \kappa_4]. \tag{4.13c}$$

We have assumed for proper modes, i.e., modes that decay exponentially as $|x| \rightarrow \infty$, that $\mathrm{Re}\{p\}$ and $\mathrm{Re}\{q\} > 0$. Generally, the solutions are complex with complex eigenvalues [47] and the fields in the substrate are damped sinusoidal functions as shown in Figure 4.3. The solutions for the transverse field functions $u(x, y)$ are obtained in both the channel region ($|y| < W/2$) and the wing region ($|y| > W/2$) by matching the fields and their derivatives at the three interfaces located at $x = 0$, d_2, and d_3; W is the channel width. The lateral field functions $v(y)$ are computed after the complex effective index profile $n_{\mathrm{eff}}(y)$ is determined. In the analysis of the transverse mode functions, we assume the absorption losses in the p-cladding, active layer, and n-cladding layers have values $\alpha_1 = 5\ \mathrm{cm}^{-1}$, $\alpha_2 = 100\ \mathrm{cm}^{-1}$, and $\alpha_3 = 5\ \mathrm{cm}^{-1}$. (The actual value of loss (or gain) in the active layer has an insignificant effect on the shape of the complex field profiles. However, the lateral distribution of the gain in the active layer above threshold is very important in determining the optoelectronic characteristics of the device.) The Al content of the active layer is $\sim$7% and corresponds to a lasing wavelength of $\lambda_0 = 0.83\ \mu$m. The Al content of the cladding layers is 33% while the Al content of the substrate (normally 0% for CSP lasers), is varied from 0% to 40%. This variation allows for a study of the coupling of light from the lasing region to the substrate. Because the mole fraction of AlAs in the substrate is variable, the value of the substrate absorption coefficient (at $\lambda = 0.83\ \mu$m) ranges from about 10,000 cm^{-1} (p-type GaAs substrate) or 5000 cm^{-1} (n-type GaAs substrate) [26] down to 10 cm^{-1} as the mole fraction of AlAs in the substrate increases beyond 0.10. A material absorption value of 10 cm^{-1} is nominally assigned to the cladding layers and the AlGaAs substrates to account for free-carrier losses.

4.2.2. *Near- and Far-Field Patterns*

Solving the above equations for $u(x, y)$ and $v(y)$ provides the near-field patterns of the laser. The far-field intensity pattern is found from the two-

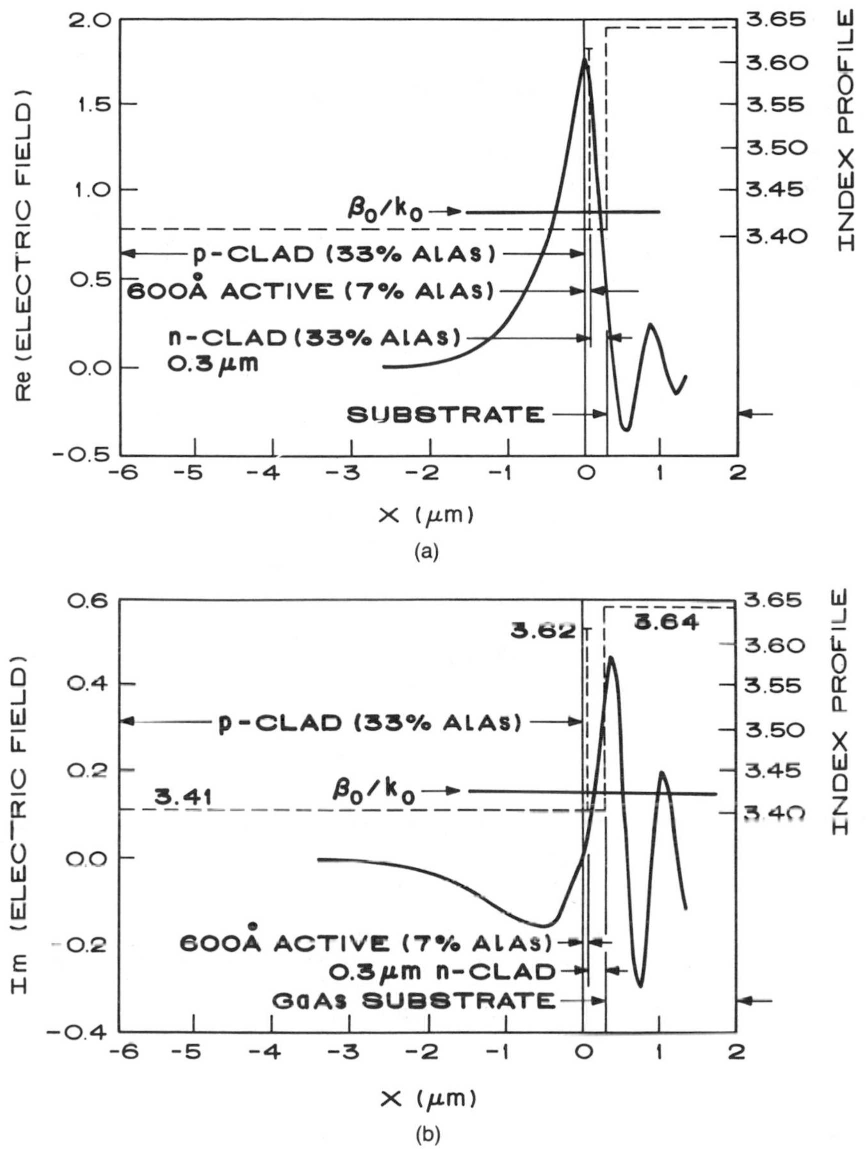

FIGURE 4.3. The (a) real and (b) imaginary part of the electric field distribution for the transverse profile shown in Figure 4.2 with $d_2 = 600$ Å and $\alpha_s = 5000\ \text{cm}^{-1}$.

dimensional Fourier transform of the aperture field distribution and the obliquity factor $g(\theta)$ [48]. The two-dimensional Fourier transform of the aperture field at the laser facet lying in the $z = 0$ plane is

$$F(k_x, k_y) = E_0 \int_{-\infty}^{\infty} \int_{-\infty}^{\infty} u(x, y)v(y) \exp[ik_x x + ik_y y]\, dx\, dy, \qquad (4.14)$$

where $k_x = k_0 \cos\varphi \sin\theta$ and $k_y = k_0 \sin\varphi \sin\theta$, and r, θ, and φ are the observation point coordinates in the far field defined by $z = r\cos\theta$, $x = r\sin\theta\cos\varphi$, and $y = r\sin\theta\sin\varphi$. The transverse radiation pattern is obtained by placing $\varphi = 0$ and varying θ. The lateral pattern is obtained by placing $\varphi = \pi/2$ and varying θ.

If we focus our attention on the transverse patterns by placing $\varphi = 0$, then

$$F(k_x, 0) = E_0 \int_{-\infty}^{\infty}\int_{-\infty}^{\infty} u(x, y)v(y)\exp[ik_x x]\,dx\,dy \tag{4.15}$$

represents an "average transform" of the transverse field function $u(x, y)$ where $v(y)$ is a weighting term. In the special case of a square channel CSP, the function $u(x, y)$ can be written as

$$u(x, y) = \begin{cases} u_{\text{ch}}(x), & |y| < W/2, \\ u_{\text{w}}(x), & |y| > W/2. \end{cases} \tag{4.16}$$

For this case, eq. (4.15) reduces to

$$F(k_x, 0) = F_{\text{ch}}(\theta)\zeta_{\text{ch}} + F_{\text{w}}(\theta)\zeta_{\text{w}}, \tag{4.17}$$

where

$$F_i(\theta) = E_0 \int_{-\infty}^{\infty} u_i(x)\exp[jk_x x]\,dx, \qquad i = \text{ch, w}, \tag{4.18}$$

and

$$\zeta_{\text{ch}} = \frac{\int_{-W/2}^{W/2} v(y)\,dy}{\int_{-\infty}^{\infty} v(y)\,dy}, \tag{4.19a}$$

$$\zeta_{\text{w}} = 2\frac{\int_{W/2}^{\infty} v(y)\,dy}{\int_{-\infty}^{\infty} v(y)\,dy}. \tag{4.19b}$$

Thus the transverse far-field radiation pattern perpendicular to the junction $I_\perp(\theta) = |g(\theta)F(k_x, 0)|^2$ for the square channel CSP laser can be written as

$$I_\perp(\theta) = I_{\text{ch}}(\theta)|\zeta_{\text{ch}}|^2 + I_{\text{w}}(\theta)|\zeta_{\text{w}}|^2 + 2|g(\theta)|^2\,\text{Re}[F_{\text{ch}}(\theta)F_{\text{w}}(\theta)\zeta_{\text{ch}}\zeta_{\text{w}}], \tag{4.20}$$

where $I_{\text{ch}}(\theta) = |g(\theta)F_{\text{ch}}(\theta)|^2$ and $I_{\text{w}}(\theta) = |g(\theta)F_{\text{w}}(\theta)|^2$. ζ_{ch}, the fraction of the

TABLE 4.1. Three models of CSP lasers.

	CSP	CSPNL	CSPNIC
d_2	600 Å	600 Å	600 Å
α_4	10,000 cm^{-1} (5000 cm^{-1})	0 cm^{-1}	10,000 cm^{-1} (5000 cm^{-1})
Δn	6.59×10^{-3} (6.82×10^{-3})	7.14×10^{-3}	2.55×10^{-3} (1.25×10^{-3})
$\Delta\alpha/k_0$	4.75×10^{-3} (5.07×10^{-3})	5.37×10^{-3}	2.57×10^{-3} (2.08×10^{-3})
$(\beta/k_0)_{\text{ch}}$	3.42315 (3.42315)	3.42315	3.42315 (3.42315)
$(\beta/k_0)_{\text{w}}$	3.41656 (3.41633)	3.41601	3.42061 (3.42190)
FWHP of $I_{\text{ch}}(\theta)$	25.18° (25.17°)	25.21°	24.88° (24.73°)
FWHP of $I_{\text{w}}(\theta)$	32.12° (32.21°)	32.40°	30.14° (28.17°)

lateral field confined to the channel, is almost unity. For the conventional CSP laser described in Table 4.1, $|\zeta_{ch}|^2$ ranges from 0.97 to 0.99 for channel widths of 4 6 μm. Therefore the transverse far-field pattern of the laser will be almost totally shaped by the Fourier transform of the transverse field in the channel region.

4.3. Lateral Optical Confinement

Waveguiding in CSP lasers appears paradoxical at first glance: Based on intuition for bound modes in dielectric waveguides, we expect that the CSP geometry has a larger effective index (real part) in the region outside the channel since a significant portion of the perpendicular field distribution there "averages in" the high index of the GaAs substrate. In addition, we expect that the large imaginary component of the effective index outside the channel region is due to high absorption (because $\alpha_4 = 5000$–$10{,}000\ \text{cm}^{-1}$ at a lasing wavelength of 0.83 μ [26]) in the GaAs substrate. However, analysis of the CSP structure shows that

(1) the effective index (real part) is higher inside the channel than outside (producing a positive index step and corresponding bound lateral modes); and
(2) the mode loss outside the channel region *increases* as the substrate absorption is *decreased.*

The reason for these apparent contradictions is that the transverse field in the regions outside the channel is not a conventional bound mode, but a complex field which radiates some power into the substrate. The conventional bound mode of a passive dielectric waveguide has decaying exponential field solutions in the first and last (the outermost, semi-infinite) layers. A leaky mode [49] has sinusoidal solutions with exponential growth in one or both of the outermost layers. When the outermost layers of an otherwise leaky waveguide have sufficient loss, the field solution is proper (referred to here as a "bound leaky mode") because the fields exponentially decay, albeit the decay is due to the absorption. Conventional bound modes have normalized transverse propagation constants β_0/k_0 that are greater than the refractive indices of the outermost layers. The complex transverse fields ("bound leaky modes") outside the channel have normalized propagation constants less than the refractive index of one (or both) outermost layers.

In Figure 4.3, the index of refraction profile of the layers outside the channel region is shown superimposed on a plot of the (a) real and (b) imaginary parts of the electric field for the fundamental mode. The magnitude (3.4163) of the normalized longitudinal propagation constant β_0/k_0 (at $y > W/2$) for the fundamental mode, also shown in Figure 4.3, corresponds to sinusoidal solutions in both the active layer and the substrate. The oscillatory behavior of the fields for $x > 0.36$ is characteristic of a complex field which radiates some power into the substrate. Even in the channel region, the mode perpendicular to the junction is also, strictly speaking, a complex field, but the electric field is so isolated from the substrate by the thick *n*-clad region that the amplitude of the field oscillations in the substrate are negligible (see Figure 4.3).

Figure 4.4 shows the (a) near-field intensities and (b) near-field phases for the fields perpendicular to the junction in both the regions inside and outside the channel for a CSP structure with $d_2 = 600$ Å and $\alpha_4 = 10{,}000\ \text{cm}^{-1}$. Note that the wavefront is tilted at 1.0° for $-1.5 < x < 0.0\ \mu$ and is tilted at about 20.2° for $x > 0.3\ \mu$ for the mode in the regions outside the channel. Since the direction of wave propagation is perpendicular to the wavefront, the wave outside the channel is therefore tilted away from the *z*-axis of the waveguide and is radiating some energy into the substrate. As the wavefront tilt increases, the guide wavelength λ_z increases (see Figure 4.5). Correspondingly, the propagation constant ($\beta = 2\pi/\lambda_z$) and the effective index ($n_{\text{eff}} = \beta/k_0$) decrease.

If we assume that the guide wavelength (λ_z) does not change for the fundamental field distribution perpendicular to the junction whether it is inside or outside the channel, we can calculate the average tilt angle of the field distribution from (see Figure 4.5)

$$\theta = \cos^{-1}\left[\frac{\lambda_p}{\lambda_z}\right] = \cos^{-1}\left[\frac{\beta_0(y > W/2)}{\beta_0(y = 0)}\right]. \tag{4.21}$$

For a structure with $d_2 = 600$ Å and $\alpha_4 = 10{,}000\ \text{cm}^{-1}$, β_0/k_0 (at $y = 0$) = 3.42315, and β_0/k_0 (at $y > W/2$) = 3.41656 which results in a tilt angle of 3.6° using eq. (4.21).

From curves such as Figure 4.4(b), a plane wave equivalent tilt Θ can be calculated between any two points x_1 and x_2 with corresponding phases (in degrees) $\Phi(x_1)$ and $\Phi(x_2)$ from

$$\Theta = \tan^{-1}\frac{[\Phi(x_1) - \Phi(x_2)]\lambda_0}{360 n_{\text{eff}}[x_1 - x_2]}. \tag{4.22}$$

Indicated in Figure 4.4(b) are plane wave equivalent tilts between the region of nonzero intensity ($-1.0 < x < 1.0$) of 8.5°, between the $1/e^2$ points in intensity of 2.5°, and between the half power points of 1.6°. Another indication of average tilt of the wavefront can be obtained from the calculated shift of the peak of the far-field intensity pattern (corresponding to the "wing" region, $|y| > W/2$) which is offset from 0° because of radiation into the substrate [13], [30–32]. For the mode perpendicular to the junction outside the channel region, the calculated shift in the far-field peak is 4.0° in air (Figure 4.4(c)), and therefore about 1.2° in the structure.

As a further illustration of the guiding mechanism, three structures are considered in Table 4.1. All three models have the same dimensions: a 600 Å active layer, 1.5-μm channel depth, and a 0.3-μm thick *n*-clad in the "wing" regions ($|y| > W/2$). The first model is that of a real CSP with losses of 10,000 cm^{-1} (5000 cm^{-1}) in the GaAs substrate. The second model (CSPNL) is the same as the first model, except that there are *no losses* in the GaAs substrate.

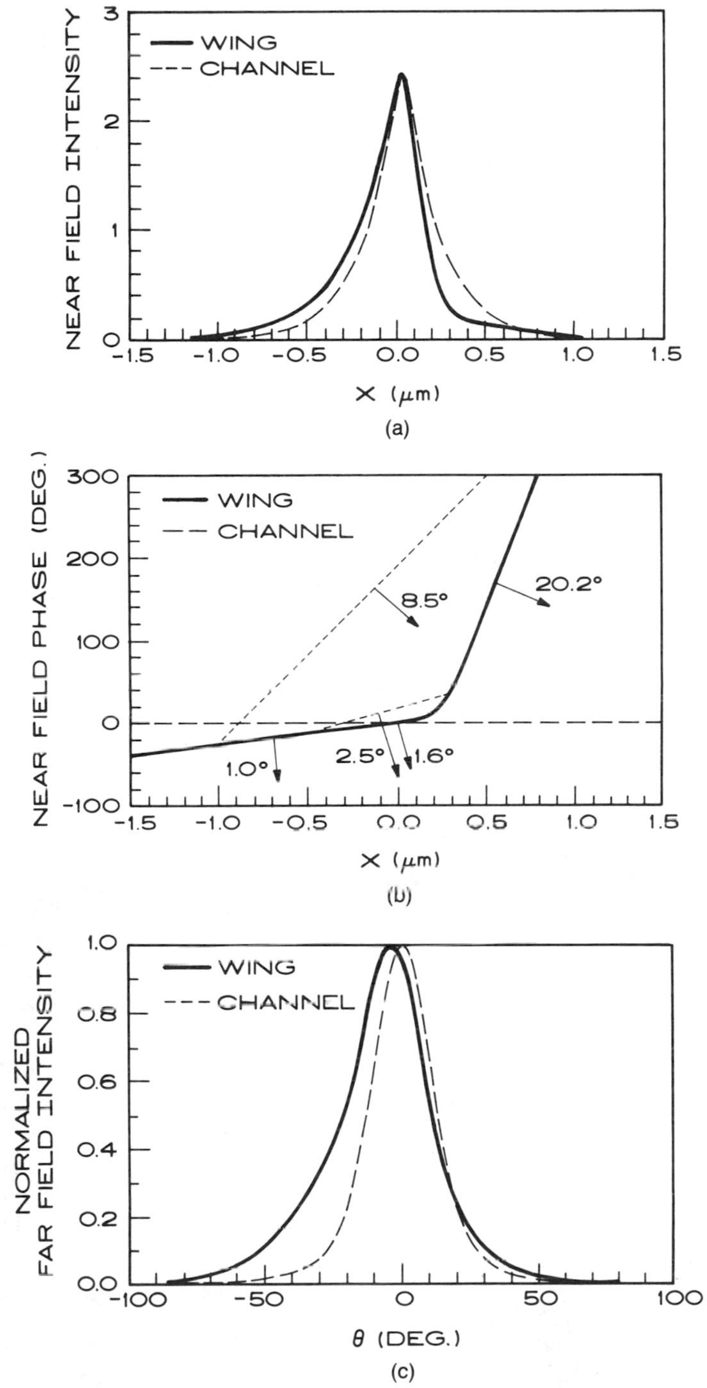

FIGURE 4.4. The transverse (a) near-field intensity, (b) near-field phase, and (c) far-field intensity $I_{ch}(\theta)$ and $I_w(\theta)$ for a conventional CSP laser with an active layer thickness of 600 Å in the region inside (– – –) and outside (——) the channel.

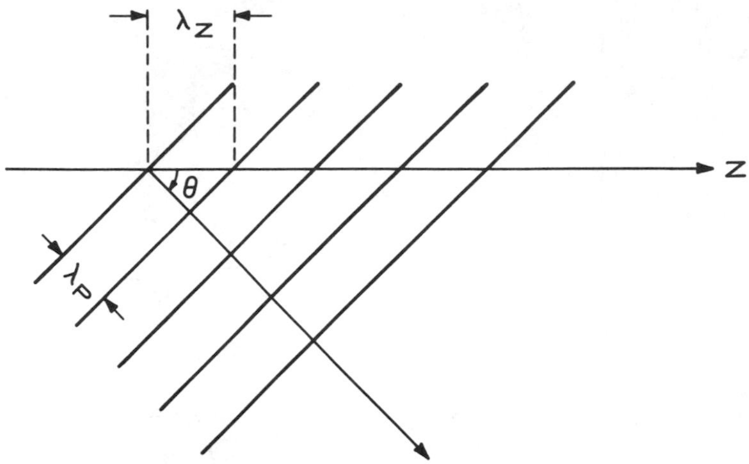

FIGURE 4.5. Relationship of the guide wavelength λ_z to the wavelength in the direction of propagation λ_p for a wave propagating at an angle θ with respect to the guide axis.

The third model (CSPNIC) is the same as the first model except that the "substrate" has the same AlAs composition as the n-clad layer (no $i$ndex $c$hange between the n-clad and the substrate). The last two models are not physical, but are pedagogically chosen to show the various effects of complex field distributions and substrate absorption on the complex effective index that provides lateral confinement in a CSP laser. The parameters listed in Table 4.1 are the real and imaginary parts of the lateral complex effective index step, and for both inside and outside the channel, the normalized longitudinal propagation constant and FWHP of the transverse beam divergence. The lateral optical confinement is due to the difference Δn and $\Delta\alpha$ in the real and imaginary parts of the complex effective index (γ_0/k_0) between the channel and wing region.

The near-field intensities, phases, and far-field intensities in the channel region are nearly identical for all three cases. However, the near-field intensities in the wing region, plotted in Figure 4.6(a) differ, primarily for $x > 0.36$, where the bound mode (CSPNIC) is the strongest damped, the CSP leaky mode is moderately damped, and the CSPNL leaky mode is undamped. The near-field phases in the wing region are plotted in Figure 4.6(b). They are similar for the CSP and CSPNL cases, with the CSPNL case having slightly more tilt for $x > 0.36$. The near-field phase of the CSPNIC has considerably

FIGURE 4.6. The transverse (a) near-field intensity, (b) near-field phase, and (c) far-field intensity in the region outside the channel region $[I_w(\theta)]$ for a conventional CSP laser (——), a CSP laser with no absorption in the GaAs substrate (---), and a structure with the same index in the substrate as the channel region, but with high losses (10,000 cm^{-1}) in the substrate (...). All structures have an active layer thickness of 600 Å. ▷

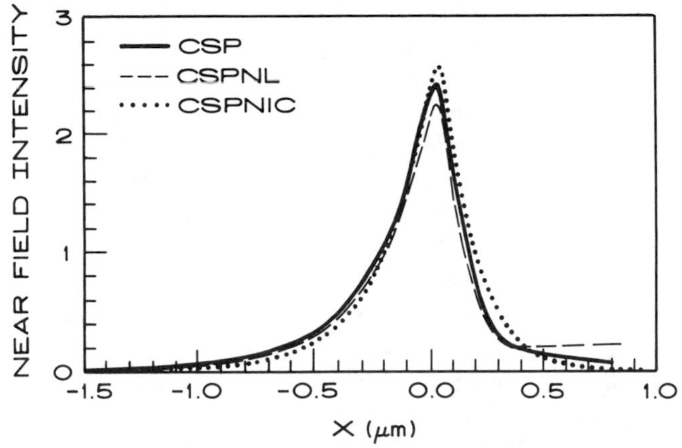
NEAR FIELD INTENSITY
CSP
CSPNL
CSPNIC
3
2
1
0
-1.5
-1.0
-0.5
0.0
0.5
1.0
X (μm)

(a)

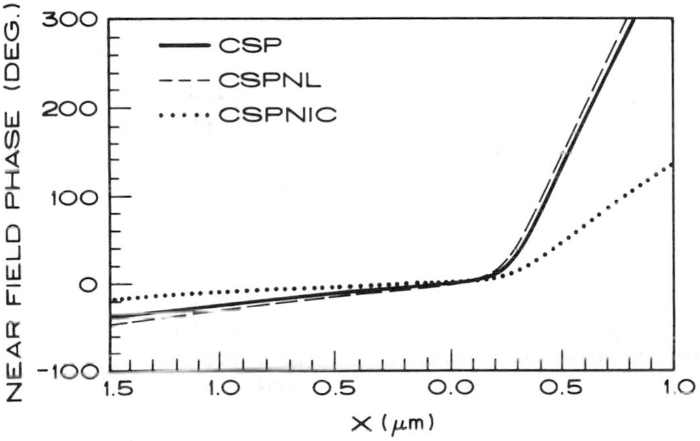
NEAR FIELD PHASE (DEG.)
CSP
CSPNL
CSPNIC
300
200
100
0
-100
1.5
1.0
0.5
0.0
0.5
1.0
X (μm)

(b)

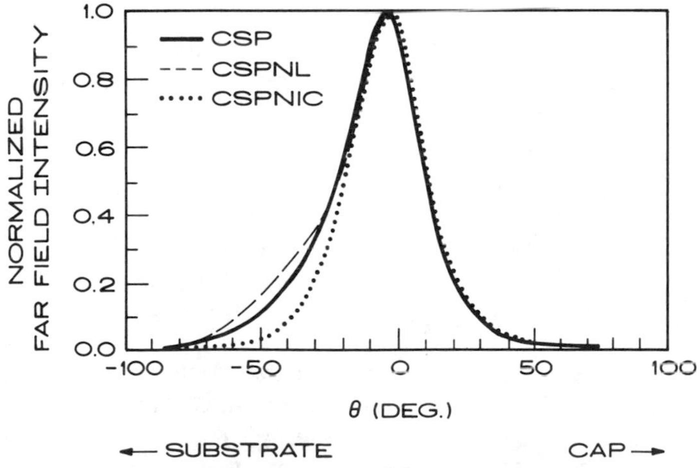
NORMALIZED FAR FIELD INTENSITY
CSP
CSPNL
CSPNIC
1.0
0.8
0.6
0.4
0.2
0.0
-100
-50
0
50
100
θ (DEG.)
SUBSTRATE
CAP

(c)

less tilt than the other cases. The far-field intensities corresponding to the near-field in the wing region are shown in Figure 4.6(c). The far-field peaks are all shifted from 0° to about 4°. Because the far-field pattern in the transverse direction obtained from either eq. (4.15) or (4.17) is due to the field in the channel and wing regions, the asymmetries in the far fields produced by the wing regions will always add a slight asymmetry to the far field produced by CSP lasers. Other causes of asymmetries in semiconductor far fields are thin cladding layers [30–32], material composition variations [13], and asymmetries in the device geometry [12], [14]. Although these far-field equations also hold for other laser geometries such as the ridge guide, the transverse fields inside and outside an ideal ridge structure both have symmetric transforms, and therefore ideal ridge (and most other) laser structures should have symmetric transverse far-field patterns.

Of the three models, the first two (CSP and CSPNL) give almost identical results, with the "no loss" model having a slightly larger complex effective index step. The third model does show that only high losses outside the channel region are sufficient to provide a lateral positive index, in agreement with an earlier qualitative calculation [11]. The index step of the third model is less than half the complex effective index step of the other two models for a substrate loss of 10,000 cm^{-1}. If the substrate absorption is 5000 cm^{-1}, the result is a positive index step of only 1.25×10^{-3} for the CSPNIC model, which is comparable to the amount of gain-induced index depression [46] expected at threshold.

These results are reasonable since from Figure 4.6(a) the CSPNL structure has a larger fraction of the mode intensity in the substrate, has slightly more tilt in the near-field phase, and has slightly more off-axis tilt in the far field than the CSP case. As the substrate absorption is decreased, more energy is radiated into the substrate resulting in an increased average wavefront tilt, and therefore both the positive index step and the mode absorption (imaginary part of the effective index) in the region outside the channel increases. Another way to view the increasing mode loss with decreasing substrate absorption is that the "skin depth" of the "bound leaky mode" increases as the substrate becomes less like a perfect conductor. Table 4.1 shows that the mode loss and the lateral index step only increase by about 13% and 8% as the substrate absorption decreases from 10,000 cm^{-1} to 0. This increase, although slight, is unexpected from an earlier explanation [11] of CSP waveguiding which predicts that the magnitude of the complex lateral effective index step approaches zero as the substrate absorption decreases towards zero.

4.3.1. *CSP Design Curves*

Figure 4.7(a) is a plot of the real part of the effective index for the mode perpendicular to the junction as a function of the *n*-clad thickness (d_3 in Figure 4.2) for active layer thicknesses of 300, 400, 600, 800, and 1000 Å. For a CSP laser to have stable near- and far-field patterns with increasing drive current, the lateral index step should be large compared to index variations due to

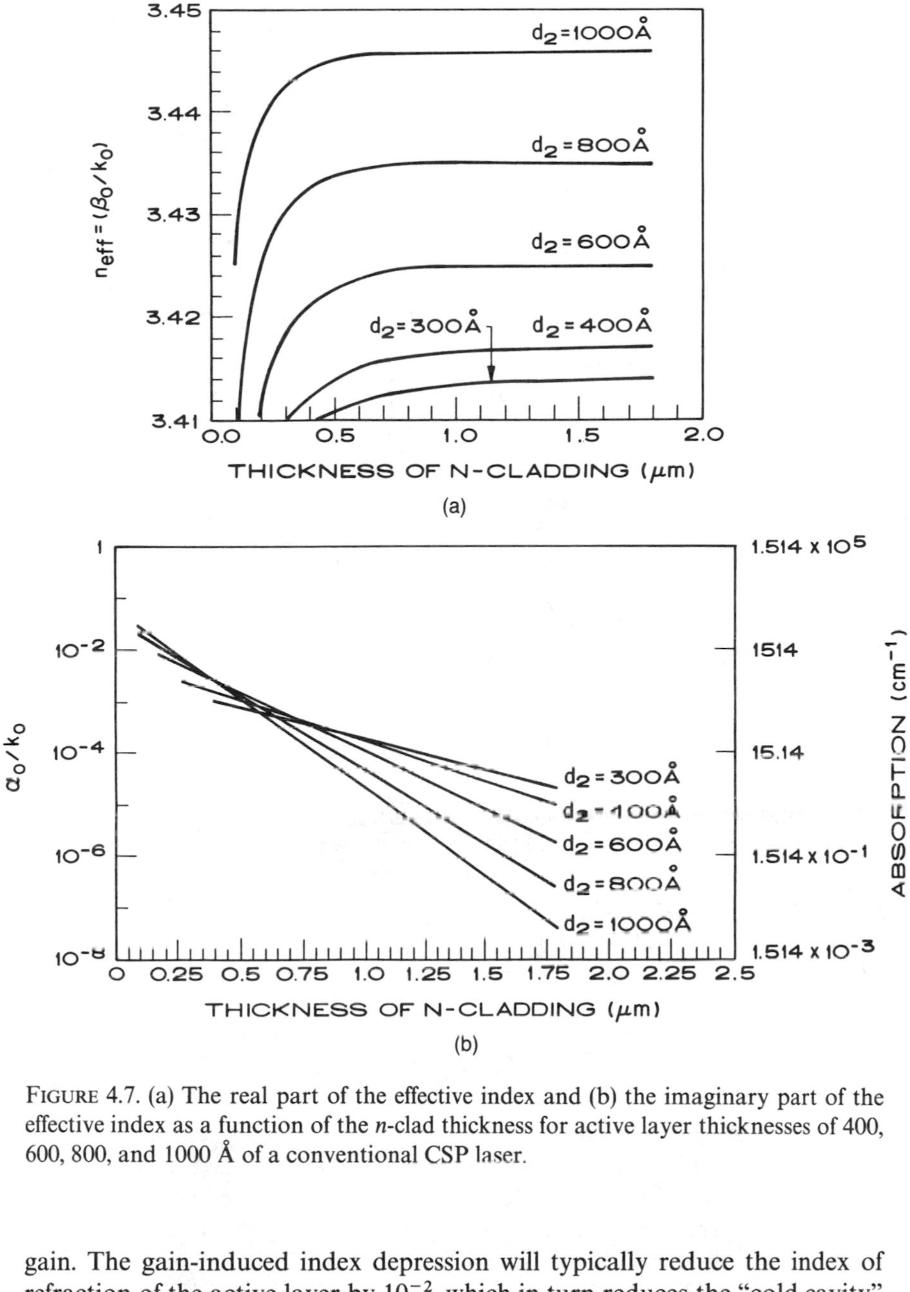

FIGURE 4.7. (a) The real part of the effective index and (b) the imaginary part of the effective index as a function of the *n*-clad thickness for active layer thicknesses of 400, 600, 800, and 1000 Å of a conventional CSP laser.

gain. The gain-induced index depression will typically reduce the index of refraction of the active layer by 10^{-2}, which in turn reduces the "cold cavity" effective index by about 10^{-3}. For this reason, a "cold cavity" lateral index step greater than 3×10^{-3} is desired to provide lateral mode stability. From Figure 4.7, this is obtained for *n*-clad thicknesses (outside of the channel) of 0.3 μm or less for channel depths greater than 1.0 μm.

Figure 4.7(b) is a plot of the imaginary part of the effective index, which

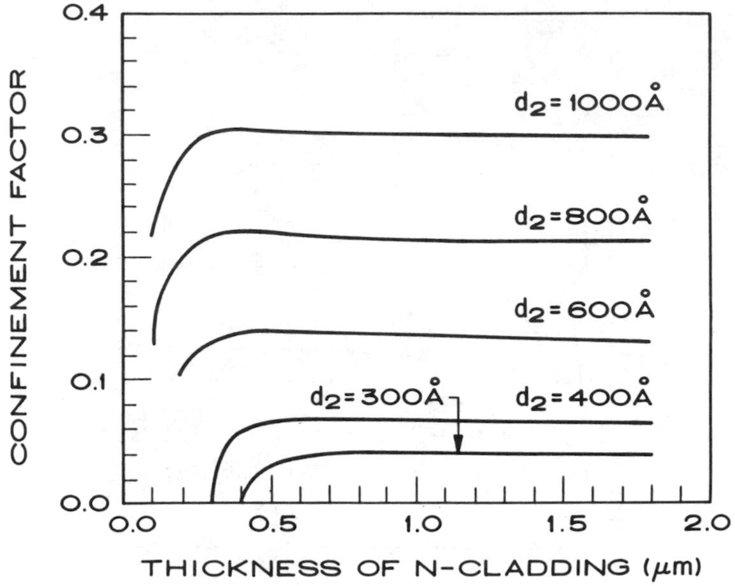

FIGURE 4.8. The active layer confinement factor as a function of the n-clad thickness for active layer thicknesses of 400, 600, 800, and 1000 Å of a conventional CSP laser.

increases with decreasing cladding thickness. The resulting magnitudes of the real and imaginary parts of the lateral effective index step for a CSP are always the same order of magnitude. Although a large imaginary component of the effective index stabilizes single laser diodes, it is detrimental to the operation of coherent CSP arrays in the fundamental array mode [50], [51].

Figure 4.8 is a plot of the active layer confinement factor (eq. (4.9b)) as a function of the n-clad thickness (d_3) for active layer thicknesses (d_2) of 300, 400, 600, 800, and 1000 Å.

4.3.2. *AlGaAs Substrates for CSP Lasers*

The use of an AlGaAs substrate or a very thick AlGaAs buffer layer is often desirable: Since AlGaAs has less meltback during LPE growth than GaAs, an AlGaAs layer may be preferred for fabricating gratings for DFB–CSP lasers [24]. The reduced meltback can also aid in maintaining the dimensions of the channel and the n-clad thickness. Due to a shorter diffusion length of minority carriers [52], [53], an AlGaAs buffer layer can be a more effective current blocking layer than conventional GaAs blocking layers [4]. Some surface emission devices such as the CSP–LOC–DBR [25] can be made more efficient if the lasing light is extracted through a transparent substrate.

Plots of the real and imaginary parts of the effective index for a CSP laser with a 600 Å active layer, and for substrate compositions of 0, 10, 20, 30, and

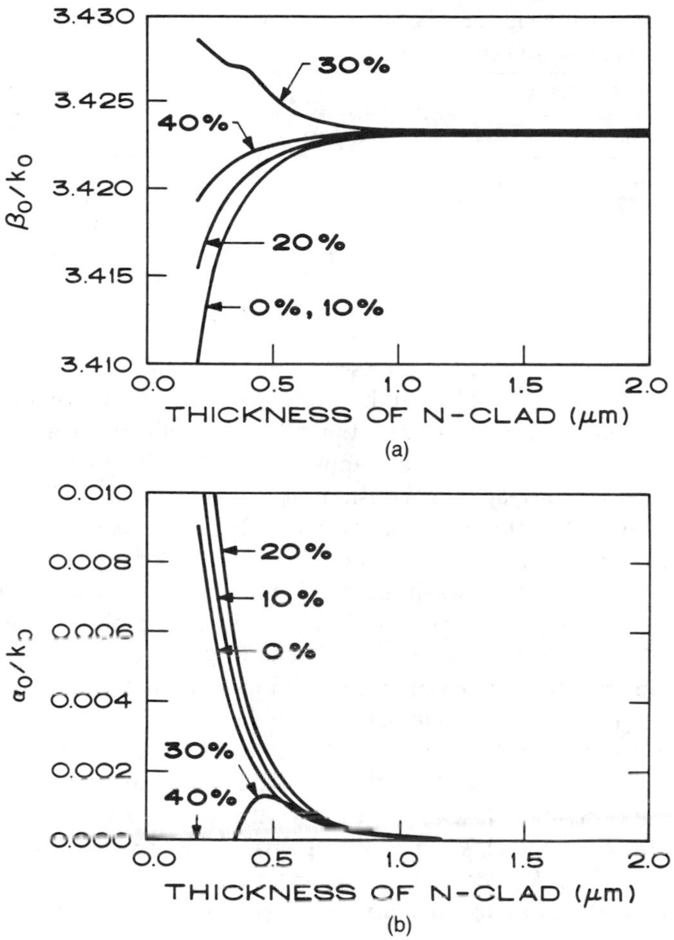

FIGURE 4.9. (a) The real part of the effective index and (b) the imaginary part of the effective index as a function of the *n*-clad thickness for an active layer thickness of 600 Å for CSP lasers with mole fractions of AlAs of 0, 10, 20, 30, and 40%.

40% AlAs are shown in Figure 4.9. The behavior of the curves for 10% and 20% AlAs have the same explanation as the effective index curves for CSP lasers on GaAs substrates: the real part of the lateral effective index is smaller outside the channel region than inside because of the wavefront tilt. Although this tilt is the dominant mechanism responsible for the smaller effective index outside the channel, there is less of an index step as the Al increases from 0% to 20% in the substrate. The reason for this is that λ_z (see Figure 4.5) in the region outside the channel is reduced as the index of the substrate decreases—the same behavior we expect for a conventional bound mode—causing the real part of the effective index to increase.

For a fixed *n*-clad thickness, for example, 0.3 μm, and for mole fractions of

TABLE 4.2. Confinement factors of CSP lasers with 0%, 10%, and 20% AlAs substrates ($d_2 = 600$ Å, $d_3 = 0.3\ \mu m$).

	0.0 AlAs		0.1 AlAs		0.2 AlAs	
Γ_{layer}	Channel	Wing	Channel	Wing	Channel	Wing
Γ_1	0.431	0.558	0.431	0.476	0.431	0.290
Γ_2	0.139	0.138	0.139	0.120	0.139	0.083
Γ_3	0.431	0.248	0.430	0.223	0.430	0.184
Γ_4	1.6×10^{-5}	0.055	5.8×10^{-5}	0.182	1.8×10^{-3}	0.443

AlAs below about 30%, the mode losses increase with increasing mole fraction of AlAs in the substrate. This can be explained by increased coupling of the mode energy to the substrate (see Γ_4, the substrate confinement factor in Table 4.2), as the index of the substrate approaches the value of the normalized transverse mode propagation constant β_0/k_0. This increased coupling of energy to the substrate is also apparent in Figure 4.10(a) which shows a reduced peak mode intensity with increased mode intensity in the substrate; in Figure 4.10(b) which shows the near-field phase; and in Figure 4.10(c) which shows an increasing tilt of the far-field patterns as the mole fraction of AlAs increases. Because of the increased field penetration (increased "skin depth") into the substrate with increasing mole fraction of AlAs, the peak temperature due to local heating at the channel shoulders [54], [55] may be reduced by spreading the absorption of the optical field further into the substrate.

As the mole fraction of AlAs in the substrate approaches 30%, the index in the substrate becomes close to the value of the normalized transverse mode propagation constant β_0/k_0 in both regions. Table 4.3 lists the applicable parameters for a CSP laser with an AlAs mole fraction of 0.30 in the substrate. For this case, the effective index for the region outside the channel is greater than the index of the substrate, corresponding to a conventional bound field distribution. Inside the channel, the effective index is less than the index of the substrate, resulting in a complex transverse field distribution ("bound leaky mode") in the channel region, exactly opposite to a conventional CSP laser. Because the effective index of the (complex) transverse field distribution in the channel region is very close to the value of the index of the substrate, the mode energy couples strongly to the substrate. This strong coupling is analogous to impedance matching in a transmission line, and is similar to the effect of enhanced coupling of mode energy to the substrate by nonuniformities in the Al composition in the channel region of a CSP laser [13]. Although this strong coupling is not apparent in the near-field intensity (Figure 4.11(a)), the near-field phase (Figure 4.11(b)) shows a wavefront tilt of 2.8° for $x > 1.5\ \mu m$. The substrate confinement factor in the channel region (Table 4.3) for this case is

FIGURE 4.10. The transverse (a) near-field intensity, (b) near-field phase, and (c) far-field intensity pattern $I_w(\theta)$ in the region outside the channel region for CSP lasers with mole fractions of AlAs in the substrate of 0(——), 10(– – –) and 20% (. . .). ▷

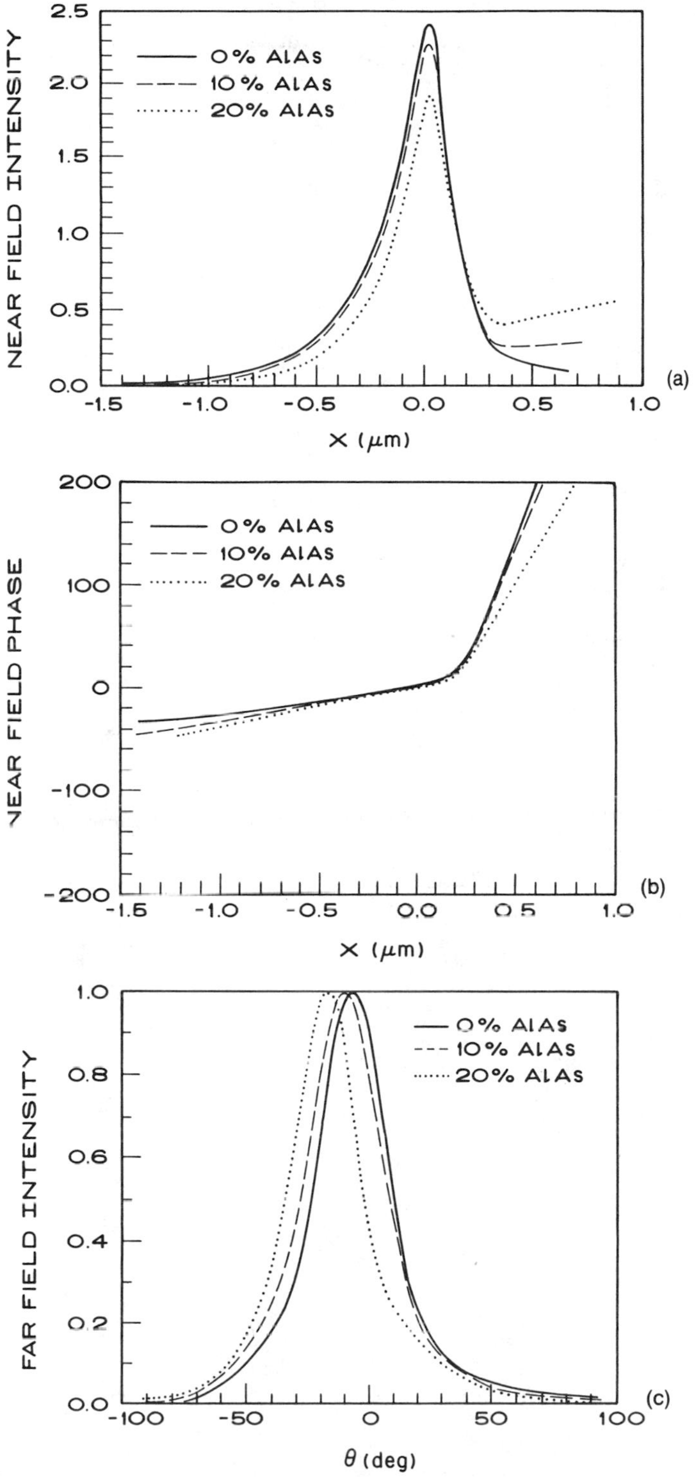
NEAR FIELD INTENSITY
0% AlAs
10% AlAs
20% AlAs
X (μm)
(a)
NEAR FIELD PHASE
0% AlAs
10% AlAs
20% AlAs
X (μm)
(b)
FAR FIELD INTENSITY
0% AlAs
10% AlAs
20% AlAs
θ (deg)
(c)

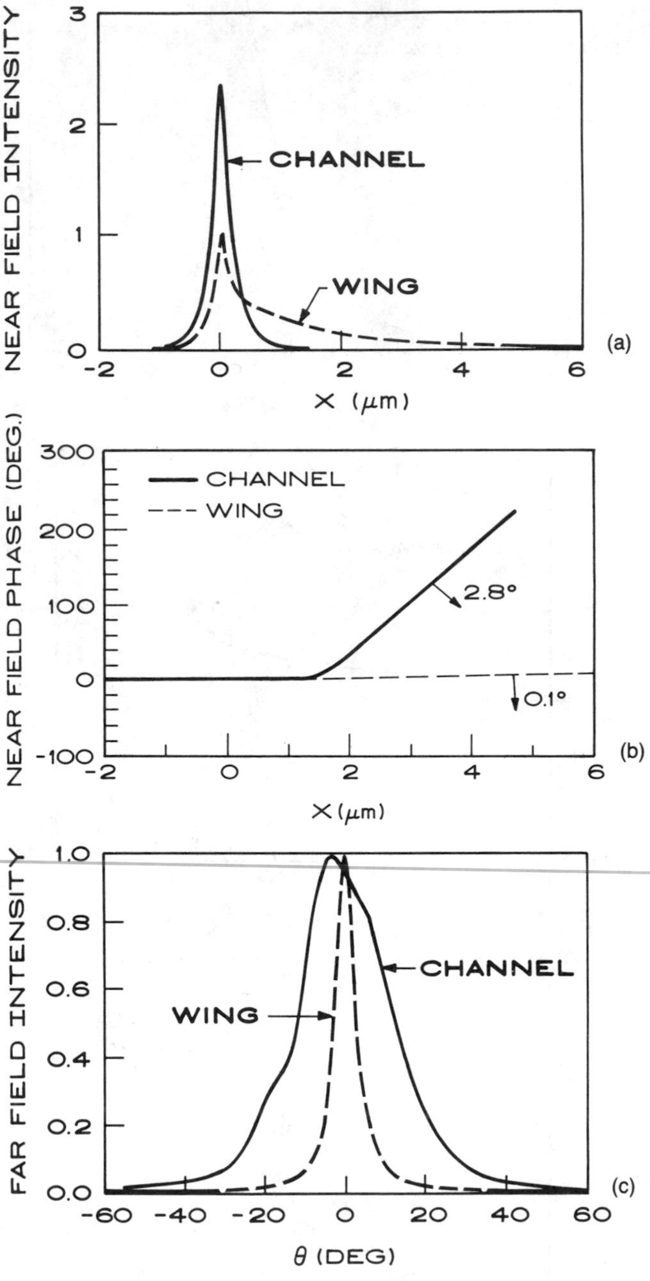

FIGURE 4.11. The transverse (a) near-field intensities, (b) near-field phases, and (c) far-field intensity patterns $I_{ch}(\theta)$ and $I_w(\theta)$ in the region inside (——) and outside (– – –) the channel for CSP laser with a mole fraction of AlAs in the substrate of 30%.

TABLE 4.3. CSP laser with 30% AlAs substrate.

Structure	Wing parameters (bound mode)	Channel parameters (leaky mode)
$d_2 = 600$ Å	$\beta/k_0 = 3.4273$	$\beta/k_0 = 3.4232$
$n_4 = 3.4270$	$\Gamma_1 = 0.16$	$\Gamma_1 = 0.43$
$\alpha_4 = 10\ \text{cm}^{-1}$	$\Gamma_2 = 0.06$	$\Gamma_2 = 0.14$
$\Delta n = -4.17 \times 10^{-3}$	$\Gamma_3 = 0.18$	$\Gamma_3 = 0.43$
$\Delta\alpha/k_0 = 2.32 \times 10^{-6}$	$\Gamma_4 = 0.60$	$\Gamma_4 = 0.0036$

two orders of magnitude larger than Γ_4 for the channel regions of any of the other cases considered (Tables 4.2 and 4.3). In addition, the far-field pattern (Figure 4.11(c)) corresponding to the channel region is characteristic of strong coupling of mode energy to the substrate—the pattern is asymmetric and tilted 2.5° towards the substrate. For the 30% case, Figure 4.9(b) is an example of a resonant region (for *n*-clad thicknesses $0.4 < d_3 < 1.0$) in which the "average wavefront tilt" argument no longer properly predicts the index step. Again, the assumption that λ_z (Figure 4.5) remains unchanged by the tilt is no longer the case. In fact, in this case the change in λ_z more than offsets the wavefront tilt. λ_z decreases (the effective index increases) because n_4 is greater than n_3 and the substrate confinement factor is very large (increasing from 0.19 to 0.95 as d_3 decreases from 1.0 to 0.4). Figure 4.12, which plots the active layer confinement factor for the cases shown in Figure 4.9, shows a large decrease in the active layer confinement factor over the same range in d_3.

As the *n*-clad thickness decreases even further below about 0.35 μm), the mode becomes a conventional bound mode and the effective index of the mode continues to increase because an increasing fraction of the mode energy is in the (slightly) higher index substrate. In Figure 4.9(b) the losses for the 30% case increases as the *n*-clad thickness decreases, until the mode changes from a complex mode into a conventional bound mode and the losses approach zero (background).

Table 4.4 lists the applicable parameters for a CSP laser which has 0.40 mole fraction of AlAs in the substrate. In this case, neither the transverse field inside the channel nor outside the channel are "bound leaky modes." The effective index curve decreases with decreasing *n*-clad thickness because the mode energy outside the channel is "averaging in" the lower clad index—just as in any ridge guide structure. The near- and far-fields for this case are shown in Figure 4.13(a) and (b). Both inside and outside the channel, the modes are bound modes with flat phase fronts.

4.3.3. *Analysis of Related Structures*

In the Introduction, we mentioned two primary variations of the CSP structure. One contains an AlGaAs layer between the substrate and *n*-clad region

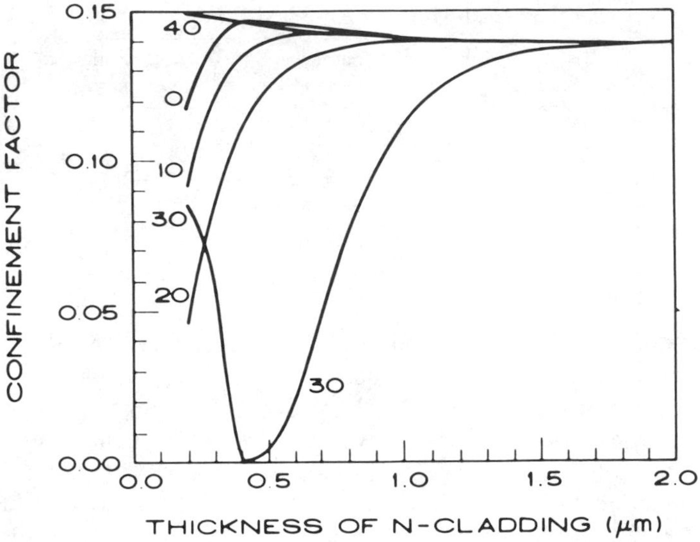

FIGURE 4.12. The active layer confinement factor as a function of the *n*-clad thickness for an active layer thickness of 600 Å for CSP lasers with mole fractions of AlAs of 0, 10, 20, 30, and 40%.

which can serve to provide current blocking and/or prevent meltback during LPE growth over the channel. The other is grown by either MOCVD or MBE and is known by several names including self-aligned stripe (SAS) [20], [22], [23]. The conventional SAS structure contains a GaAs current blocking layer (with the same dopant polarity as the substrate) about 0.2–0.4 μm above the active layer. A segment of the blocking layer is removed and a second growth of material with the same composition as the cladding layer buries the GaAs blocking layers. The resulting structure is comparable to a CSP laser in that the opening in the blocking layer is equivalent to the channel width and the GaAs blocking layer is equivalent to the substrate. One feature of the SAS structure is that since the blocking layer is a finite thickness (typically 1 μm) compared to the infinitely thick (> 50 μm) substrate, the calculated differential

TABLE 4.4. CSP laser with 40% AlAs substrate.

Structure	Wing parameters (bound mode)	Channel parameters (bound mode)
$d_2 = 600$ Å	$\beta/k_0 = 3.4209$	$\beta/k_0 = 3.4231$
$n_4 = 3.364$	$\Gamma_1 = 0.51$	$\Gamma_1 = 0.43$
$\alpha_4 = 10\ \text{cm}^{-1}$	$\Gamma_2 = 0.15$	$\Gamma_2 = 0.14$
$\Delta n = 2.21 \times 10^{-3}$	$\Gamma_3 = 0.32$	$\Gamma_3 = 0.43$
$\Delta\alpha/k_0 = 0.0$	$\Gamma_4 = 0.032$	$\Gamma_4 = 1.15 \times 10^{-5}$

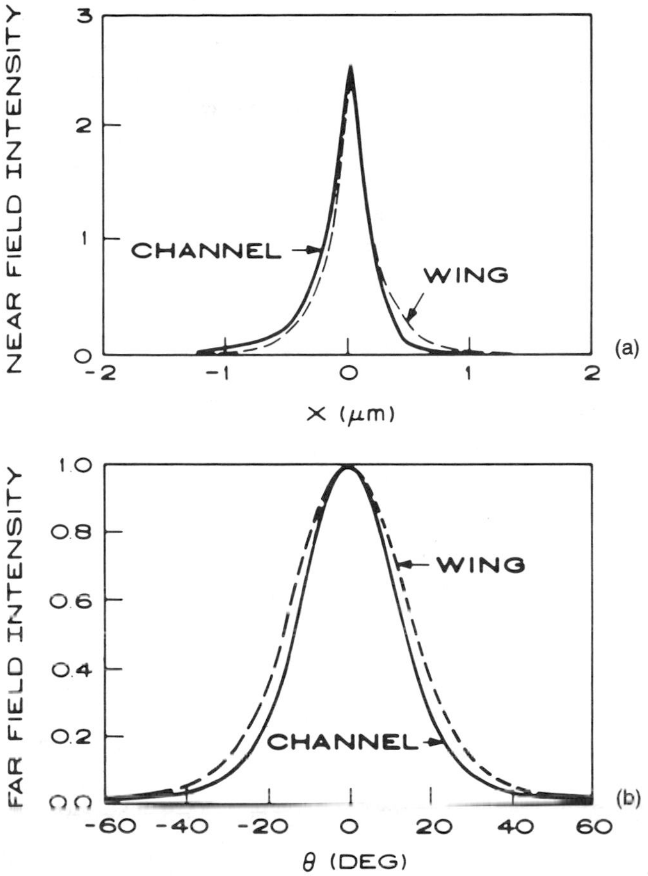

FIGURE 4.13. The transverse (a) near-field intensity and (b) far-field intensity pattern $I_{ch}(\theta)$ and $I_w(\theta)$ for a CSP laser with a mole fraction of AlAs in the substrate of 40% in the region inside (——) and outside (- - -) the channel.

quantum efficiency is somewhat larger than for a conventional CSP. For 4-μm wide openings in the blocking layer or channel, the increase in differential quantum efficiency can be as much as 65%.

The calculations shown here for CSP lasers suggest that AlGaAs layers can be used for antimeltback and for current blocking in related structures such as the SAS or V-channeled substrate inner stripe (VSIS) [4] device. However, for both of these CSP-related structures, the value of the effective index outside the channel region is very sensitive to laser wavelength, layer composition, and layer thickness. For some parameter values, the effective index can even oscillate about the effective index of the channel region for small changes in a specific parameter (e.g., 1000–2000 Å in layer thickness). The sensitivity of the effective index for these structures can be thought of as due to the various layers behaving as a dielectric stack which changes from high reflectivity to low reflectivity as the index or layer thickness is changed. To obtain an accurate value of the effective index in channeled structures with one or more layers on either side of the channel requires the consideration of additional higher-order modes [56], [57].

4.4. Transverse Asymmetries

This section analyzes the effects of observed nonuniformities in the Al composition in the channel regions. We approximate the nonuniform Al composition using a five-layer dielectric waveguide model. Analytic expressions for the five-layer waveguide are obtained [58] using a simple extension of the four-layer model discussed above, or general algorithms for calculating complex modes in plane-layered, complex-dielectric structures [59], [60] may be used.

4.4.1. *Higher AlAs Concentration at the Bottom of the Channel*

An experimental technique which can be used for the analysis of surface composition on the cleaved facets of semiconductor lasers is Auger Electron Spectroscopy (AES). Figure 4.14(b) shows Auger spectra (using a primary electron beam with a resolution of about 1000 Å) that indicate the surface composition on a cleaved facet of a CSP-type laser at the points $x = 0.3$ μm and $x = 1.2$ μm (see Figure 4.14(a)). The magnitudes of the Ga, As, and Al lines shown on the spectra reflect the concentration of these constituents at the two points and indicate that the Al concentration at the bottom of the channel is about twice that just below the active region. Note that the change in the magnitude of the Al line is tracked by a corresponding change in the magnitude of the Ga line while the As line has remained essentially unchanged. Examining a random sampling of CSP lasers has established that the magnitude of the concentration variations indicated in Figure 4.14(b) ranges from zero to about a factor of two.

Aluminum composition grading has been observed in channels etched in both the V-groove (Figure 4.15(a)) and the dovetail (Figure 4.15(b)) directions. The Al concentration variation we are reporting can be seen qualitatively in the scanning electron micrograph (SEM) of the cleaved facet shown for the square-shaped channel in Figure 4.15(b). Here, part of the backscattered electron signal is due to the average atomic number Z of the surface under examination. Thus, brighter regions in the micrograph are regions of material with a higher atomic number compared to neighboring darker regions, so that regions of higher Al concentration will be darker than those of lower Al concentration. The difference in atomic number between Al and Ga is 2.5. In the SEM of Figure 4.15(b), the laser structure is clearly defined. The 0.9 μm

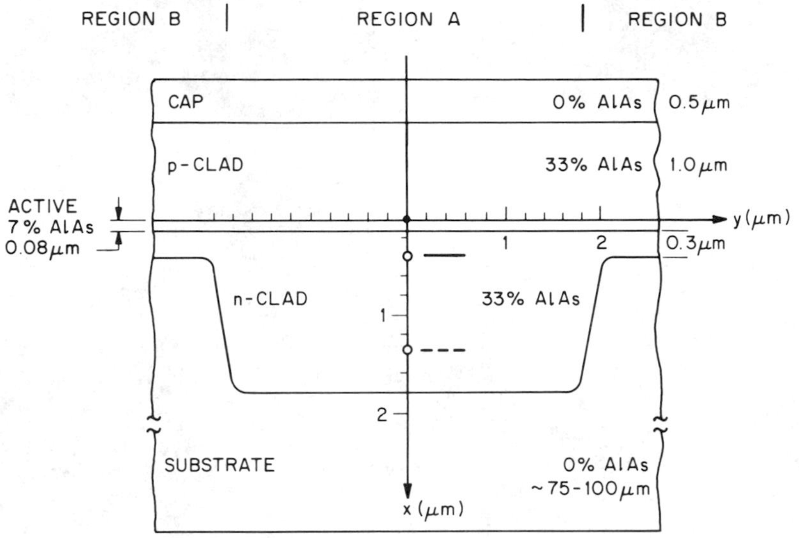

(a)

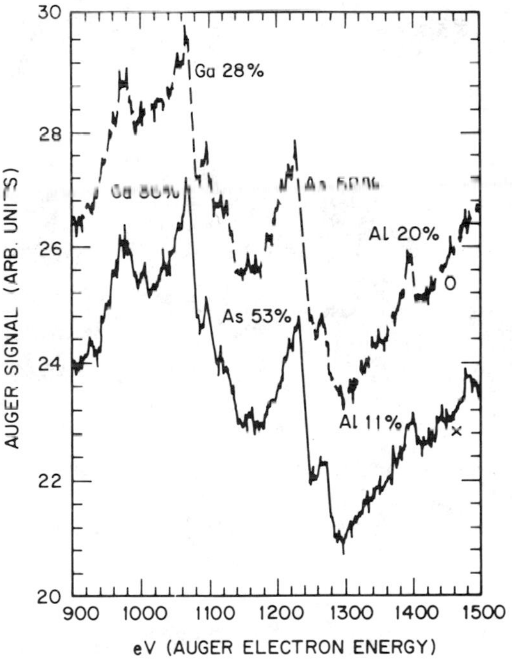

(b)

FIGURE 4.14. (a) Geometry of a typical CSP-type laser; $x = 0$ is the top of the active layer and $x = 1.8$ μm is the bottom of the channel. (b) Auger analysis of a cleaved facet of a CSP-type laser showing a higher Al composition near the bottom of the channel ($x = 1.4$ μm, dashed line) than near the top of the channel ($x = 0.4$ μm, solid line).

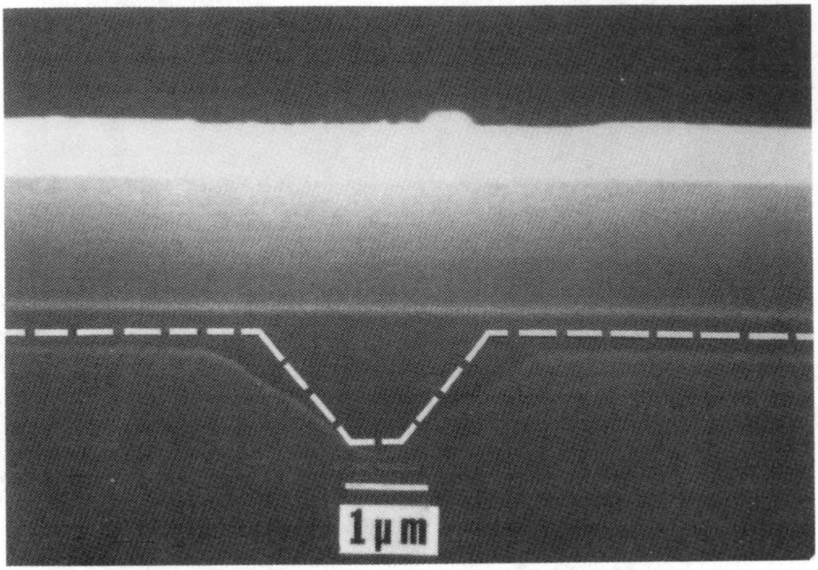

(a)

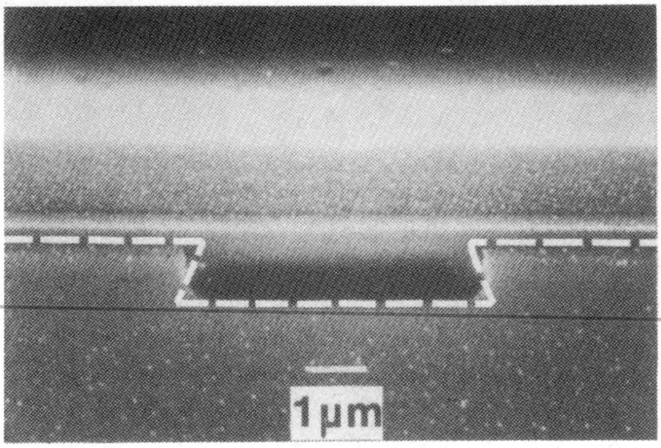

(b)

FIGURE 4.15. Scanning electron micrographs of a cross section of a CSP-type laser with the channel etched in (a) the V-groove direction and (b) the dovetail direction. The dashed black lines indicates the channel profiles before growth.

cap layer and 0.1 μm active layer are delineated by the bright horizontal images (0% and 7% AlAs content, respectively), while the two cladding layers (both nominally 33% AlAs) show up as darker regions on either side of the active layer. Within the channel shown in Figure 4.15(b), the dark region indicates a significantly higher AlAs content at the channel bottom, while the lighter region indicates a significantly lower AlAs content adjacent to the active layer—in agreement with the Auger analysis data shown in Figure 4.14(b).

A nonuniform Al/Ga ratio within the channel will effect the dielectric profile perpendicular to the junction. Figure 4.16 contains a series of possible index profiles together with their electric field distributions for (a) a uniform Al/Ga ratio in the channel (CSP); (b) a higher Al/Ga ratio at the channel bottom (CSP–LOC); and (c) a lower Al/Ga ratio at the channel bottom ESL–CSP). Graded index profiles are also possible. However, their characteristics are qualitatively similar to the abrupt step profiles discussed in this paper. The Al concentrations and corresponding index values for the three structures shown in Figure 4.16 are tabulated in Tables 4.1, 4.2, and 4.3. The loss of each epilayer is taken to be 10 cm^{-1}. The value used for the substrate loss α_s (at $\lambda = 0.83\ \mu m$), for all calculations is 5000 cm^{-1} [26]. However, there is insignificant difference in all calculated results as α_s varies from 1000 to 10,000 cm^{-1}, in agreement with earlier reports [3], [15].

The electric field distribution perpendicular to the p–n junction in the channel, $E_y(x)$, is obtained by solving the one-dimensional wave equation (4.8a) assuming region A of Figure 4.14(a) is infinitely wide with the usual boundary conditions.

The real part of the complex index of refraction $n^*(x)$, shown in Figure 4.16, is the real part of the square root of the complex relative electric permittivity. The effective index of region A (Figure 4.14(a)) is

$$n^{A_{\text{eff}}} = \text{Im}(\gamma/k_0) \tag{4.23}$$

and the effective attenuation coefficient of region A is

$$\alpha = \text{Re}(\gamma) \tag{4.24}$$

The theoretical calculations of the transverse electric field and near-field intensity distributions are both normalized according to eq. (4.9c), and the transverse far-field intensity distribution is obtained from eq. (4.15). In this section, we calculate and plot only the contribution to the transverse far-field due to the channel region and ignore the contribution from the regions outside the channel. We make this approximation because:

(1) the contribution from the regions outside the channel is negligible for practical channel widths; and
(2) the far-field distribution, considering only the channel region, would be symmetric if the composition in the channel region was homogeneous.

The data in Figures 4.14 and 4.15 indicate cases in which the Al concentration is highest in the bottom of the channels. In these cases, the resulting perpendicular index profile is no longer that of a simple double heterostructure (Figure 4.16(a)), but that of a large optical cavity (Figure 4.16(b)). The various parameters used in calculating the index and near-field profiles in Figure 4.16 are listed in Tables 4.5, 4.6, and 4.7. The LOC index profile was originally introduced intentionally [28] to lower the optical power density by

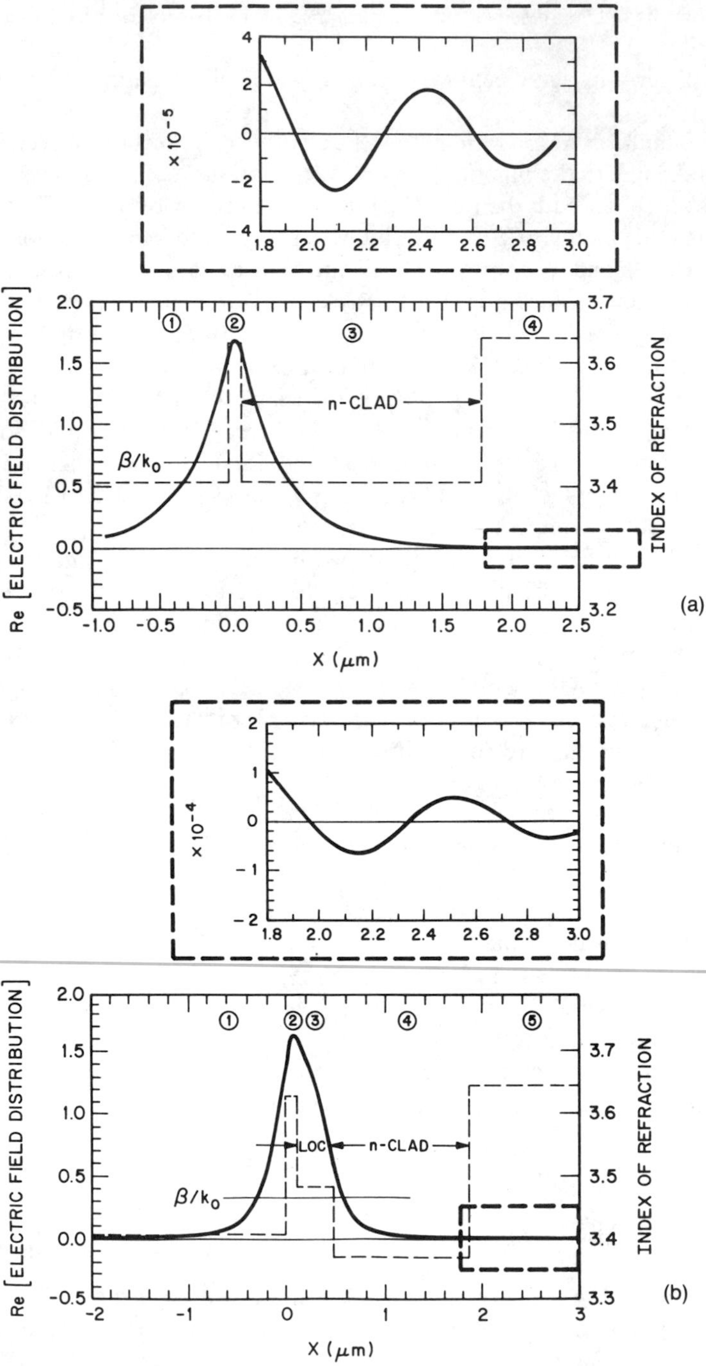

FIGURE 4.16. Index profiles (– – –) and corresponding electric field distributions (——) for (a) a conventional CSP laser; (b) a CSP–LOC laser; and (c) an ESL–CSP laser. The layer compositions, thicknesses, and effective index for each structure are listed in Tables 4.5–4.7. The dashed rectangles in (a) and (b) show the field distributions on expanded scales for $x > 1.8\ \mu m$.

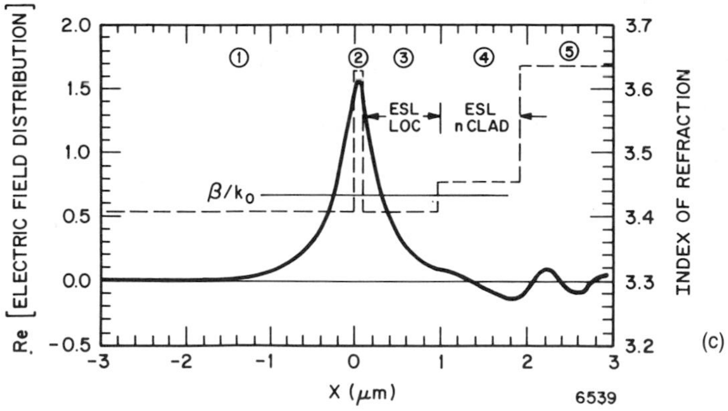

FIGURE 4.16 (*continued*)

TABLE 4.5. CSP structure.

Layer	Thickness (μm)	% AlAs	Index ($\lambda = 0.83$ μm)	Γ_{layer}	Re$\{n^*_{eff}\}$	Re$\{n_e^{ff}\}$
1. *p*-Clad	>1.0	33	3.40657	0.389		
2. Active	0.08	7	3.62805	0.222	3.43401	6.6×10^{-5}
3. *n*-Clad	1.8	33	3.40657	0.389		
4. Substrate	~75	0	3.64	3.4×10^{-6}		

TABLE 4.6. CSP–LOC structure.

Layer	Thickness (μm)	% AlAs	Index ($\lambda = 0.83$ μm)	Γ_{layer}	Re$\{n^*_{eff}\}$	Im$\{n^*_{eff}\}$
1. *p*-Clad	>1.0	33	3.40657	0.19043		
2. Active	0.08	7	3.62805	0.19205	3.43401	
3. LOC	0.4	22	3.48276	0.59205	3.46932	6.6×10^{-5}
4. *n*-Clad	1.8	40	3.364	0.02546		
5. Substrate	~75	0	3.64	2.9×10^{-9}		

TABLE 4.7. ESL–CSP structure.

Layer	Thickness (μm)	% AlAs	Index ($\lambda = 0.83$ μm)	Γ_{layer}	Re$\{n^*_{eff}\}$	Im$\{n^*_{eff}\}$
1. *p*-Clad	>1.0	33	3.40657	0.331		
2. Active	0.08	7	3.62805	0.190		
3. LOC	0.9	33	3.40657	0.346	3.43405	5.11×10^{-4}
4. *n*-Clad	1.8	26.3	3.45245	0.125		
5. Substrate	~75	0	3.64	0.008		

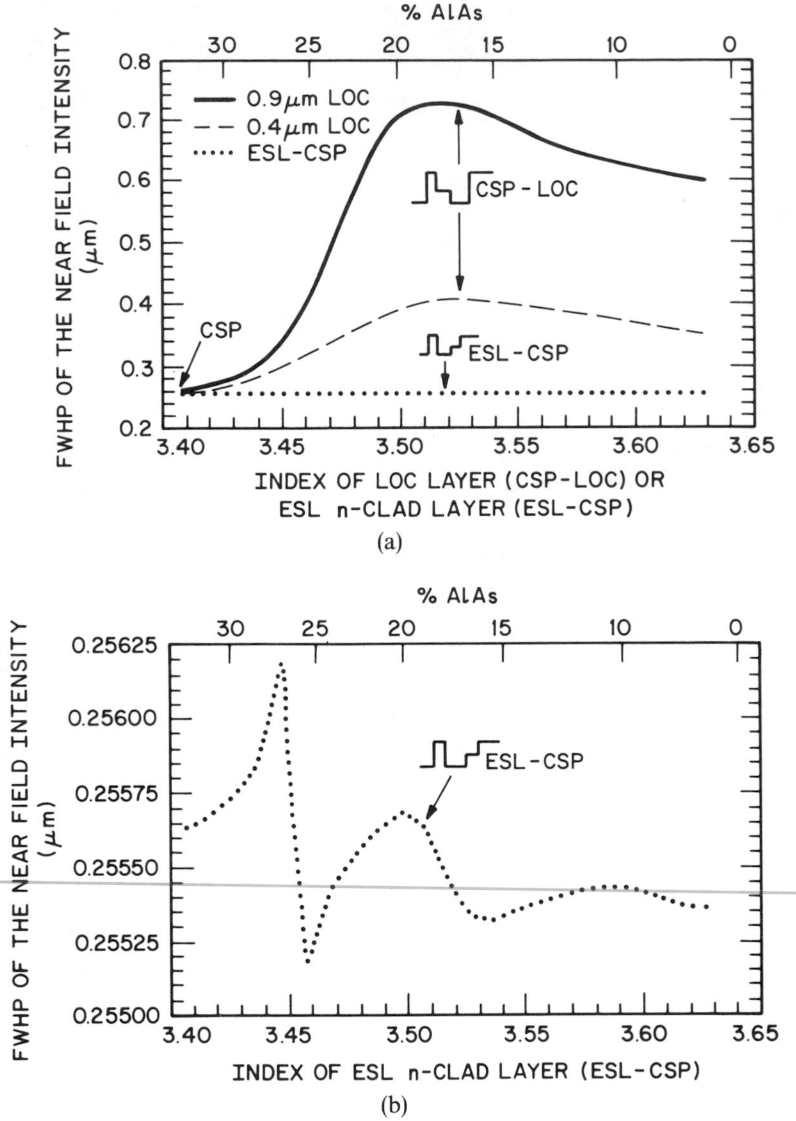

FIGURE 4.17. (a) Calculated near-field FWHP as a function of the % AlAs (or index of refraction at $\lambda = 0.83$ μm) of a 0.4 μm (---) and 0.9 μm (——) thick LOC layer (CSP–LOC geometry) or of the 0.9 μm (...) thick *n*-clad layer (ESL–CSP geometry). The *n*-clad layer (layer 4, Figure 4.16(b)) for the CSP–LOC layer has an AlAs mole fraction of 0.33, and the LOC layer (layer 3, Figure 4.16(c)) for the ESL–CSP has an AlAs mole fraction of 0.33. (b) The calculated near-field FWHP as a function of AlAs or index of refraction for the ESL–CSP structure on an expanded scale. The common point to all three curves (at an index value = 3.40657) corresponds to the conventional CSP laser described in Table 4.5.

producing a wider perpendicular near-field distribution than that of a conventional DH configuration. In Figure 4.17(a) the FWHP of the near-field intensity is plotted as a function of AlAs composition (or index at $\lambda = 0.83$ μm) of the LOC layer assuming the AlAs mole fraction of the *p*- and *n*-clad layer is 0.33 and that of the active layer is 0.07. The CSP–LOC FWHP near-field intensity of the fundamental mode reaches a maximum of about 0.4 μm (for a 0.4-μm thick LOC layer) and 0.72 μm (for a 0.9-μm thick LOC layer) at AlAs compositions of the LOC layer of about 18%. These curves are not meant to be design curves for LOC structures. They characterize patterns of behavior and do not consider, for example, higher-order modes which are possible in very thick (>0.5 μm) LOC layers [28], [29].

Parenthetically, if the Al content is greater at the bottom of the channel than in the *p-clad* layer, it is possible for a LOC layer of a CSP–LOC to actually decrease the FWHP of the near-field intensity. This is shown in Figure 4.18(a), which is a plot similar to that in Figure 4.17(a) except that the *n*-clad region has an AlAs mole fraction of 45%: the FWHP of the near-field intensity initially decreases as the Al is increased in the LOC layer. In both Figures 4.17 and 4.18, as the AlAs composition of the LOC layer is decreased below 33%, the CSP–LOC FWHP near-field spot size is always greater than that for a conventional CSP laser.

Figure 4.19 contains experimentally observed photographs of of the near fields of three CSP lasers operating at 20 mW cw, all from the same wafer, showing considerable variation in the perpendicular spot size. The near fields in Figure 4.19(b) and (c) show some asymmetry in the direction of the channel, which is not characteristic of a conventional CSP DH with equal AlAs compositions in the *p*- and *n*-clad layers. These experimental results are in qualitative agreement with the calculated near fields shown in Figure 4.20(a) for a conventional CSP laser, a CSP–LOC laser (LOC composition = 22% AlAs, *n*-clad composition = 33% AlAs), and an ESL–CSP laser (ESL–LOC layer composition = 33% AlAs, ESL *n*-clad layer composition = 26%). The far fields corresponding to the near-field intensity and phase distributions of Figure 4.20(a) and (b) are shown in Figure 4.20(c).

An additional consequence of Al composition changes in the channel region is a variation in the FWHP perpendicular beam divergence. The FWHP of the far-field radiation lobe of the laser corresponding to Figure 4.19(a) is 35°, while that of the laser corresponding to Figure 4.19(c) is 28°. While this range in FWHP perpendicular beam divergence could be explained by a variation in active layer thickness from 650 Å to 1000 Å for a conventional CSP laser, an alternative explanation is a nonuniform Al composition in the channel region as shown in Figures 4.18(b) and 4.21. In Figure 4.21 the FWHP perpendicular beam divergences are plotted (for the same parameters used in Figure 4.17 to plot the FWHP of the near field) as the Al composition of the LOC layer (CSP–LOC structures) or *n*-clad layer (ESL–CSP structures) varies from 33% to 7%. Figure 4.21 shows that nonuniformities in the Al composition in the channel region can cause a range in FWHP perpendicular

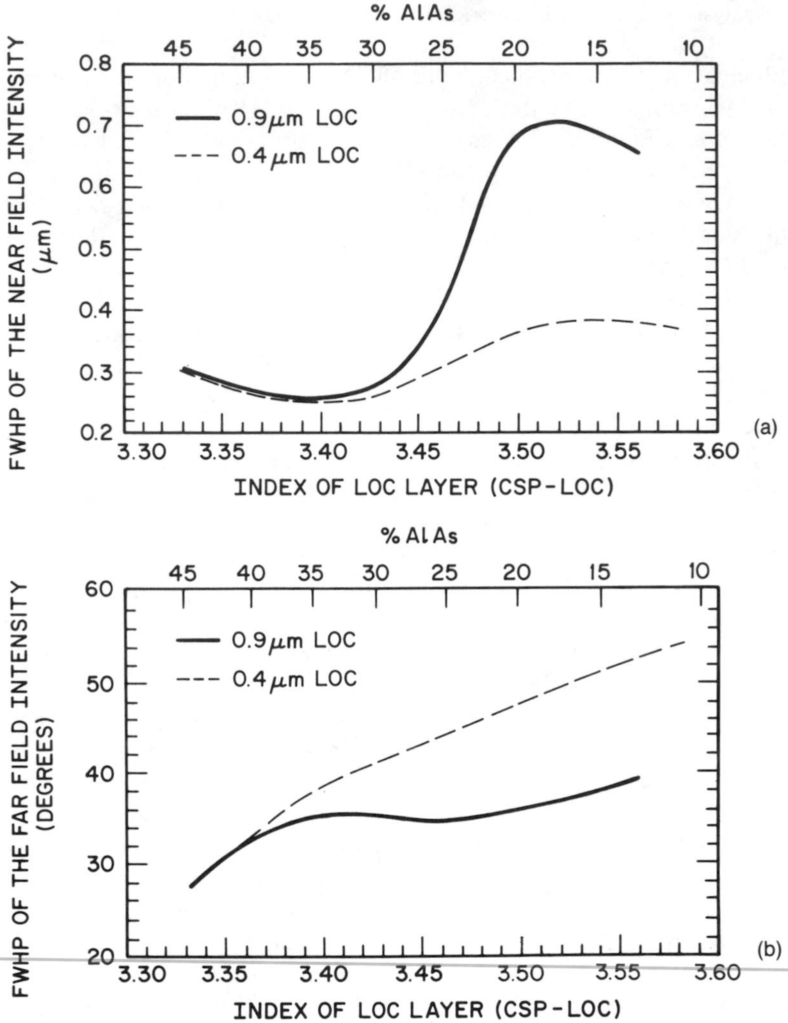

FIGURE 4.18. Calculated (a) near-field FWHP, and (b) far-field FWHP as a function of the % AlAs (or index of refraction at $\lambda = 0.83$ μm) of a 0.4 μm (---) and a 0.9 μm (——) thick LOC layer of a CSP-LOC with an n-clad AlAs mole fraction of 0.45.

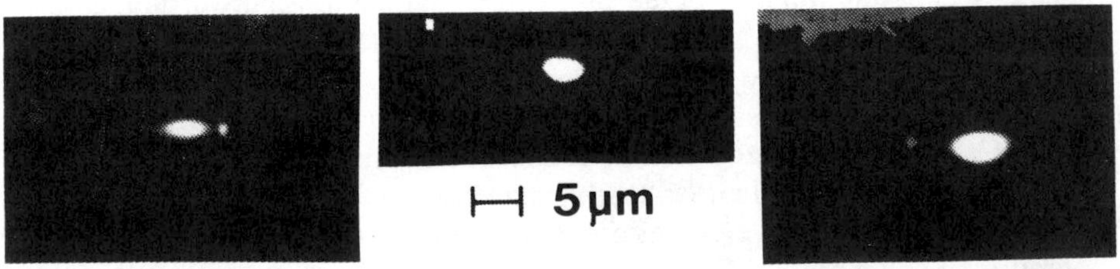

(a) (b) (c)

FIGURE 4.19. Near-field micrographs of CSP-type lasers operating at the same power output showing significantly different perpendicular near fields.

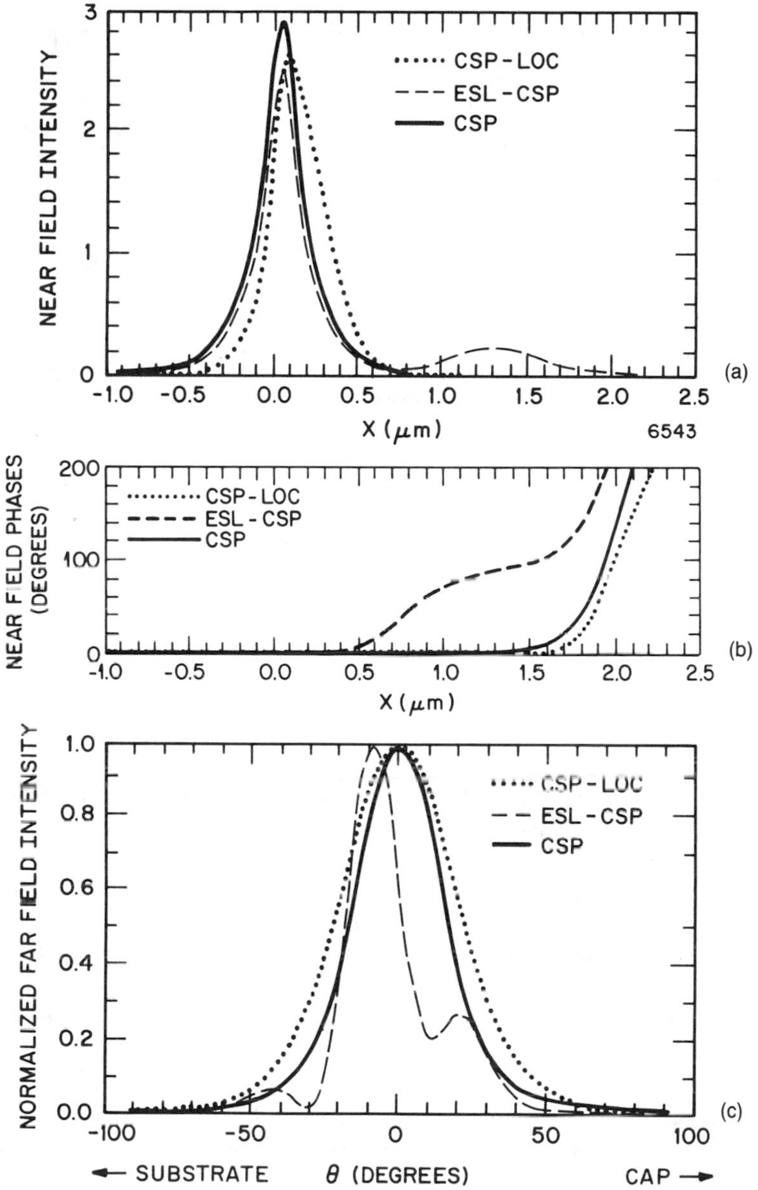

FIGURE 4.20. Calculated (a) near-field intensities, (b) near-field phases, and (c) far-field intensities $I_{ch}(\theta)$ perpendicular to the junction for a conventional CSP laser (——), a CSP–LOC laser (. . .), and an ESL–CSP laser (– –) for the parameters listed in Tables 4.4–4.7.

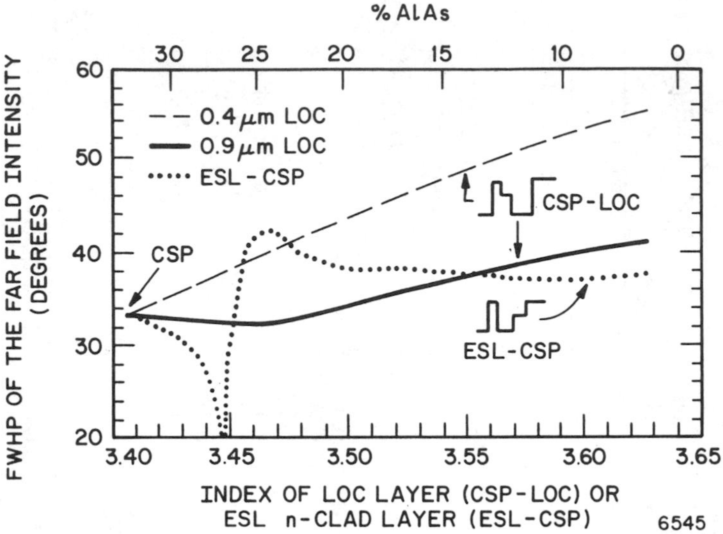

FIGURE 4.21. Calculated far-field FWHP as a function of the % AlAs (or index of refraction at $\lambda = 0.83$ μm) of a 0.4 μm (---) and 0.9 μm (——) thick LOC layer (CSP–LOC geometry) or of the 0.9 μm (...) thick *n*-clad layer (ESL–CSP geometry). As in Figure 4.17, the *n*-clad layer for the CSP–LOC layer has an AlAs mole fraction of 0.33, and the LOC layer (layer 3) for the ESL–CSP has an AlAs mole fraction of 0.33. The common point to all three curves (at an index value = 3.40657) corresponds to the conventional CSP laser described in Table 4.5.

beam divergences from about 20° to more than 50° for a fixed active layer thickness of 800 Å. These calculations indicate that observed variations in FWHP perpendicular beam divergences can result not only from variations in active layer thicknesses, but also from variations in Al compositions in the LOC or *n*-clad layers of any semiconductor laser.

Finally, the larger perpendicular near-field spot sizes shown in Figure 4.19 (and calculated in Figures 4.17 and 4.18) result in a decreased optical power density which might be expected to result in longer operating life at moderate to high power. This expectation was found experimentally for the three lasers (all from the same wafer) whose near-field photos are shown in Figure 4.19. Operating at 20 mW cw and at 30 °C, the first unit (6a) lasted 50 hours, the second unit (6b) about 1000 hours, while the third (6c) was still operating after 17,000 hours.

4.4.2. *Lower AlAs Concentration at the Bottom of the Channel*

Shown in Figure 4.22 is a schematic drawing of the channel region of a CSP laser together with Al concentrations as given by AES data. Here, the Al

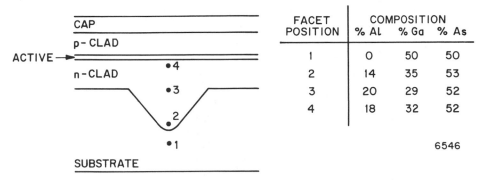

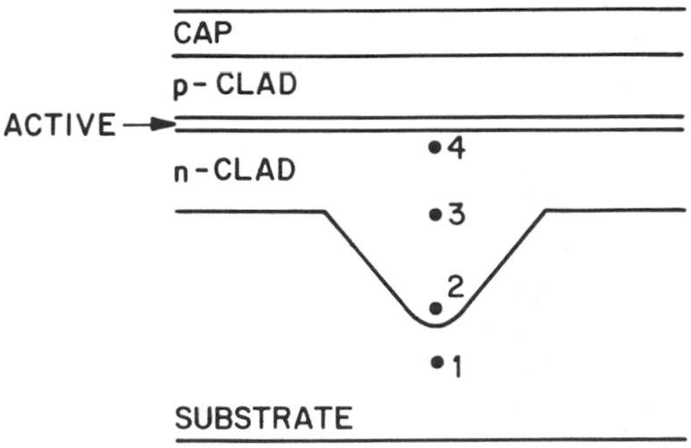

FACET POSITION	COMPOSITION % AL	% Ga	% As
1	0	50	50
2	14	35	53
3	20	29	52
4	18	32	52

6546

FIGURE 4.22. Composition measured by Auger analysis at four positions along a cleaved facet of a CSP-type laser showing a lower Al composition near the bottom of the channel than near the top of the channel.

content is lower at the bottom of the channel. The resulting index profile (Figure 4.16(c)) can be thought of as "pulling" some of the mode power into the substrate, increasing the mode loss. This mechanism, which we call Enhanced Substrate Loss (ESL), can also be explained by realizing that all of the modes supported by the index profiles shown in Figure 4.16 are complex modes [49] (i.e., the longitudinal and transverse wave vectors have a real and imaginary component), because the field solutions to the electromagnetic wave equation (4.8a) are "sinusoidal" in the substrate. Usually, the n-clad region separating the active region (and LOC layer, if present) from the substrate is thick enough ($> 1\ \mu$m) that field penetration of the laser mode into the substrate is negligible (see Figure 4.16(a), (b)). However, if the mole fraction of AlAs is lower at the bottom of the channel than at the top, the higher-index portion (layer 4 of Figure 4.16(c)) in the channel acts like an antireflection coating [61] between the high-aluminum, low-index portion (layer 3) and the no-aluminum, very high-index substrate (layer 5), thereby coupling or redistributing a larger fraction of the mode power into the substrate. In Figure 4.23 the ratio of energy confined in the laser substrate to the total mode energy (Γ_s, the substrate confinement factor) is plotted as a function of the mole fraction of AlAs (or index of refraction at $\lambda = 0.83\ \mu$m) of the bottom half of the channel region (layer 4) for the index profile shown in Figure 4.16(c). The peaks in this plot correspond to index values that optimize the coupling of light into the substrate for the 0.9 μm thickness of the index "matching" layer (layer 4). In the vicinity of these peaks, the mode has significant energy directed both along the waveguide axis *and* into the substrate, and the corresponding far-field pattern is no longer peaked perpendicular to the facet, but is tilted towards the substrate on the order of a few degrees as shown by the calculated far fields—see Figure 4.20(c) and insets in Figure 4.23. This far-field tilt is consistent with the asymmetric near-field phase of the ESL–CSP laser (Figure 4.20(b)) which, unlike that of the CSP–

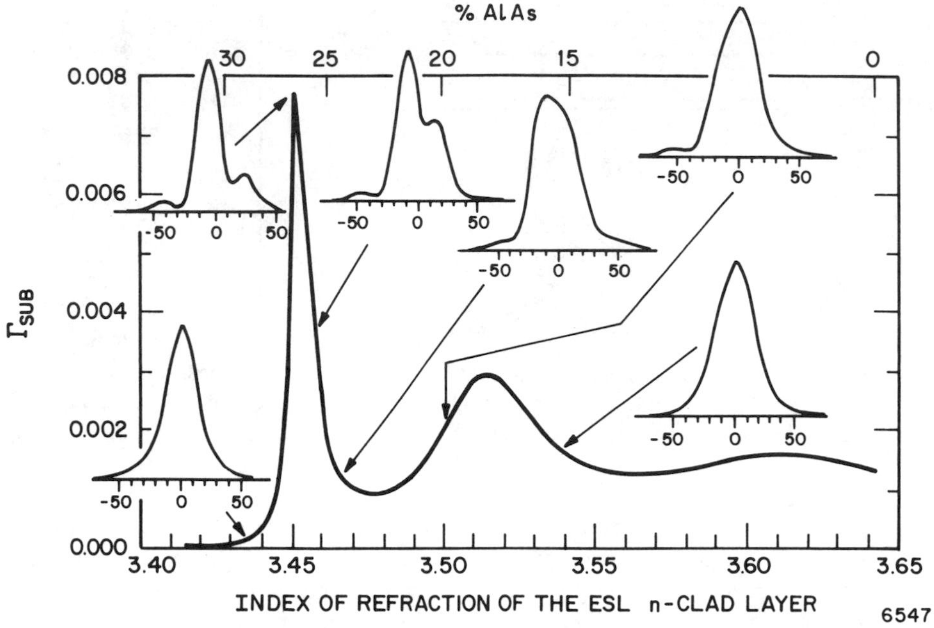

FIGURE 4.23. The substrate confinement factor Γ_s as a function of the % AlAs (or index of refraction at $\lambda = 0.83$ μm) of the n-clad layer at the bottom of the channel of an ESL–CSP laser. The inset far-field intensity $I_{ch}(\theta)$ versus angle patterns show a large variation in asymmetry as a function of % AlAs of the n-clad layer.

LOC or CSP laser, is not flat over the region ($0.4 < x < 2.0$) where the optical field has a significant amplitude. The asymmetric far-field pattern for the ESL–CSP laser in Figure 4.20(c) is caused by the asymmetry of the near-field phase variation in Figure 4.20(b).

For both the ESL–CSP and CSP–LOC lasers, the thicknesses and refractive indices of the LOC and n-clad layers chosen for the theoretical models are not unique. The cases illustrated in this chapter correspond to fixed layer thicknesses with a varying index of refraction in one layer (Figures 4.17, 4.18, and 4.23). However, a curve similar to Figure 4.23 is obtained by assuming an index n_4 ($n_3 < n_4 < n_5$) for the n-clad layer and varying the thickness of the n-clad layer.

Experimentally observed far-field intensity patterns of several ESL–CSP lasers (with V-groove channels) from the same wafer are shown in Figure 4.24. The experimental asymmetric profiles agree qualitatively with the theoretical profiles shown in Figure 4.23. In general, the asymmetric far-field profile for a given ESL–CSP laser does not change as a function of power even at power levels above 100 mW. The shift of the peak of the experimental far fields from the normal to the facet ranges from less than 2° to about 9°, in agreement with the calculated shifts (up to 8°) shown in Figures 4.20(c) and 4.23. The expe-

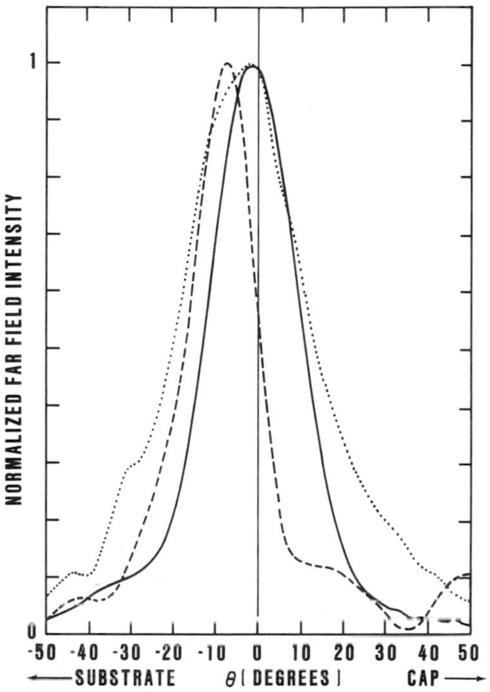

FIGURE 4.24. Experimentally measured far-field intensity patterns of four ESL–CSP lasers.

rimental range in FWHP of the far-field patterns range from 18° to 33° compared to a theoretical range of 20–42° for the parameters assumed in Figure 4.21. Distinct sidelobes appear in the experimental far field (dashed line) shown in Figure 4.24 and in some of the calculated far fields shown in Figures 4.23 and 4.20(c). These sidelobes can be explained as the interference pattern in the far field between two peaks of the near-field intensity. For example, the near-field distribution of the ESL–CSP laser shown in Figure 4.20(a) can be approximated by two point sources separated by 1.3 μm. From Figure 4.20(b) the smaller peak is shifted in phase by an average of about 90° relative to the main peak. The interference pattern from this simple two-point approximation has a central peak at 9.2° and sidelobes at 28.6° and $-52.9°$, in general agreement with the ESL–CSP far-field pattern shown in Figure 4.20(c).

Asymmetries and radiation pattern shifts in the perpendicular far field have been observed in many types of AlGaAs semiconductor lasers and we suggest that nonuniform Al compositions are often the explanation.

4.4.3. *Discussion of AlAs Concentration Variations*

The cause of a nonuniform Al/Ga ratio occurring within the channel of a CSP laser during the LPE growth of the first cladding layer is not fully understood. However, we know

(1) that the layer growth must be faster at the bottom of the channel than at the top in order to result in flat growth profiles;
(2) that there must be a lateral component of growth within the channel as well as a perpendicular component;
(3) that the wall and bottom curvatures of the channel present crystallographic planes to the growth nucleation that are different from those above the channel where the growth is planar [62]; and
(4) that there can be varying degrees of meltback at the channel walls and shoulders as shown in Figure 4.15.

Any of these conditions can readily affect the composition of the ternary compound that initially nucleates and freezes out from the AlGaAs melt, and can alter this composition as the growth proceeds and overall growth conditions change. Furthermore, if the Al/Ga ratio changes during the initial growth, then changes in the local melt composition may occur which might further change the Al/Ga ratio later along in the growth. From the AlAs–GaAs phase diagram [63], very small changes in Al in the melt produce very significant changes in the Al content in the grown material. The basis for a nonuniform Al/Ga ratio within the channel of a CSP laser discussed above is also consistent with the variability of this effect from channel to channel since it would be the local conditions around each channel (shape, initial freeze-out rate, local melt composition) that would determine the magnitude of the effect. The lasers discussed in this chapter were grown at 800 °C with about 4 °C of supersaturation and a cooling rate of 0.75 °C/min. The *p*- and *n*-dopants were Ge and Sn.

While the CSP–LOC structure produced by higher Al concentrations at the bottom of the channels has generally familiar and well-understood consequences, the ESL–CSP structure produced by lower Al concentrations at the bottom of the channel has some peculiar properties. First, the internal losses of the perpendicular component of the ESL mode can be almost an order of magnitude higher than that of the conventional CSP or CSP–LOC, as shown in Figure 4.25, which could noticeably increase the threshold current and decrease both efficiency and lifetime. (However, we have not seen a clear correlation between increases in threshold current and decreases in differential quantum efficiency with asymmetries in the far-field pattern of our lasers.) Second, the near-field and far-field FWHP plots summarized in Figures 4.17, 4.18, 4.20, and 4.21 are not those intuitively expected from plane wave, uniform intensity diffraction theory which equates the far-field beam divergence θ_B to λ/D, where D is the near-field aperture: The ESL–CSP structure can have a narrower FWHP perpendicular near-field intensity *and* a narrower FWHP perpendicular beam divergence than a CSP or CSP–LOC structure. Additionally, while the FWHP of the perpendicular near-field spot sizes for the CSP–LOC structures are two to three times larger than that for a conventional CSP (Figure 4.17), the CSP–LOC FWHP beam divergences are larger by typically 10° (Figure 4.21). Finally, both the CSP–LOC FWHP beam divergences (Figures 4.18(b) and 4.21) and the FWHP near-field intensities

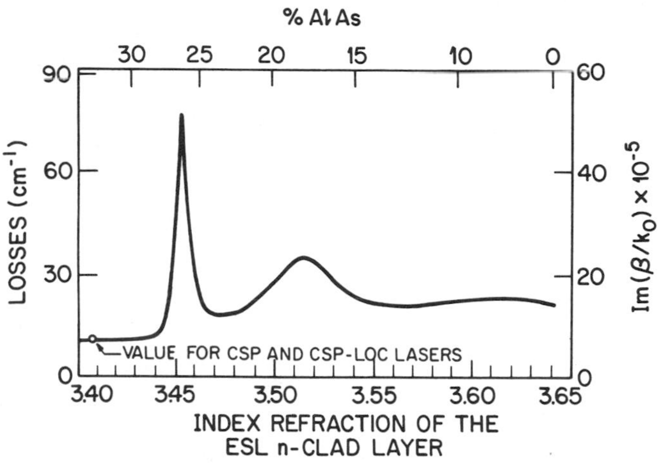

FIGURE 4.25. The perpendicular mode loss in region A of Figure 4.1(a) as a function of the % AlAs (or index of refraction at $\lambda = 0.83$ μm) of the n-clad layer at the bottom of the channel of an ESL–CSP laser. The right-hand ordinate is the imaginary part of the effective index.

(Figures 17(a) and 18(a)) simultaneously increase over some ranges of AlAs variation. These apparent contradictions of the expected relationship computed from variables that are Fourier transform pairs result because the FWHP, by itself, of the various near-field intensities can be a misleading and inappropriate representation of the near-field aperture. For example, the ESL–CSP structure has a very narrow FWHP beam divergence of about 24° (Figure 4.20(c)) and a narrow FWHP near-field spot size of about 0.26 μm. However, because of an irregular distribution of mode power, the effective near-field aperture (Figure 4.20(a)) is actually about 1.5 μm. The reason for these apparent contradictions is primarily because the FWHP of the near-field intensities contain varying fractions of the mode power. Such apparent contradictions would be largely eliminated[1] by a definition of near-field aperture and far-field beam divergence corresponding to near- and far-field widths enclosing the same percentage (e.g., 80%) of mode power, instead of the experimentally convenient measurement of FWHP.

The effect of Al variations on lateral mode confinement has not been investigated in detail. However, since only the fundamental lateral mode is observed experimentally, the difference between the magnitude of the complex effective index in the channel (region A of Figure 4.14(a)) and outside the channel (region B of Figure 4.14(a)), $\Delta n_{\text{eff}}(=|n^{*A}_{\text{eff}}| - |n^{*B}_{\text{eff}}|)$, must be much less than 0.1 for channel widths ranging from 4 μm to 7 μm. Figure 4.26 shows

[1] The far-field beam divergence from a fixed aperture D will, however, have slight to moderate variations as either the near-field intensity distribution or the near-field phase is changed.

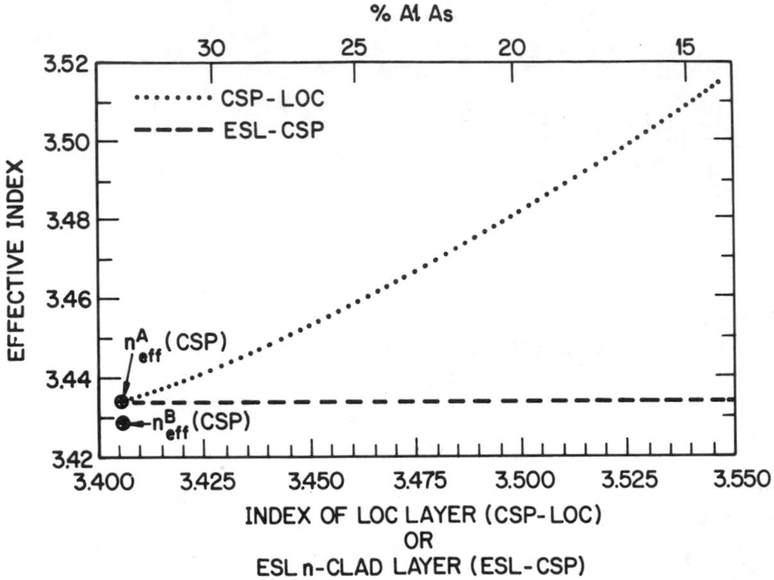

FIGURE 4.26. The calculated effective index in the channel as a function of the % AlAs (or index of refraction at $\lambda = 0.83\ \mu m$) of a 0.4 μm (...) thick LOC layer (CSP–LOC geometry) or of a 0.9 μm (---) thick *n*-clad layer (ESL–CSP geometry). The *n*-clad layer for the CSP–LOC layer has an AlAs mole fraction of 0.33, and the LOC layer (layer 3) for the ESL–CSP has an AlAs mole fraction of 0.33. The common point to both curves (at an index value = 3.40657) corresponds to the conventional CSP laser described in Table 4.5. The effective index of the ESL–CSP laser changes by less than 4×10^{-4} over the index range of the *n*-clad layer shown. Also shown are typical design values for the effective index in the channel $n_{eff}^{A(CSP)}$ and outside the channel $n_{eff}^{B(CSP)}$ for a conventional CSP laser.

that while the effective index in the channel region (n_{eff}^{*A}) for the ESL–CSP structure changes negligibly with changes in Al concentration ($\sim 4 \times 10^{-4}$), the effective index in the channel region of the CSP–LOC structure changes significantly ($\sim (2\text{–}6) \times 10^{-2}$). Thus, the observed absence of higher-order lateral modes, for the CSP–LOC structure, suggests that Al composition changes may occur in the *n*-clad layer *outside* the channel that follow, at least partially, those inside the channel, and result in a corresponding increase in n_{eff}^{*B}.

4.5. Above-Threshold Analysis

4.5.1. *Field Analysis*

The above-threshold analysis of the optical fields in CSP–DH lasers is obtained by using the effective index method to solve Maxwell's equations for

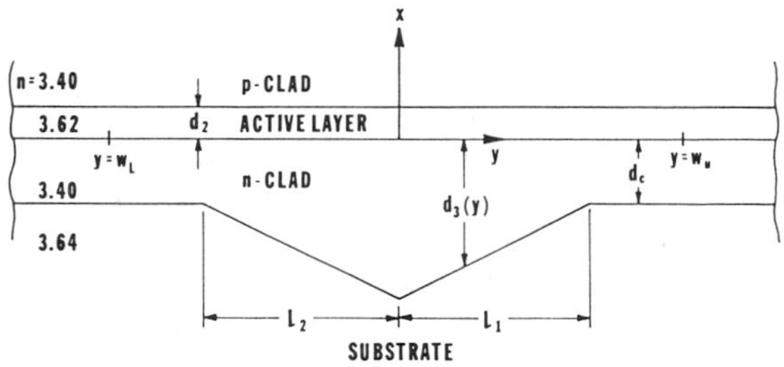

(a)

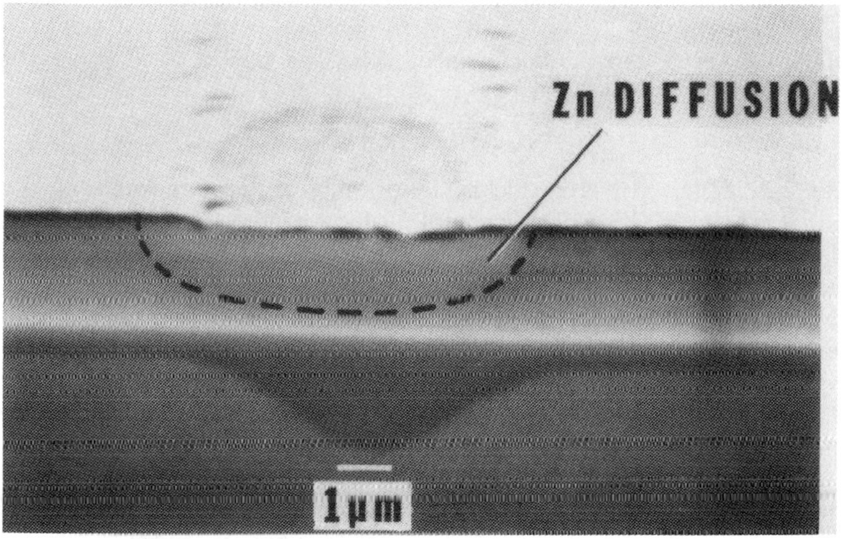

(b)

FIGURE 4.27. The CSP–DH laser structure; (a) idealized model, and (b) scanning electron micrograph illustrating misalignment of the Zn diffusion with the channel.

the basic asymmetric structure shown in Figure 4.27(a). Asymmetry of the channel, apparent in Figure 4.15(a), can occur during etching of the channel prior to growth or during growth by meltback. Zinc diffused regions, used to confine current, may be misaligned with respect to the channel as shown in Figure 4.27(b).

From Section 4.2, the effective index method first requires solving for the transverse fields (perpendicular to the active layer) using simple multilayer waveguide theory. From the solution to the layered waveguide field at specific lateral points, the lateral effective index along the y direction is established (see Figure 4.29(a)). When the various layers have loss or gain, the effective index is complex. The absorption in the active layer is accounted for in the expression of the gain profile which we discuss later. The GaAs substrate has a very large absorption coefficient ($\sim$5000–10,000 cm^{-1}) while those of the p- and n-cladding layers are on the order of 10 cm^{-1}.

For a lasing mode the eigenvalue γ must satisfy the oscillation condition $\mathrm{Re}\{\gamma\} = -G/2$, where $G = -\ln(R_1 R_2)/(2L)$ is the modal gain, R_1 and R_2 are the facet reflectivities, and L is the length of the laser. Earlier we defined $\alpha = \mathrm{Re}[\gamma]$, and in this section we put $G/2 = \alpha$, where G represents modal gain. The magnitude of G is equal to the emission losses at the laser facets. The differential equation for the lateral field $v(y)$, as given in eq. (4.10), is modified as follows:

$$\frac{\partial^2 v}{\partial y^2} + [\gamma^2 - \gamma_0^2 + k_0^2 \Gamma_c(y) \kappa_v(y, N_{ph})] v = 0, \tag{4.25}$$

where $\Gamma_c(y)$ is the complex confinement factor

$$\Gamma_c = \frac{\int_0^{d_2} u^2(x, y)\, dx}{\int_{-\infty}^{\infty} u^2(x, y)\, dx}, \tag{4.26}$$

where d_2 is the active layer thickness. For an ideal structure with no loss or gain, the value of Γ_c is real. The complex effective index n_{eff} is found from eq. (4.11). The value $\kappa_v(y, N_{ph})$ is the gain (or carrier) dependent portion of the dielectric constant of the active layer and N_{ph} is the photon density in the active layer. At threshold, $N_{ph} = 0$, and $\kappa_v(y, N_{ph})$ is determined from the carrier injection and ambipolar diffusion processes in the active layer. Above threshold, $N_{ph} > 0$, and $\kappa_v(y, N_{ph})$ is then a function of $v(y)$ so that eq. (4.25) is nonlinear. In terms of the active layer gain $g(y)$ and index changes due to carrier injection $\delta n(y)$, $\kappa_{v2}(y, N_{ph})$ can be expressed as [64], [65]

$$\kappa_{v2}(y, N_{ph}) = 2n_2 \delta n(y) + \frac{i n_2 g(y)}{k_0}. \tag{4.27}$$

The value $\delta n(y)$ is linearly related to the carrier or gain profile as $\delta n(y) = Rg(y)/k_0$ where R lies between -1 and -4 for laser structures of the type discussed here [46]. This relates the index change $\delta n(y)$ at a point in the active layer to the gain $g(y)$ at that position using a linear relationship. A more accurate description of $\delta n(y)$ in terms of g is a functional dependence where R/k_0 is the first term of a Taylor series of the carrier-dependent refractive index change.

The solution of eq. (4.25) is obtained by numerical methods using a multipoint differential equation solver with prescribed boundary conditions [60]. In the structure of Figure 27(a), the complex effective index of refraction is constant for both $y > w_u$ and $y < w_l$. If the optical fields and injected carrier profiles are neglected in the far lateral positions $y > w_u$ and $y < w_l$, the solution of eq. (4.25) can be written as

$$v(y) = \begin{cases} v(w_u) \exp[p(w_u - y)], & y > w_u, \\ v(w_l) \exp[q(w_l + y)], & y < w_l, \end{cases} \tag{4.28}$$

where $v(w_u)$ and $v(w_l)$ are the boundary solutions. The complex values p and

q are the evanescent field decay coefficients and satisfy the conditions

$$-p^2 = \gamma^2 - \gamma_0^2(w_u) + k_0^2\Gamma_c(y)\kappa_v(y, 0), \tag{4.29a}$$

$$-q^2 = \gamma^2 - \gamma_0^2(w_l) + k_0^2\Gamma_c(y)\kappa_v(y, 0). \tag{4.29b}$$

We have assumed that at the points $y = w_u$ and $y = w_l$, the optical fields are sufficiently small so that hole-burning effects can be neglected.

Above threshold, the high optical fields in the region below the stripe contact become so intense that electron/hole pairs are depleted at a fast rate. The rate of recombination is proportional to the product of the photon density $N_{ph}(x, y)$ and the optical gain coefficient $g(y)$. Thus, stimulated recombination acts as a sink for carriers while the applied current density is the source. The resulting carrier distribution must be found from solutions of the diffusion equation.

4.5.2. *Carrier Diffusion*

To determine the source and sink terms in the diffusion equation, we must describe the current injection into the active layer as well as the optical fields inside the laser cavity. Current injection into the active layer depends on the stripe contact width S, and the resistivities and thicknesses of the various layers. In addition, if Zn is diffused through the cap and p-clad layers above the channel, the effects of the Zn diffused region must be included in the model. Computations of the current distribution in the active layer of various CSP–DH laser structures have been made by using a finite-element code. Models were developed for a self-consistent calculation for the voltage drop across the active layer caused by the electron/hole diffusion along the lateral direction [66]. These calculations have been made for devices with Zn diffusion in order to determine the effects of current confinement due to various diffusion depths. For very shallow depths the current spreads in the active layer; however, for deep diffusion depths, the current is almost constant under the stripe contact [67]. In fact, calculations on Zn diffused geometries show that the current spreading can be expressed by the well-known expression [68]

$$J_x(y) = \begin{cases} J_0, & |y| \le S/2, \\ J_0/[1 + (|y| - S/2)/y_0]^2, & |y| > S/2, \end{cases} \tag{4.30}$$

where J_0 is the current density under the stripe. The value of y_0 is primarily a function of Zn diffusion depth and stripe contact width, and the resistivities of the p-GaAs cap and p-AlGaAs cladding layers. For typical structures where the Zn diffusion front extends approximately halfway through the p-clad layer, the value of y_0 lies in the range of a few tenths to several microns. The large values of y_0 occur for small stripe widths. Nevertheless, the exact shape of the current injection profile plays a secondary role in the shape of the carrier profile in the active layer because the ambipolar diffusion process causes carrier redistribution.

The optical hole burning process plays a major role in the nonlinear nature of the electrical and optical properties of the laser. The field inside the cavity can be considered as a superposition of two traveling waves. Because the waves are propagating in opposite directions, there is a quasi-standing wave pattern in the axial direction. Spatial hole burning would occur along the axial direction if carrier diffusion lengths were small. However, since the carrier diffusion length $L_D \gg \lambda_g/2$ where λ_g $(=\text{Im}\{2\pi/\gamma)$ is the guide wavelength, we can neglect this axial variation of the carrier density [69]. The photon density at a point in the axial direction is the sum of the photon densities in the forward and backward traveling waves. The photon density in the active layer is given as the fraction $\Gamma_2(y)$ of optical power in the transverse direction overlapping the active layer. The real function $\Gamma_2(y)$ (eq. (4.9b)) is different from the complex confinement factor $\Gamma_c(y)$ given in eq. (4.26) that arises from the normalization of the complex-valued wave functions. The total photon density in the active layer becomes

$$N_{ph} = P_0 n_{eff} \Gamma(y)|v(y)|^2[\exp(Gz) + R_1 \exp(G\{2L - z\})]/[hvcd_2\langle|v(y)|^2\rangle], \tag{4.31}$$

where P_0 is power (related to the electric field amplitude E_0 of eq. (4.3)) in the forward traveling wave at $z = 0$ (mirror 2), n_{eff} is the effective modal refractive index, c is the velocity of light, and R_1 is the reflectivity of mirror 1 located at $z = L$, and G is the modal gain coefficient. The term $\langle|v(y)|^2\rangle$ is the integral of the field intensity over the lateral dimension. The z dependence of N_{ph} is nearly uniform over the cavity length so that the two exponential terms can be averaged over the length of the cavity [69], [70], [71]. The resulting expression for the stimulated recombination term R_{st} in the diffusion equation becomes

$$\begin{aligned} R_{st} = 2P_1\Gamma(y)|v(y)|^2 \times \{1 - R_1 - (R_1R_2)^{1/2} \\ + (R_1/R_2)^{1/2}\}g(y)/[hv\, d_2\langle|v(y)|^2\rangle(1 - R_1)\ln(1/R_1R_2)], \end{aligned} \tag{4.32}$$

where the emission from the front facet $P_1 = P_0(1 - R_1)\exp(GL)$, and P_0 is the power inside the laser at $z = 0$.

The necessary source and sink terms in the diffusion equation are now completely defined. Letting $N(y)$ denote the electron-hole pair density, the diffusion equation becomes

$$\frac{D_e \partial^2 N(y)}{\partial y^2} - \frac{N(y)}{\tau_s} - BN^2(y) = -\frac{J_x(y)}{[qd_2]} + R_{st}, \tag{4.33}$$

where D_e $(=L_D^2/\tau_s)$ is the effective diffusion coefficient, τ_s is the spontaneous carrier lifetime, and B is the bimolecular recombination term. The gain coefficient is expressed as

$$g(y) = aN(y) - b. \tag{4.34}$$

In GaAs active layers, we use $a = 2.5 \times 10^{-16}$ cm^2 and $b = 190$ cm^{-1} [72].

Substitution of the gain expression into eq. (4.33) gives a differential equation for the carrier density where the term P_1 is the emission power out of the

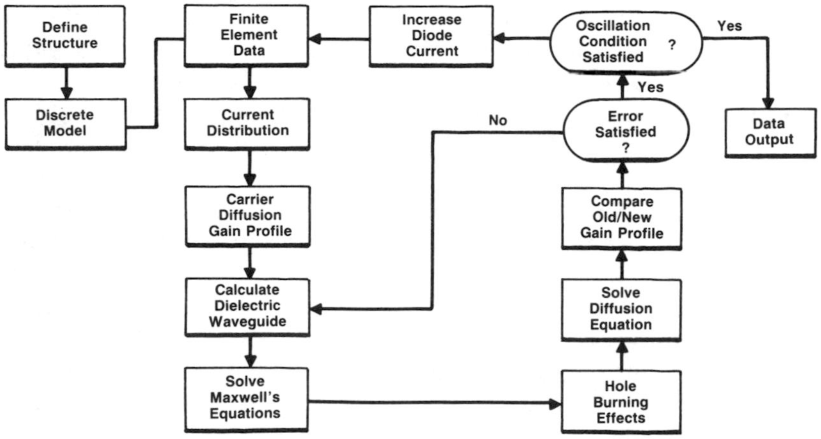

FIGURE 4.28. Flow chart summarizing the numerical calculations for laser oscillation above threshold.

front facet, assumed as a parameter, and the wave function $v(y)$ is a solution to eq (4.25). Of course, self-consistency requires the simultaneous solution of both the wave equation which has a dielectric constant that is carrier dependent and the carrier diffusion equation which has an optical field dependence. The flow chart in Figure 4.28 illustrates our method of solution. In Appendix A, we discuss the numerical formulation of the differential equations which are solved by a multipoint boundary-value routine.

4.5.3. *Discussions of Lateral Asymmetries*

The basic laser structure is illustrated in Figure 4.27, and Table 4.5 gives the parameters used in the calculations. In the following discussion, we consider a symmetrical CSP–DH laser where the Zn diffusion is symmetrically located over the channel. For the asymmetrical structures, we first assume that the slope of the walls of the channel are different, with the Zn diffused region symmetrical about $y = 0$. Next we study the effects of a misaligned (relative to the channel) Zn diffused region over a symmetrical channel. In the cases shown in Figures 4.29–4.31, and 4.32(a), the width S of the Zn diffused region is 4 μm, the same as the width of the symmetric channel. In Figure 4.32(b), $S = 6$ μm which is larger than the channel. In Figures 4.29–4.32, we show (a) the effective index, (b) the gain distribution in the active layer, (c) the lateral near-field intensity, and (d) the lateral far-field radiation patterns at various emission powers. At threshold, the gain in the active layer (Figures 4.29–4.32) is distributed according to the diffusion of carriers in the absence of hole burning.

For the symmetric waveguide, the dimensions are $l_1 = l_2 = 2$ μm while for the asymmetric case (Figure 4.30), $l_1 = 2$ μm, and $l_2 = 1.7$ μm. This range of values is not a typical for our fabrication process. Differences in lengths l_1 and l_2 arise during fabrication as discussed earlier and can be associated with, for example, the crystal orientation so that both chemical etch rates and meltback rates can be different for the two sloping walls of the channel. Other effects of asymmetries can arise due to the asymmetrical growth of AlGaAs in the sloping channels.

For a perfect CSP laser, both the effective index (Figure 4.29(a)) and the carriers (Figure 4.29(b)) are distributed symmetrically about the center of the waveguide for all power levels. However, for $l_1 < l_2$, the effective index at threshold (Figure 4.30(a)) is asymmetric with a corresponding shift of the peak near-field intensity (Figure 4.30(c)) along the positive y-axis. Although the gain distribution (Figure 4.30(b)) appears symmetric at threshold, the gain profile is redistributed asymmetrically with increasing power with the highest peak of the gain distribution occurring on the negative y-axis. This redistribution of the gain profile with drive for the asymmetric case is a result of the near-field intensity distribution (Figure 4.30(c)) preferentially saturating the gain (Figure 4.30(b)) along the positive y-axis. The asymmetry of lateral gain in turn forces the effective index profile to become even more asymmetric (the lateral effective index step increases more along the negative y-axis than along the positive y-axis) because of the gain-induced refractive index depression effect—which causes an increase in the shift of the optical field in the positive y direction with drive (as much as 0.25 μm at 50 mW as shown in Figure 4.30(c)). As this interplay between the effective index profile, gain distribution, and near-field intensity continues, the central peak of the near-field distribution increasingly shifts along the positive y-axis while the magnitude of the gain distribution increases much more along the negative y-axis than along the positive y-axis.

For a strongly asymmetric CSP structure, the interplay between the effective index profile, gain distribution, and near-field intensity can limit the output power to a few tens of milliwatts: In Figure 4.31, the 4-μm wide Zn diffused region of a symmetric channel CSP laser is displaced 0.5 μm along the negative y direction. For this case, the slight asymmetry in the effective index (Figure 4.31(a)) near threshold is drastically enhanced as the peak of the near-field distribution (Figure 4.31(c)) increasingly shifts along the positive y-axis, increasing the gain (Figure 4.31(b)) almost exclusively along the negative y-axis. As a result, the power in such a device saturates since increased drive current reduces the interaction of the lasing mode with the gain and prevents saturation of the spontaneous emission. For this theoretical case, the total emission power of the fundamental mode could not be increased above approximately 30 mW at current levels above 125 mA as shown in Figure 4.33.

If the misalignment of the Zn diffusion is less, or the width of the Zn diffused region is larger than the channel width, the device behavior is similar to the case of a CSP laser with an asymmetric channel (Figure 4.30). The gain distribution for a 4-μm wide diffusion front displaced 0.2 μm is shown in Figure 4.32(a) while that of a 6-μm wide diffusion front displaced 0.5 μm is shown in Figure 4.32(b). In both cases the symmetric channel is 4 μm wide.

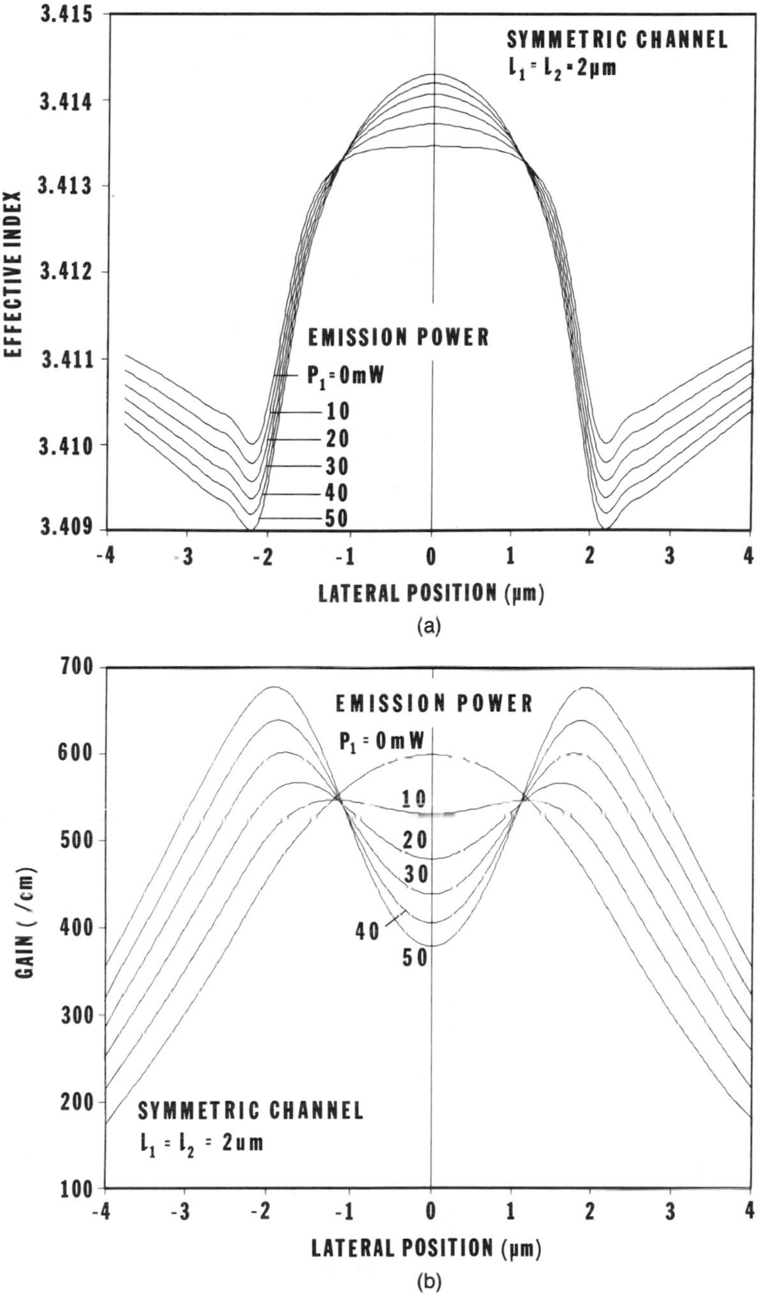

FIGURE 4.29. Calculated (a) lateral effective index profiles, (b) lateral gain profiles in the active layer, (c) lateral near-field intensity distributions, and (d) lateral far-field intensity patterns for a symmetric CSP–DH structure ($l_1 = 2.0\ \mu$m and $l_2 = 2.0\ \mu$m) with a 4-μm wide Zn diffusion front for output powers of 0, 10, 20, 30, 40, and 50 mW.

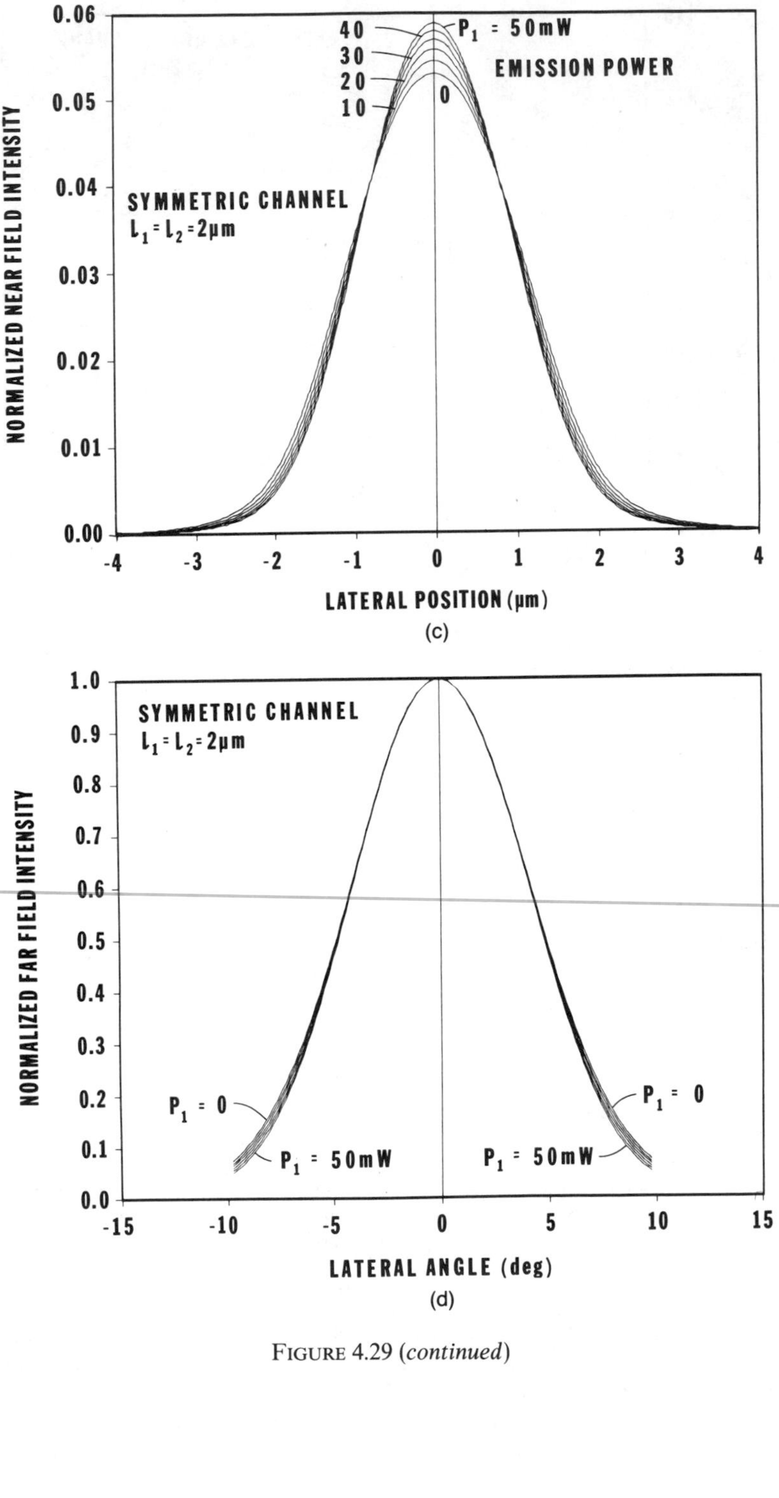

FIGURE 4.29 (*continued*)

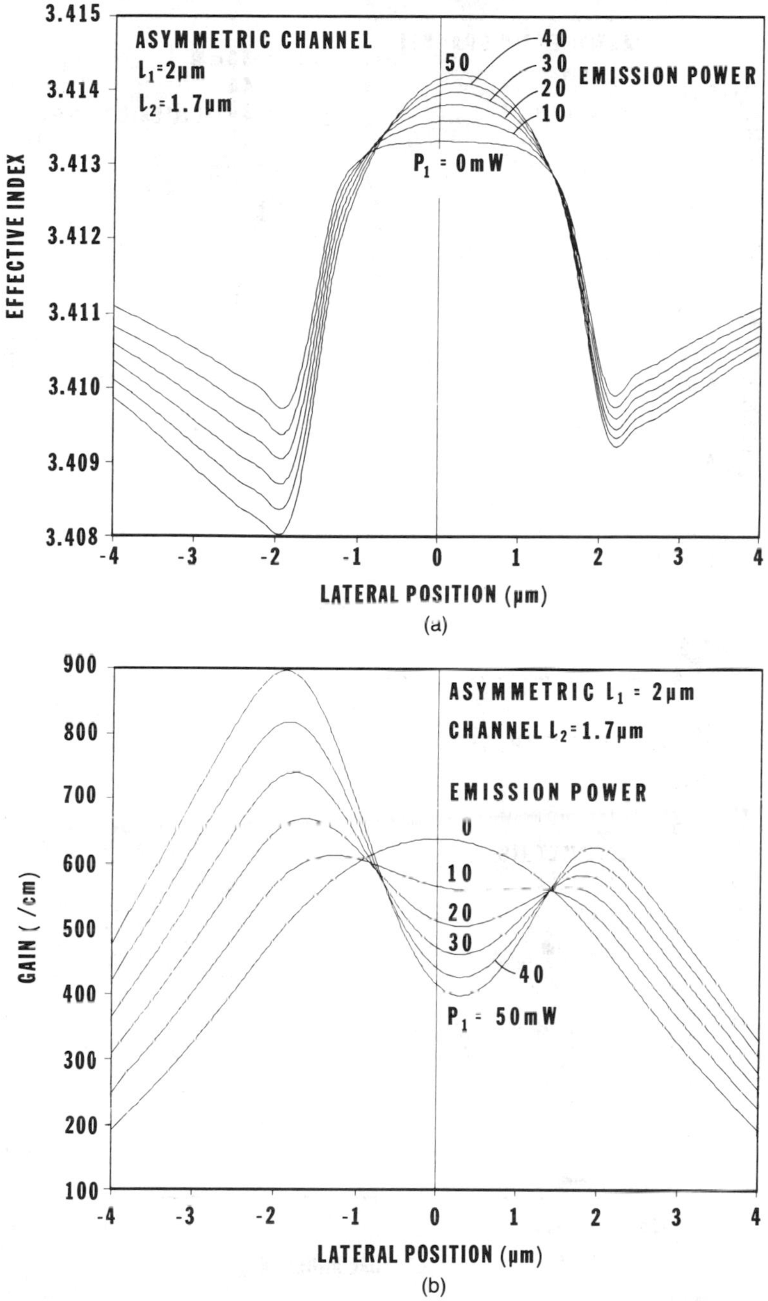

FIGURE 4.30. Calculated (a) lateral effective index profiles, (b) lateral gain profiles in the active layer, (c) lateral near-field intensity distributions, and (d) lateral far-field intensity patterns for an asymmetric CSP–DH structure ($l_1 = 2.0\ \mu$m and $l_2 = 1.7\ \mu$m) with a 4-μm wide Zn diffusion front for output powers of 0, 10, 20, 30, 40, and 50 mW.

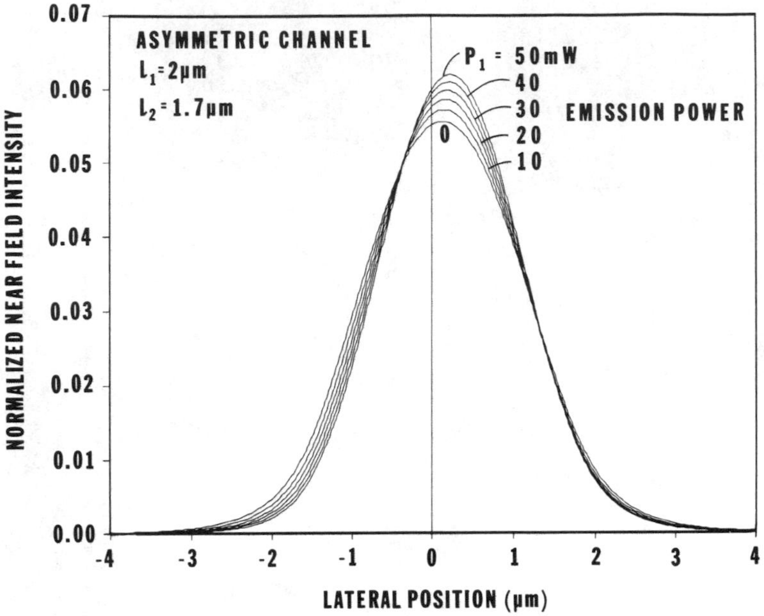

(c)

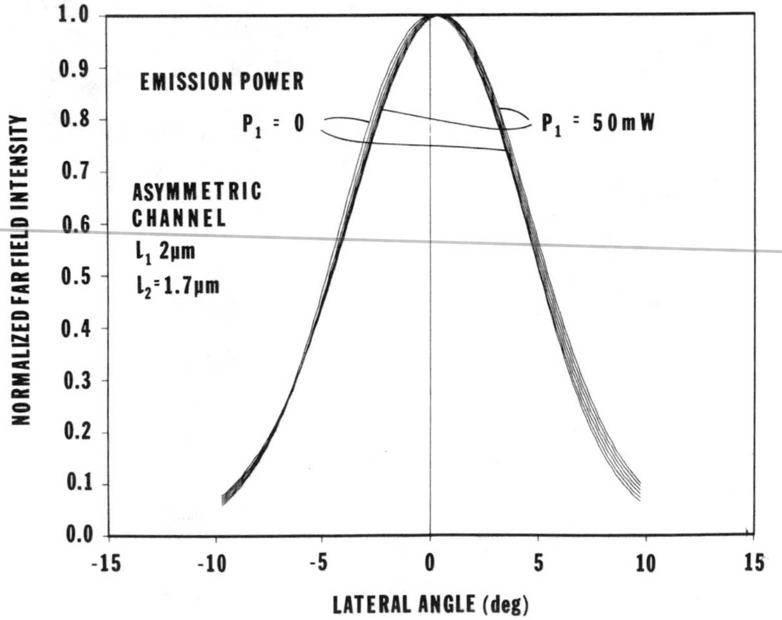

(d)

FIGURE 4.30 (*continued*)

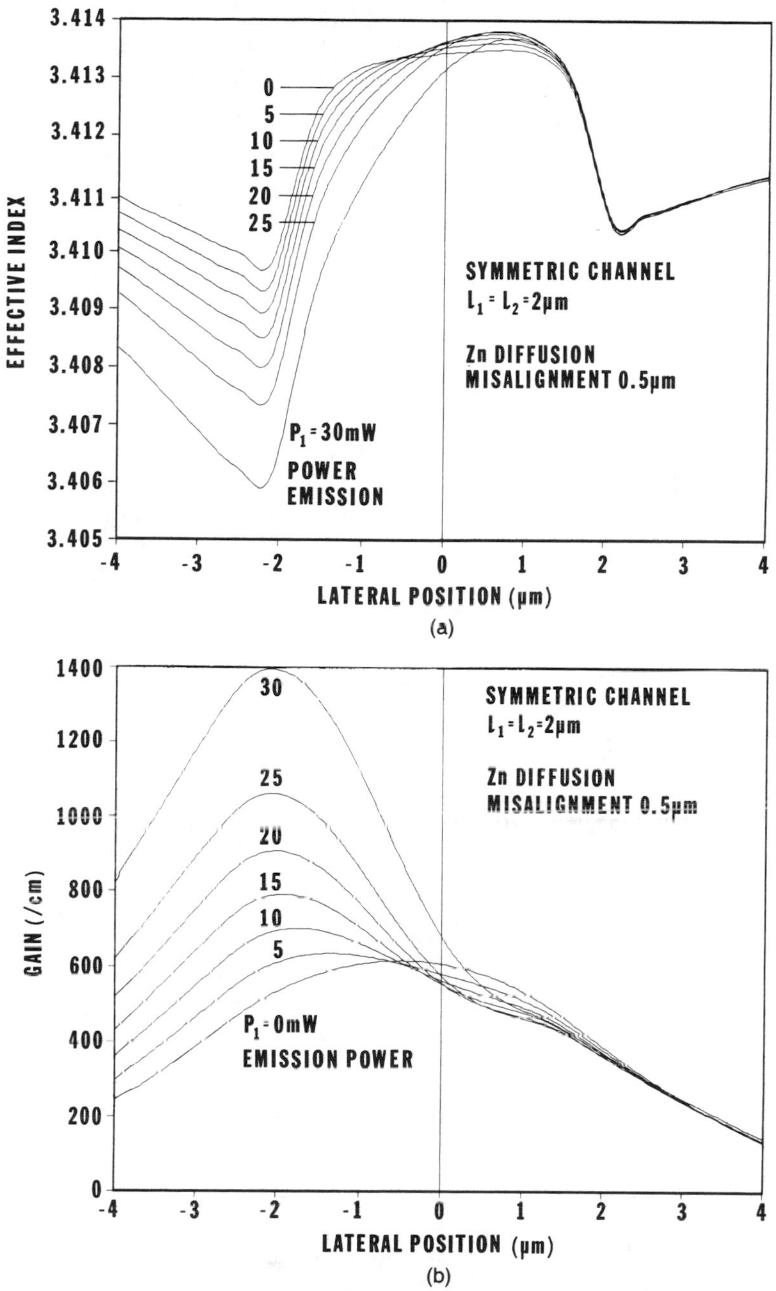

FIGURE 4.31. Calculated (a) lateral effective index profiles, (b) lateral gain profiles in the active layer, (c) lateral near-field intensity distributions, and (d) lateral far-field intensity patterns for a symmetric CSP–DH structure ($l_1 = 2.0$ μm and $l_2 = 2.0$ μm) with a misaligned 4-μm wide Zn diffusion front for output powers of 0, 5, 10, 15, 20, 25, and, 30 mW. The Zn diffusion is centered at $y = -0.5$ μm.

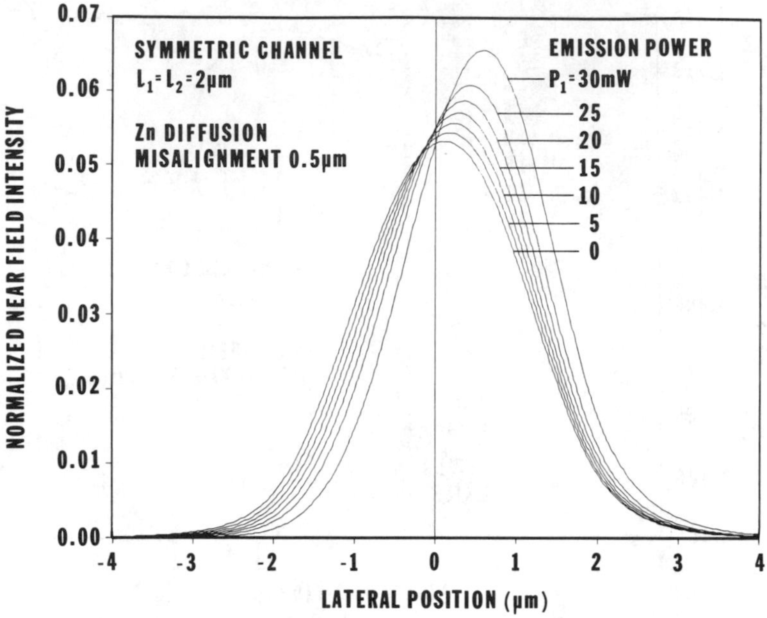

(c)

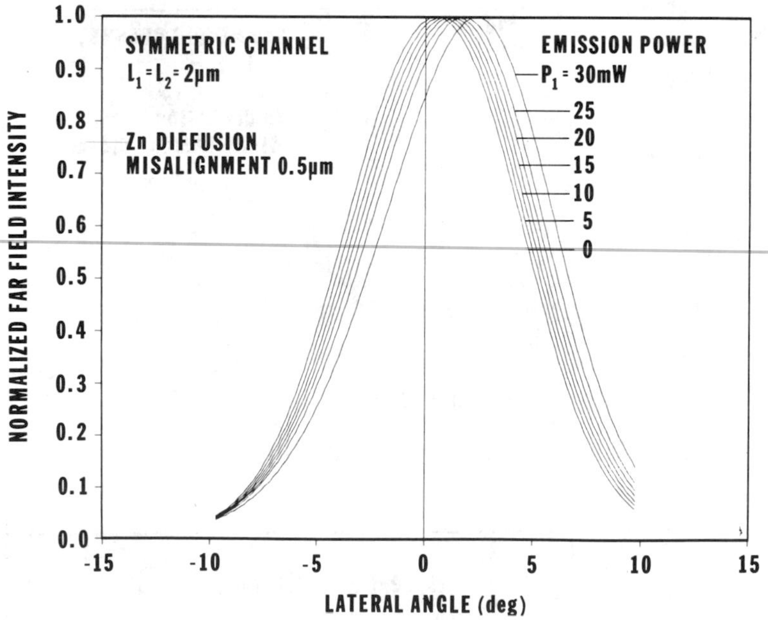

(d)

FIGURE 4.31 (*continued*)

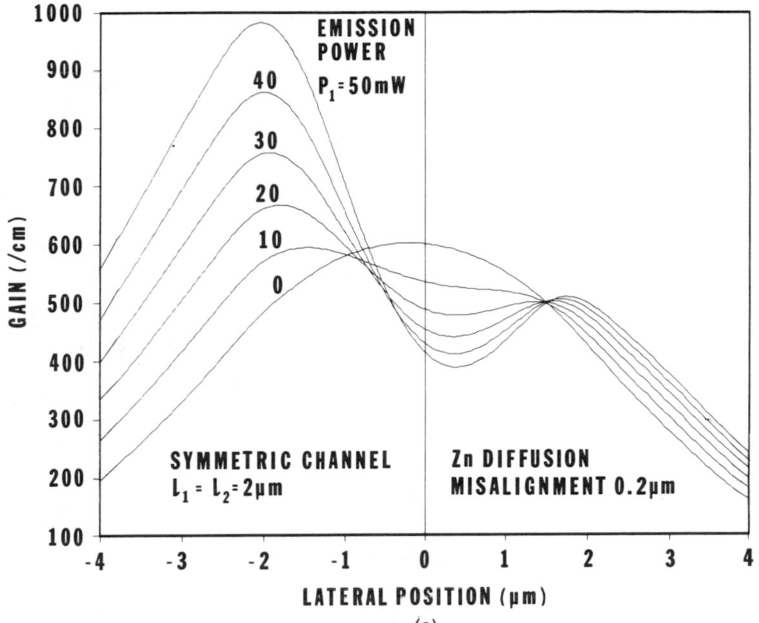

(a)

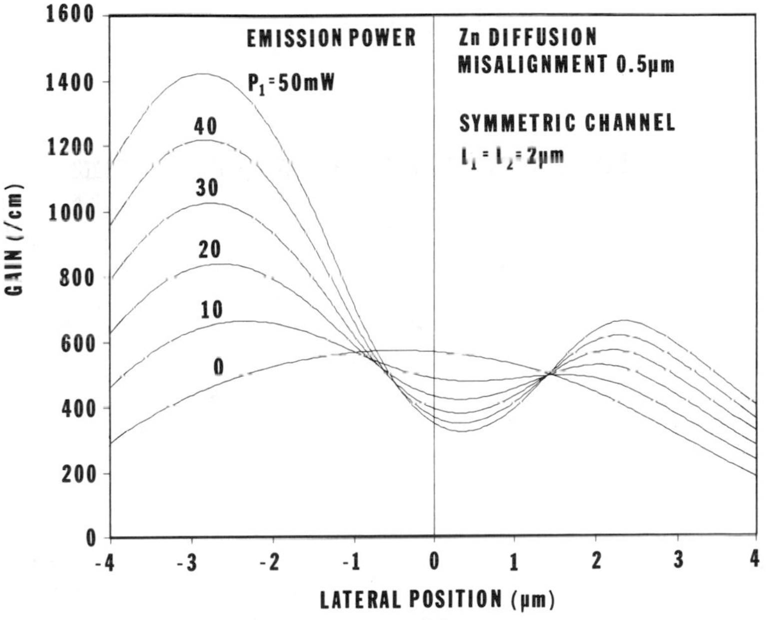

(b)

FIGURE 4.32. Calculated lateral gain profiles in the active layer for a symmetric CSP–DH structure for output powers of 0, 10, 20, 30, 40, and 50 mW with (a) a misaligned 4-μm wide Zn diffusion front centered at $y = -0.2$ μm, and (b) a misaligned 6-μm wide Zn diffusion front centered at $y = -0.5$ μm.

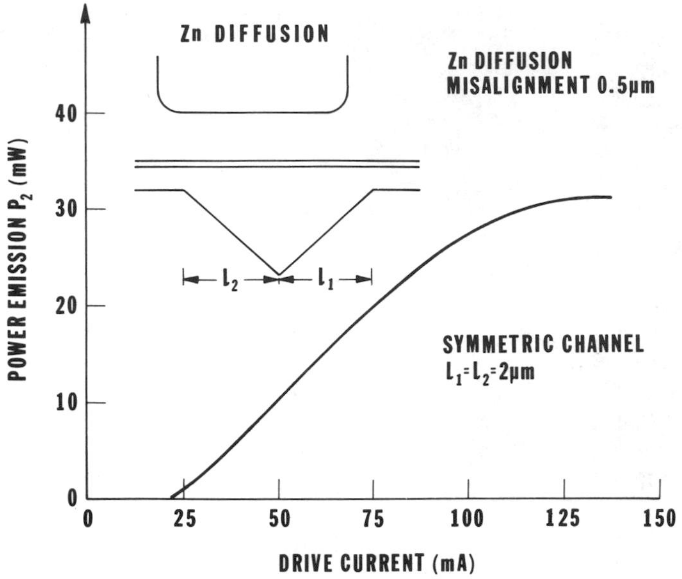

FIGURE 4.33. Calculated power versus current curves for a CSP–DH structure with a symmetric ($l_1 = 2.0$ μm and $l_2 = 2.0$ μm) channel, and a 4-μm wide diffusion front which is offset from the channel by 0.5 μm.

Not only are the gain distributions shown in Figure 4.32 similar to the gain distribution shown in Figure 4.30(b), the effective index, near-field distribution, and far-field pattern curves for the two cases shown in Figure 4.32 are very similar to the corresponding curves shown in Figure 4.30. The near-field shifts corresponding to the cases shown in Figure 4.32(a) and (b) were both about 0.25 μm at 50 mW (the same as the asymmetric channel case shown in Figure 4.29(c)) and the far-field peak shifts were both about 0.8° at 50 mW (slightly more than the asymmetric channel case shown in Figure 4.29(d)). For the strongly asymmetric CSP structure of Figure 4.31, the near-field peak shifts about 0.6 μm and the far-field peak moves about 2.5° as the output power increases from threshold to 30 mW. These near- and far-field shifts

---

FIGURE 4.34. (a) Calculated power versus current curves for a CSP–DH structure with: (1) a symmetric ($l_1 = 2.0$ μm and $l_2 = 2.0$ μm) channel, symmetric 4-μm wide diffusion front; (2) an asymmetric ($l_1 = 2.0$ μm and $l_2 = 1.7$ μm) channel, symmetric 4-μm wide diffusion front; (3) a symmetric ($l_1 = 2.0$ μm and $l_2 = 2.0$ μm) channel, misaligned 4-μm wide diffusion front centered at $y = -0.2$ μm; and (4) a symmetric ($l_1 = 2.0$ μm and $l_2 = 2.0$ μm) channel, misaligned 6-μm wide diffusion front centered at $y = -0.5$ μm. The device geometry is shown in Figure 4.28 and the numerical parameters given in Table 4.8. (b) Experimentally measured power versus current curve for a CSP–DH laser. ▷

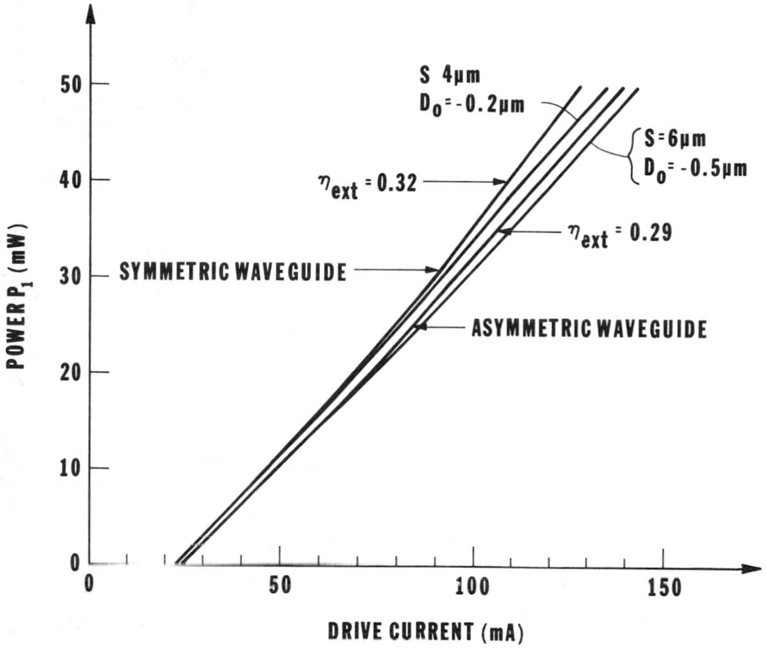
S 4μm
D_0=-0.2μm
S=6μm
D_0=-0.5μm
η_{ext} = 0.32
η_{ext} = 0.29
SYMMETRIC WAVEGUIDE
ASYMMETRIC WAVEGUIDE
POWER P_1 (mW)
50
40
30
20
10
0
0
50
100
150
DRIVE CURRENT (mA)

(a)

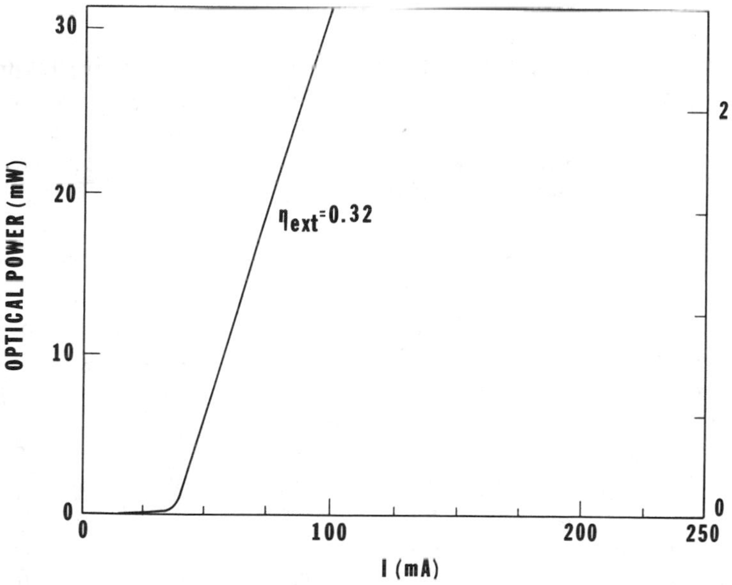
η_{ext}=0.32
OPTICAL POWER (mW)
30
20
10
0
2
0
0
100
200
250
I (mA)

(b)

TABLE 4.8. CSP modeling parameters.

	Cladding layers	Active region	Substrate
Mole fraction AlAs	0.30	0.07	0.0
Refractive index	3.40	3.62	3.64
Absorption coefficient	10 cm^{-1}	—	5000 cm^{-1}

$\lambda = 0.83\ \mu m$ (lasing wavelength);
$d_2 = 0.06\ \mu m$ (active layer thickness);
$B = 10^{-12}\ cm^3/sec$ (bimolecular coefficient);
$L_D = 3\ \mu m$ (carrier diffusion length);
$\tau_s = 3 \times 10^{-9}$ sec (carrier lifetime),
$R_1 = 0.32$, $R_2 = 0.85$ (facet reflection coefficients);
$d_c = 0.4\ \mu m$ (n-clad layer thickness outside of the channel);
$y_0 = 0.5\ \mu m$ (current decay parameter);
$S = 4\ \mu m$ (stripe width);
$L = 250\ \mu m$ (device length).

occur with asymmetries in the CSP structure because the gain-induced refractive index perturbation has a magnitude comparable to the "built-in" effective index profile. If the "built-in" effective index profile was large compared to the gain-induced index perturbations, the near- and far-fields should be stable with drive.

Because the gain in a CSP laser is predominantly above the channel, we expect that the lateral effective index step decreases with increasing drive. However, calculations indicate that the opposite occurs: the carriers are depleted in the center of the channel (Figures 4.29(b)–4.31(b), 4.32) by the optical field (which increases the effective index in the center of the channel), the gain is largest near the channel edges (which decreases the effective index outside of the channel), and the lateral effective index step is increased. This explanation is also consistent with astigmatism and phase-front measurements made on CSP lasers [73] which show no measurable change with increasing power.

For both the symmetric and asymmetric cases, the FWHP of the far-field pattern increases with increasing output power, although in the symmetric case, the increase is comparable to experimental error (a few tenths of a degree). The increase in the FWHP of the far-field pattern is due to narrowing of the near-field intensity (Figures 4.29(c)–4.31(c)) as the lateral effective index step (Figures 4.29(a)–4.31(a)) increases with increasing power output. Experimental measurements showing increases in FWHP of the far-field pattern of about 10% (over a power range of 100 mW) with power for a CSP laser appear in Figure 3 of reference [7].

In Figure 4.34(a) the calculated values of the optical emission from the front facet (front-facet reflectivity $R_1 = 0.32$, rear-facet reflectivity $R_2 = 0.85$) for the four cases described in Figures 4.29–4.32 as a function of the device drive current whose distribution is given by eq. (4.30) is shown. In the computations, we have assumed the internal quantum efficiency is unity and there is no loss of carriers transversing heterojunctions. Therefore, the current axis can be scaled to account for known leakage in realistic devices. A complete list of the parameters used to calculate the power as a function of drive current for the cases shown in Figure 4.34 is given in Table 4.8. Both the threshold currents and slope efficiencies are different for the different cases. For the symmetric CSP, $I_{th} = 22$ mA, while it is approximately 24 mA for the asymmetric channel case. Note that increasing the Zn diffusion front from 4 μm to 6 μm results in only a slight decrease in the external differential quantum efficiency. Figure 4.34(b) shows an experimental P–I curve of a typical CSP laser with minimal asymmetries and with geometry, composition, and facet coatings comparable to the numerical examples. Experimentally, thresholds as low as 37 mA were measured.

4.5.4. *Performance Considerations*

As discussed in Section 4.3, while the large imaginary part of the effective index outside the channel region of a CSP laser helps stabilize the mode, it also increases the threshold current and decreases the differential quantum efficiency. Using the computational procedure described above, the power versus

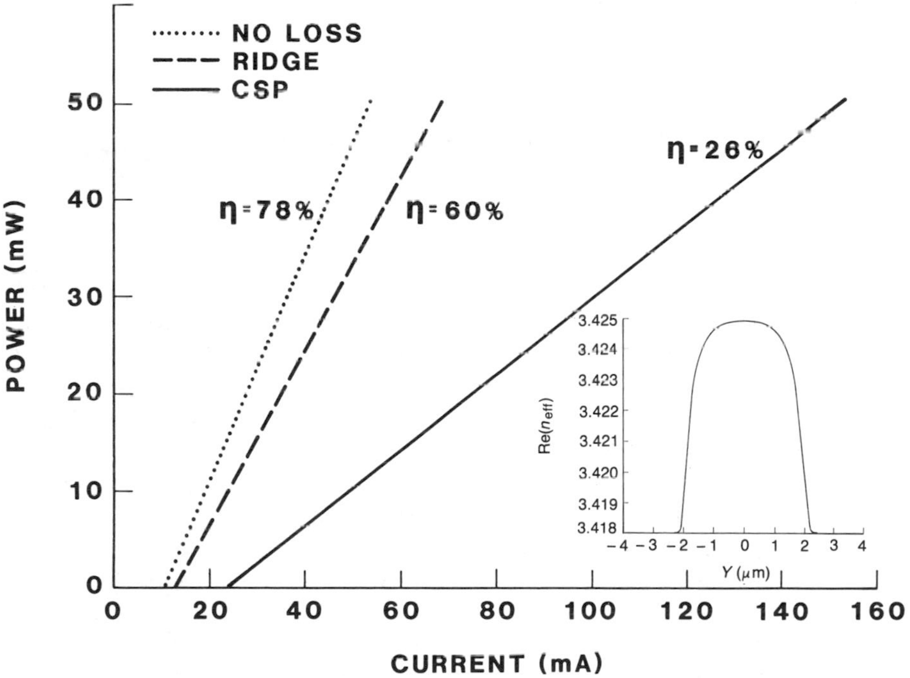

FIGURE 4.35. Calculated power versus current curves for a conventional CSP laser (——), a conventional ridge guide laser (– – –), and a lossless ridge guide laser (. . .).

current curves for a CSP-type laser, a ridge guide laser with cladding layer losses of 10 cm^{-1}, and a lossless ridge guide laser (all with active layer thicknesses of 600 Å) are calculated and the results are plotted in Figure 4.35. All three structures have the same lateral index step (6.822×10^{-3}) and profile (see inset of Figure 4.35) with a channel or ridge width of 4 μm. The large differences in threshold current and differential quantum efficiency are due to the reduced losses outside the ridge of the two ridge structures compared to the high losses outside the channel for the CSP device.

The value of the injected current at which the first-order lateral mode reaches threshold was also calculated for the three structures of Figure 4.35. While the first-order mode of the ridge structures reached threshold at power levels in the 10–20 mW range, the first-order mode of the CSP structure was only approaching transparency at 200 mW. Since the channel and ridge widths are equal in these calculations, we must note that 4 μm is a relatively narrow channel for a CSP laser, while 4 μm is relatively wide for a ridge guide. As the CSP channel is widened from 8 μm to 12 μm, the power level at which the first higher-order lateral mode reaches threshold is reduced from about 100 mW to 30 mW. As the width of the ridge is reduced to 2 μm, the first higher-order lateral mode is calculated to reach threshold at about 50 mW. Such calculations indicate that the onset of higher-order modes is not a significant effect in CSP lasers of common channel widths (less than 8 μm). Rather, variations and irregularities in the output characteristics of CSP lasers are more likely due to compositional variations [13] and geometrical asymmetries [12], [14].

For the devices of Figure 4.35, $\Delta\alpha_0/k_0 = 0$ for the ridge guide structures and 5.074×10^{-3} for the GaAs substrate CSP laser.

4.6. Conclusions

We have presented a physical explanation of the guiding mechanism in CSP lasers: the lateral index step is determined by (1) the amount of wavefront tilt, and (2) the magnitude of the guide wavelength λ_z. In most cases, the wavefront tilt is the dominant mechanism affecting the lateral index step, and slight changes in the magnitude of λ_z explain minor changes in the real part of the lateral index step with changes in the AlAs composition ($0 < x < 0.3$) of the substrate. However, when the normalized transverse propagation constant is very close to the refractive index of the substrate (as in the example of a substrate with an AlAs composition equal to 0.3), a resonant condition occurs and the dominatc mechanism governing the value of the effective index is the magnitude of the guide wavelength λ_z. Therefore, CSP lasers are index guided even when there is no substrate absorption.

Because of the contribution of the tilted wavefront in the calculation of the far-field radiation pattern, all CSP lasers will have a slight asymmetry in their transverse far-field pattern.

Design curves which provide the complex lateral effective index step as a function of n-clad thickness with the active layer thickness as a parameter are presented. We also show that the CSP guiding mechanism provides a positive lateral real index step for substrates with mole fractions of AlAs ranging from 0 to values approaching 0.30.

The net result of the effect described here is that due to compositional changes in the n-clad layer within the channel, a CSP–LOC or ESL–CSP can be inadvertently grown instead of a conventional CSP laser, all three of which have significantly different characteristics. We also give a theoretical explanation for asymmetric perpendicular far-field patterns which have been experimentally observed in CSP and other AlGaAs laser structures.

A comprehensive model of the CSP–DH laser structure fabricated with sloping V-groove channels is developed. The laser characteristics are computed using a self-consistent calculation of the optical field and of the electron/hole distribution in the active layer of the laser. These detailed computations show how small asymmetries due to device fabrication can significantly affect the optical and electrical properties of the CSP–DH laser.

Many applications of semiconductor lasers require both near- and far-field stability with drive current. Although our analysis shows that the lateral effective index step increases with power, narrowing the near-field distribution and broadening the far-field pattern of even a perfect CSP laser, only asymmetries in the device cause far-field shifts and measurable broadening of the far-field pattern. The asymmetries studied in this paper show how the lateral effective index profile can be deformed to produce near-field movements greater than 0.5 μm and far-field pattern shifts of about 2.5°.

The most serious asymmetry introduced here occurs because of Zn diffusion misalignment. The effect of misalignments on the performance can be minimized by making the Zn diffusion region wider than the channel at the expense of only a slight decrease of efficiency. Additionally, these calculations show the necessity of precisely positioning the Zn diffusion region even when the diffusion front is wider than the channel. CSP–DH lasers fabricated with blocking layers [4], [5], [8], [20] have self-aligned current confinement and are not vulnerable to Zn diffusion misalignment. However, nonuniformities in the blocking layers will have the same effect as Zn diffusion misalignment.

The effects of asymmetries of the type discussed here are particularly pronounced when the CSP–DH type lasers are designed to have a weak lateral index guiding mechanism. Weak index guiding produces optical modes with broad lateral near fields that have a low peak optical intensity for a given power level. Such low peak optical intensity levels are desirable for avoiding catastrophic optical damage at the facets of semiconductor lasers. Strongly index-guided structures, while less sensitive to power-dependent field shifts, require small spot sizes (with corresponding higher peak power densities) to maintain single lateral mode operation. High-power lasers with weak index guided structures require device geometries with minimal asymmetries in order to produce lasers with characteristics that are insensitive to drive current.

Acknowledgments

The authors would like to acknowledge Nancy Dinkel and Bernard Goldstein for the expert growth and fabrication of the CSP lasers reported herein; excellent computational support by Valerie Mason; and extensive and numerous technical discussions on the topic of CSP lasers with Nils Carlson, John Connolly, Michael Ettenberg, Bernard Goldstein, Jacob Hammer, and Valerie Mason.

References

[1] M. Sakamoto, D.F. Welch, G.L. Harnagel, W. Streifer, H. Kung, and D.R. Scifres (1988), Ultrahigh power 38 W continuous-wave monolithic laser diode arrays, *Appl. Phys. Lett.*, **52**, 26, 2220–2221.

[2] G.A. Evans, N.W. Carlson, J.M. Hammer, M. Lurie, J.K. Butler, S.L. Palfrey, R. Amantea, L.A. Carr, F.Z. Hawrylo, E.A. James, C.J. Kaiser, J.B. Kirk, and W.F. Reichert (1989), Two-dimensional coherent laser arrays using grating surface emission, *IEEE J. Quantum Electronics*, **QE-25**, 6, 1525–1538.

[3] K. Aiki, M. Nakamura, T. Kuroda, J. Umeda, R. Ito, Naoki Chinone, and M. Maeda (1978), Transverse mode stabilized $Al_xGa_{1-x}As$ injection lasers with channeled-substrate-planar structure, *IEEE J. Quantum Electronics*, **QE-14**, 2, 89–94.

[4] S. Yamamoto, H. Hayashi, S. Yano, T. Sakurai, and T. Hijikata (1982), Visible GaAlAs V-channeled substrate inner stripe laser with stabilized mode using *p*-GaAs substrate, *Appl. Phys. Lett.*, **40**, 5, 372–374.

[5] S. Yamamoto, N. Miyauchi, S. Maei, T. Morimoto, O. Yamamoto, S. Yano, and T. Hijikata (1985), High output power characteristics in broad-channeled substrate inner stripe lasers, *Appl. Phys. Lett.*, **46**, 4, 319–321.

[6] K. Hamada, M. Wada, H. Shimizu, M. Kume, F. Susa, T. Shibutani, N. Yoshikawa, K. Itoh, G. Kano, and I. Teramoto (1985), A 0.2 W continuous wave laser with buried twin-ridge substrate structure, *IEEE J. Quantum Electronics*, **QE-21**, 6, 623–628.

[7] B. Goldstein, M. Ettenberg, N.A. Dinkel, and J.K. Butler (1985), A high-power channeled-substrate-planar AlGaAs laser, *Appl. Phys. Lett.* **47**, 655–657.

[8] T. Shibutani, M. Kume, K. Hamada, H. Shimizu, K. Itoh, G. Kano, and I. Teramoto (1987), A novel high-power laser structure with current-blocked regions near cavity facets, *IEEE J. Quantum Electronics*, **QE-23**, 6, 760–764.

[9] T. Kuroda, M. Nakamura, K. Aiki, and J. Umeda (1978), Channeled-substrate-planar structure $Al_xGa_{1-x}As$ lasers: an analytical waveguide study, *Appl. Optics*, **17**, 20, 3264–3267.

[10] K.A. Shore (1981), Above-threshold analysis of channeled-substrate-planar (CSP) laser, *IEE Proc.* Part I, 9–15.

[11] S. Wang, C. Chen, A.S. Liao, and L. Figueroa (1981), Control of mode behavior in semiconductor lasers, *IEEE J. Quantum Electronics*, **QE-17**, 4, 453–468.

[12] J.K. Butler, G.A. Evans, and B. Goldstein (1987), Analysis and performance of channeled-substrate-planar double-heterojunction lasers with geometrical asymmetries, *IEEE J. Quantum Electronics*, **QE-23**, 11, 1890–1899.

[13] G.A. Evans, B. Goldstein, and J.K. Butler (1987), Observations and consequences of non-uniform aluminum concentrations in the channel regions of AlGaAs channeled-substrate-planar lasers, *IEEE J. Quantum Electronics*, **QE-23**, 11, 1900–1908.

[14] T. Ohtoshi, K. Yamaguchi, C. Nagaoka, T. Uda, Y. Murayama, and N. Chinone (1987), A two-dimensional device simulator of semiconductor lasers, *Solid-State Electronics*, **30**, 6, 627–638.

[15] G.A. Evans, J.K. Butler, and V.M. Masin (1988), Lateral optical confinement of channeled-substrate-planar lasers with GaAs/AlGaAs substrates, *IEEE J. Quantum Electronics*, **QE-24**, 5, 737–749.

[16] T. Kadowaki, T. Aoyagi, S. Hinata, N. Kaneno, Y. Seiwa, K. Ikeda, and W. Susaki (1986), Long-lived phase-locked laser arrays mounted on an Si-submount with Au–Si solder with a junction-down configuration, *IEEE International Semiconductor Laser Conference*, Kanazawa, Japan, Conference Program and Abstrate of Papers, pp. 84–85.

[17] B. Goldstein, N.W. Carlson, G.A. Evans, N. Dinkel, and V.J. Masin (1987), Performance of a channeled-substrate-planar high-power phase-locked array operating in the diffraction limit, *Electronics Lett.*, **23**, 21, 1136–1137.

[18] M. Taneya, M. Matsumoto, S. Matsui, S. Yano, and T. Hijikata (1985), 0° phase mode operation in phased-array laser diode with symmetrically branching waveguide, *Appl. Phys. Lett.*, **47**, 4, 341–343.

[19] D.F. Welch, P. Cross, D. Scifres, W. Streifer, and R.D. Burnham (1986), In-phase emission from index-guided laser array up to 400 mW, *Electronics Lett.*, **22**, 6, 293–294.

[20] K. Uomi, S. Nakatsuka, T. Ohtoshi, Y. Ono, N. Chinone, and T. Kajimura (1984), High-power operation of index-guided visible GaAs/GaAlAs multi-quantum well lasers, *Appl. Phys. Lett.* **45** (8), 818–821.

[21] J. J. Yang, C.S. Hong, J. Niesen, and L. Figueroa (1985), High-power single longitudinal mode operation of inverted channel substrate planar lasers, *J. Appl. Phys.*, **58**, 4480–4482.

[22] H. Tanaka, M. Mushiage, Y. Ishida (1985), Single-longitudinal-mode self-aligned (AlGa)As double-heterostructure lasers fabricated by molecular beam epitaxy, *Japan. J. Appl. Phys.* **24**, 2, L89–L90.

[23] K. Yagi, H. Yamauchi, and T. Niina (1986), High external differential quantum efficiency (80%) SCH lasers grown by MBE, *IEEE International Semiconductor Laser Conference*, Kanazawa, Japan, Conference Program and Abstract of Papers, pp. 158–159.

[24] B. Goldstein, G.A. Evans, N. Dinkel, J. Kirk, and J. Connolly (1988), An efficient AlGaAs channeled-substrate-planar distributed feedback laser, *Appl. Phys. Lett.*, **53**, 7, 550–552.

[25] G.A. Evans, J.M. Hammer, N.W. Carlson, F.R. Elia, E.A. James, and J.B. Kirk (1986), Surface-emitting second-order distributed Bragg reflector laser with dynamic wavelength stabilization and far-field angle of 0.25°, *Appl. Phys. Lett.*, **49**, 6, 314–315.

[26] H.C. Casey, Jr., D.D. Sell, and K.W. Wecht (1975), Concentration dependence of the absorption coefficient for *n*- and *p*-type GaAs between 1.3 and 1.6 eV, *J. Appl. Phys.*, **46**, 1, 250–257.

[27] J. Connolly, T.R. Stewart, D.B. Gilbert, S.E. Slavin, D.B. Carlin, and M. Ettenberg (1988), High-power 0.87 μm channeled substrate planar lasers for spaceborn communications, in *SPIE Proceedings*, Vol. 885, Los Angeles, CA.

[28] H. Kressel and J.K. Butler (1977), *Semiconductor Lasers and Heterojunction LEDs*, Academic Press, New York, Chap. 7, Sec. 5, pp. 230–234.

[29] T. Hayakawa, T. Suyama, H. Hayashi, S. Yamamoto, S. Yano, and T. Hijikata (1983), Mode characteristics of large-optical-cavity V-channeled substrate inner stripe injection lasers, *IEEE J. Quantum Electronics*, **QE-19**, 10, 1530–1536.

[30] J.K. Butler, H. Kressel, and I. Ladany (1975), Internal optical losses in very thin CW heterojunction laser diodes," *IEEE J. Quantum Electronics*, **QE-11**, 7, 402–408.

[31] W. Streifer, Robert D. Burnham, and D.R. Scifres (1976), Substrate radiation losses in GaAs heterostructure lasers, *IEEE J. Quantum Electronics*, **QE-12**, 3, 177–182.

[32] D.R. Scifres, W. Streifer, and R.D. Burnham (1976), Leaky wave room-temperature double heterostructure GaAs : GaAlAs diode laser, *Appl. Phys. Lett.*, **29**, 1, 23–25.

[33] J. Buus (1979), A model for the static properties of DH lasers, *IEEE J. Quantum Electronics*, **QE-15**, 734–739.

[34] R. Lang (1979), Lateral transverse mode instability and its stabilization in stripe geometry injection lasers, *IEEE J. Quantum Electronics*, **QE-15**, 718–726.

[35] W. Streifer, D.R. Scifres, and R.D. Burnham (1980), Above-threshold analysis of double-heterojunction diode lasers with laterally tapered active regions, *Appl. Phys. Lett.*, **37**, 877–879.

[36] W. Streifer, D.R. Scifres, and R.D. Burnham (1981), Channelled substrate non-planar laser analysis, Part II: Lasers with tapered active regions, *IEEE J. Quantum Electronics*, **QE-17**, 1521–1530.

[37] W. Streifer, D.R. Scifres, and R.D. Burnham (1981), Channelled substrate non-planar laser analysis, Part I: for modulation, *IEEE J. Quantum Electronics*, **QE-17**, 736–744.

[38] C.B. Su (1981), An analytical solution of kinks and nonlinearities driven by near-field displacement instabilities in stripe geometry diode lasers, *J. Appl. Phys.*, **52**, 2665–2673.

[39] M. Ueno, R. Lang, S. Matsumoto, H. Kawano, T. Furuse, and I. Sakuma (1982), Optimum designs for InGaAsP/InP (1 = 1.3 μm) planoconvex waveguide lasers under lasing conditions, *IEE Proc.*, **129**, I, 218–228.

[40] K.A. Shore (1983), Above-threshold current leakage effects in stripe-geometry injection lasers, *Opt. Quantum Electronics*, **15**, 371–379.

[41] G.P. Agrawal (1984), Lateral analysis of quasi-index guided injection lasers: transitions from gain to index guiding, *IEEE J. Lightwave Technology*, **LT-2**, 537–543.

[42] J. Buus (1985), Principles of semiconductor laser modelling, *IEE Proc. J.*, **132**, 42–51.

[43] E.A.J. Marcatili (1969), Dielectric rectangular waveguide and directional coupler for integrated optics, *Bell System Tech. J.*, **48**, 2071–2102.

[44] G.B. Hocker and W.K. Burns (1977), Mode dispersion in diffused channel waveguides by the effective index method, *Appl. Opt.*, **16**, 113–118.

[45] W. Streifer and E. Kapon (1979), Applications of the equivalent-index method to DH diode lasers, *Appl. Opt.*, **18**, 3724–3725.

[46] J. Manning and R. Olshansky (1983), The carrier-induced index change in AlGaAs and 1.3 μm InGaAs diode lasers, *IEEE J. Quantum Electronics*, **QE-19**, 10, 1525–1530.

[47] A. Sommerfeld (1949), *Partial Differential Equations in Physics*, Academic Press, New York, p. 195.

[48] L. Lewin (1975), Obliquity-factor correction to solid-state radiation patterns," *J. Appl. Phys.*, **46**, 5, 2323–2324.

[49] T. Tamir and F.Y. Kou (1986), Varieties of leaky waves and their excitation along multilayered structures, *IEEE J. Quantum Electronics*, **QE-22**, 4, 544–551.

[50] L. Figueroa, T.L. Holcomb, K. Burghard, D. Bullock, C.B. Morrison, L.M. Zinkiewicz, and G.A. Evans (1986), Modeling of the optical characteristics for twinchannel laser (TCL) structures, *IEEE J. Quantum Electronics*, **QE-22**, 2141-2149.

[51] N.W. Carlson, V.J. Masin, M. Lurie, B. Goldstein, and G.A. Evans (1987), Measurement of the coherence of a single-mode phase-locked diode laser Array, *Appl. Phys. Lett.*, **51**, 9, 643–645.

[52] H. Neumann and U. Flohrer (1974), Electron mobility in Al_xGa_{1-x}, *Phys. Stat. Sol.* (a), **25**, 2, K145–K147.

[53] S. Adachi (1985), GaAs, AlAs, and $Al_xGa_{1-x}As$: material parameters for use in research and device parameters, *J. Appl. Phys.*, **58**, 3, R1–R29.

[54] S. Todoroki, M. Sawai, and K. Aiki (1985), Temperature distribution along the striped active region in high power GaAlAs visible lasers, *J. Appl. Phys.*, **58**, 1124–28.

[55] S. Todoroki (1986), Influence of local heating on current-optical output power characteristics in $Ga_{1-x}Al_xAs$ lasers, *J. Appl. Phys.*, **60**, 61–65.

[56] S. Chinn and R. Spiers (1982), Calculation of separated multiclad-layer stripe geometry laser modes, *IEEE J. Quantum Electronics*, **QE-18**, 6, 984

[57] M. Amann (1986), Rigorous waveguiding analysis of the separated multiclad-layer stripe-geometry laser, *IEEE J. Quantum Electronics*, **QE-22**, 10, 1992.

[58] H. Kressel and J.K. Butler (1977), *Semiconductor Lasers and Heterojunction LEDs*, Academic Press, New York, Chap. 5, p. 172.

[59] R.B. Smith and G.L. Mitchel, Calculation of complex propagating modes in arbitrary, plane-layered, complex dielectric structures. I. Analytic formulation. II. Fortran program MODEIG, EE Technical Report, No. 206, University of Washington, Seattle, WA.

[60] V. Asder, J. Christiansen, and R.D. Russel (1979), COLSYS—a collocation code for boundary value problems, in *Codes for Boundary Value Problems*, edited by B. Childs et al., Lecture Notes in Computer Science, Vol. 76, Springer-Verlag, New York.

[61] M. Born and E. Wolf (1975), *Principles of Optics*, Pergamon Press, New York, Chap. 1, pp. 61–66.

[62] J.W. Cahn and D.W. Hoffman (1974), A vector thermodynamics for anisotropic surfaces—II. Curved and faceted surfaces, *Acta Metallurgica*, **22**, 1205–1214.

[63] H. Kressel and J.K. Butler (1977), *Semiconductor Lasers and Heterojunction LEDs*, Academic Press, New York, Chap. 11, p. 372.

[64] J.K. Butler and D. Botez (1982), Mode characteristics of nonplanar double-heterojunction and large-optical-cavity laser structures, *IEEE J. Quantum Electronics*, **QE-18**, 952–961.

[65] J.K. Butler and D. Botez (1984), Lateral mode discrimination and control in high-power single-mode diode lasers of the large-optical-cavity (LOC) type, *IEEE J. Quantum Electronics*, **QE-20**, 879–891.

[66] W.B. Joyce (1982), Role of the conductivity of the confining layers in DH-laser spatial hole burning effects, *IEEE J. Quantum Electronics*, **QE-18**, 2005–2009.

[67] R. Papannareddy, W.E. Ferguson, Jr., and J.K. Butler (1988), Four models of lateral current spreading in double-heterostructure stripe-geometry lasers, *IEEE J. Quantum Electronics*, **QE-24**, 1, 60–65.

[68] H. Yonezu, I. Sakuma, K. Kobayashi, T. Kamejima, M. Ueno, and Y. Nannichi (1973), A GaAs/AlGaAs double heterostructure planar stripe laser, *Japan J. Appl. Phys.*, **12**, 1585–1592.

[69] J.K. Butler, G.A. Evans, and N.W. Carlson (1989), Nonlinear characterization of modal gain and effective index saturation in channelled-substrate-planar double-heterojunction lasers, *IEEE J. Quantum Electronics*, **QE-25**, 7, 1646–1652.

[70] S. Hasuo and T. Ohmi (1974), Spatial distribution of the light intensity in the injection lasers, *Japan J. Appl. Phys.*, **13**, 1429–1434.

[71] M.J. Adams (1977), Longitudinal field control in stripe-geometry lasers, *Electronics Lett.* **13**, 236–237.

[72] F. Stern (1976), Calculated spectral dependence of gain in excited GaAs, *J. Appl. Phys.*, **47**, 5382–5386.

[73] N.W. Carlson, V.J. Masin, G.A. Evans, B. Goldstein, and J.K. Butler (1986), Phase front measurements of high-power diode lasers, Paper TuQ6, *Digest of Technical Papers, Conference on Lasers and Electro-Optics*, San Francisco, CA.

Appendix A. Lateral Field Analysis

The solution of the lateral fields using the multipoint boundary-value solver COLSYS [60] is formulated here. The differential equation for the lateral field function $v(y)$ is

$$\frac{d^2v}{dy^2} + [\gamma^2 - \gamma_0^2(y) + k_0^2\Gamma_c(y)\kappa_v(y)]v = 0, \qquad (4.A.1)$$

where $v(y)$ is the complex-valued function $v = v_r + jv_i$. The functions $\gamma_0(y)$, the transverse complex effective index, and $\Gamma_c(y)$, the complex confinement factor of the active layer, are determined from solutions of the transverse fields as dictated by the effective index method. The relative dielectric constant $\kappa_v = 2n_2\delta n_2(y) + jn_2 g(y)/k_0$, where $g(y)$ is the gain of the active layer and is determined from the electron/hole pair density in the active layer. We assume the carrier-induced index variation n_2 is linearly related to the gain. In the V-groove CSP–DH laser structures (see Figure 4.27(a)), the functional dependence of $\gamma_0(y)$ and $\Gamma_c(y)$ are constants at lateral positions where $y > w_u$ and $y < w_l$. However, the dielectric constant κ_v of the active layer can have variations outside the V-groove regions depending on the width of the stripe contact. κ_v will settle to a constant value at several diffusion lengths from the point at $y = S/2$, the stripe contact edge.

In our formulation it is assumed that the coefficient of v is constant for $y > w_u$ and for $y < -w_l$. Therefore, $v(y)$ has a simple exponential dependence in the two regions as given by eq. (4.28). Hence, the field solutions for $w_l < y < w_u$, solved by COLSYS, are matched to the evanescent fields in the far lateral positions. Putting, $\gamma = \alpha + j\beta$, $\gamma_0 = \alpha_0 + j\beta_0$, $\Gamma_c = \Gamma_{cr} + j\Gamma_{ci}$, and $\kappa_v = \kappa_{vr} + j\kappa_{vi}$, the set of differential equations must be separated into their real and imaginary parts. Before specifying the equations, it is appropriate to normalize the complex wave function $v(y)$ as follows:

$$I(y) = k_0 \int_{-\infty}^{y} v^2(y')\,dy' \tag{4.A.2}$$

so that $I(\infty) = 1$. Since v is complex, the above represents two real equations. Since $\gamma = \alpha + j\beta$ is unknown, two new differential equations are added so that COLSYS can solve for the eigenvalues α, β. At this point, we can write the following system of equations that govern the above assumptions:

$$\frac{d\alpha}{dy} = 0, \tag{4.A.3a}$$

$$\frac{d\beta}{dy} = 0, \tag{4.A.3b}$$

$$\frac{dI_r}{dy} = k_0(v_r^2 - v_i^2), \tag{4.A.3c}$$

$$\frac{dI_i}{dy} = 2k_0 v_r v_i, \tag{4.A.3d}$$

$$\frac{d^2 v_r}{dy^2} = -\mathrm{Re}\{[\gamma^2 - \gamma_0^2 + k_0^2\Gamma_c\kappa_v]v\}, \tag{4.A.3e}$$

$$\frac{d^2 v_i}{dy^2} = -\mathrm{Im}\{[\gamma^2 - \gamma_0^2 + k_0^2\Gamma_c\kappa_v]v\}. \tag{4.A.3f}$$

In the above system, there are four first-order equations and two second-order ones which requires six boundary conditions. Three required at $y = w_l$, are

$$\frac{dv_r}{dy} - q_r v_r + q_i v_i = 0, \tag{4.A.4a}$$

$$\frac{dv_i}{dy} - q_i v_r - q_r v_i = 0, \tag{4.A.4b}$$

$$2(q_r I_r - q_i I_i) - v_r^2 + v_i^2 = 0, \tag{4.A.4c}$$

$$2(q_i I_r + q_r I_i) - v_r v_i = 0, \tag{4.A.4d}$$

and at $y = w_u$

$$\frac{dv_r}{dy} + p_r v_r - p_i v_i = 0, \tag{4.A.5a}$$

$$\frac{dv_i}{dy} + p_i v_r + p_r v_i = 0, \tag{4.A.5b}$$

$$2(p_r I_r - p_i I_i) + v_r^2 - v_i^2 - 2p_r = 0, \tag{4.A.5c}$$

$$p_i I_r + p_r I_i + v_r v_i - p_i = 0. \tag{4.A.5d}$$

The complex quantities p and q are related to α and β via eq. (4.29). Separate routines required for COLSYS must compute the Jacobian of the differential equations (4.A.3) and of the boundary conditions (4.A.5). Since the set of differential equations are nonlinear, COLSYS can be started only by supplying initial trial solutions. Therefore, starting eigenvalues α and β and approximate field solutions v_r and v_i must be supplied. The value of β is related to the modal propagation constant while α is related to the modal gain. In our initial guesses, we assume $v_i = 0$. The initial guess for v_r is associated with the desired modal solution. For the fundamental mode, v_r has a single antinode, whereas the first higher-order mode has two antinodes with a 180° phase shift between them.

Appendix B. Solution of the Diffusion Equation

The carrier distribution in the active layer is determined from a solution of the diffusion equation

$$L_D^2 \frac{d^2N}{dy^2} - [1 + B\tau_s N(y) + aR_0\Gamma(y)|v(y)|^2]N(y) = -\frac{\tau_s J(y)}{qd_2} + R_0 b\Gamma(y)|v(y)|^2, \tag{4.B.1}$$

where a and b are defined in eq. (4.34) and

$$R_0 = \frac{2P_1\tau_s(1 - R_1 - \sqrt{R_1R_2} + \sqrt{R_1/R_2})}{hvd_2(1 - R_1) < |v(y)|^2 > \ln(1/R_1R_2)}. \tag{4.B.2}$$

To simplify the computations, the field $v(y)$ is assumed to be negligibly small in the regions $y < w_l$ and $y > w_u$. Further, we assume $B\tau_s N(y) \ll 1$ in the far lateral regions so that the resulting diffusion equation is linear. Hence, eq. (4.B.1) reduces to

$$L_D^2 \frac{d^2N}{dy^2} - N(y) = -\frac{\tau_s J_0}{qd_2} \frac{1}{\left(1 + \dfrac{|y| - S/2}{y_0}\right)^2}. \tag{4.B.3}$$

In the regions $y > w_u$ and $y < w_l$, the solutions of eq. (4.B.3) must be matched to the numerical solution of eq. (4.B.1) in the interior region $w_l < y < w_u$. The solution of eq. (4.B.3) involves both the homogeneous and particular solutions. In the interior region eq. (4.B.1) is solved using the routine COLSYS with the boundary conditions at $y = w_l$ and $y = w_u$ as follows:

$$L_D \frac{dN}{dy}\bigg|_{y=w_u} + N(w_u) - \frac{\tau_s J_0 y_0^2}{q d_2 L_D^2} \int_0^\infty \frac{e^{-t}}{(t + t_u)^2}\, dt = 0, \tag{4.B.4}$$

$$L_D \frac{dN}{dy}\bigg|_{y=w_l} - N(w_l) - \frac{\tau_s J_0 y_0^2}{q d_2 L_D^2} \int_0^\infty \frac{e^{-t}}{(t + t_l)^2}\, dt = 0, \tag{4.B.5}$$

where $t_u = (y_0 + w_u - S/2)/L_D$ and $t_l = (y_0 - w_l - S/2)/L_D$.

5
Correlation Theory of Electromagnetic Radiation Using Multipole Expansions

NICHOLAS GEORGE* and AVSHALOM GAMLIEL*

5.0. Introduction

In electromagnetic theory a wide variety of important problems can be treated in spherical coordinates using multipole expansions. It is also reasonable to expect that these expansions would provide a useful formalism for the solution of problems in statistical electromagnetic theory. The main theme of this chapter is the extension of multipole expansions into the realm of correlation theory. In the initial sections of this chapter, we present the basic elements of the Debye potentials [1], including a detailed discussion of uniqueness. This formulation of the Debye potentials follows the presentation in the literature by Bouwkamp and Casimir [2] as well as that in the treatise by Papas [3]. The elegant formalisms of vector spherical harmonics and dyads are also employed [4]. While this use of vector spherical harmonics and dyads requires some patience in becoming accustomed to the notation [5], we believe that it is particularly worthwhile in several instances.

In our treatment of statistical problems in electromagnetic theory, we write general dyadic cross-spectral densities of the field [6]. Thinking of the nonion form, one will realize that this expectation provides a concise representation for the nine possible second-order moments of the vector field. In Section 5.4, treating the far-zone region, we present expressions for the Poynting vector, the cross-spectral dyadic for the electric field, and a representation of the Stokes parameters. As an important illustrative example, we derive an expression for the total radiated power.

In the same way that multipole expansions provide a compact scalar representation for the vector electromagnetic field, the dyadic cross-spectral densities can logically be represented in terms of the corresponding scalar cross-spectral densities as shown in Section 5.5.1. The scalar cross-spectral densities required for representing the dyadic cross-spectral densities are obtained in Section 5.5.2 using symmetrical expansions of the multipole moments, and in Section 5.5.3 using symmetrical expansions of source terms for the Debye potentials.

* The Institute of Optics, University of Rochester, Rochester, NY 14627, USA.

5.1. Electromagnetic Waves in Spherical Polar Coordinates

5.1.1. *Maxwell's Field Equations*

We consider a region of space in which the time-dependent form of Maxwell's equations can be written in the MKSA system of units as

$$\nabla \times \mathbf{E}^{(\mathbf{R})}(\mathbf{r}, t) = -\frac{\partial}{\partial t}\mathbf{B}^{(\mathbf{R})}(\mathbf{r}, t), \tag{5.1}$$

$$\nabla \times \mathbf{H}^{(\mathbf{R})}(\mathbf{r}, t) = \mathbf{J}(\mathbf{r}, t) + \frac{\partial}{\partial t}\mathbf{D}^{(\mathbf{R})}(\mathbf{r}, t). \tag{5.2}$$

Here the field vectors are the electric field $\mathbf{E}^{(\mathbf{R})}(\mathbf{r}, t)$, the magnetic field $\mathbf{H}^{(\mathbf{R})}(\mathbf{r}, t)$, the electric displacement $\mathbf{D}^{(\mathbf{R})}(\mathbf{r}, t)$, and the magnetic induction $\mathbf{B}^{(\mathbf{R})}(\mathbf{r}, t)$. The superscript $^{(\mathbf{R})}$ is used to indicate that the vectors are real-valued functions of position and time, and the source current density $\mathbf{J}^{(\mathbf{R})}(\mathbf{r}, t)$ is measured in current per unit area. In the region that we are considering the well-known divergence equations are

$$\nabla \cdot \mathbf{D}^{(\mathbf{R})}(\mathbf{r}, t) = \rho(\mathbf{r}, t), \tag{5.3}$$

$$\nabla \cdot \mathbf{B}^{(\mathbf{R})}(\mathbf{r}, t) = 0, \tag{5.4}$$

in which $\rho(\mathbf{r}, t)$ is the charge per unit volume.

In this chapter space-frequency relations are used throughout, and hence we rewrite Maxwell's equations above in the space-frequency domain by taking the temporal Fourier transform of eqs. (5.1) and (5.2). We illustrate the Fourier transform convention by using the electric field $\mathbf{E}^{(\mathbf{R})}(\mathbf{r}, t)$ and its transform $\mathbf{E}(\mathbf{r}; \nu)$:

$$\mathbf{E}(\mathbf{r}; \nu) = \int_{-\infty}^{\infty} \mathbf{E}^{(\mathbf{R})}(\mathbf{r}, t)e^{i2\pi\nu t}\, dt. \tag{5.5}$$

We assert the existence of this transform including the statistical case in which $\mathbf{E}^{(\mathbf{R})}(\mathbf{r}, t)$ is not mean square integrable [7].

The resulting space-frequency representation of Maxwell's equations is given by

$$\nabla \times \mathbf{E}(\mathbf{r}; \nu) = i2\pi\nu\mathbf{B}(\mathbf{r}; \nu), \tag{5.6}$$

$$\nabla \times \mathbf{H}(\mathbf{r}; \nu) = \mathbf{J}(\mathbf{r}; \nu) - i2\pi\nu\mathbf{D}(\mathbf{r}; \nu). \tag{5.7}$$

Here and after we will use the terminology electric field and magnetic field when referring to these Fourier transform representations.

The medium is taken to be linear, time-invariant, isotropic, and homogeneous so that we may characterize it by the scalar dielectric constant $\varepsilon(\nu)$ and the magnetic permeability $\mu(\nu)$, which interrelate the field vectors as follows:

$$\mathbf{D}(\mathbf{r}; \nu) = \varepsilon(\nu)\mathbf{E}(\mathbf{r}; \nu), \tag{5.8}$$

$$\mathbf{B}(\mathbf{r}; \nu) = \mu(\nu)\mathbf{H}(\mathbf{r}; \nu). \tag{5.9}$$

In a source-free region by eqs. (5.3), (5.4), (5.6), and (5.7) we can show that both the electric field and the magnetic field satisfy the vector Helmholtz equations, namely

$$(\nabla^2 + k^2)\mathbf{E}(\mathbf{r}; \nu) = 0, \tag{5.10}$$

$$(\nabla^2 + k^2)\mathbf{H}(\mathbf{r}; \nu) = 0, \tag{5.11}$$

where the wave number k is given by

$$k = 2\pi\nu\sqrt{\mu\varepsilon}. \tag{5.12}$$

Now let us consider a more complicated case in which the source current $\mathbf{J}(\mathbf{r}; \nu)$ is different from zero throughout some finite domain and eqs. (5.6) and (5.7) are the appropriate form of Maxwell's equations. Our immediate objective is to express the radial components of the electric and magnetic fields, E_r, H_r, in terms of source currents. These two radial components will be seen to play a central role in finding an explicit solution for the fields outside the source distribution; this will be clear from the uniqueness theorem that is derived in Section 5.1.2.

Taking the curl of eqs. (5.6) and (5.7), using eqs. (5.8) and (5.9), and forming the scalar product $(\mathbf{r}\cdot)$ yield the following equations:

$$\mathbf{r}\cdot\nabla \times \nabla \times \mathbf{H}(\mathbf{r}; \nu) - k^2\mathbf{r}\cdot\mathbf{H}(\mathbf{r}; \nu) = \mathbf{r}\cdot\nabla \times \mathbf{J}(\mathbf{r}; \nu), \tag{5.13}$$

$$\mathbf{r}\cdot\nabla \times \nabla \times \mathbf{E}(\mathbf{r}; \nu) - k^2\mathbf{r}\cdot\mathbf{E}(\mathbf{r}; \nu) = i2\pi\nu\mu\mathbf{r}\cdot\mathbf{J}(\mathbf{r}; \nu). \tag{5.14}$$

Using the vector identity

$$\mathbf{r}\cdot(\nabla \times \nabla \times \mathbf{F}) = -\nabla^2(\mathbf{r}\cdot\mathbf{F}) + 2\nabla\cdot\mathbf{F} + \mathbf{r}\cdot\nabla(\nabla\cdot\mathbf{F}), \tag{5.15}$$

where $\mathbf{F}$ is any vector, we rewrite eqs. (5.13) and (5.14) to obtain the following differential equations for $\mathbf{r}\cdot\mathbf{E}$ and $\mathbf{r}\cdot\mathbf{H}$:

$$(\nabla^2 + k^2)[\mathbf{r}\cdot\mathbf{H}(\mathbf{r}; \nu)] = -\mathbf{r}\cdot\nabla \times \mathbf{J}(\mathbf{r}; \nu), \tag{5.16}$$

$$(\nabla^2 + k^2)\left[\mathbf{r}\cdot\mathbf{E}(\mathbf{r}; \nu) + \frac{i}{2\pi\nu\varepsilon}\mathbf{r}\cdot\mathbf{J}(\mathbf{r}; \nu)\right] = \frac{1}{i2\pi\nu\varepsilon}\mathbf{r}\cdot\nabla \times \nabla \times \mathbf{J}(\mathbf{r}; \nu). \tag{5.17}$$

Integral solutions of each of these equations can be written using the free-space Green's function $G(\mathbf{r}, \mathbf{r}'; \nu)$ given by

$$G(\mathbf{r}, \mathbf{r}'; \nu) = \frac{1}{4\pi}\frac{e^{ik|\mathbf{r}-\mathbf{r}|}}{|\mathbf{r} - \mathbf{r}'|}. \tag{5.18}$$

The resulting expressions for the radial components of the electric and magnetic fields are as follows:

$$\mathbf{r}\cdot\mathbf{H}(\mathbf{r}) = \int G(\mathbf{r}, \mathbf{r}')\mathbf{r}'\cdot\nabla' \times \mathbf{J}(\mathbf{r}')\, d^3r', \tag{5.19}$$

$$\mathbf{r}\cdot\mathbf{E}(\mathbf{r}) = \frac{1}{i2\pi\nu\varepsilon}\mathbf{r}\cdot\mathbf{J}(\mathbf{r}) - \frac{1}{i2\pi\nu\varepsilon}\int G(\mathbf{r}, \mathbf{r}')\mathbf{r}'\cdot\nabla' \times \nabla' \times \mathbf{J}(\mathbf{r}')\, d^3r', \tag{5.20}$$

where ∇' denotes operation with respect to the coordinates of $\mathbf{r}'$ and we omitted the explicit dependence of the current and the Green's function on ν. Using a well-known vector identity, eqs. (5.19) and (5.20) can be integrated by parts, and noting that the source current $\mathbf{J}(\mathbf{r})$ vanishes outside a bounded region we find $\mathbf{r}\cdot\mathbf{E}$ and $\mathbf{r}\cdot\mathbf{H}$

$$\mathbf{r}\cdot\mathbf{H}(\mathbf{r}) = \int \mathbf{J}(\mathbf{r}')\cdot\nabla' \times [\mathbf{r}'G(\mathbf{r},\mathbf{r}')]\, d^3r', \tag{5.21}$$

and

$$\mathbf{r}\cdot\mathbf{E}(\mathbf{r}) = \frac{1}{i2\pi\nu\varepsilon}\mathbf{r}\cdot\mathbf{J}(\mathbf{r}) - \frac{1}{i2\pi\nu\varepsilon}\int \mathbf{J}(\mathbf{r}')\cdot\nabla' \times \nabla' \times [\mathbf{r}'G(\mathbf{r},\mathbf{r}')]\, d^3r'. \tag{5.22}$$

After obtaining a general solution for the radial components of the field we show that these solutions are sufficient for a complete and unique specification of all field components.

5.1.2. *A Uniqueness Theorem in Spherical Polar Coordinates*

In this section we present an important theorem concerning the form of the solution to Maxwell's equations.[1] Consider a source region located near the center of the coordinate system and enclosed by a sphere S of radius R_0, with the major interest being to describe the electromagnetic field in the region outside the sphere. We will show that a determination of the radial components of the electromagnetic fields $E_r(\mathbf{r};\nu)$ and $H_r(\mathbf{r};\nu)$ on S is adequate to establish the remaining components of the fields.

In the derivation we consider a source-free region V between the inner sphere of radius R_0 and an outer sphere of radius R_1. In this region Maxwell's equations are valid, the electromagnetic fields $\mathbf{E}(\mathbf{r};\nu)$ and $\mathbf{H}(\mathbf{r};\nu)$ are assumed to exist without singularities, and their first derivatives are well behaved.

Theorem. *Let* $\mathbf{E}$, $\mathbf{H}$ *and* $\mathbf{E}'$, $\mathbf{H}'$ *be solution pairs of Maxwell's equations in the region* V. *If we assert that* $E_r = E'_r$ *and* $H_r = H'_r$ *on* S *it then follows that the fields are identical, i.e.,* $\mathbf{E} = \mathbf{E}'$ *and* $\mathbf{H} = \mathbf{H}'$ *in* V.

First, in terms of field differences $\Delta\mathbf{E}$, $\Delta\mathbf{H}$ is defined by

$$\Delta\mathbf{E} = \mathbf{E}(\mathbf{r};\nu) - \mathbf{E}'(\mathbf{r};\nu), \tag{5.23}$$

and

$$\Delta\mathbf{H} = \mathbf{H}(\mathbf{r};\nu) - \mathbf{H}'(\mathbf{r};\nu), \tag{5.24}$$

we substitute in Maxwell's equations, eqs. (5.1)–(5.4), to obtain the following

[1] The theorem and the general method of proof follow that in Bouwkamp and Casimir [2].

expressions in spherical coordinates:

$$i2\pi\nu\varepsilon r\Delta E_\theta = \frac{\partial}{\partial r}(r\Delta H_\varphi), \tag{5.25}$$

$$-i2\pi\nu\varepsilon r\Delta E_\varphi = \frac{\partial}{\partial r}(r\Delta H_\theta), \tag{5.26}$$

$$-i2\pi\nu\mu r\Delta H_\theta = \frac{\partial}{\partial r}(r\Delta E_\varphi), \tag{5.27}$$

$$i2\pi\nu\mu r\Delta H_\varphi = \frac{\partial}{\partial r}(r\Delta E_\theta), \tag{5.28}$$

$$\frac{\partial}{\partial\theta}(\Delta H_\varphi \sin\theta) = \frac{\partial}{\partial\varphi}\Delta H_\theta, \tag{5.29}$$

$$\frac{\partial}{\partial\theta}(\Delta E_\varphi \sin\theta) = \frac{\partial}{\partial\varphi}\Delta E_\theta. \tag{5.30}$$

Equations (5.25)–(5.28) are combined and solved for rE_θ, rH_φ and rE_φ, rH_θ, and expressed in traveling wave form, namely,

$$r\Delta E_\theta = A_1(\theta, \varphi)e^{ikr} + B_1(\theta, \varphi)e^{-ikr}, \tag{5.31}$$

$$r\Delta H_\Phi = g[A_1(\theta, \varphi)e^{ikr} - B_1(\theta, \varphi)e^{-ikr}], \tag{5.32}$$

$$r\Delta H_\theta = g[A_2(\theta, \varphi)e^{ikr} - B_2(\theta, \varphi)e^{-ikr}], \tag{5.33}$$

$$r\Delta E_\Phi = -A_2(\theta, \varphi)e^{ikr} - B_2(\theta, \varphi)e^{-ikr}, \tag{5.34}$$

where the wave admittance g is defined by $g = \sqrt{\varepsilon/\mu}$. In writing these solutions the complex-valued integration constants $A_1(\theta, \varphi)$, $B_1(\theta, \varphi)$ and $A_2(\theta, \varphi)$, $B_2(\theta, \varphi)$ correspond, respectively, to E_θ, H_φ and E_φ, H_θ field pairs. Equations (5.32) and (5.33) are substituted into eq. (5.29) and correspondingly eqs. (5.31) and (5.34) into eq. (5.30). The resulting pair of differential equations for the integration constants is

$$\frac{\partial}{\partial\theta}[\sin\theta(A_1 e^{ikr} - B_1 e^{-ikr})] = \frac{\partial}{\partial\varphi}(A_2 e^{ikr} - B_2 e^{-ikr}) \tag{5.35}$$

and

$$-\frac{\partial}{\partial\theta}[\sin\theta(A_2 e^{ikr} + B_2 e^{-ikr})] = \frac{\partial}{\partial\varphi}(A_1 e^{ikr} + B_1 e^{-ikr}), \tag{5.36}$$

where the explicit dependence of the integration constants on θ and φ is omitted for brevity.

Considering a variety of problems, e.g., outgoing waves only, we can assert the independence of A_1, A_2 from B_1, B_2. Hence it follows that we can write

$$\frac{\partial}{\partial\theta}[A_1(\theta, \varphi) \sin \theta] = \frac{\partial}{\partial\varphi} A_2(\theta, \varphi), \tag{5.37}$$

$$\frac{\partial}{\partial\theta}[A_2(\theta, \varphi) \sin \theta] = -\frac{\partial}{\partial\varphi} A_1(\theta, \varphi). \tag{5.38}$$

Corresponding equations can be written for B_1, B_2, but from this point these analogous expressions will be omitted.

In order to apply methods from the theory of complex variables, we would like to establish an analytic function in terms of the field coefficients A_1 and A_2. Both the proper grouping of A_1, A_2 and a coordinate mapping of θ, φ must be considered. We define the real and imaginary parts of A_1 and A_2 by

$$A_1(\theta, \varphi) = u_1(\theta, \varphi) + iv_1(\theta, \varphi), \tag{5.39}$$

$$A_2(\theta, \varphi) = u_2(\theta, \varphi) + iv_2(\theta, \varphi). \tag{5.40}$$

Substituting eqs. (5.39) and (5.40) into eqs. (5.37) and (5.38) and separating the real and imaginary parts we obtain

$$\frac{\partial}{\partial\theta}[u_1 \sin \theta] - \frac{\partial}{\partial\varphi} u_2, \tag{5.41}$$

$$\frac{\partial}{\partial\theta}[v_1 \sin \theta] = \frac{\partial}{\partial\varphi} v_2, \tag{5.42}$$

$$\frac{\partial}{\partial\theta}[u_2 \sin \theta] = -\frac{\partial}{\partial\varphi} u_1, \tag{5.43}$$

$$\frac{\partial}{\partial\theta}[v_2 \sin \theta] = -\frac{\partial}{\partial\varphi} v_1. \tag{5.44}$$

In order to construct the Cauchy–Riemann equations, consider the transformation defined by letting the complex variable z be

$$z = \varphi + i\psi(\theta). \tag{5.45}$$

Using this transformation in eqs. (5.41)–(5.44) we observe that an analytic function $F(z)$ satisfying the Cauchy–Riemann conditions is obtained, with $F(z)$ given by

$$F(z) = [A_1 + iA_2] \sin \theta, \tag{5.46}$$

where A_1 and A_2 depend implicitly on the complex variable z. Moreover, the transformation function $\psi(\theta)$ satisfies the differential equation

$$\frac{\partial\psi}{\partial\theta} = \frac{1}{\sin \theta}. \tag{5.47}$$

The function $\psi(\theta)$ itself is therefore given by

$$\psi(\theta) = \log\left[\tan\frac{\theta}{2}\right]. \tag{5.48}$$

Since $|A_1|$ and $|A_2|$ must be finite it follows that $|F(z)|$ is a bounded function independent of z. From eq. (5.46) we also see that $F(z) = 0$ for $\theta = 0, \pi$.

The theorem of Liouville[1,2] states that a function which is analytic and regular in every finite region of the complex z-plane, and is uniformly bounded, is a constant. From this it follows that $F(z)$ is identically zero. Hence the uniqueness theorem is proved.

5.1.3. *Debye Potentials*

An elegant formulation of solutions of Maxwell's equations in potential form was given by Debye [1] for the source-free region. The electric and magnetic field vectors are expressed in the form

$$\mathbf{E}(\mathbf{r}; \nu) = \nabla \times \nabla \times [\mathbf{r}\Pi_1(\mathbf{r}; \nu)] + i\frac{k}{g}\nabla \times [\mathbf{r}\Pi_2(\mathbf{r}; \nu)], \tag{5.49}$$

$$\mathbf{H}(\mathbf{r}; \nu) = \nabla \times \nabla \times [\mathbf{r}\Pi_2(\mathbf{r}; \nu)] - ikg\nabla \times [\mathbf{r}\Pi_1(\mathbf{r}; \nu)]. \tag{5.50}$$

Outside the source distribution, substituting eqs. (5.49) and (5.50) into eqs. (5.6) and (5.7) with an appropriate choice of gauge, one can verify that the scalar potential functions $\Pi_1(\mathbf{r}; \nu)$ and $\Pi_2(\mathbf{r}; \nu)$ are solutions of the scalar Helmholtz equation, i.e.,

$$(\nabla^2 + k^2)\Pi_1(\mathbf{r}; \nu) = 0 \tag{5.51}$$

and

$$(\nabla^2 + k^2)\Pi_2(\mathbf{r}; \nu) = 0, \tag{5.52}$$

where the wave number k is given by eq. (5.12).

It is interesting to discuss the form of eqs. (5.49) and (5.50) in light of the uniqueness theorem of the preceding section. First, consider the case for which $\Pi_2(\mathbf{r}; \nu)$ is set to zero. By computing $\mathbf{r} \cdot \mathbf{H}$ from eq. (5.50) we see that $H_r = 0$. However, by eq. (5.49), in general, the term $\nabla \times \nabla \times [\mathbf{r}\Pi_1(\mathbf{r})]$ leads to all components of $\mathbf{E}$, and in particular E_r is different from zero. This form of solution is commonly termed an E-type wave.

From the symmetry of eqs. (5.49) and (5.50), by analogy, it is clear that the terminology H-type wave is appropriate for fields generated by the potential $\Pi_2(\mathbf{r}; \nu)$ alone. That is, the potential $\Pi_2(\mathbf{r}; \nu)$ gives rise to all components of $\mathbf{H}$. However, the corresponding radial electric field contributed by $\Pi_2(\mathbf{r}; \nu)$ is identically zero, i.e.,

$$\mathbf{r} \cdot \nabla \times [\mathbf{r}\Pi_2(\mathbf{r}; \nu)] \equiv 0. \tag{5.53}$$

Since by eqs. (5.51) and (5.52) the Debye potentials are solutions of the scalar Helmholtz equation, we present a brief summary of their solution in spherical harmonics, so that the notation we are using is evident. For a comprehensive review of functional notations used in the literature on light scattering one should consult Kerker [9]. Our notation follows that in Hill [5], with but minor variations from that in Bouwkamp and Casimir [2] or in Papas [3].

---

[1] See, for example, Copson [8, Sec. 4.34].

By the method of separation of variables, we can readily establish the form of the general solution for eqs. (5.51) and (5.52) which consists of the product of spherical Hankel functions of the first and second kinds with the spherical harmonics. The spherical harmonics contain a product of the associated Legendre polynomials, in general, of the first and second kinds, with the azimuthal functions $e^{\pm im\varphi}$. Here, for the sake of definiteness, we consider the region V outside a sphere S of radius R_0 as in Section 5.1.2. Hence we drop the associated Legendre polynomials of the second kind, Q_n^m, since they have a singularity at $\theta = 0$, and we omit terms in the Hankel functions of the second kind, $h_n^{(2)}(kr)$, in order to restrict the problem to outgoing waves. The appropriate form for either Debye potential $\Pi(\mathbf{r}; v)$ is given by

$$\Pi(\mathbf{r}) = \sum_{n=0}^{\infty} \sum_{m=-n}^{n} \alpha_n^m h_n^{(1)}(kr) Y_n^m(\theta, \varphi), \tag{5.54}$$

in which α_n^m is an expansion coefficient representing a_n^m, b_n^m according to whether we are expanding Π_1 or Π_2. The Hankel function of the first kind, $h_n^{(1)}(kr)$, can be written in terms of the spherical Bessel functions or the half-order cylindrical Hankel functions of the first kind, namely,

$$h_n^{(1)}(kr) = j_n(kr) + iy_n(kr), \tag{5.55}$$

$$h_n^{(1)}(kr) = \sqrt{\frac{\pi}{2kr}} H_{n+1/2}^{(1)}(kr). \tag{5.56}$$

The spherical harmonics, $Y_n^m(\theta, \varphi)$, can be written in the form

$$Y_n^m(\theta, \varphi) = \frac{1}{\sqrt{2\pi}} \Theta_n^m(\theta) e^{im\varphi}, \tag{5.57}$$

in which

$$\Theta_n^m(\theta) = (-1)^m \sqrt{\frac{2n+1}{2} \frac{(n-m)!}{(n+m)!}} P_n^m(\mu). \tag{5.58}$$

The associated Legendre polynomial of the first kind, $P_n^m(\mu)$, can be computed by

$$P_n^m(\mu) = \frac{(1-\mu^2)^{m/2}}{2^n n!} \frac{d^{n+m}}{d\mu^{n+m}} (\mu^2 - 1)^n, \tag{5.59}$$

where

$$\mu = \cos\theta. \tag{5.60}$$

For a comprehensive treatment of the properties of spherical harmonics, the reader is referred to Hobson [10].

The orthogonality relation for the spherical harmonics, $Y_n^m(\theta, \varphi)$, in eq. (5.57) can be expressed in the form

$$\int_{(4\pi)} Y_n^m(\theta, \varphi) [Y_{n'}^{m'}(\theta, \varphi)]^* \, d\Omega = \delta_{nn'} \delta_{mm'}, \tag{5.61}$$

where δ_{ij} is the Kronecker symbol, $d\Omega$ is the solid angle, and the asterisk denotes complex conjugation. Particularly useful formulas for the associated Legendre polynomials are also included. The first is an orthogonality relationship, namely,

$$\int_{-1}^{1} \frac{P_n^m(\mu) P_n^{m'}(\mu)}{1-\mu^2} \, d\mu = 0 \qquad (m \neq m'). \tag{5.62}$$

The second is the normalization relation for the defining form of P_n^m in eq. (5.59), i.e.,

$$\int_{-1}^{1} [P_n^m(\mu)]^2 \, d\mu = \frac{2}{2n+1} \frac{(n+m)!}{(n-m)!}. \tag{5.63}$$

For the region outside the source domain we write general expressions for the Debye potentials $\Pi_1(\mathbf{r})$ and $\Pi_2(\mathbf{r})$ of eqs. (5.49) and (5.50) in the form

$$\Pi_1(\mathbf{r}; \nu) = \sum_{n=0}^{\infty} \sum_{m=-n}^{n} a_n^m \Pi_n^m(\mathbf{r}), \tag{5.64}$$

$$\Pi_2(\mathbf{r}; \nu) = \sum_{n=0}^{\infty} \sum_{m=-n}^{n} b_n^m \Pi_n^m(\mathbf{r}). \tag{5.65}$$

From eq. (5.54) we write the following form of $\Pi_n^m(\mathbf{r})$

$$\Pi_n^m(\mathbf{r}) = Y_n^m(\theta, \varphi) h_n(kr). \tag{5.66}$$

Here $h_n(kr)$ is used to denote $h_n^{(1)}(kr)$. There is also a temporal frequency, ν, dependence implicit in $\Pi_n^m(\mathbf{r})$ as well as in the expansion coefficients a_n^m, b_n^m for the electric and magnetic potentials, respectively.

At this point we would like to relate the coefficients a_n^m, b_n^m to the source current $\mathbf{J}(\mathbf{r}; \nu)$ in the region $r < R_0$. From eqs. (5.21) and (5.22) it is clear that we should form the scalar product $\mathbf{r} \cdot \mathbf{E}$ and $\mathbf{r} \cdot \mathbf{H}$ using the notation of eqs. (5.64) and (5.65). Combining eqs. (5.64) and (5.65) with eqs. (5.49) and (5.50) and forming the scalar product immediately give the following result:

$$\mathbf{r} \cdot \mathbf{E}(\mathbf{r}) = \sum_{n=0}^{\infty} \sum_{m=-n}^{n} a_n^m \mathbf{r} \cdot [\nabla \times \nabla \times \mathbf{r} \Pi_n^m(\mathbf{r})] \tag{5.67}$$

and

$$\mathbf{r} \cdot \mathbf{H}(\mathbf{r}) = \sum_{n=0}^{\infty} \sum_{m=-n}^{n} b_n^m \mathbf{r} \cdot [\nabla \times \nabla \times \mathbf{r} \Pi_n^m(\mathbf{r})]. \tag{5.68}$$

One can verify the following vector identity:

$$\mathbf{r} \cdot \{\nabla \times \nabla \times [r \Pi_n^m(\mathbf{r})]\} = n(n+1) \Pi_n^m(\mathbf{r}), \tag{5.69}$$

which is substituted into eqs. (5.67) and (5.68) with the result

$$\mathbf{r} \cdot \mathbf{E}(\mathbf{r}) = \sum_{n=0}^{\infty} \sum_{m=-n}^{n} n(n+1) a_n^m \Pi_n^m(\mathbf{r}) \tag{5.70}$$

and

$$\mathbf{r}\cdot\mathbf{H}(\mathbf{r}) = \sum_{n=0}^{\infty}\sum_{m=-n}^{n} n(n+1)b_n^m\Pi_n^m(\mathbf{r}). \tag{5.71}$$

In order to express the expansion coefficients, a_n^m, b_n^m, using eqs. (5.21) and (5.22), we start with an orthogonal expansion of the Green's function in eq. (5.18). In the literature we find the following useful expansion:

$$\frac{e^{ik|\mathbf{r}-\mathbf{r}'|}}{4\pi|\mathbf{r}-\mathbf{r}'|} = \frac{ik}{4\pi}\sum_{n=0}^{\infty}\sum_{m=-n}^{n} h_n(kr)Y_n^m(\theta,\varphi)j_n(kr')[Y_n^m(\theta',\varphi')]^*, \tag{5.72}$$

where $\mathbf{r}$ is a field point and $\mathbf{r}'$ is the source point [3, Sec. 4.4]. Equation (5.72) is substituted into eq. (5.21) and the summation and integration are interchanged to yield

$$\mathbf{r}\cdot\mathbf{H}(\mathbf{r}) = \frac{ik}{4\pi}\sum_{n=0}^{\infty}\sum_{m=-n}^{n} h_n(kr)Y_n^m(\theta,\varphi)\int \mathbf{J}(\mathbf{r}')\cdot\nabla'\times[\mathbf{r}'j_n(kr')Y_n^m(\theta',\varphi')]^*\,d^3r'. \tag{5.73}$$

Similarly, by eqs. (5.72) and (5.22) and noting that $\mathbf{J}(\mathbf{r}) = 0$ outside the source domain, we can write

$$\mathbf{r}\cdot\mathbf{E}(\mathbf{r}) = -\frac{1}{4\pi g}\sum_{n=0}^{\infty}\sum_{m=-n}^{n} h_n(kr)Y_n^m(\theta,\varphi)\int \mathbf{J}(\mathbf{r}')\cdot\nabla'\times\nabla'\times[\mathbf{r}'j_n(kr')Y_n^m(\theta',\varphi')]^*\,d^3r'. \tag{5.74}$$

TABLE 5.1. Spherical polar components of the electromagnetic fields. The explicit expression for $\partial Y_n^m/\partial\theta$ is given in eq. (5.82). The expansion coefficients, a_n^m, b_n^m, are given in terms of the source current $\mathbf{J}(\mathbf{r})$ by eqs. (5.75) and (5.76), respectively.

Field component	Π_1 (Electric multipole, a_n^m)	Π_2 (Magnetic multipole, b_n^m)
$\mathbf{E}(\mathbf{r};\nu)$		
$\mathbf{e}_r$	$\frac{n(n+1)}{r}h_n(kr)Y_n^m$	0
$\mathbf{e}_\theta$	$\frac{1}{r}\frac{d}{dr}[rh_n(kr)]\frac{\partial}{\partial\theta}Y_n^m$	$-\frac{k}{g}\frac{m}{\sin\theta}h_n(kr)Y_n^m$
$\mathbf{e}_\varphi$	$\frac{im}{r\sin\theta}\frac{d}{dr}[rh_n(kr)]Y_n^m$	$-i\frac{k}{g}h_n(kr)\frac{\partial}{\partial\theta}Y_n^m$
$\mathbf{H}(\mathbf{r};\nu)$		
$\mathbf{e}_r$	0	$\frac{n(n+1)}{r}h_n(kr)Y_n^m$
$\mathbf{e}_\theta$	$kg\frac{m}{\sin\theta}h_n(kr)Y_n^m$	$\frac{1}{r}\frac{d}{dr}[rh_n(kr)]\frac{\partial}{\partial\theta}Y_n^m$
$\mathbf{e}_\varphi$	$ikgh_n(kr)\frac{\partial}{\partial\theta}Y_n^m$	$\frac{im}{r\sin\theta}\frac{d}{dr}[rh_n(kr)]Y_n^m$

Comparing eq. (5.71) to (5.73) and eq. (5.70) to eq. (5.74), we obtain the desired expression for the expansion coefficients in terms of the source current

$$a_n^m = -\frac{1}{4\pi g}\frac{1}{n(n+1)}\int \mathbf{J}(\mathbf{r}')\cdot\nabla'\times\nabla'\times[\mathbf{r}' j_n(kr')Y_n^m(\theta',\varphi')]^*\,d^3r', \quad (5.75)$$

$$b_n^m = \frac{ik}{4\pi}\frac{1}{n(n+1)}\int \mathbf{J}(\mathbf{r}')\cdot\nabla'\times[\mathbf{r}' j_n(kr')Y_n^m(\theta',\varphi')]^*\,d^3r'. \quad (5.76)$$

While our interest here has been primarily to show eqs. (5.75) and (5.76), still for completeness we include in Table 5.1 equations for the separate field components of **E** and **H**.

Later in Section 5.2.1, independently, we derive expressions for the fields **E** and **H** using vector spherical harmonics.

We point out that the expansion coefficients, a_n^m, b_n^m, can be considered independent variables in the sense that any number of them may be taken as nonzero, and a complete vector electromagnetic field satisfying Maxwell's equations is given by eqs. (5.49) and (5.50). Implicitly, a_n^m, b_n^m depend on the primary distribution of the source current $\mathbf{J}(\mathbf{r}; \nu)$ as in eqs. (5.75) and (5.76).

5.1.4. *Vector Spherical Harmonics*

In the study of electromagnetic theory we often encounter situations where the physical configuration requires solutions of differential equations in spherical polar coordinates. In particular, solutions of Maxwell's equations can frequently be expressed in terms of series expansions involving the surface spherical harmonics $Y_n^m(\theta, \varphi)$. In previous sections the multipole expansion, as well as a number of related results, have been expressed in spherical harmonics. In some cases fairly intricate algebraic manipulations (not presented in the text) have been involved in going from one expression to the next. In this section we introduce the vector spherical harmonics which are found to be particularly helpful in multipole theory. There are two factors contributing to their usefulness. One is that multiple curl, div, and other vector operations are conveniently handled, by taking advantage of important commutation relationships. Second, the vector orthogonalization properties provide a compact formalism for presenting a great many equations. Vector spherical harmonics, the origin and functional basis and applications to atomic and nuclear physics, are given by Blatt and Weisskopf [4]. We make use of the identities and notation contained in two articles by Hill [5] and Hill and Landshoff [11].

Consider the following differential operation on the scalar spherical harmonics $i\mathbf{r}\times\nabla Y_n^m(\theta, \varphi)$. It is clear that this operation generates a vector having components in the plane perpendicular to the polar vector **r**. Let Λ denote the differential operator

$$\Lambda = -i\mathbf{r}\times\nabla, \quad (5.77)$$

which when operating on the surface spherical harmonics, $Y_n^m(\theta, \varphi)$, generates the vector spherical harmonics $\mathbf{X}_n^m(\theta, \varphi)$. Here, $\mathbf{X}_n^m(\theta, \varphi)$ is defined by the relation

$$\Lambda Y_n^m(\theta, \varphi) = \sqrt{n(n+1)}\,\mathbf{X}_n^m(\theta, \varphi). \tag{5.78}$$

The formal definition of the vector spherical harmonics, $\mathbf{X}_n^m(\theta, \varphi)$, $\mathbf{V}_n^m(\theta, \varphi)$, $\mathbf{W}_n^m(\theta, \varphi)$, expressed in spherical coordinates are

$$\mathbf{X}_n^m = 0\mathbf{e}_r - \frac{mY_n^m}{\sqrt{n(n+1)}\sin\theta}\mathbf{e}_\theta - \frac{i}{\sqrt{n(n+1)}}\frac{\partial Y_n^m}{\partial\theta}\mathbf{e}_\varphi, \tag{5.79}$$

$$\mathbf{V}_n^m = -\sqrt{\frac{n+1}{2n+1}}Y_n^m\mathbf{e}_r + \frac{1}{\sqrt{(n+1)(2n+1)}}\frac{\partial Y_n^m}{\partial\theta}\mathbf{e}_\theta + \frac{imY_n^m}{\sqrt{(n+1)(2n+1)}\sin\theta}\mathbf{e}_\varphi, \tag{5.80}$$

$$\mathbf{W}_n^m = \sqrt{\frac{n}{2n+1}}Y_n^m\mathbf{e}_r + \sqrt{\frac{1}{n(2n+1)}}\frac{\partial Y_n^m}{\partial\theta}\mathbf{e}_\theta + \frac{imY_n^m}{\sqrt{n(2n+1)}\sin\theta}\mathbf{e}_\varphi. \tag{5.81}$$

Here $\partial Y_n^m/\partial\theta$ is given by

$$\frac{\partial}{\partial\theta}Y_n^m = \tfrac{1}{2}\sqrt{(n-m)(n+m+1)}\,Y_n^{m+1}\mathbf{e}^{-i\varphi} - \tfrac{1}{2}\sqrt{(n+m)(n-m+1)}\,Y_n^{m-1}\mathbf{e}^{i\varphi}. \tag{5.82}$$

We observe that the vector spherical harmonics, $\mathbf{X}_n^m(\theta, \varphi)$, do not have a component in the $\mathbf{e}_r$ direction, as is expected from eqs. (5.77) and (5.78). $\mathbf{V}_n^m$ and $\mathbf{W}_n^m$ are seen to have components along the three unit vectors of the spherical coordinate system.

An important property of the vector spherical harmonics is their mutual orthogonality with respect to integration over all solid angles. Using the orthogonality conditions for the surface spherical harmonics, eq. (5.61), and for the associated Legendre polynomials, eq. (5.62) in eqs. (5.79)–(5.81), we obtain the following orthogonality conditions:

$$\int_{(4\pi)} \mathbf{X}_n^m \cdot [\mathbf{X}_{n'}^{m'}]^*\, d\Omega = \delta_{nn'}\delta_{mm'}, \tag{5.83}$$

$$\int_{(4\pi)} \mathbf{V}_n^m \cdot [\mathbf{V}_{n'}^{m'}]^*\, d\Omega = \delta_{nn'}\delta_{mm'}, \tag{5.84}$$

$$\int_{(4\pi)} \mathbf{W}_n^m \cdot [\mathbf{W}_{n'}^{m'}]^*\, d\Omega = \delta_{nn'}\delta_{mm'}, \tag{5.85}$$

$$\int_{(4\pi)} \mathbf{V}_n^m \cdot [\mathbf{X}_{n'}^{m'}]^*\, d\Omega = 0, \tag{5.86}$$

$$\int_{(4\pi)} \mathbf{X}_n^m \cdot [\mathbf{W}_{n'}^{m'}]^*\, d\Omega = 0, \tag{5.87}$$

$$\int_{(4\pi)} \mathbf{W}_n^m \cdot [\mathbf{V}_{n'}^{m'}]^*\, d\Omega = 0. \tag{5.88}$$

A part of the usefulness of the vector spherical harmonics rests upon having an adequate listing of identities. For this the reader is referred to the literature, especially the article by Hill [5]. To illustrate a few representative identitie we present the expression for $\nabla Y_n^m(\theta, \varphi)$ and a selected list of vector operations on $\mathbf{X}_n^m(\theta, \varphi)$

$$\nabla Y_n^m = \frac{n}{r}\sqrt{\frac{n+1}{2n+1}}\mathbf{V}_n^m + \frac{n+1}{r}\sqrt{\frac{n}{2n+1}}\mathbf{W}_n^m, \tag{5.89}$$

$$-i\nabla \times [F(r)\mathbf{X}_n^m] = \sqrt{\frac{n}{2n+1}}\left[\frac{dF}{dr} - \frac{n}{r}F\right]\mathbf{V}_n^m + \sqrt{\frac{n+1}{2n+1}}\left[\frac{dF}{dr} + \frac{n+1}{r}F\right]\mathbf{W}_n^m, \tag{5.90}$$

$$\nabla^2[F(r)\mathbf{X}_n^m] = \mathbf{X}_n^m\left[\frac{\partial^2}{\partial r^2} + \frac{2}{r}\frac{\partial}{\partial r} - \frac{n(n+1)}{r^2}\right]F(r), \tag{5.91}$$

$$-i\mathbf{r} \times \mathbf{X}_n^m = \sqrt{\frac{n}{2n+1}}\mathbf{V}_n^m + \sqrt{\frac{n+1}{2n+1}}\mathbf{W}_n^m. \tag{5.92}$$

This completes our description of the vector spherical harmonics, $\mathbf{X}_n^m(\theta, \varphi)$, $\mathbf{V}_n^m(\theta, \varphi)$, $\mathbf{W}_n^m(\theta, \varphi)$. In the next section we apply these vector functions to the problem of expressing the fields $\mathbf{E}(\mathbf{r}; \nu)$ and $\mathbf{H}(\mathbf{r}; \nu)$ as well as the current distribution $\mathbf{J}(\mathbf{r}; \nu)$.

5.2. Expansion of the Sources and the Radiation Field in Vector Spherical Harmonics

5.2.1. *The Radiation Field*

In this section our purpose is to obtain general expressions for the radiation fields $\mathbf{E}(\mathbf{r})$, $\mathbf{H}(\mathbf{r})$. The earlier representation in eqs. (5.49) and (5.50) containing differential operators is evaluated. This gives us an explicit representation in terms of vector spherical harmonics.

In Section 5.1.4 the vector spherical harmonics were described together with the introduction of the operator Λ. This operator provides systematic and convenient means for carrying out the curl operation. At this point we would like to obtain expressions for both the curl and the curl–curl operations on $[\mathbf{r}\Pi_{1,2}(\mathbf{r})]$.

The curl of $[\mathbf{r}\Pi_{1,2}(\mathbf{r})]$ can be expanded using a well-known vector identity, and $\nabla \times \mathbf{r} = \mathbf{0}$, to yield

$$\nabla \times [\mathbf{r}\Pi_{1,2}(\mathbf{r})] = -\mathbf{r} \times \nabla\Pi_{1,2}(\mathbf{r}). \tag{5.93}$$

With the previously adopted definition for Λ in eq. (5.77),

$$\Lambda = -i\mathbf{r} \times \nabla, \tag{5.94}$$

we rewrite eq. (5.93) in the form

$$\nabla \times [\mathbf{r}\Pi_{1,2}(\mathbf{r})] = -i\Lambda\Pi_{1,2}(\mathbf{r}). \tag{5.95}$$

Substituting eqs. (5.64) and (5.65) into eq. (5.95), noting that Λ commutes with the spherical Hankel functions, and using the identity eq. (5.78), we obtain the following forms:

$$\Lambda\Pi_1(\mathbf{r}) = \sum_{n,m} a_n^m h_n(kr)\sqrt{n(n+1)}\mathbf{X}_n^m, \tag{5.96}$$

$$\Lambda\Pi_2(\mathbf{r}) = \sum_{n,m} b_n^m h_n(kr)\sqrt{n(n+1)}\mathbf{X}_n^m, \tag{5.97}$$

with sums taken over m, n as in eq. (5.74).

It is convenient to rewrite eqs. (5.96) and (5.97) in terms of the familiar curl operator giving the desired results

$$\nabla \times [\mathbf{r}\Pi_1(\mathbf{r})] = -i \sum_{n,m} a_n^m \sqrt{n(n+1)}\, h_n(kr)\mathbf{X}_n^m \tag{5.98}$$

and

$$\nabla \times [\mathbf{r}\Pi_2(\mathbf{r})] = -i \sum_{n,m} b_n^m \sqrt{n(n+1)}\, h_n(kr)\mathbf{X}_n^m. \tag{5.99}$$

Now consider the following curl–curl operation which we write with eq. (5.94)

$$\nabla \times \nabla \times [\mathbf{r}\Pi_n^m(\mathbf{r})] = -i\nabla \times \Lambda\Pi_n^m(\mathbf{r}). \tag{5.100}$$

From the definition of the function $\Pi_n^m(\mathbf{r})$ in eq. (5.66) and from eq. (5.100), it follows that

$$\nabla \times \nabla \times [\mathbf{r}\Pi_n^m(\mathbf{r})] = -i\nabla \times [F_n(kr)\mathbf{X}_n^m], \tag{5.101}$$

where

$$F_n(kr) = \begin{cases} j_n(kr), & r < R_0, \\ h_n(kr), & r > R_0. \end{cases} \tag{5.102}$$

Here R_0 is the radius of a sphere S enclosing the entire source distribution.

In the evaluation of eq. (5.101) the following identity is important:

$$-i\nabla \times [F(r)\mathbf{X}_n^m]$$

$$= \sqrt{\frac{n}{2n+1}}\left[\frac{dF}{dr} - \frac{n}{r}F\right]\mathbf{V}_n^m + \sqrt{\frac{n+1}{2n+1}}\left[\frac{dF}{dr} + \frac{n+1}{r}F\right]\mathbf{W}_n^m. \tag{5.103}$$

Setting $F(r) = F_n(kr)$ in the last equation and using the recurrence relations [12]

$$\frac{d}{dz}F_n(z) - \frac{n}{z}F_n(z) = -F_{n+1}(z) \tag{5.104}$$

and

$$\frac{d}{dz}F_n(z) + \frac{n+1}{z}F_n(z) = F_{n-1}(z). \tag{5.105}$$

Equation (5.100) can be expressed in the form

$$\nabla \times \nabla \times [\mathbf{r}\Pi_n^m(\mathbf{r})]$$

$$= k\sqrt{n(n+1)}\left[\sqrt{\frac{n+1}{2n+1}}h_{n-1}(kr)\mathbf{W}_n^m - \sqrt{\frac{n}{2n+1}}h_{n+1}(kr)\mathbf{V}_n^m\right], \tag{5.106}$$

when $r > R_0$, and

$$\nabla \times \nabla \times [\mathbf{r}\Pi_n^m(\mathbf{r})]$$

$$= k\sqrt{n(n+1)}\left[\sqrt{\frac{n+1}{2n+1}}j_{n-1}(kr)\mathbf{W}_n^m - \sqrt{\frac{n}{2n+1}}j_{n+1}(kr)\mathbf{V}_n^m\right], \quad (5.107)$$

when $r < R_0$. Finally, eqs. (5.98) and (5.106) are substituted into eqs. (5.49) and (5.50). We then obtain general vector solutions for the radiation field

$$\mathbf{E}(\mathbf{r}) = k\sum_{n,m}\sqrt{n(n+1)}$$

$$\times\left\{a_n^m\left[\sqrt{\frac{n+1}{2n+1}}h_{n-1}(kr)\mathbf{W}_n^m - \sqrt{\frac{n}{2n+1}}h_{n+1}(kr)\mathbf{V}_n^m\right] + \frac{b_n^m}{g}h_n(kr)\mathbf{X}_n^m\right\} \quad (5.108)$$

and

$$\mathbf{H}(\mathbf{r}) = k\sum_{n,m}\sqrt{n(n+1)}$$

$$\times\left\{b_n^m\left[\sqrt{\frac{n+1}{2n+1}}h_{n-1}(kr)\mathbf{W}_n^m - \sqrt{\frac{n}{2n+1}}h_{n+1}(kr)\mathbf{V}_n^m\right] - g a_n^m h_n(kr)\mathbf{X}_n^m\right\}. \quad (5.109)$$

In the following sections the expansions of the field into vector spherical harmonics in eqs. (5.108) and (5.109) are applied in deriving expressions for statistical moments of interest.

5.2.2. *A Uniqueness Theorem*

The advantages of expressing the fields **E** and **H** in vector spherical harmonics become apparent in any radiation problem which can be posed in spherical polar coordinates. As an illustrative example, we derive a theorem that stipulates the specification of *one* of the tangential fields on the surface of the sphere of radius R_0, for the unique determination of all of the components of the electromagnetic field in the region outside the sphere.

Theorem. *Let a source current distribution* $\mathbf{J}(\mathbf{r}; \nu)$ *be defined throughout a domain D that is completely enclosed by a sphere S of radius* R_0. *If we specify either the tangential electric field* $\mathbf{E}_t$ *or the tangential magnetic field* $\mathbf{H}_t$ *on the surface of the sphere, then the electromagnetic fields* $\mathbf{E}(\mathbf{r}; \nu)$ *and* $\mathbf{H}(\mathbf{r}; \nu)$ *are uniquely determined for all points outside the sphere* ($|\mathbf{r}| > R_0$).

To prove the theorem we first consider the case where the tangential electric field $\mathbf{E}_t$ is specified on S. Starting from the expansion of the electric field, eq.

(5.108), and using the orthogonality of the vector spherical harmonics, eqs. (5.83)–(5.88), it is straightforward to show that the following three relations hold:

$$\int_S \mathbf{E}(\mathbf{r})\cdot[\mathbf{W}_n^m(\theta,\varphi)]^*\,d\Omega = k(n+1)\sqrt{\frac{n}{2n+1}}\,h_{n-1}(kr)a_n^m, \tag{5.110}$$

$$\int_S \mathbf{E}(\mathbf{r})\cdot[\mathbf{V}_n^m(\theta,\varphi)]^*\,d\Omega = -kn\sqrt{\frac{n+1}{2n+1}}\,h_{n+1}(kr)a_n^m, \tag{5.111}$$

$$\int_S \mathbf{E}(\mathbf{r})\cdot[\mathbf{X}_n^m(\theta,\varphi)]^*\,d\Omega = \frac{k}{g}\sqrt{n(n+1)}\,h_n(kr)b_n^m. \tag{5.112}$$

Recalling that by eq. (5.79) the vector spherical harmonics $\mathbf{X}_n^m(\theta, \varphi)$ have no radial component, we can readily solve eq. (5.112) for b_n^m, namely,

$$b_n^m = \frac{g}{kh_n(kR_0)\sqrt{n(n+1)}}\int_S \mathbf{E}_t(R_0,\theta,\varphi)\cdot[\mathbf{X}_n^m(\theta,\varphi)]^*\,d\Omega. \tag{5.113}$$

Equation (5.113) implies that by specifying the tangential electric field $\mathbf{E}_t$ we have uniquely determined the set of multipole coefficients $\{b_n^m\}$, which in turn lead to the complete specification of the scalar potential $\Pi_2(\mathbf{r})$, as is evident from eq. (5.65).

It is a slightly more complicated task to achieve a similar relation for the coefficients $\{a_n^m\}$, due to the fact that both the spherical harmonics $\mathbf{W}_n^m(\theta, \varphi)$ and $\mathbf{V}_n^m(\theta, \varphi)$ contain a component in the $\mathbf{e}_r$ direction. Comparing the two radial components in eqs. (5.80) and (5.81) it is easy to see that the linear combination $\sqrt{n}\mathbf{V}_n^m(\theta, \varphi) + \sqrt{n+1}\mathbf{W}_n^m(\theta, \varphi)$ has no contribution in the radial direction and therefore

$$\mathbf{E}(\mathbf{r})\cdot[\sqrt{n}\mathbf{V}_n^m + \sqrt{n+1}\mathbf{W}_n^m] = \mathbf{E}_t(\mathbf{r})\cdot[\sqrt{n}\mathbf{V}_n^m + \sqrt{n+1}\mathbf{W}_n^m]. \tag{5.114}$$

Using eq. (5.114) in the integrations of eqs. (5.110) and (5.111), we obtain an explicit expression for the coefficients $\{a_n^m\}$, i.e.,

$$a_n^m = \frac{1}{k}\sqrt{\frac{2n+1}{n(n+1)}}\left\{\frac{\int_D \mathbf{E}_t(\mathbf{r})\cdot[\sqrt{n}\mathbf{V}_n^m + \sqrt{n+1}\mathbf{W}_n^m]^*\,d\Omega}{(n+1)h_{n-1}(kR_0) - nh_{n+1}(kR_0)}\right\}. \tag{5.115}$$

We have thus shown that the specification of $\mathbf{E}_t$ on the sphere S leads to the unique determination of the coefficients $\{a_n^m\}$ and $\{b_n^m\}$, which by eqs. (5.64) and (5.65) give scalar potentials $\Pi_1(\mathbf{r})$ and $\Pi_2(\mathbf{r})$. From the similarity in eqs. (5.108) and (5.109) it is clear that relations equivalent to eqs. (5.113) and (5.115) can be derived from $\mathbf{H}_t$, which completes the proof.

5.2.3. *Expansion of the Current Distribution*

In this section the representations for the expansion coefficients a_n^m and b_n^m in terms of volume integrals are to be expressed in an interesting alternative form.

We expand the current distribution in terms of vector spherical harmonics. In this expansion we define spherical projection functions, $f_n^m(r)$, $g_n^m(r)$, $h_n^m(r)$, which measure the overlap integral of the current density with each of the three vector spherical harmonics. This representation for the source distribution gives an insightful understanding of the radiation field in terms of source multipoles.

The current density $\mathbf{J}(\mathbf{r})$ is expanded in terms of $\mathbf{j}_n^m(\mathbf{r})$ as follows:

$$\mathbf{J}(\mathbf{r}) = \sum_{n=0}^{\infty} \sum_{m=-n}^{n} \mathbf{j}_n^m(\mathbf{r}), \tag{5.116}$$

in which $\mathbf{j}_n^m(\mathbf{r})$ is given by

$$\mathbf{j}_n^m(\mathbf{r}) = \frac{1}{r}[f_n^m(r)\mathbf{X}_n^m + g_n^m(r)\mathbf{V}_n^m + h_n^m(r)\mathbf{W}_n^m]. \tag{5.117}$$

The spherical projection functions, $f_n^m(r)$, $g_n^m(r)$, $h_n^m(r)$, are introduced together with the explicit factor $1/r$, as in Blatt and Weisskopf [4]. In general, the expansion in eqs. (5.116) and (5.117) is valid because of the completeness of the vector spherical harmonics. The orthogonality relationships, eqs. (5.83)–(5.88), lead immediately to the following expressions for the spherical projection functions:

$$\frac{1}{r} f_n^m(r) = \int_{(4\pi)} (\mathbf{X}_n^m)^* \cdot \mathbf{J}(\mathbf{r})\, d\Omega, \tag{5.118}$$

$$\frac{1}{r} g_n^m(r) = \int_{(4\pi)} (\mathbf{V}_n^m)^* \cdot \mathbf{J}(\mathbf{r})\, d\Omega, \tag{5.119}$$

$$\frac{1}{r} h_n^m(r) = \int_{(4\pi)} (\mathbf{W}_n^m)^* \cdot \mathbf{J}(\mathbf{r})\, d\Omega. \tag{5.120}$$

It is important to stress that the spherical projection function $h_n^m(r)$ should not be confused with the Hankel function $h_n(r)$.

We would now like to express the coefficients of the electric and magnetic multipoles in terms of the spherical projection functions. Combining eqs. (5.75) and (5.107) for a_n^m, and correspondingly eqs. (5.76) and (5.98) for b_n^m, gives the following intermediate results:

$$a_n^m = -\frac{k}{4\pi g} \frac{1}{\sqrt{n(n+1)(2n+1)}} \int \mathbf{J}(\mathbf{r}') \cdot [\mathbf{W}_n^m j_{n-1}(kr')\sqrt{n+1} - \mathbf{V}_n^m j_{n+1}(kr')\sqrt{n}]^*\, d^3r', \tag{5.121}$$

$$b_n^m = \frac{k}{4\pi} \frac{1}{\sqrt{n(n+1)(2n+1)}} \int \mathbf{J}(\mathbf{r}') \cdot [\mathbf{X}_n^m j_n(kr')]^*\, d^3r'. \tag{5.122}$$

The desired result for a_n^m and b_n^m in terms of the spherical projection functions is obtained by substituting eq. (5.117) into eqs. (5.121) and (5.122), respectively, and using the orthogonality relations in eqs. (5.83)–(5.88), namely,

$$a_n^m = -\frac{k}{4\pi g} \frac{1}{\sqrt{n(n+1)(2n+1)}}$$

$$\times \left\{ \sqrt{n+1} \int h_n^m(r) j_{n-1}(kr) r \, dr - \sqrt{n} \int g_n^m(r) j_{n+1}(kr) r \, dr \right\}, \tag{5.123}$$

$$b_n^m = \frac{k}{4\pi} \frac{1}{\sqrt{n(n+1)}} \int f_n^m(r) j_n(kr) r \, dr. \tag{5.124}$$

Noting in eqs. (5.123) and (5.124) that the radial dependence is in the form of Bessel transforms of the spherical projection functions, we define $\eta_n^m(k)$, $\gamma_n^m(k)$, and $\varphi_n^m(k)$ as follows:

$$\eta_n^m(k) = \frac{1}{\sqrt{n(2n+1)}} \int h_n^m(r) j_{n-1}(kr) r \, dr, \tag{5.125}$$

$$\gamma_n^m(k) = \frac{1}{\sqrt{(n+1)(2n+1)}} \int g_n^m(r) j_{n+1}(kr) r \, dr, \tag{5.126}$$

$$\varphi_n^m(k) = \frac{1}{\sqrt{n(n+1)}} \int f_n^m(r) j_n(kr) r \, dr. \tag{5.127}$$

Using the transform representation we write a compact form for the multipole coefficients

$$a_n^m = \frac{k}{4\pi g} [\gamma_n^m(k) - \eta_n^m(k)] \tag{5.128}$$

and

$$b_n^m = \frac{k}{4\pi} \varphi_n^m(k). \tag{5.129}$$

Equations (5.128) and (5.129) give us the desired expressions for the coefficients of the electric and magnetic multipoles in terms of expressions of the distribution of the source current. Comparing eqs. (5.129) and (5.122) we note that the Bessel transform term $\varphi_n^m(k)$ does not contain any contribution from radial components of the current distribution; i.e., it is entirely determined by source currents flowing in the $\mathbf{e}_\varphi$ and $\mathbf{e}_\theta$ directions. As can be seen by comparing eqs. (5.128) and (5.121) the Bessel transforms $\gamma_n^m(k)$ and $\eta_n^m(k)$ correspond, respectively, to the vector spherical harmonics $\mathbf{V}_n^m$ and $\mathbf{W}_n^m$. Hence, the electric multipole coefficient a_n^m has a value depending upon both the longitudinal and the transverse components of the current in the source distribution.

5.2.4. *The Far Zone*

The form in which we expressed the electromagnetic fields in the previous sections is a complete and general solution of Maxwell's equations. This solution is valid outside the source domain; extension to include the source region, inside S, is given in Section 5.5.3 and in the literature.[1] In particular, the solution is valid in the far zone, which includes all points of space for which the parameters $kr \gg 1$ and $r/R_0 \gg 1$. Specifically, we require that points in the far zone are sufficiently distanced from the source compared with the two characteristic quantities: the wavelength λ and the source size R_0.

The emphasis on evaluating the far-zone case has both physical and mathematical rationale. Physically, we are interested in practical situations where the radiated fields are calculated for positions of space in which the presence of a detector will not significantly perturb their properties. Mathematically, radiation problems tend to simplify for far-zone calculations due to asymptotic expansions of the appropriate Green's functions. This constitutes a considerable computational convenience that will become apparent in the present section.

The radiation fields, eqs. (5.108) and (5.109), contain the spherical Hankel functions $h_n(kr)$, which contain the only dependence on $r = |\mathbf{r}|$ in these expressions. The asymptotic expansion of the Hankel functions [13] has the form

$$h_n(kr) \sim (-i)^{n+1} \frac{e^{ikr}}{kr} \qquad (kr \to \infty). \tag{5.130}$$

In order to obtain the far-zone expressions for the electromagnetic fields at an arbitrary (θ, φ) it is therefore sufficient to substitute eq. (5.130) into eqs. (5.108) and (5.109), namely,

$$\mathbf{E}^{(\infty)}(\mathbf{r}) = \frac{e^{ikr}}{r} \sum_{n,m} (-i)^n \sqrt{n(n+1)} \times \left\{ a_n^m \left[\sqrt{\frac{n+1}{2n+1}} \mathbf{W}_n^m + \sqrt{\frac{n}{2n+1}} \mathbf{V}_n^m \right] - \frac{ib_n^m}{g} \mathbf{X}_n^m \right\}, \tag{5.131}$$

$$\mathbf{H}^{(\infty)}(\mathbf{r}) = \frac{e^{ikr}}{r} \sum_{n,m} (-i)^n \sqrt{n(n+1)} \times \left\{ a_n^m ig \mathbf{X}_n^m + b_n^m \left[\sqrt{\frac{n+1}{2n+1}} \mathbf{W}_n^m + \sqrt{\frac{n}{2n+1}} \mathbf{V}_n^m \right] \right\}. \tag{5.132}$$

The structure of eqs. (5.131) and (5.132) provides an additional physical insight. To understand its significance, consider an observer in the far zone. He sees the current distribution as a small, point-like source and therefore expects the fields to have the general form of an outgoing spherical wave together with some amplitude and phase factors. Indeed, on inspecting the last two equations it is evident that both consist of a superposition of spherical waves with amplitudes and phases determined by the coefficients of the various multipoles existing in the source current distribution.

[1] See papers by Nisbet [14], [15] for continuation of the region of validity into the source distribution.

It is also instructive to consider the decomposition of the fields in the far zone into components parallel to unit vectors in the spherical coordinate system. Starting with the electric field $\mathbf{E}^{(\infty)}(\mathbf{r})$ in eq. (5.131), we substitute for the vector spherical harmonics in terms of their components in directions $\mathbf{e}_r$, $\mathbf{e}_\theta$, $\mathbf{e}_\varphi$, eqs. (5.79)–(5.81). The resulting components for the electric field are

$$\mathbf{E}^{(\infty)}(\mathbf{r})\cdot\mathbf{e}_r = 0, \tag{5.133}$$

$$\mathbf{E}^{(\infty)}(\mathbf{r})\cdot\mathbf{e}_\theta = \frac{e^{ikr}}{r}\sum_{n,m}(-i)^n\left\{\frac{1}{g}\frac{imb_n^m}{\sin\theta}Y_n^m + a_n^m\frac{\partial Y_n^m}{\partial\theta}\right\}, \tag{5.134}$$

$$\mathbf{E}^{(\infty)}(\mathbf{r})\cdot\mathbf{e}_\varphi = \frac{e^{ikr}}{r}\sum_{n,m}(-i)^n\left\{a_n^m\frac{im}{\sin\theta}Y_n^m - b_n^m\frac{1}{g}\frac{\partial Y_n^m}{\partial\theta}\right\}, \tag{5.135}$$

and for the magnetic field

$$\mathbf{H}^{(\infty)}(\mathbf{r})\cdot\mathbf{e}_r = 0, \tag{5.136}$$

$$\mathbf{H}^{(\infty)}(\mathbf{r})\cdot\mathbf{e}_\theta = \frac{e^{ikr}}{r}\sum_{n,m}(-i)^n\left\{b_n^m\frac{\partial Y_n^m}{\partial\theta} - a_n^m\frac{igm}{\sin\theta}Y_n^m\right\}, \tag{5.137}$$

$$\mathbf{H}^{(\infty)}(\mathbf{r})\cdot\mathbf{e}_\varphi = \frac{e^{ikr}}{r}\sum_{n,m}(-i)^n\left\{b_n^m\frac{im}{\sin\theta}Y_n^m + a_n^m g\frac{\partial Y_n^m}{\partial\theta}\right\}. \tag{5.138}$$

Parenthetically, we would like to make a remark about converting these far-zone solutions to the cgs system of units. If, as is customary, in eqs. (5.49) and (5.50) the terms $2\pi\nu\mu$ and $2\pi\nu\varepsilon$ are replaced by k/g and kg, respectively. Then in eqs. (5.134)–(5.138) the cgs form is obtained by setting $g = 1$. In the above equations a_n^m, given by eq. (5.121), must be modified; replacing $4\pi g$ in the denominator by c and, correspondingly, b_n^m in eq. (5.122) is modified by replacing 4π by c.

The vanishing of the radial components, eqs. (5.133) and (5.136), is consistent with the transverse nature of the electromagnetic fields. Nevertheless, we emphasize that these results are correct to first order in $1/r$ as a consequence of the asymptotic formula (5.130).

5.3. Second-Order Statistics of the Electromagnetic Field

5.3.1. *Dyadic Operators and the Notion Form*

A natural representation of both scalar $(\cdot)$ and vector $(\times)$ products is the dyadic operator. We will find this notation particularly convenient in the representation of second-order statistics of the electromagnetic radiation. We, therefore, review briefly the definition of a dyad and trace a few of its elementary properties, without pretending to present the subject exhaustively [16].

We introduce the notion of a dyadic, starting with the well-known equation from vector analysis in which an arbitrary vector **u** is represented by its components along the x-, y-, z-axes, as follows:

$$\mathbf{u} = \mathbf{e}_x(\mathbf{e}_x \cdot \mathbf{u}) + \mathbf{e}_y(\mathbf{e}_y \cdot \mathbf{u}) + \mathbf{e}_z(\mathbf{e}_z \cdot \mathbf{u}). \tag{5.139}$$

We can conveniently symbolize this sum by regrouping terms in the form

$$\mathbf{u} = (\mathbf{e}_x\mathbf{e}_x + \mathbf{e}_y\mathbf{e}_y + \mathbf{e}_z\mathbf{e}_z) \cdot \mathbf{u}. \tag{5.140}$$

In the first factor of eq. (5.140) we see pairs of vectors grouped without indication of scalar product or cross product between them. This provides an example of a dyadic which is generalized in the next paragraph.

By definition the terminology dyad is used to refer to the grouping of two vectors **uv**, without any indicated operation; and the dyadic polynomial is written in the form

$$\mathfrak{D} = \sum_n \mathbf{u}_n\mathbf{v}_n. \tag{5.141}$$

The old English font is used to denote a dyadic. The operation $\mathfrak{D} \cdot \mathbf{r}$ is defined using eq. (5.141) by

$$\mathfrak{D} \cdot \mathbf{r} = \left(\sum_n \mathbf{u}_n\mathbf{v}_n\right) \cdot \mathbf{r}, \tag{5.142}$$

$$\mathfrak{D} \cdot \mathbf{r} = \sum_n \mathbf{u}_n(\mathbf{v}_n \cdot \mathbf{r}). \tag{5.143}$$

Similarly, we define the operation $\mathbf{r} \cdot \mathfrak{D}$ by

$$\mathbf{r} \cdot \mathfrak{D} = \sum_n (\mathbf{r} \cdot \mathbf{u}_n)\mathbf{v}_n. \tag{5.144}$$

In general, the vectors resulting from eqs. (5.143) and (5.144) are not equal; and usually, the order of vectors in a dyad must be retained. Separately, one can establish that the combination of vectors in a dyad is distributive.

From eq. (5.141) it follows for any vectors **u** and **v** that the dyadic $\mathfrak{D}$ can be expressed in Cartesian coordinates as a sum of nine terms, i.e., the nonion form

$$\begin{aligned}\mathfrak{D} = {} & \mathbf{e}_x\mathbf{e}_x u_x v_x + \mathbf{e}_x\mathbf{e}_y u_x v_y + \mathbf{e}_x\mathbf{e}_z u_x v_z \\ & + \mathbf{e}_y\mathbf{e}_x u_y v_x + \mathbf{e}_y\mathbf{e}_y u_y v_y + \mathbf{e}_y\mathbf{e}_z u_y v_z \\ & + \mathbf{e}_z\mathbf{e}_x u_z v_x + \mathbf{e}_z\mathbf{e}_y u_z v_y + \mathbf{e}_z\mathbf{e}_z u_z v_z,\end{aligned} \tag{5.145}$$

in which u_x, v_x ... are the Cartesian components of the vectors **u** and **v**, respectively. In view of eq. (5.145) it is useful to define the matrix of a dyad **uv** by

$$(\mathbf{uv})_\mathrm{M} = \begin{bmatrix} u_x v_x & u_x v_y & u_x v_z \\ u_y v_x & u_y v_y & u_y v_z \\ u_z v_x & u_z y_y & u_z v_z \end{bmatrix}. \tag{5.146}$$

An analogous representation to eq. (5.145) can be written in any orthogonal curvilinear coordinate system. Furthermore, by eq. (5.145), we remark that equality of two dyads is true if and only if equality holds on a term by term basis in the nonion form.

An important quantity associated with the dyadic $\mathfrak{D}$ in eq. (5.141) is the scalar invariant, denoted by $\mathfrak{D}_S$, defined by summing the dot products of individual dyads $\mathbf{u}_n \cdot \mathbf{v}_n$

$$\mathfrak{D}_s = \sum_n \mathbf{u}_n \cdot \mathbf{v}_n. \tag{5.147}$$

Using the matrix representation for the dyadic polynomial in eqs. (5.141) and (5.147), we see that the scalar of the dyadic $\mathfrak{D}_S$ is given by the trace of the matrix representation, i.e.,

$$\mathfrak{D}_s = \mathrm{Tr}[\mathfrak{D}]_M \tag{5.148}$$

or

$$\mathfrak{D}_s = \sum_n \mathrm{Tr}[\mathbf{u}_n \mathbf{v}_n]_M. \tag{5.149}$$

Similarly, the vector of the dyadic $\mathfrak{D}$ in eq. (5.141) is defined as the sum of the cross products

$$\mathfrak{D}_\times = \sum_n \mathbf{u}_n \times \mathbf{v}_n. \tag{5.150}$$

The discussion of dyads is concluded with the remark that scalar and cross products for the differential operator ∇ can be written by direct analogy with eqs. (5.146) and (5.147).

5.3.2. *Energy Density of the Electromagnetic Field*

Consider now a sphere S of radius R_0 enclosing a current distribution $\mathbf{J}(\mathbf{r}; \nu)$, as in Sections 5.1 and 5.2. It is well known that the time-average energy density of the electromagnetic field at a point $\mathbf{r}$ consists of two components [17]. One is the energy density per unit volume in the electric field, u_E, given by

$$u_E = \tfrac{1}{4}\varepsilon \mathbf{E}^*(\mathbf{r}; \nu) \cdot \mathbf{E}(\mathbf{r}; \nu). \tag{5.151}$$

Similarly, the time-average energy density per unit volume in the magnetic field, u_M, is given by

$$u_M = \tfrac{1}{4}\mu \mathbf{H}^*(\mathbf{r}; \nu) \cdot \mathbf{H}(\mathbf{r}; \nu). \tag{5.152}$$

Hence, by eqs. (5.151) and (5.152) the total time-average energy density per unit volume is the sum $u_E + u_M$, i.e.

$$u(\mathbf{r}) = \tfrac{1}{4}[\varepsilon \mathbf{E}^*(\mathbf{r}; \nu) \cdot \mathbf{E}(\mathbf{r}; \nu) + \mu \mathbf{H}^*(\mathbf{r}; \nu) \cdot \mathbf{H}(\mathbf{r}; \nu)]. \tag{5.153}$$

In the formulation of problems in statistical electromagnetic theory it is customary to compute second-order moments. For calculations of energy density we are interested in second- and fourth-order moments of the electric field and the magnetic field. We will assume throughout that the fluctuations may be characterized by stationary ensembles. Moreover, in order to obtain information about spatial variation in addition to energy density, it is common to calculate the second-order moment at two states defined by coordinates $(\mathbf{r}_1; \nu)$ and $(\mathbf{r}_2; \nu)$. Hence, for stochastic processes by analogy to eq. (5.153), the second-order moment of interest denoted by $U(\mathbf{r}_1, \mathbf{r}_2; \nu)$ is written as

$$U(\mathbf{r}_1, \mathbf{r}_2; \nu) = \tfrac{1}{4}\varepsilon\langle \mathbf{E}^*(\mathbf{r}_1; \nu)\cdot\mathbf{E}(\mathbf{r}_2; \nu)\rangle + \tfrac{1}{4}\mu\langle \mathbf{H}^*(\mathbf{r}_1; \nu)\cdot\mathbf{H}(\mathbf{r}_2; \nu)\rangle. \quad (5.154)$$

The angle brackets in eq. (5.154) are used to denote an ensemble average taken over the stochastic process, and the constitutive parameters μ and ε are assumed deterministic.

A fairly general fourth-order moment of interest,

$$\langle \mathbf{E}(\mathbf{r}_1, \nu_1)\cdot\mathbf{E}(\mathbf{r}_1, \nu_1)\mathbf{E}^*(\mathbf{r}_2, \nu_2)\cdot\mathbf{E}(\mathbf{r}_2, \nu_2)\rangle, \quad (5.155)$$

arises in the treatment of speckle problems in which the expectation can be taken over both the medium and the sources. In evaluation of this moment, assuming Gaussian statistics, we are lead naturally to consider the following second-order moment

$$\langle \mathbf{E}^*(\mathbf{r}_1; \nu_1)\cdot\mathbf{E}(\mathbf{r}_2; \nu_2)\rangle. \quad (5.156)$$

In this case we note that the two states are $(\mathbf{r}_1; \nu_1)$ and $(\mathbf{r}_2; \nu_2)$.

In this analysis we prefer the dyad operator formalism for the second-order moment following the notation which was introduced by Carter and Wolf [6]. For stochastic processes the dyadic generalization of eq. (5.154) is given by

$$\mathfrak{U}(\mathbf{r}_1, \mathbf{r}_2; \nu) = \tfrac{1}{4}\varepsilon\langle \mathbf{E}^*(\mathbf{r}_1; \nu)\mathbf{E}(\mathbf{r}_2; \nu)\rangle + \tfrac{1}{4}\mu\langle \mathbf{H}^*(\mathbf{r}_1; \nu)\mathbf{H}(\mathbf{r}_2; \nu)\rangle. \quad (5.157)$$

We define two cross-spectral dyads: $\mathfrak{W}_{\mathrm{ee}}(\mathbf{r}_1, \mathbf{r}_2; \nu)$ and $\mathfrak{W}_{\mathrm{hh}}(\mathbf{r}_1, \mathbf{r}_2; \nu)$ for the electric and magnetic field densities, respectively, i.e.,

$$\mathfrak{W}_{\mathrm{ee}}(\mathbf{r}_1, \mathbf{r}_2; \nu) = \langle \mathbf{E}^*(\mathbf{r}_1; \nu)\mathbf{E}(\mathbf{r}_2; \nu)\rangle \quad (5.158)$$

and

$$\mathfrak{W}_{\mathrm{hh}}(\mathbf{r}_1, \mathbf{r}_2; \nu) = \langle \mathbf{H}^*(\mathbf{r}_1; \nu)\mathbf{H}(\mathbf{r}_2; \nu)\rangle. \quad (5.159)$$

We note that eqs. (5.158) and (5.159) each contains nine scalar cross-spectral densities.

By taking the scalar of the dyads in eq. (5.157) we obtain a representation for the cross-spectral density $U(\mathbf{r}_1, \mathbf{r}_2; \nu)$ in eq. (5.154), i.e.,

$$U(\mathbf{r}_1, \mathbf{r}_2; \nu) = \mathfrak{U}_{\mathrm{s}}(\mathbf{r}_1, \mathbf{r}_2; \nu). \quad (5.160)$$

Using eqs. (5.158) and (5.159), and the matrix representation for the dyad $\mathfrak{U}$ as in eq. (5.146), we write the general result

$$U(\mathbf{r}_1, \mathbf{r}_2; \nu) = \frac{\varepsilon}{4}\mathrm{Tr}[\mathfrak{W}_{\mathrm{ee}}(\mathbf{r}_1, \mathbf{r}_2; \nu)]_{\mathrm{M}} + \frac{\mu}{4}\mathrm{Tr}[\mathfrak{W}_{\mathrm{hh}}(\mathbf{r}_1, \mathbf{r}_2; \nu)]_{\mathrm{M}}. \quad (5.161)$$

The expression for the energy density in the electromagnetic field is obtained by setting $\mathbf{r}_1 = \mathbf{r}_2 = \mathbf{r}$ in eq. (5.161). This ensemble average energy density is consequently given by

$$U(\mathbf{r}, \mathbf{r}; \nu) = \frac{\varepsilon}{4} \operatorname{Tr}[\mathfrak{W}_{\mathrm{ee}}(\mathbf{r}, \mathbf{r}; \nu)]_{\mathrm{M}} + \frac{\mu}{4} \operatorname{Tr}[\mathfrak{W}_{\mathrm{hh}}(\mathbf{r}, \mathbf{r}; \nu)]_{\mathrm{M}}. \tag{5.162}$$

Since the trace of a matrix remains invariant to a change of the coordinate system, we can immediately write, for example, $\operatorname{Tr}(\mathfrak{W}_{\mathrm{ee}})_{\mathrm{M}}$ in the spherical polar coordinate system as

$$\begin{aligned} \operatorname{Tr}[\mathfrak{W}_{\mathrm{ee}}(\mathbf{r}_1, \mathbf{r}_2; \nu)] &= \langle E_{\mathrm{r}}^*(\mathbf{r}_1; \nu) E_{\mathrm{r}}(\mathbf{r}_2; \nu)\rangle + \langle E_\theta^*(\mathbf{r}_1; \nu) E_\theta(\mathbf{r}_2; \nu)\rangle \\ &\quad + \langle E_\Phi^*(\mathbf{r}_1; \nu) E_\Phi(\mathbf{r}_2; \nu)\rangle, \end{aligned} \tag{5.163}$$

from which the special case $\mathbf{r}_1 = \mathbf{r}_2 = \mathbf{r}$ follows immediately.

5.3.3. *The Dyadic Cross-Spectral Densities of the Field and the Multipole Coupling Coefficients*

The representation of the cross-spectral density $U(\mathbf{r}_1, \mathbf{r}_2; \nu)$ in eq. (5.157) requires expressions for the dyadic cross-spectral densities $\mathfrak{W}_{\mathrm{ee}}$ and $\mathfrak{W}_{\mathrm{hh}}$. In this section we derive an explicit expression for $\mathfrak{W}_{\mathrm{ee}}$ in terms of the expansion of the electric field in vector spherical harmonics, eq. (5.108).

It is useful to introduce the following notation for the expected values of pairs of expansion coefficients. The definitions of $A(\mathrm{N})$, $B(\mathrm{N})$, $C(\mathrm{N})$, $C^\dagger(\mathrm{N})$ are given by

$$A(\mathrm{N}) = \langle a_n^{m*} a_{n'}^{m'}\rangle, \tag{5.164}$$

$$B(\mathrm{N}) = \langle b_n^{m*} b_{n'}^{m'}\rangle, \tag{5.165}$$

$$C(\mathrm{N}) = \langle a_n^{m*} b_{n'}^{m'}\rangle, \tag{5.166}$$

$$C^\dagger(\mathrm{N}) = \langle b_n^{m*} a_{n'}^{m'}\rangle. \tag{5.167}$$

Here the multiple indices n, m, n', m' are indicated by N, i.e.,

$$\mathrm{N} = (n, m; n', m'). \tag{5.168}$$

Physically, we observe that the $A(\mathrm{N})$ coefficients represent cross-coupling among E-type multipoles, and similarly $B(\mathrm{N})$ represent that for the H-type multipoles. The coefficients $C(\mathrm{N})$ and $C^\dagger(\mathrm{N})$ represent cross-coupling between different E- and H-type multipoles.

Substituting eq. (5.108) into eq. (5.158), expanding, and taking the ensemble average lead to the desired expression for the dyadic $\mathfrak{W}_{\mathrm{ee}}$. In taking the ensemble average we note that the random process is completely characterized by the expansion coefficients a_n^m and b_n^m, and the expectation brackets commute with the Hankel functions and the vector spherical harmonics so that the result is given by

$$\begin{aligned}
&\mathfrak{W}_{ee}(\mathbf{r}_1, \mathbf{r}_2; \nu)\\
&= k^2 \sum_{\mathbf{N}} \sqrt{n(n+1)n'(n'+1)}\\
&\times \Bigg\{ \frac{A(\mathbf{N})}{\sqrt{(2n+1)(2n'+1)}} \Big[\sqrt{(n+1)(n'+1)}\, h^*_{n-1}(kr_1) h_{n'-1}(kr_2) \mathbf{W}^{m^*}_n(1) \mathbf{W}^{m'}_{n'}(2)\\
&\qquad + \sqrt{nn'}\, h^*_{n+1}(kr_1) h_{n'+1}(kr_2) \mathbf{V}^{m^*}_n(1) \mathbf{V}^{m'}_{n'}(2)\\
&\qquad - \sqrt{n'(n+1)}\, h^*_{n-1}(kr_1) h_{n'+1}(kr_2) \mathbf{W}^{m^*}_n(1) \mathbf{V}^{m'}_{n'}(2)\\
&\qquad - \sqrt{n(n'+1)}\, h^*_{n+1}(kr_1) h_{n'-1}(kr_2) \mathbf{V}^{m^*}_n(1) \mathbf{W}^{m'}_{n'}(2)\\
&+ \frac{1}{g^2} B(\mathbf{N}) h^*_n(kr_1) h_{n'}(kr_2) \mathbf{X}^{m^*}_n(1) \mathbf{X}^{m'}_{n'}(2)\\
&+ \frac{1}{g} \frac{C(\mathbf{N})}{\sqrt{2n+1}} \mathbf{X}^{m'}_{n'}(2) h_{n'}(kr_2)\\
&\times \big[\sqrt{n+1}\, h_{n-1}(kr_1) \mathbf{W}^m_n(1) - \sqrt{n}\, h_{n+1}(kr_1) \mathbf{V}^m_n(1)\big]^*\\
&+ \frac{1}{g} \frac{C^\dagger(\mathbf{N})}{\sqrt{2n'+1}} \mathbf{X}^{m^*}_n(1) h^*_n(kr_1)\\
&\times \big[\sqrt{n'+1}\, h_{n'-1}(kr_2) \mathbf{W}^{m'}_{n'}(2) - \sqrt{n'}\, h_{n'+1}(kr_2) \mathbf{V}^{m'}_{n'}(2)\big] \Bigg\}
\end{aligned} \tag{5.169}$$

In eq. (5.169) for $\mathfrak{W}_{ee}$, abbreviations have been used for angular coordinates of the vector spherical harmonics, i.e.,

$$(1) \;\Leftrightarrow\; (\theta_1, \varphi_1), \tag{5.170}$$

$$(2) \;\Leftrightarrow\; (\theta_2, \varphi_2). \tag{5.171}$$

Equation (5.169) for $\mathfrak{W}_{ee}$ gives us the desired dyadic polynomial form in terms of vector spherical harmonics. Similarly, eq. (5.109) can be substituted into eq. (5.159) and a corresponding form can be obtained for $\mathfrak{W}_{hh}$. We can also obtain a dyadic $\mathfrak{W}_{eh}$ starting with the definition

$$\mathfrak{W}_{eh}(\mathbf{r}_1, \mathbf{r}_2; \nu) = \langle \mathbf{E}^*(\mathbf{r}_1; \nu) \mathbf{H}(\mathbf{r}_2; \nu) \rangle \tag{5.172}$$

and using eqs. (5.108) and (5.109).

Returning now to the problem in which the expansion coefficients a^m_n and b^m_n are given by integrals over the source current, eqs. (5.121) and (5.122), we observe that the coupling coefficients in eqs. (5.164)–(5.167) can be formally expressed in terms of the source distribution $\mathbf{J}(\mathbf{r}; \nu)$. In the deterministic case $\mathfrak{B}_{ee}$ is obtained from eq. (5.169) simply by dropping the expectation brackets, and of course this carries over to the coupling coefficients $A(\mathbf{N})$, $B(\mathbf{N})$, $C(\mathbf{N})$, $C^\dagger(\mathbf{N})$.

Alternatively, we can express the coupling coefficients in terms of the Bessel transforms using eqs. (5.128) and (5.129) in eqs. (5.164)–(5.167). The result of a straightforward calculation is the following simple forms for the multipole coupling coefficients:

$$A(\mathbf{N}) = \left(\frac{k}{4\pi g}\right)^2 \langle \gamma_n^{m*}(k)\gamma_{n'}^{m'}(k) + \eta_n^{m*}(k)\eta_{n'}^{m'}(k) - \gamma_n^{m*}(k)\eta_{n'}^{m'}(k) - \eta_n^{m*}(k)\gamma_{n'}^{m'}(k)\rangle, \tag{5.173}$$

$$B(\mathbf{N}) = \left(\frac{k}{4\pi}\right)^2 \langle \varphi_n^{m*}(k)\varphi_{n'}^{m'}(k)\rangle, \tag{5.174}$$

$$C(\mathbf{N}) = \frac{1}{g}\left(\frac{k}{4\pi}\right)^2 \langle [\gamma_n^{m}(k) - \eta_n^{m}(k)]^*\varphi_{n'}^{m'}(k)\rangle, \tag{5.175}$$

$$C^{\dagger}(\mathbf{N}) = \frac{1}{g}\left(\frac{k}{4\pi}\right)^2 \langle \varphi_n^{m*}(k)[\gamma_{n'}^{m'}(k) - \eta_{n'}^{m'}(k)]\rangle \tag{5.176}$$

5.4. Statistical Representation in the Far Zone

5.4.1. *The Poynting Vector*

In this section we start with the general form for the Poynting vector

$$\mathbf{S}(\mathbf{r}; \nu) = \tfrac{1}{2}\Re[\mathbf{E}^*(\mathbf{r}; \nu) \times \mathbf{H}(\mathbf{r}; \nu)], \tag{5.177}$$

where $\Re$ denotes the real part. By eq. (5.172) for the cross-spectral dyadic $\mathfrak{W}_{\text{eh}}$ we can express the ensemble average of the Poynting vector $\langle \mathbf{S}(\mathbf{r}; \nu)\rangle$ as

$$\langle \mathbf{S}(\mathbf{r}; \nu)\rangle = \tfrac{1}{2}\Re[\mathfrak{W}_{\text{eh}}(\mathbf{r}, \mathbf{r}; \nu)]_{\times}. \tag{5.178}$$

The subscript $\times$ in eq. (5.178) denotes the vector product of the dyad as in eq. (5.150).

In the far zone, by eqs. (5.133) and (5.136), there are only transverse components of the electric and magnetic fields (of order $1/r$); hence the Poynting vector $\mathbf{S}^{(\infty)}(\mathbf{r}; \nu)$ has only a radial component. Using the definition for the vector product, eq. (5.150) in eq. (5.178), we find the following relation for the Poynting vector

$$\langle \mathbf{S}^{(\infty)}(\mathbf{r}; \nu)\rangle = \tfrac{1}{2}\Re\langle [E_\theta^{(\infty)*}(\mathbf{r}; \nu)H_\Phi^{(\infty)}(\mathbf{r}; \nu) - E_\Phi^{(\infty)*}(\mathbf{r}; \nu)H_\theta^{(\infty)}(\mathbf{r}; \nu)]\rangle \mathbf{e}_r. \tag{5.179}$$

It is also straightforward to show that eqs. (5.134)–(5.137) give the well-known ratios between transverse components of the electric and magnetic field vectors, namely,

$$\frac{H_\theta}{E_\varphi} = g \tag{5.180}$$

and

$$\frac{H_\theta}{E_\varphi} = -g. \tag{5.181}$$

Here the transverse wave admittance is $g = \sqrt{\varepsilon/\mu}$.

Substituting eqs. (5.180) and (5.181) into eq. (5.179) we obtain

$$\langle \mathbf{S}^{(\infty)}(\mathbf{r}; \nu) \rangle = \frac{g}{2} \langle |E_\theta^{(\infty)}(\mathbf{r}; \nu)|^2 + |E_\varphi^{(\infty)}(\mathbf{r}; \nu)|^2 \rangle \mathbf{e}_r. \tag{5.182}$$

Evaluating eq. (5.162) for $U^{(\infty)}(\mathbf{r}, \mathbf{r}; \nu)$ by eqs. (5.180) and (5.181) we reach the well-known result that in the far zone the Poynting vector and the energy density are related by

$$\langle \mathbf{S}^{(\infty)}(\mathbf{r}; \nu) \rangle = cU^{(\infty)}(\mathbf{r}, \mathbf{r}; \nu)\mathbf{e}_r, \tag{5.183}$$

where c is the speed of light. Also, as a consequence of eqs. (5.180) and (5.181), we obtain the well-known equality between the energy stored in the electric field and that which is stored in the magnetic field, namely,

$$U_{\mathrm{E}}^{(\infty)}(\mathbf{r}; \nu) = U_{\mathrm{M}}^{(\infty)}(\mathbf{r}; \nu) = \frac{\varepsilon}{4}[|E_\theta^{(\infty)}(\mathbf{r}; \nu)|^2 + |E_\Phi^{(\infty)}(\mathbf{r}; \nu)|^2]. \tag{5.184}$$

Finally, eqs. (5.134) and (5.135) are substituted into eq. (5.182) giving the expected value for the Poynting vector in the far zone

$$\begin{aligned}\langle \mathbf{S}^{(\infty)}(\mathbf{r}; \nu) \rangle = {} & \frac{1}{2gr^2} \mathbf{e}_r \sum_N (-i)^{n'-n} \\ & \times \Bigg\{ [g^2 A(\mathbf{N}) + B(\mathbf{N})] \left[\frac{mm'}{\sin^2\theta} Y_n^{m*} Y_{n'}^{m'} + \left(\frac{\partial Y_n^{m*}}{\partial\theta} \right) \left(\frac{\partial Y_{n'}^{m'}}{\partial\theta} \right) \right] \\ & + \frac{ig}{\sin\theta} [C(\mathbf{N}) - C^\dagger(\mathbf{N})] \left[m Y_n^{m*} \frac{\partial Y_{n'}^{m'}}{\partial\theta} + m' Y_{n'}^{m'} \frac{\partial Y_n^{m*}}{\partial\theta} \right] \Bigg\}.\end{aligned} \tag{5.185}$$

5.4.2. *Cross-Spectral Dyadic of the Electric Field*

From eq. (5.158) we write the electric cross-spectral density dyad for the far zone, $\mathfrak{W}_{\mathrm{ee}}^{(\infty)}(\mathbf{r}_1, \mathbf{r}_2; \nu)$

$$\mathfrak{W}_{\mathrm{ee}}^{(\infty)}(\mathbf{r}_1, \mathbf{r}_2; \nu) = \langle \mathbf{E}^{(\infty)*}(\mathbf{r}_1; \nu) \mathbf{E}^{(\infty)}(\mathbf{r}_2; \nu) \rangle. \tag{5.186}$$

By eq. (5.131) the appropriate electric field vector expressed in vector spherical harmonics is

$$\begin{aligned}\mathbf{E}^{(\infty)}(\mathbf{r}) = {} & \frac{e^{ikr}}{r} \sum_{n,m} (-i)^n \sqrt{n(n+1)} \\ & \times \left\{ \frac{a_n^m}{\sqrt{2n+1}} [\sqrt{n+1}\mathbf{W}_n^m + \sqrt{n}\mathbf{V}_n^m] - i\frac{b_n^m}{g} \mathbf{X}_n^m \right\}.\end{aligned} \tag{5.187}$$

Substituting eq. (5.187) into eq. (5.186) we obtain

$$\mathfrak{W}_{ee}^{(\infty)}(\mathbf{r}_1, \mathbf{r}_2; \nu)$$

$$= \frac{e^{ik(r_2 - r_1)}}{r_1 r_2} \sum_{N} \sqrt{\frac{n(n+1)n'(n'+1)}{(2n+1)(2n'+1)}}(-i)^{n'-n}$$

$$\times \Big\{ A(\mathbf{N})[\sqrt{(n+1)(n'+1)}\,\mathbf{W}_n^{m*}(1)\mathbf{W}_{n'}^{m'}(2) + \sqrt{nn'}\,\mathbf{V}_n^{m*}(1)\mathbf{V}_{n'}^{m'}(2)$$

$$+ \sqrt{n(n'+1)}\,\mathbf{V}_n^{m*}(1)\mathbf{W}_{n'}^{m'}(2) + \sqrt{n'(n+1)}\,\mathbf{W}_n^{m*}(1)\mathbf{V}_{n'}^{m'}(2)]$$

$$+ \frac{1}{g^2} B(\mathbf{N})\sqrt{(2n+1)(2n'+1)}\,\mathbf{X}_n^{m*}(1)\mathbf{X}_{n'}^{m'}(2)$$

$$- \frac{i}{g} C(\mathbf{N})\sqrt{2n'+1}\,\mathbf{X}_{n'}^{m'}(2)[\sqrt{n+1}\,\mathbf{W}_n^{m*}(1) + \sqrt{n}\,\mathbf{V}_n^{m*}(1)]$$

$$+ \frac{i}{g} C^{\dagger}(\mathbf{N})\sqrt{2n+1}\,\mathbf{X}_n^{m*}(1)[\sqrt{n'+1}\,\mathbf{W}_{n'}^{m'}(2) + \sqrt{n}\,\mathbf{V}_{n'}^{m'}(2)]\Big\}. \tag{5.188}$$

This representation is particularly convenient in performing integrations over a large sphere due to the well-known orthogonality relations of the vector spherical harmonics. On the other hand, it is often useful to express the electric field in spherical polar coordinates. In this later case the matrix form of the dyad $\mathfrak{W}_{ee}^{(\infty)}(\mathbf{r}_1, \mathbf{r}_2; \nu)$ can be written as

$$[\mathfrak{W}_{ee}^{(\infty)}(\mathbf{r}_1, \mathbf{r}_2; \nu)]_M = \left\langle \begin{bmatrix} 0 \\ E_\theta^{(\infty)*}(\mathbf{r}_1, \nu) \\ E_\varphi^{(\infty)*}(\mathbf{r}_1, \nu) \end{bmatrix} [0 \quad E_\theta^{(\infty)}(\mathbf{r}_2, \nu) \quad E_\varphi^{(\infty)}(\mathbf{r}_2, \nu)] \right\rangle, \tag{5.189}$$

in which E_θ and E_φ are given by eqs. (5.134) and (5.135), respectively.

On forming the product of the two matrices in eq. (5.189) the resultant matrix is seen to be of rank 2. This matrix contains correlations involving only transverse components of the field. In the literature on electromagnetic theory it is well known as the coherence matrix [18]. Let this matrix be denoted by the symbol

$$[\mathfrak{J}]_M = \begin{bmatrix} \langle E_\theta^{(\infty)*}(\mathbf{r}; \nu)E_\theta^{(\infty)}(\mathbf{r}; \nu)\rangle & \langle E_\theta^{(\infty)*}(\mathbf{r}; \nu)E_\Phi^{(\infty)}(\mathbf{r}; \nu)\rangle \\ \langle E_\Phi^{(\infty)*}(\mathbf{r}; \nu)E_\theta^{(\infty)}(\mathbf{r}; \nu)\rangle & \langle E_\Phi^{(\infty)*}(\mathbf{r}; \nu)E_\Phi^{(\infty)}(\mathbf{r}; \nu)\rangle \end{bmatrix}. \tag{5.190}$$

It is observed that the coherence matrix is Hermitian. Other properties of this matrix are treated extensively in the literature. The matrix elements are

$$\langle E_\theta^{(\infty)*}(\mathbf{r}; \nu)E_\theta^{(\infty)}(\mathbf{r}; \nu)\rangle$$

$$= \frac{1}{g^2 r^2} \sum_{N} (-i)^{n'-n}$$

$$\times \Big\{ g^2 A(\mathbf{N})\left(\frac{\partial Y_n^{m*}}{\partial\theta}\right)\left(\frac{\partial Y_{n'}^{m'}}{\partial\theta}\right) + \mathbf{B}(\mathbf{N})\frac{mm'}{\sin^2\theta} Y_n^{m*} Y_{n'}^{m'}$$

$$+ ig\left[C(\mathbf{N})\left(\frac{\partial Y_n^{m*}}{\partial\theta}\right)\frac{m'}{\sin\theta} Y_{n'}^{m'} - C^{\dagger}(\mathbf{N})\frac{m}{\sin\theta} Y_n^{m*}\left(\frac{\partial Y_{n'}^{m'}}{\partial\theta}\right)\right]\Big\}, \tag{5.191}$$

$$\langle E_\varphi^{(\infty)*}(\mathbf{r}; \nu) E_\varphi^{(\infty)}(\mathbf{r}; \nu)\rangle$$

$$= \frac{1}{g^2 r^2} \sum_N (-i)^{n'-n}$$

$$\times \left\{ g^2 A(\mathbf{N}) \frac{mm'}{\sin^2\theta} Y_n^{m*} Y_{n'}^{m'} + B(\mathbf{N}) \left(\frac{\partial Y_n^{m*}}{\partial\theta}\right)\left(\frac{\partial Y_{n'}^{m'}}{\partial\theta}\right)\right.$$

$$\left. + \frac{ig}{\sin\theta}\left[C(\mathbf{N}) m Y_n^{m*} \left(\frac{\partial Y_{n'}^{m'}}{\partial\theta}\right) - C^\dagger(\mathbf{N}) \left(\frac{\partial Y_n^{m*}}{\partial\theta}\right) m' Y_{n'}^{m'}\right]\right\}, \tag{5.192}$$

$$\langle E_\theta^{(\infty)*}(\mathbf{r}; \nu) E_\varphi^{(\infty)}(\mathbf{r}; \nu)\rangle$$

$$= \frac{1}{g^2 r^2} \sum_N (-i)^{n'-n}$$

$$\times \left\{ ig^2 A(\mathbf{N}) \left(\frac{\partial Y_n^{m*}}{\partial\theta}\right) \frac{m'}{\sin\theta} Y_{n'}^{m'} + iB(\mathbf{N}) \frac{m}{\sin\theta} Y_n^{m*} \left(\frac{\partial Y_{n'}^{m'}}{\partial\theta}\right)\right.$$

$$\left. + g\left[C^\dagger(\mathbf{N}) \frac{mm'}{\sin^2\theta} Y_n^{m*} Y_{n'}^{m'} - C(\mathbf{N}) \left(\frac{\partial Y_n^{m*}}{\partial\theta}\right)\left(\frac{\partial Y_{n'}^{m'}}{\partial\theta}\right)\right]\right\}, \tag{5.193}$$

$$\langle E_\theta^{(\infty)*}(\mathbf{r}; \nu) E_\Phi^{(\infty)}(\mathbf{r}; \nu)\rangle = [\langle E_\Phi^{(\infty)*}(\mathbf{r}; \nu) E_\theta^{(\infty)}(\mathbf{r}; \nu)\rangle]^*. \tag{5.194}$$

5.4.3. *The Stokes Parameters*

The Stokes parameters represent another well-known method for characterizing the polarization properties of an electromagnetic field. Their relationship to the coherence matrix has been discussed extensively in the literature. Using the conventional notation we write the defining equations for the Stokes parameters s_0, s_1, s_2, s_3:

$$s_0 = \langle E_\theta^{(\infty)*}(\mathbf{r}; \nu) E_\theta^{(\infty)}(\mathbf{r}; \nu)\rangle + \langle E_\varphi^{(\infty)*}(\mathbf{r}; \nu) E_\varphi^{(\infty)}(\mathbf{r}; \nu)\rangle, \tag{5.195}$$

$$s_1 = \langle E_\theta^{(\infty)*}(\mathbf{r}; \nu) E_\theta^{(\infty)}(\mathbf{r}; \nu)\rangle - \langle E_\varphi^{(\infty)*}(\mathbf{r}; \nu) E_\varphi^{(\infty)}(\mathbf{r}; \nu)\rangle, \tag{5.196}$$

$$s_2 = \langle E_\theta^{(\infty)*}(\mathbf{r}; \nu) E_\varphi^{(\infty)}(\mathbf{r}; \nu)\rangle + \langle E_\varphi^{(\infty)*}(\mathbf{r}; \nu) E_\theta^{(\infty)}(\mathbf{r}; \nu)\rangle, \tag{5.197}$$

$$s_3 = i[\langle E_\varphi^{(\infty)*}(\mathbf{r}; \nu) E_\theta^{(\infty)}(\mathbf{r}; \nu)\rangle - \langle E_\theta^{(\infty)*}(\mathbf{r}; \nu) E_\varphi^{(\infty)}(\mathbf{r}; \nu)\rangle]. \tag{5.198}$$

One useful application of the Stokes parameters is in characterizing propagation through a polarizing medium. We find the output Stokes parameters as a matrix product of the 4×4 matrix that characterizes the medium and the 4×1 matrix for the input Stokes parameters. Here, for the sake of completeness, we derive explicit expressions for the Stokes parameters using eqs. (5.191)–(5.194) in eqs. (5.195)–(5.198)

$$s_0 = \frac{1}{g^2 r^2} \sum_N (-i)^{n'-n} \times \left\{ [g^2 A(\mathbf{N}) + B(\mathbf{N})] \left[\left(\frac{\partial Y_n^{m*}}{\partial \theta} \right) \left(\frac{\partial Y_{n'}^{m'}}{\partial \theta} \right) + \frac{mm'}{\sin^2 \theta} Y_n^{m*} Y_{n'}^{m'} \right] + ig[C(\mathbf{N}) - C^\dagger(\mathbf{N})] \left[\left(\frac{\partial Y_n^{m*}}{\partial \theta} \right) \frac{m'}{\sin \theta} Y_{n'}^{m'} + \left(\frac{\partial Y_{n'}^{m'}}{\partial \theta} \right) \frac{m}{\sin \theta} Y_n^{m*} \right] \right\}, \tag{5.199}$$

$$s_1 = \frac{1}{g^2 r^2} \sum_N (-i)^{n'-n} \times \left\{ [g^2 A(\mathbf{N}) - B(\mathbf{N})] \left[\left(\frac{\partial Y_n^{m*}}{\partial \theta} \right) \left(\frac{\partial Y_{n'}^{m'}}{\partial \theta} \right) - \frac{mm'}{\sin^2 \theta} Y_n^{m*} Y_{n'}^{m'} \right] + ig[C(\mathbf{N}) + C^\dagger(\mathbf{N})] \left[\left(\frac{\partial Y_n^{m*}}{\partial \theta} \right) \frac{m'}{\sin \theta} Y_{n'}^{m'} + \left(\frac{\partial Y_{n'}^{m'}}{\partial \theta} \right) \frac{m}{\sin \theta} Y_n^{m*} \right] \right\}, \tag{5.200}$$

$$s_2 = \frac{1}{g^2 r^2} \sum_N (-i)^{n'-n} \times \left\{ [ig^2 A(\mathbf{N}) - B(\mathbf{N})] \left[\left(\frac{\partial Y_n^{m*}}{\partial \theta} \right) \frac{m'}{\sin \theta} Y_{n'}^{m'} - \frac{m}{\sin \theta} Y_n^{m*} \left(\frac{\partial Y_{n'}^{m'}}{\partial \theta} \right) \right] - g[C(\mathbf{N}) + C^\dagger(\mathbf{N})] \left[\left(\frac{\partial Y_n^{m*}}{\partial \theta} \right) \left(\frac{\partial Y_{n'}^{m'}}{\partial \theta} \right) - \frac{mm'}{\sin^2 \theta} Y_n^{m*} Y_{n'}^{m'} \right] \right\}, \tag{5.201}$$

$$s_3 = \frac{1}{g^2 r^2} \sum_N (-i)^{n'-n} \times \left\{ [g^2 A(\mathbf{N}) + B(\mathbf{N})] \left[\frac{m}{\sin \theta} Y_n^{m*} \left(\frac{\partial Y_{n'}^{m'}}{\partial \theta} \right) + \frac{m'}{\sin \theta} Y_{n'}^{m'} \left(\frac{\partial Y_n^{m*}}{\partial \theta} \right) \right] + ig[C(\mathbf{N}) + C^\dagger(\mathbf{N})] \left[\frac{mm'}{\sin^2 \theta} Y_n^{m*} Y_{n'}^{m'} + \left(\frac{\partial Y_n^{m*}}{\partial \theta} \right) \left(\frac{\partial Y_{n'}^{m'}}{\partial \theta} \right) \right] \right\}. \tag{5.202}$$

5.4.4. *Total Radiated Power*

In order to obtain an expression for the total radiated power per unit frequency interval, $P(\nu)$, which flows out of a large sphere, we calculate the radial component of the Poynting vector and integrate it over the closed surface. The sphere is taken large enough so that we are well into the far zone; the resulting integral for $P(\nu)$ is given by

$$P(\nu) = \int_{(4\pi)} \langle S^{(\infty)}(\mathbf{r}; \nu) \rangle \cdot \mathbf{e}_r r^2 \, d\Omega, \tag{5.203}$$

where the solid angle $d\Omega$ is equal to

$$d\Omega = \sin\theta \, d\theta \, d\varphi, \tag{5.204}$$

and (4π) denotes integration over all solid angles. In principle we could use eq. (5.185) to express the expected value of the Poynting vector. It is however simpler to combine eqs. (5.182) and (5.203). We then obtain an expression for the total radiated power as a function of the far-zone components of the electric field

$$P(\nu) = \frac{g}{2} \int_{(4\pi)} \langle |E_\theta^{(\infty)}(\mathbf{r}; \nu)|^2 + |E_\varphi^{(\infty)}(\mathbf{r}; \nu)|^2 \rangle r^2 \, d\Omega. \tag{5.205}$$

Substituting eqs. (5.134) and (5.135) into eq. (5.205), we are able to integrate this expression readily using the orthogonality conditions for the vector spherical harmonics, eqs. (5.83)–(5.88). The result for the total radiated power is found to be

$$P(\nu) = \frac{g}{2} \sum_{n,m} n(n+1)[\langle |a_n^m|^2 \rangle + \langle |b_n^m|^2 \rangle]. \tag{5.206}$$

This expression for the radiated power is established without making any assumption about the spatial correlations of source currents. The cross terms of the form $\langle a_n^{m*} b_{n'}^{m'} \rangle$ are not present in eq. (5.206) due to the orthogonality of the vector spherical harmonics. We remark that the multipoles in the Debye expansion contribute to the total radiated power in a simple additive fashion, as if each multipole radiated independently of the others.

5.5. Expansion of the Multipole Moments

5.5.1. *A Representation of the Dyadic Cross-Spectral Densities in Terms of Scalar Cross-Spectral Densities*

For radiation problems in statistical electromagnetics, the second-order moments of the Debye potentials play a central role. Our earlier expressions, e.g., the dyad $\mathfrak{W}_{ee}$ in eqs. (5.158) and (5.169), have been expressed in vector spherical harmonics, although we see that in the final form the coefficients $\{a_n^m\}$ and $\{b_n^m\}$, or rather their expectations, eqs. (5.162)–(5.165), completely define $\mathfrak{W}_{ee}$. As an alternative form we present a description of the dyad $\mathfrak{W}_{ee}$ using moments of scalar potentials $\Pi_1(\mathbf{r})$ and $\Pi_2(\mathbf{r})$. This gives us the advantage of having a description of the statistics of vector electromagnetic fields entirely in terms of scalar cross-spectral densities.

From the electric and magnetic Debye potentials we form four scalar cross-spectral densities denoted by P_{11}, P_{22}, P_{12}, P_{21}, and defined by the equations

$$P_{11}(\mathbf{r}_1, \mathbf{r}_2; \nu) = \langle \Pi_1^*(\mathbf{r}_1; \nu)\Pi_1(\mathbf{r}_2; \nu)\rangle, \tag{5.207}$$

$$P_{22}(\mathbf{r}_1, \mathbf{r}_2; \nu) = \langle \Pi_2^*(\mathbf{r}_1; \nu)\Pi(\mathbf{r}_2; \nu)\rangle, \tag{5.208}$$

$$P_{12}(\mathbf{r}_1, \mathbf{r}_2; \nu) = \Pi_1^*(\mathbf{r}_1; \nu)\Pi_2(\mathbf{r}_2; \nu)\rangle, \tag{5.209}$$

$$P_{21}(\mathbf{r}_1, \mathbf{r}_2; \nu) = \langle \Pi_2^*(\mathbf{r}_1; \nu)\Pi_1(\mathbf{r}_2; \nu)\rangle. \tag{5.210}$$

Only the first three need to be calculated, and the fourth follows by taking the conjugate complex of P_{12} and permuting $\mathbf{r}_1$ with $\mathbf{r}_2$. Equation (5.49) is substituted into eq. (5.158) to yield the following preliminary form for the dyad $\mathfrak{W}_{ee}$ in terms of the Debye potentials:

$$\mathfrak{W}_{ee}(\mathbf{r}_1, \mathbf{r}_2; \nu) = \left\langle \left[\nabla_1 \times \nabla_1 \times \mathbf{r}_1\Pi_1(\mathbf{r}_1) + \frac{ik}{g}\nabla_1 \times \mathbf{r}_1\Pi_2(\mathbf{r}_1) \right]^* \right.$$
$$\left. \left[\nabla_2 \times \nabla_2 \times \mathbf{r}_2\Pi_1(\mathbf{r}_2) + \frac{ik}{g}\nabla_2 \times \mathbf{r}_2\Pi_2(\mathbf{r}_2) \right] \right\rangle. \tag{5.211}$$

Expanding eq. (5.211) and substituting the defining statistical moments, eqs. (5.207)–(5.210), yield the expression

$$\begin{aligned}\mathfrak{W}_{ee}(\mathbf{r}_1, \mathbf{r}_2; \nu) = {} & (\nabla_1 \times \nabla_1 \times \mathbf{r}_1)(\nabla_2 \times \nabla_2 \times \mathbf{r}_2)P_{11}(\mathbf{r}_1, \mathbf{r}_2; \nu) \\ & + \left[\frac{k}{g}\right]^2 (\nabla_1 \times \mathbf{r}_1)(\nabla_2 \times \mathbf{r}_2)P_{22}(\mathbf{r}_1, \mathbf{r}_2; \nu) \\ & + i\frac{k}{g}(\nabla_1 \times \nabla_1 \times \mathbf{r}_1)(\nabla_2 \times \mathbf{r}_2)P_{12}(\mathbf{r}_1, \mathbf{r}_2; \nu) \\ & - i\frac{k}{g}(\nabla_1 \times \mathbf{r}_1)(\nabla_2 \times \nabla_2 \times \mathbf{r}_2)P_{21}(\mathbf{r}_1, \mathbf{r}_2; \nu).\end{aligned} \tag{5.212}$$

Equation (5.212) can be written in a more compact form using a combination of curl and the Λ operator in eq. (5.95), i.e.,

$$\begin{aligned}\mathfrak{W}_{ee}(\mathbf{r}_1, \mathbf{r}_2; \nu) = {} & (\nabla_1 \times \Lambda_1)(\nabla_1 \times \Lambda_2)P_{11}(\mathbf{r}_1, \mathbf{r}_2; \nu) + \left[\frac{k}{g}\right]^2 \Lambda_1\Lambda_2 P_{22}(\mathbf{r}_1, \mathbf{r}_2; \nu) \\ & + i\frac{k}{g}(\nabla_1 \times \mathbf{r}_1)\Lambda_2 P_{12}(\mathbf{r}_1, \mathbf{r}_2; \nu) - i\frac{k}{g}\Lambda_1(\nabla_2 \times \mathbf{r}_2)P_{21}(\mathbf{r}_1, \mathbf{r}_2; \nu).\end{aligned} \tag{5.213}$$

In a similar manner, by eqs. (5.50) and (5.159), we obtain the corresponding form for $\mathfrak{W}_{hh}$

$$\begin{aligned}\mathfrak{W}_{hh}(\mathbf{r}_1, \mathbf{r}_2; \nu) = {} & (\nabla_1 \times \Lambda_1)(\nabla_2 \times \Lambda_2)P_{22}(\mathbf{r}_1, \mathbf{r}_2; \nu) + (kg)^2\Lambda_1\Lambda_2 P_{11}(\mathbf{r}_1, \mathbf{r}_2; \nu) \\ & + ikg\Lambda_1(\nabla_2 \times \Lambda_2)P_{12}(\mathbf{r}_1, \mathbf{r}_2; \nu) \\ & - ikg\Lambda_2(\nabla_1 \times \Lambda_1)P_{21}(\mathbf{r}_1, \mathbf{r}_2; \nu).\end{aligned} \tag{5.214}$$

Finally, an expression for $\mathfrak{W}_{\text{eh}}$ is obtained by substituting eqs. (5.49) and (5.50) into eq. (5.172)

$$\mathfrak{W}_{\text{eh}}(\mathbf{r}_1, \mathbf{r}_2; \nu) = (\nabla_1 \times \Lambda_1)(\nabla_2 \times \Lambda_2)P_{12}(\mathbf{r}_1, \mathbf{r}_2; \nu) - k^2\Lambda_1\Lambda_2 P_{21}(\mathbf{r}_1, \mathbf{r}_2; \nu)$$
$$- i\frac{k}{g}\Lambda_1(\nabla_2 \times \Lambda_2)P_{22}(\mathbf{r}_1, \mathbf{r}_2; \nu)$$
$$- ikg(\nabla_1 \times \Lambda_1)\Lambda_2 P_{11}(\mathbf{r}_1, \mathbf{r}_2; \nu). \tag{5.215}$$

Equations (5.213)–(5.215) illustrate the statistical representation of second-order moments for the vector electromagnetic fields in terms of essentially three scalar cross-spectral densities. It is also clear from these equations that the formulation of all possible second-order moments for the vector fields is considerably simpler than we might expect by counting all permutations of the scalar field components. The mathematical properties of these cross-spectral densities formed by the Debye potentials will be studied in the following sections.

5.5.2. *Moments of Debye Potentials; Symmetrical Expansions*

The second-order moments in eqs. (5.207)–(5.210) are expressed in terms of the expansion coefficients $\{a_n^m\}$, $\{b_n^m\}$ for the electric and magnetic potentials, respectively, in the form

$$P_{11}(\mathbf{r}_1, \mathbf{r}_2) = \sum_{n,m}\sum_{n',m'} \langle a_n^{m*} a_{n'}^{m'} \rangle \Pi_n^{m*}(\mathbf{r}_1)\Pi_{n'}^{m'}(\mathbf{r}_2), \tag{5.216}$$

$$P_{22}(\mathbf{r}_1, \mathbf{r}_2) = \sum_{n,m}\sum_{n',m'} \langle b_n^{m*} b_{n'}^{m'} \rangle \Pi_n^{m*}(\mathbf{r}_1)\Pi_{n'}^{m'}(\mathbf{r}_2), \tag{5.217}$$

$$P_{12}(\mathbf{r}_1, \mathbf{r}_2) = \sum_{n,m}\sum_{n',m'} \langle a_n^{m*} b_{n'}^{m'} \rangle \Pi_n^{m*}(\mathbf{r}_1)\Pi_{n'}^{m'}(\mathbf{r}_2), \tag{5.218}$$

$$P_{21}(\mathbf{r}_1, \mathbf{r}_2) = \sum_{n,m}\sum_{n',m'} \langle b_n^{m*} a_{n'}^{m'} \rangle \Pi_n^{m*}(\mathbf{r}_1)\Pi_{n'}^{m'}(\mathbf{r}_2), \tag{5.219}$$

Here the explicit dependence of the correlations on ν has been omitted for brevity. In eqs. (5.216) and (5.217) we observe that the double summations are, in general, symmetric forms. It is interesting at the outset to consider an illustrative source distribution that results in a simple functional form for the statistical moments. As explained in Section 5.1.3 the expansion coefficients $\{a_n^m\}$, $\{b_n^m\}$ can be specified independently. Hence we may assert that these coefficients are chosen to be statistically uncorrelated, i.e.,

$$\langle a_n^{m*} b_{n'}^{m'} \rangle = \langle a_n^{m*} \rangle \langle b_{n'}^{m'} \rangle. \tag{5.220}$$

Moreover, we assert that the ensemble averages of $\{a_n^m\}$, $\{b_n^m\}$ have zero means such that

$$\langle a_n^m \rangle = \langle b_n^m \rangle = 0 \tag{5.221}$$

together with

$$\langle a_n^{m*} a_{n'}^{m'} \rangle = \langle a_n^{m*} a_n^m \rangle \delta_{nn'} \delta_{mm'}, \tag{5.222}$$

$$\langle b_n^{m*} b_{n'}^{m'} \rangle = \langle b_n^{m*} b_n^m \rangle \delta_{nn'} \delta_{mm'}. \tag{5.223}$$

A direct consequence of the present choice of statistics is the collapse of the double sums in eqs. (5.216)–(5.219) into a single sum to yield

$$P_{11}(\mathbf{r}_1, \mathbf{r}_2) = \sum_{n,m} \langle a_n^{m*} a_n^m \rangle \Pi_n^{m*}(\mathbf{r}_1) \Pi_n^m(\mathbf{r}_2), \tag{5.224}$$

$$P_{22}(\mathbf{r}_1, \mathbf{r}_2) = \sum_{n,m} \langle b_n^{m*} b_n^m \rangle \Pi_n^{m*}(\mathbf{r}_1) \Pi_n^m(\mathbf{r}_2), \tag{5.225}$$

$$P_{12}(\mathbf{r}_1, \mathbf{r}_2) = P_{21}(\mathbf{r}_1, \mathbf{r}_2) = 0. \tag{5.226}$$

It is clear from eqs. (5.224)–(5.226) that the dyadic cross-spectral densities, together with expressions derived from them, are significantly simplified when the expansion coefficients $\{a_n^m\}$, $\{b_n^m\}$ are chosen to be statistically independent. For example, the expression for the average far-zone Poynting vector, eq. (5.185), takes the form

$$\langle \mathbf{S}^{(\infty)}(\mathbf{r}; \nu) \rangle = \frac{1}{2gr^2} \mathbf{e}_r \sum_{n,m} [g^2 \langle |a_n^m|^2 \rangle + \langle |b_n^m|^2 \rangle] \left[\frac{m^2}{\sin^2 \theta} |Y_n^m|^2 + \left| \frac{\partial Y_n^m}{\partial \theta} \right|^2 \right]. \tag{5.227}$$

5.5.3. *Generalized Orthogonal Representation of the Moments*

In Section 5.5.1 we represented the dyadic cross-spectral densities of the electromagnetic field in terms of statistical moments of the Debye potentials. So far, we expressed the Debye potentials in terms of the expansions of Bouwkamp and Casimir [2], which emphasized the decomposition of a radiating source into separate multipole contributions. In this section we formulate the Debye potentials as solutions of inhomogeneous Helmholtz equations. An immediate advantage of this representation is the resulting simplified formalism for the statistical moments of the Debye potentials. In addition, this formulation extends the region of validity of eqs. (5.49) and (5.50) to include the entire source domain.

In this section for the inclusion of sources in eqs. (5.49) and (5.50), i.e., so that $\nabla \cdot \mathbf{E}(\mathbf{r}) = \rho(\mathbf{r})/\varepsilon$, we generalize eqs. (5.51) and (5.52). It has been shown by Nisbet [14], [15] that a suitable gauge transformation can convert eqs. (5.51) and (5.52) into inhomogeneous Helmholtz equations, namely,

$$(\nabla^2 + k^2)\Pi_1(\mathbf{r}) = -\rho_1(\mathbf{r}), \tag{5.228}$$

$$(\nabla^2 + k^2)\Pi_2(\mathbf{r}) = -\rho_2(\mathbf{r}). \tag{5.229}$$

Here the source terms are given by

$$\rho_1(\mathbf{r}) = -\frac{1}{r}\left[\frac{\mathbf{J}(\mathbf{r}) \cdot \mathbf{e}_r}{ikg} + \frac{\partial l(\mathbf{r})}{\partial r}\right], \tag{5.230}$$

$$\rho_2(\mathbf{r}) = ikg\frac{G(\mathbf{r})}{r}, \tag{5.231}$$

and the functions $G(\mathbf{r})$ and $l(\mathbf{r})$ are related to the source current distribution through the coupled differential equations

$$-\frac{1}{r\sin\theta}\frac{\partial G}{\partial\varphi}+\frac{1}{r}\frac{\partial l}{\partial\theta}=-\frac{\mathbf{J}(\mathbf{r})\cdot\mathbf{e}_\theta}{ik}, \tag{5.232}$$

$$\frac{1}{r}\frac{\partial G}{\partial\theta}+\frac{1}{r\sin\theta}\frac{\partial l}{\partial\varphi}=-\frac{\mathbf{J}(\mathbf{r})\cdot\mathbf{e}_\varphi}{ik}. \tag{5.233}$$

Equations (5.228) and (5.229) are valid throughout the whole space and, in particular, at all points inside the source domain. The general solution for the scalar potentials in these equations is given by

$$\Pi_j(\mathbf{r})=\int_{\text{sources}}\rho_j(\mathbf{r}')\frac{e^{ik|\mathbf{r}-\mathbf{r}'|}}{4\pi|\mathbf{r}-\mathbf{r}'|}\,d^3r' \qquad (j=1,2). \tag{5.234}$$

We can therefore use the Debye potentials $\Pi_1(\mathbf{r})$, $\Pi_2(\mathbf{r})$ derived from eq. (5.234), together with eqs. (5.49) and (5.50), to express the electromagnetic fields $\mathbf{E}$ and $\mathbf{H}$ both inside and outside the source domain.

When the source distribution is a stationary random process, both scalar source terms $\rho_j(\mathbf{r};\nu)\,(j=1,2)$, also represent stationary random processes and we may characterize them by the scalar cross-spectral density functions

$$W_{\rho_1}(\mathbf{r}_1,\mathbf{r}_2;\nu)=\langle\rho_1^*(\mathbf{r}_1;\nu)\rho_1(\mathbf{r}_2;\nu)\rangle, \tag{5.235}$$

$$W_{\rho_2}(\mathbf{r}_1,\mathbf{r}_2;\nu)=\langle\rho_2^*(\mathbf{r}_1;\nu)\rho_2(\mathbf{r}_2;\nu)\rangle. \tag{5.236}$$

It can be verified that both cross-spectral density functions are Hermitian-symmetric, i.e., that

$$W_{\rho_j}(\mathbf{r}_1,\mathbf{r}_2;\nu)=W_{\rho_j}^*(\mathbf{r}_2,\mathbf{r}_1;\nu) \qquad (j=1,2) \tag{5.237}$$

and they are also continuous functions of the position variables $\mathbf{r}_1$ and $\mathbf{r}_2$. Since we assume that both source functions $\rho_j(\mathbf{r};\nu)\,(j=1,2)$, are well behaved, then in every finite domain D the scalar cross-spectral densities are L_2 functions,[1] explicitly,

$$\int_D\int_D|W_{\rho_j}(\mathbf{r}_1,\mathbf{r}_2;\nu)|^2\,d^3r_1\,d^3r_2<\infty \qquad (j=1,2). \tag{5.238}$$

An additional property of the cross-spectral densities defined by eqs. (5.235) and (5.236) is that they are *nonnegative definite* functions (See Appendix A in Ref. [19]), which means they satisfy the inequality

$$\int_D\int_D W_{\rho_j}(\mathbf{r}_1,\mathbf{r}_2;\nu)f^*(\mathbf{r}_1)f(\mathbf{r}_2)\,d^3r_1\,d^3r_2\geq 0 \qquad (j=1,2) \tag{5.239}$$

with any function $f(\mathbf{r})$ which is square integrable over the domain D.

[1] For a definition of L_2 functions in this connection, see Tricomi [20].

The significance of properties of the cross-spectral densities just cited becomes evident when we consider either cross-spectral density as the kernel of a Fredholm integral equation

$$\int_D W_\rho(\mathbf{r}_1, \mathbf{r}_2; \nu)\Phi_n(\mathbf{r}_1; \nu)\, d^3r_1 = \lambda_n(\nu)\Phi_n(\mathbf{r}_2; \nu). \tag{5.240}$$

Here $\Phi_n(\mathbf{r}; \nu)$ are the eigenfunctions, $\lambda_n(\nu)$ are the eigenvalues of the kernel $W_\rho(\mathbf{r}_1, \mathbf{r}_2; \nu)$, and the index n has the same dimensionality of the space which is spanned by $\mathbf{r}_1$ and $\mathbf{r}_2$. Mercer's theorem states that when the kernel is a continuous, Hermitian-symmetric, nonnegative function which does not vanish identically, it may be represented in terms of the eigenvalues and eigenfunctions of eq. (5.240) in the form[1]

$$W_\rho(\mathbf{r}_1, \mathbf{r}_2; \nu) = \sum_{n=0}^{\infty} \lambda_n(\nu)\Phi_n^*(\mathbf{r}_1; \nu)\Phi_n(\mathbf{r}_2; \nu) \tag{5.241}$$

and this series converges absolutely and uniformly for all $\mathbf{r}_1$ and $\mathbf{r}_2$ in the source domain.

The theory of integral equations supplements the Mercer expansion with a number of useful theorems regarding the eigenvalues and eigenvectors appearing in eq. (5.240). It can be shown that all eigenvalues are real and nonnegative. Furthermore, if the kernel does not vanish identically, there exists at least one nonzero eigenvalue. The eigenfunctions of the kernel that correspond to different eigenvalues are orthogonal, and eigenfunctions that correspond to a degenerate eigenvalue can be made mutually orthonormal. The set $\Phi_n(\mathbf{r}; \nu)$ may or may not be complete.

For statistical applications the Mercer expansion, eq. (5.241), is also useful in representing an ensemble of functions $U_\rho(\mathbf{r}; \nu)$ as a linear superposition of eigenfunctions, i.e.,

$$U_\rho(\mathbf{r}; \nu) = \sum_n u_n \Phi_n(\mathbf{r}; \nu), \tag{5.242}$$

where u_n are expansion coefficients that may depend on frequency ν. It can be shown (see, also, Lumley [23]) that the coefficients u_n can always be chosen uncorrelated; in particular, we may set

$$\langle u_n^* u_m \rangle = \lambda_n \delta_{nm}. \tag{5.243}$$

We stress that this choice of coefficients can be made for all current distributions and does not have the physical implications of Section 5.5.2. With this choice of coefficients, it is easy to see by eqs. (5.242) and (5.243) that the cross-spectral densities W_ρ are expressed in terms of the ensemble average [19] and [22]:

$$W_\rho(\mathbf{r}_1, \mathbf{r}_2; \nu) = \langle U_\rho^*(\mathbf{r}_1; \nu) U_\rho(\mathbf{r}_2; \nu) \rangle. \tag{5.244}$$

Equations (5.241)–(5.243) show that it is possible to construct an ensemble of uncorrelated functions which represent the cross-spectral density in a

[1] Smithies [21, p. 127].

simple form. Physically, we may regard eqs. (5.242)–(5.244) as giving an expression for the source distribution in terms of a sum of uncorrelated spatial distributions that yield an identical cross-spectral density.

We shall now apply the preceding formulation to obtain a representation for second-order moments of the Debye potentials. Starting from the expansion, eq. (5.241), we introduce the sets of eigenvectors and eigenfunctions

$$W_{\rho_j}(\mathbf{r}_1, \mathbf{r}_2; \nu) = \sum_n \lambda_n^{(j)}(\nu)\Phi_n^{(j)*}(\mathbf{r}_1; \nu)\Phi_n^{(j)}(\mathbf{r}_2; \nu) \qquad (j = 1, 2). \qquad (5.245)$$

By substituting eq. (5.245) into eq. (5.234), and using eq. (5.207), we obtain at once

$$P_{11}(\mathbf{r}_1, \mathbf{r}_2; \nu) = \sum_n \lambda_n^{(1)}(\nu)\Psi_n^{(1)*}(\mathbf{r}_1; \nu)\Psi_n^{(1)}(\mathbf{r}_2; \nu), \qquad (5.246)$$

and

$$P_{22}(\mathbf{r}_1, \mathbf{r}_2; \nu) = \sum_n \lambda_n^{(2)}(\nu)\Psi_n^{(2)*}(\mathbf{r}_1; \nu)\Psi_n^{(2)}(\mathbf{r}_2; \nu). \qquad (5.247)$$

Here the expansion functions Ψ_n are given by

$$\Psi_n^{(j)}(\mathbf{r}; \nu) = \int \Phi_n^{(j)}(\mathbf{r}'; \nu)\frac{e^{ik|\mathbf{r}-\mathbf{r}'|}}{|\mathbf{r} - \mathbf{r}'|}\, d^3r' \qquad (j = 1, 2). \qquad (5.248)$$

It is emphasized that, in general, the set $\{\Psi_n(\mathbf{r}; \nu)\}$ is not orthogonal, and the expansions in eqs. (5.246) and (5.247) are not the Mercer expansions for P_{11} and P_{22}. However, if we compare the expressions in eqs. (5.216) and (5.217) with eqs. (5.246) and (5.247), respectively, we recognize the advantage of the later form, because the summation is performed over three indices rather than four, without resorting to the assumptions associated with eqs. (5.216) and (5.217). In addition, expansions (5.246) and (5.247) are valid throughout the whole space, whereas eqs. (5.216) and (5.217) may take a different form according to whether the observation point is inside or outside the sphere enclosing the source distribution.

5.6. Summary

The general theme of this chapter is to connect the subjects of multipole expansions and statistical electromagnetic theory. In this summary we include a section-by-section review so that the reader may more easily gain insight as to the organization of the material.

In Section 5.1, after a brief review of the space-frequency representation of Maxwell's equations followed by a statement of a uniqueness theorem in spherical coordinates, we define the Debye potentials $\Pi_1(\mathbf{r}, \nu)$, $\Pi_2(\mathbf{r}, \nu)$ corresponding to E- and H-type waves, respectively. From the uniqueness theorem it is clear that the radial field components are adequate to specify completely the radiation field, hence it is natural to proceed with the calculation of the radial components of the electric and magnetic fields as in eqs. (5.67) and (5.68), or as evaluated explicitly in terms of the source currents in eqs. (5.73) and (5.74). The multipole expansion coefficients, a_n^m, b_n^m, are given in terms of the source current $\mathbf{J}(\mathbf{r})$ by eqs. (5.75) and (5.76). Table 5.1 contains a complete solution for the radiation field outside the source domain. This formulation of the Debye potentials follows the presentation in the literature by Bouwkamp and Casimir [2] as well as that in the treatise by Papas [3].

In Section 5.2 we derive expressions for the electric and magnetic fields in terms of vector spherical harmonics, as in eqs. (5.108) and (5.109). It is interesting and useful to expand the source current $\mathbf{J}(\mathbf{r})$ into vector spherical harmonics as defined in eqs. (5.116) and (5.117), where we introduce the spherical projection functions $f_n^m(r)$, $g_n^m(r)$, $h_n^m(r)$. Using this representation for the current $\mathbf{J}(\mathbf{r})$ we express the multipole expansion coefficients a_n^m and b_n^m in terms of Bessel transforms of the spherical projection functions, eqs. (5.128) and (5.129).

In Section 5.3 second-order statistics of the electromagnetic field are formulated. After a breif review of dyadic operators we describe the energy density of the field in terms of two cross-spectral dyadics $\mathfrak{W}_{\mathrm{ee}}$ and $\mathfrak{W}_{\mathrm{hh}}$ defined in eqs. (5.158) and (5.159), respectively. An explicit expression for $\mathfrak{W}_{\mathrm{ee}}$ in terms of vector spherical harmonics is given in eq. (5.169). As we might expect the second-order moments of the fields are functionally dependent on the expected values of products of multipole expansion coefficients. Hence it is convenient to define the coupling coefficients $A(\mathbf{N})$, $B(\mathbf{N})$, $C(\mathbf{N})$ in eqs. (5.164)–(5.166).

The correspondence of dyadic cross-spectral densities to measurable quantities in the far zone is treated in Section 5.4. We derive expressions in terms of second-order coupling coefficients for the Poynting vector, the dyadic cross-spectral density of the electric field, the Stokes parameters, and the total radiated power in eqs. (5.185), (5.188), (5.199)–(5.202), and (206), respectively.

In Section 5.5 we consider second-order moments formed by the ensemble averages of the Debye potentials Π_1 and Π_2. These moments give a description of the statistics of the vector electromagnetic field in terms of three scalar cross-spectral densities. Symmetrical expansions for the second-order moments are written for the case when the coupling coefficients are statistically uncorrelated.

In Section 5.5.3 we introduce new symmetrical expansions, eqs. (5.246) and (5.247), for moments of the Debye potentials. The expansion functions are established from a consideration of source densities that are introduced in the differential equations for the Debye potentials.

References

[1] P. Debye (1909), *Ann. Phys. (Leip.)*, **30**, 57.
[2] C.J. Bouwkamp and H.B.G. Casimir (1954), *Physica*, **20**, 539.

[3] C.H. Papas (1965), *Theory of Electromagnetic Wave Propagation*, McGraw-Hill, New York.
[4] J.M. Blatt and V.F. Weisskopf (1952), *Theoretical Nuclear Physics*. See, in particular, Appendix B.
[5] E.L. Hill (1954), *Amer. J. Phys.*, **22**, 211.
[6] W.H. Carter and E. Wolf (1987), *Phys. Rev.*, A, **36**, 1258.
[7] W.B. Davenport and W.L. Root (1958), *An Introduction to the Theory of Random Signals and Noise*, McGraw-Hill, New York.
[8] E.T. Copson (1978), *An Introduction to the Theory of Functions of a Complex Variable*, Oxford University Press, Oxford.
[9] M. Kerker (1969), *The Scattering of Light and Other Electromagnetic Radiation*, Academic Press, New York.
[10] E.W. Hobson (1939), *Spherical and Ellipsoidal Harmonics*, Wiley, New York.
[11] E.L. Hill and R. Landshoff (1938), *Rev. Mod. Phys.*, **10**, 87.
[12] M. Abramowitz and I.A. Stegun (1965), *Handbook of Mathematical Functions*, Dover, New York, Sec. 10.1.
[13] A. Erdelyi (1956), *Asymptotic Expansions*, Dover, New York, Sec. 3.6.
[14] A. Nisbet (1955), *Physica*, **21**, 799.
[15] A. Nisbet (1955), *Proc. Roy. Soc. London*, **A 231**, 250.
[16] C.E. Weatherburn (1949), *Advanced Vector Analysis*, G. Bell & Sons, London.
[17] W.R. Smythe (1950), *Static and Dynamic Electricity*, McGraw-Hill, New York.
[18] M. Born and E. Wolf (1980), *Principles of Optics*, 6th ed., Pergamon Press, Oxford.
[19] E. Wolf (1982), *J. Opt. Soc. Amer.*, **72**, 343.
[20] F.G. Tricomi (1985), *Integral Equations*, Dover, New York.
[21] F. Smithies (1958), *Integral Equations*, Cambridge University Press, Cambridge.
[22] E. Wolf (1986), *J. Opt. Soc. Amer.*, *A* **3**, 76–85.
[23] J.L. Lumley (1970), *Stochastic Tools in Turbulence*, Academic Press, New York.

6
On Fractal Electrodynamics

DWIGHT L. JAGGARD*

> Clouds are not spheres, mountains are not cones,
> coastlines are not circles and bark is not smooth,
> nor does lightning travel in a straight line.
>
> B.B. Mandelbrot
> *The Fractal Geometry of Nature*

6.1. Prologue

Geometry has played a crucial role in the formulation and understanding of mathematical methods in electromagnetic theory. It was the pioneering work of James Clerk Maxwell and Lord Rayleigh in the nineteenth century that provided the first glimpse into the world of wave interactions with regular Euclidean structures, and led to the solution of canonical scattering problems typical of the early and mid-twentieth century. These problems involved scattering by edges, spheres, and infinite cylinders and the guiding of waves by planar and cylindrical structures of regular cross section. More recently, wave interactions with wedges, ellipsoids, cones, cubes, truncated cylinders, and small groups of these objects have been described using both experimental and numerical methods to attack increasingly complex geometry. In most cases, these more complex objects are either descriptive of simple man-made objects or are elementary approximations to more complicated objects which occur in nature. However, models which accurately display the multitude of scale lengths typical of naturally occurring structures are not well suited to Euclidean description and so fall outside the realm of problems noted above. It is exactly these structures, sometimes denoted variegated, ramified, spiky, tortuous, pathological, wiggly, or wildly irregular, and their interactions with electromagnetic waves which are of interest here. We seek connections between the geometrical parameters or descriptors of these structures and meaningful physical quantities which capture their electromagnetic qualities. Applications of this work include the characterization of electromagnetic or other waves with geophysical structures and ocean surfaces, radiation from lightning and coronas, scattering and characterization of rough fibers, propagation in random media, and scattering from rough surfaces and the synthesis of a host of structures with specified scattering properties.

---

* Moore School of Electrical Engineering, University of Pennsylvania, Philadelphia, PA 19104-6390, USA.

Two challenges immediately present themselves. The first concerns the modeling of realizable multiple-scale structures while the second concerns the computation of their interaction with electromagnetic waves. Here we treat selected aspects of both challenges by a careful blend of fractal geometry and Maxwell's theory of electromagnetism which define a research area we call *fractal electrodynamics.* Although the treatment of fractal electrodynamics given here is personal and not intended to be exhaustive, this chapter is self-contained. Notions of fractal geometry and the introduction of physical fractals and their models are introduced and followed by illustrative applications developed recently at the University of Pennsylvania.

6.2. Almost Everything You Wanted to Know About Fractals

6.2.1. *What are Fractals?*

Although Euclidean or classical geometry was formulated over two thousand years ago, fractal geometry is a relative newcomer. Fractal geometry finds its roots in the work of the late nineteenth and early twentieth century mathematicians. Its expression, however, is still more recent. In the mid-1970s, Mandelbrot introduced the fractal concept and coined the term *fractal*[1] [1]. This allowed us to describe complex objects or phenomena, from electron orbits to galaxies to stock market fluctuations, with the same ease that we describe children's blocks. In addition, this concept quantified phenomena such as chaos and dynamical stability and supplied the tools necessary for research in diverse areas such as scene generation and analysis, image compression algorithms, lightning propagation, fluid flow in porous materials, and the propagation of epidemics through specified populations.

In the spirit of its originator, fractals are not rigorously defined, but apply to a variety of wiggly spiky structures. However, it will be useful to concentrate our attention here on objects that are invariant under change of scale and invariant under displacement. That is, we consider objects which are self-similar and homogeneous, at least in the mean. As will become apparent, the flexible fractal concept provides a framework for considering both order and disorder and appropriate mixtures thereof, and can apply to both random and deterministic structures.

Self-similar objects possess structure on all scales as shown schematically in Figure 6.1. That is, each time a portion of a fractal structure is magnified, it reveals another layer of finer detail which is similar, in the mean, to the unmagnified image.

Fractals characterized by scale invariance possess characteristics not com-

[1] *Fractal*, after the latin *fractus*, means broken and refers to irregular fragments.

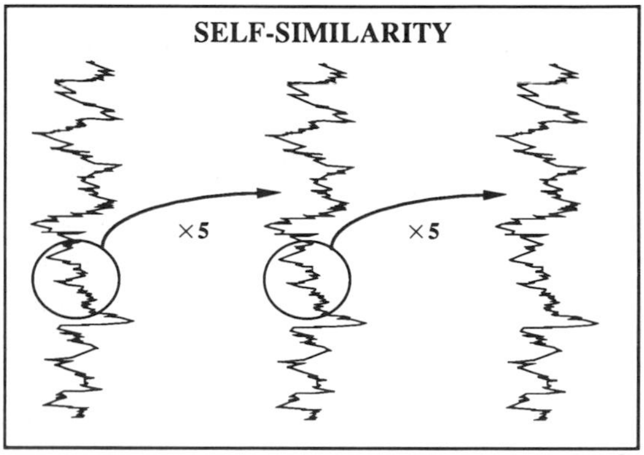

FIGURE 6.1. The self-similarity of fractals is revealed by the increasingly fine structure which becomes apparent with increased resolution. Here each portion of an irregular fractal curve, when enlarged, appears in the form of the original structure, at least in the mean

mon to Euclidean geometry. First, as we examine the finer and finer structure of a fractal, it becomes apparent that the derivatives of these structures become unbounded to maintain self-similarity. Second, it appears that it may be difficult to specify many fractals through formulas because of this increasingly fine structure. Therefore, it is expected that fractal construction is more likely to be specified by iteration than by explicit mathematical expressions. Third, statistical fractals keep in balance long-range order and short-range disorder. Finally, the usual concept of dimension has to be revisited since one-dimensional fractal curves, such as those of Figure 6.1, tend to partially fill-up areas and so occupy more space than is usual for one-dimensional lines but less than that occupied by two-dimensional surfaces. Likewise, roughened fractal surfaces, such as the fractal mountains displayed in Figure 6.2, tend to

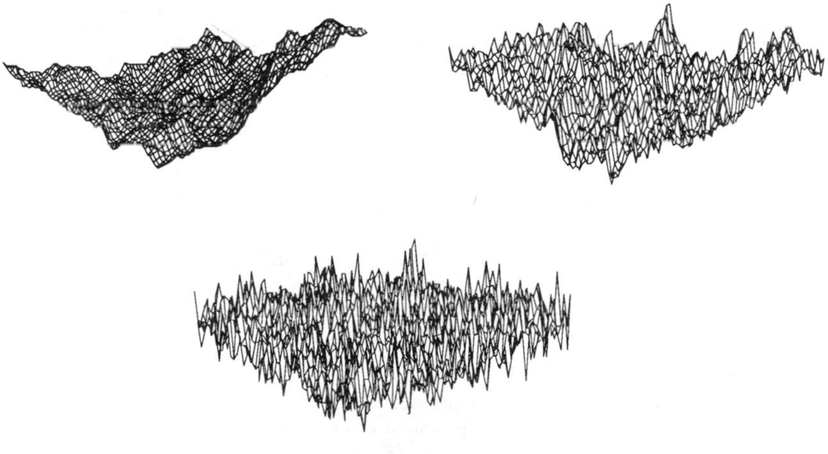

FIGURE 6.2. Gently undulating and rugged fractal mountains which partially fill a three-dimensional volume. The fractal dimension of these surfaces is bounded by the integers two and three.

fill a volume partially and so take up more space than is typical for usual two-dimensional planes but less than that taken up by three-dimensional volumes.

6.2.2. *What Is the Fractal Dimension?*

From the discussion and the figures above, it is clear that the geometrical concept of *dimension* requires modification when applied to fractals. That is, fractals seem to take up more space than their Euclidean allocation. We also note that although we have described fractals and their attributes, we have not yet quantified a measure of their fragmentation or roughness. For these purposes, a geometrically based fractional dimension is well suited and will be referred to here as the *fractal dimension* D.[1] This is counterdistinction to the geometrical or Euclidean dimension d which is restricted to integer values.

The concept of the dimension and its extension to fractals can be understood by the following heuristic argument [3]. Consider the measurement of the line segment with length L shown at the top of Figure 6.3. If we use a one-dimensional yardstick of length ε, the total number N of yardsticks contained in the line segment is simply

$$N(\varepsilon) = \left(\frac{L}{\varepsilon}\right)^1. \tag{6.1}$$

Similarly, consider the measurement of the area $A\ (= L^2)$ of the square shown in the middle of Figure 6.3. If we use a two-dimensional square yardstick of side ε, the total number N of yardsticks contained in the square is

$$N(\varepsilon) = \left(\frac{L}{\varepsilon}\right)^2. \tag{6.2}$$

Likewise, consider the measurement of the volume $V\ (= L^3)$ of the cube shown at the bottom of Figure 6.3. If we use a three-dimensional cubical yardstick of side ε, the total number N of yardsticks contained in the cube is

$$N(\varepsilon) = \left(\frac{L}{\varepsilon}\right)^3. \tag{6.3}$$

In each of the relations (6.1)–(6.3), the exponent represents the Euclidean dimension d of the problem under consideration.

This concept of dimension can be expanded by defining a generalized fractal dimension D through the relation of yardstick size ε and number of yardsticks

[1] Note that many fractal dimensions can be defined depending on the application [2]. That is, a structure may have certain geometrical properties giving rise to a geometrically defined fractal dimension, but physical characteristics of the structure such as scattering may posses other more useful characteristic fractal dimension. In this chapter we are primarily interested in geometrically based fractal dimensions found by the disk covering method (discussed later) and will relate this fractal dimension of an object to its wave interaction properties.

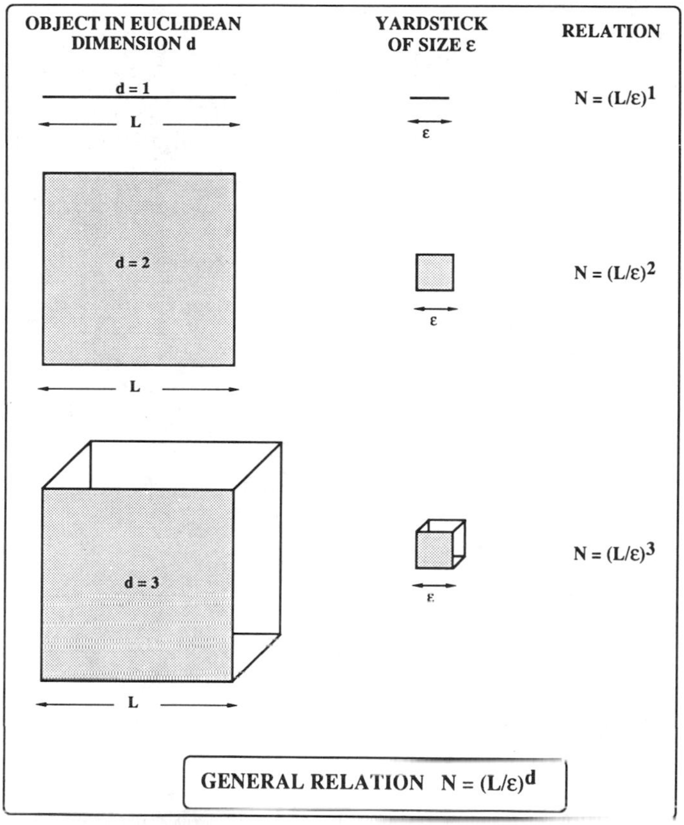

FIGURE 6.3. Use of a yardstick to obtain the dimension of Euclidean objects. Here N is the number of yardstick lengths ε in each object of side L.

N contained in an arbitrary object as

$$N(\varepsilon) \equiv C\varepsilon^{-D}, \tag{6.4}$$

where the appropriate constant C is dependent on the Euclidean dimension in which the object is embedded. Relation (6.4) is a generalization of (6.1)–(6.3) since D becomes d for simple Euclidean objects where the dimension is an integer, but may take on fractional values for other more pathological objects. Rearranging this expression yields the desired dimension directly as

$$D \equiv \frac{d[\ln N(\varepsilon)]}{d[\ln(1/\varepsilon)]} \xrightarrow[D\text{ constant}]{} \frac{[\ln N(\varepsilon)] - [\ln C]}{[\ln(1/\varepsilon)]} \xrightarrow[\lim \varepsilon\to 0]{} \frac{[\ln N(\varepsilon)]}{[\ln(1/\varepsilon)]}. \tag{6.5}$$

The fractal dimension definition (6.5) is the basis for the *disk covering method* of determining the fractal dimension. In this method, we choose appropriate disks of size ε (e.g., line segments of length ε for finding the fractal dimension of curves, circles of radius ε for finding the fractal dimension of surfaces or spheres of radius ε for finding the fractal dimension of solids) to cover or fill the object. As the yardstick size ε is varied and the number of disks N is determined, use of (6.5) yields the fractal dimension. This method is particularly suitable for discretized images, scenes or photomicrographs of small aggregates. In this case, these two-dimensional images are analyzed and classified with square disks to obtain the same result. In this form, the method is often referred to as the *box counting method.*

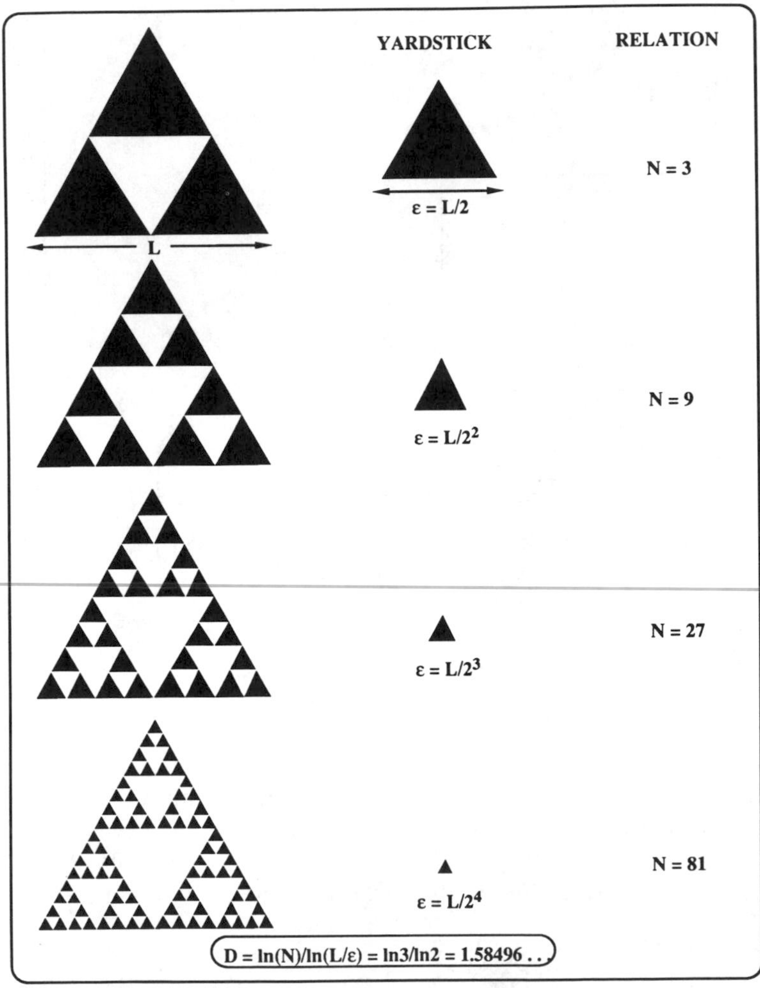

FIGURE 6.4. Determining the fractal dimension D of a Sierpiński gasket using the disk covering method with triangular yardsticks and various magnifications or resolution. As the magnification increases (top to bottom) its increasingly fine structure is apparent. Here $D = 1.58496\ldots.$

A closely related technique, the *coastline method*, comes from the problem of measuring rough fractal perimeters such as national boundaries [1]. Here the perimeter or boundary length L can become unbounded as the measuring yardstick size ε becomes smaller since the decreasing yardstick takes into account the fine detail of the structure. Since this length is the product of N and ε, we find immediately that

$$L(\varepsilon) = C\varepsilon^{1-D}. \tag{6.6}$$

Generalizations to higher Euclidean dimension yield similar results for areas and volumes so that we find

$$L_d(\varepsilon) = C\varepsilon^{d-D} \qquad (d = 1, 2, 3), \tag{6.7}$$

where L_1 [$= L(\varepsilon)$] is the length given in (6.6), L_2 [$= A(\varepsilon)$] is the desired area, or L_3 [$= V(\varepsilon)$] is the desired volume and C is an appropriate constant.

As an example of the fractal dimension D defined by the disk covering method (6.5), examine a standard non-Euclidean object, the Sierpiński gasket whose construction is given in the next section. Here it is most convenient to use a triangular yardstick for determining the fractal dimension of the Sierpiński gasket as shown in Figure 6.4. As the resolution of the observer increases or, correspondingly, as the yardstick decreases in length, the increasingly fine structure of the gasket is observed and measured (top to bottom of Figure 6.4). Using the fundamental relation (6.5), the fractal dimension is found as $D = \ln(N)/\ln(L/\varepsilon) = \ln(3)/\ln(2) = 1.58496\ldots$. The result is physically appealing in that the Sierpiński gasket does not completely cover an area, therefore its dimension must be less than two, while it certainly occupies more space than does a line, therefore its dimension must be greater than unity. The self-similarity of this gasket is clear in that each portion of the gasket, when magnified, looks like the original entire gasket.

6.2.3. *How Can I Become a Fractal Mechanic?*

If self-similarity becomes the major attribute of fractals, their geometrical construction becomes straightforward for certain families of stylized and regular fractals. Formalizing the stages of growth in Figure 6.5, we start with an initiator and a generator. Each stage of fractal growth is found by applying the generator, or its scaled replica, to the initiator. The initiator governs the gross shape of the fractal structure while the generator provides the detailed structure and ensures self-similarity and long-range correlation. At the top of Figure 6.5 is shown an initiator (left) composed of an equilateral triangle of unit length and a single generator (right) which is an inverted triangle of half-length. The generator excises the region where it acts on the initiator. Repeated scaling and application of the generator yields a mathematical fractal which becomes a diaphanous veil. The first four stages of growth are shown here.

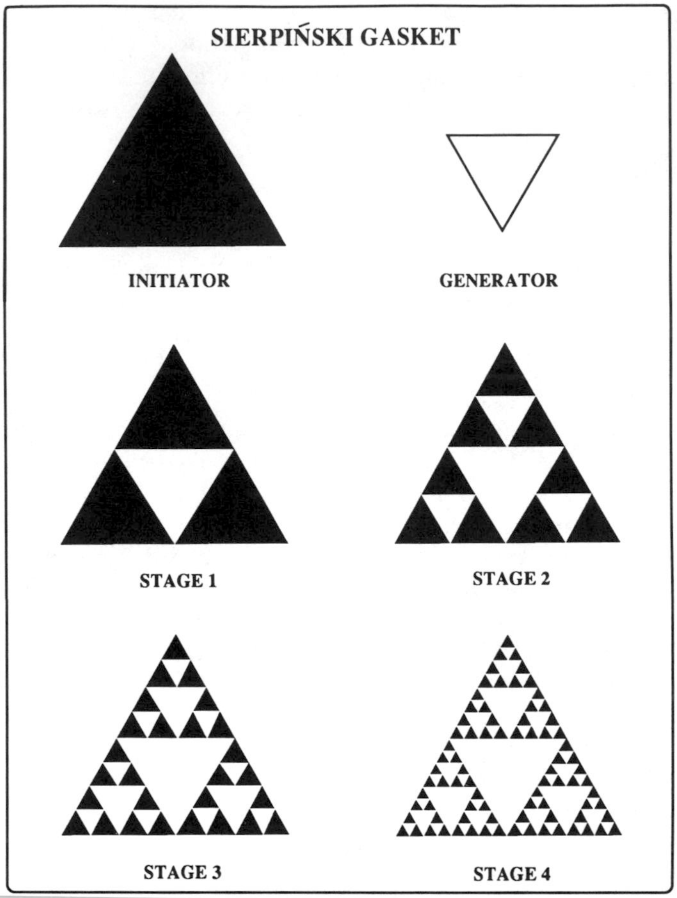

FIGURE 6.5. Growth of a Sierpiński gasket made from the repeated action of a scaled version of a generator on an initiator. Carried on indefinitely, this action produces a diaphanous veil. Three-dimensional analogues of this produce a Sierpiński sponge.

The growth of two additional well-known deterministic fractal structures, the Von Koch snowflake and the Cantor bar are shown in Figure 6.6. The Von Koch snowflake is a stylized model for rough boundaries and is characterized by a finite area but unbounded perimeter as the growth process progresses. This snowflake is formed by a triangular initiator and generator, the former of unit length and the latter of one-third length. The generator and its scaled replicas are applied to the midpoints of initiator and the previous stage of growth. Using the method of the previous subsection, the fractal dimension of its perimeter is found to be $D = \ln(4)/\ln(3) = 1.26186\ldots$.

The Cantor bar is formed by a line segment whose middle third is repeatedly removed. In this case the initiator is defined as a line segment of unit length

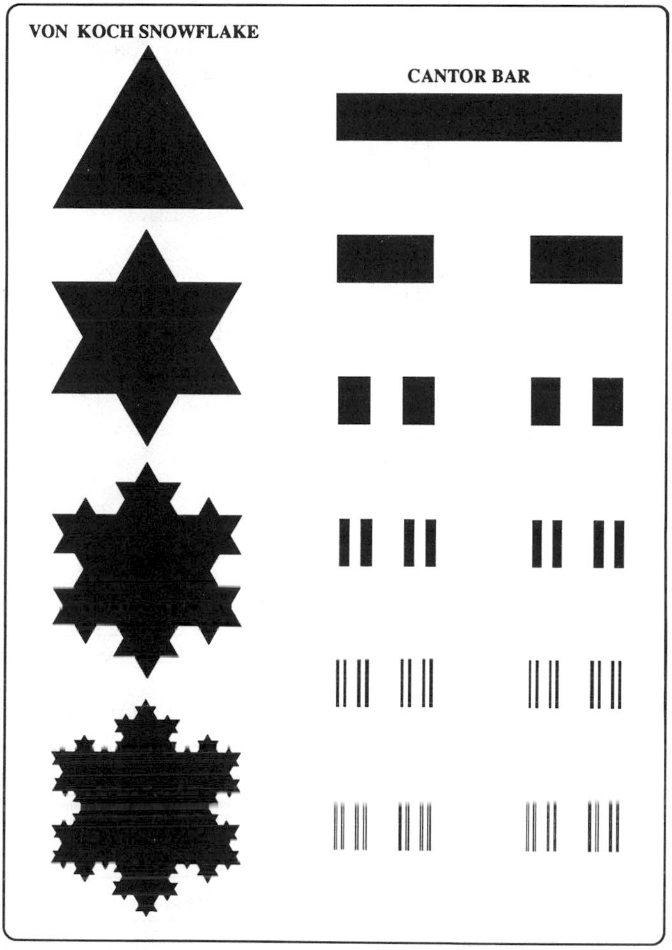

FIGURE 6.6. Growth of the Von Koch snowflake ($D = 1.26186\ldots$) and Cantor bar/dust (for Cantor dust $D = 0.630930\ldots$) fractals. The initiator for the snowflat is an equilateral triangle of unit length while the generator is one of one-third length. The initiator for the bar is a line segment of unit length and its generator is an excising line segment of length one-third. A Cantor dust is formed from a Cantor bar in the limit as the height of the bar becomes vanishingly small.

and the generator is defined as an excising line segment of length one-third. The bar is formed by repeated application of the generator and its scaled replica to the middle third of the initiator or the previous stage of growth. When the thickness of the Cantor bar becomes vanishingly small, the resultant fractal becomes Cantor dust that can be used to model the behavior of sparse intergalactic dust. Cantor bars and dust also appear in the distribution of bandgaps in almost-periodic structures. The fractal dimension of a Cantor dust is $D = \ln(2)/\ln(3) = 0.630930\ldots$. which is less than unity as expected. Finally, we note that when successive stages of a Cantor bar are stacked upon each other, we can form a Cantor bar surface which models rough but regular interfaces.

The structures described to this point are deterministic closed-boundary structures. Open tree-like structures, both deterministic and statistical, can also be defined using an extension of the previously described initiator–generator construction scheme. Here, for random fractal structures, instead of using a single generator, use several generators, one of which can be chosen (statistically) at each growth point and each stage of growth. An example of a deterministic fractal tree (left) and its statistical counterpart (right) is given in Figure 6.7. Such fractal structures can be used to model or synthesize diverse objects from vegetation to large networks [4] or antenna arrays [5].

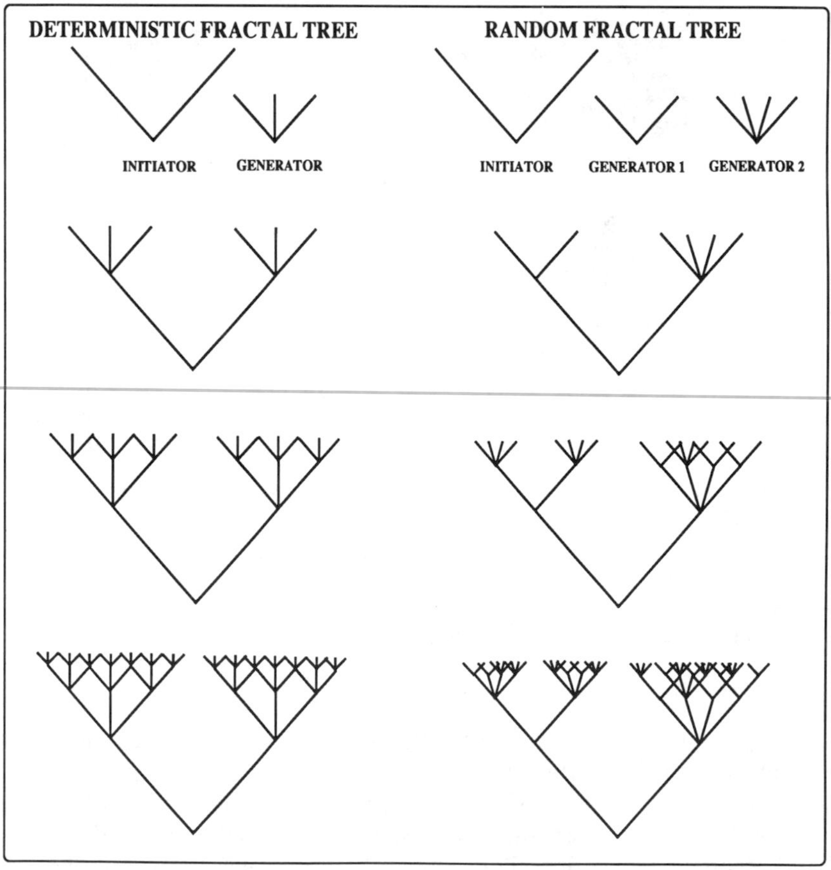

FIGURE 6.7. Fractal tree growth (top to bottom) with overlapping branches, both deterministic (left) and random (right).

TABLE 6.1. Heuristic summary of fractal and traditional or Euclidean geometrical attributes.

Fractal geometry	Euclidean geometry
Form given by rule (recipe)	Form given by formula
Structure on all scales	One or several scale lengths
Dilation symmetry (self-similarity)	No self-similarity
Fractional dimension possible	Integer dimensions
Long-range correlation	Variable correlation
Described as ramified, variegated, spiky	Described as regular
Non-differentiable boundaries	Differentiable boundaries

It is useful to contrast and compare the differences between classical or Euclidean geometry and fractal geometry. In Table 6.1 is a brief heuristic summary based on the concepts presented in this section. In keeping with the original fractal concept which is somewhat fuzzy, this table indicates typical characteristics which, although not universal, provide a useful guide in differentiating between the old and new geometries.

6.3. Physical and Mathematical Fractals

6.3.1. *The Need for Physical Fractals and Their Models*

From a physical point of view, fractals are often the result of regular but nonperiodic forces which give rise to complex structures through repetitive actions. Just as a coastline is formed through the repetitive action of wind and waves, its geometry is the result of a careful balance between periodic and aperiodic processes. This gives rise to long-range order in random fractal structures despite their short-range disorder. In this sense, natural fractal representations are neither the usual periodic functions of mathematical physics nor the random functions or correlations of statistical mechanics and signal processing. However, the stylized fractals of the previous section do not do justice to the structures observed in nature. This brings us to the connection between the previous constructions and fractals which are suitable for everyday description and electromagnetic analysis. This section presents the mathematical tools necessary for fractal electromagnetics.

The fractals of the previous section are *mathematical fractals* which are defined by continuous iteration. Here we call fractals which correspond to naturally occurring objects or physical processes *physical fractals* as opposed to these *mathematical fractals*. While mathematical fractals have self-similarity on all scales, we note that naturally occurring shapes such as geological structures, sea surfaces, vegetation, turbulence, and lightning all have characteristic inner and outer scale lengths. In other words, these structures exhibit fractal characteristics over a particular regime of scale lengths.

TABLE 6.2. Heuristic comparison of mathematical and physical fractals.

Mathematical fractals	Physical Fractals
Fractal on all scales	Fractal over a regime of scale sizes
Self-similar on all scales	Self-similar over a regime of scale sizes
Not characterized by Euclidean geometry	May be characterized by Euclidean geometry outside of fractal regime
Nondifferentiable boundaries	Boundaries with large but bounded derivatives

Even our representations of mathematical fractals have limitations in scale length which are bounded by page size and printing press resolution. Therefore, it is necessary to introduce models suitable for physical fractals to complement the models of mathematical fractals. These physical fractals possess a fractal regime in which mathematical fractals provide appropriate approximations. However, these physical fractals may also possess nonfractal regimes in which their characteristics are adequately specified through traditional means. As an example, we can consider a modest tree in which patterns of veins in leaves imitate patterns of twigs which resemble those of branches which in turn are like those of major trunks. Thus, within the scale range of centimeters to several meters, the tree appears fractal. However, for scales on the order of millimeters or less or for scales on the order of several tens of meters or more, the tree is not multiscaled and hence is not fractal.

In Table 6.2 we summarize some of the differences between mathematical and physical fractals and in the next section a sample model for physical fractals is given.

6.3.2. *Bandlimited Fractal Functions*

It is useful to introduce the notion of *bandlimited fractals* as models for physical fractals. By appropriately limiting the spatial frequencies of these fractal objects, we can model and analyze the fractal characteristics of objects over some finite spatial scale while ignoring, or treating by traditional means, the phenomena of interest in the nonfractal regime of scale lengths. From a practical point of view, this has the very important advantage of limiting or bounding the derivatives of such models so that for computational purposes common differential operators can operate on the characteristic functions.

Intuitively, we might obtain the fine structure present in a fractal function by simply adding together a series of increasingly oscillatory functions. A candidate model which is self-similar (at discrete magnifications) is the Weierstrass function introduced to be everywhere continuous but nowhere differentiable. This function and several variations have been investigated previously for applications to mathematical fractals [1], [6].

Here, in keeping with our desire to create bandlimited fractals to model physical fractals, we appropriately truncate a variant of the Weierstrass function and so introduce the *bandlimited Weierstrass function* $W(x)$ defined by the relation [7]

$$W(x) = \eta \frac{\sqrt{2}[1 - b^{(2D-4)}]^{1/2}}{[b^{(2D-4)N_1} - b^{(2D-4)(N_2+1)}]^{1/2}} \sum_{n=N_1}^{N_2} b^{n(D-2)} \cos(2\pi s b^n x + \theta_n) \quad (6.8)$$

as a function of the coordinate x. Here sb^{N_1} $(b > 1)$ is the fundamental spatial frequency, D is denoted the fractal dimension of this function and can take on values ranging from 1 to 2, s is a scaling factor, and θ_n are specified phases, often taken to be a random distribution on $[0, 2\pi]$. The number of tones or spectral lines is given by N $(= N_2 - N_1 + 1)$ and η^2 is the variance of this function. In the limit as the number of tones approaches infinity ($N_1 \to -\infty$ and $N_2 \to \infty$) and all phase terms are taken as zero ($\theta_n = 0$), eq. (6.8) satisfies the scaling or self-similarity relation $W(\beta x) = \beta^{(2-D)} W(x)$ where $\beta = b^{-m}$, m being an integer. For a limited number of tones, this function is fractal for scale sizes ranging from an inner scale $(sb^{N_2})^{-1}$ to an outer scale $(sb^{N_1})^{-1}$ and is self-similar in this regime.

The question of whether the parameter D of this function is the same fractal dimension found by the disk covering or coastline method has apparently not yet been proved. However, it is not an unreasonable assumption and in the limit as $N \to \infty$, it has been shown that D becomes an upper bound to the fractal dimension [8]. Here, as done elsewhere in the limit of an infinite number of tones [1], [6], D of relation (6.8) will be referred to as the fractal dimension.

The bandlimited Weierstrass function (6.8) is a member of the family of almost-periodic functions described by Besicovitch [9] and so possesses an autocorrelation which is again an almost-periodic function. From first principles, it can be shown that the autocorrelation of the bandlimited Weierstrass function with fractal dimension D is a bandlimited Wierstrass function of fractal dimension $2(D - 1)$ and that the maximum value of the autocorrelation is the variance η^2.

We note that for given b, D is an indicator of the relative strength or amplitude of each spatial frequency present in $W(x)$ as can be directly seen from (6.8) or its power spectral density $S(v)$. From the defining relation (6.8), the latter is found to be a function of spatial frequency v of the form

$$S(v) = \eta^2 \frac{[1 - b^{(2D-4)}]}{2[b^{(2D-4)N_1} - b^{(2D-4)(N_2+1)}]} \sum_{n=N_1}^{N_2} b^{2n(D-2)}[\delta(v - sb^n) + \delta(v + sb^n] \quad (6.9)$$

which represents a line spectrum. Here as in (6.8) it is clear that the fractal dimension D determines the spectral roll-off with increasing v. The case $D \to 2$ yields a flat spectrum and so exhibits strong fluctuations due to the strong presence of high frequencies. In the opposing limit when $D \to 1$, an undulating curve with weak fluctuations appears since only the fundamental spatial frequency component is dominant unless $b \to 1$ as well.

For the case when $b \to 1$, the spectral lines tend to merge into a continuum. In this limit, the continuous approximation $\bar{S}(\nu)$ to the power spectral density $S(\nu)$ becomes

$$\bar{S}(\nu) = \eta^2 \frac{[1 - b^{(2D-4)}]s^{-(2D-4)}}{2[b^{(2D-4)N_1} - b^{(2D-4)(N_2+1)}]\ln(b)} \nu^{(2D-5)} \qquad (1 < D < 2). \quad (6.10)$$

The geometrical spacing of the tones or spectral lines and their power-law decay as given in (6.10) are responsible for the self-similarity relation. In particular, it can be shown from Fourier transform relations (see, e.g., [10]) that if a self-similar fractal function $g(x)$ satisfies the scaling relation,

$$g(x) = \alpha g(\beta x), \quad (6.11)$$

where α and β are scaling constants, its power spectral density $G(\nu)$ satisfies the relation

$$G(\nu) = \frac{\alpha^2}{\beta} G\left(\frac{\nu}{\beta}\right). \quad (6.12)$$

Therefore, a power-law power spectral density, such as (6.10), of the form

$$G(\nu) = \Gamma \nu^{-\gamma}, \quad (6.13)$$

with constants Γ and γ, must satisfy the additional constraint, $\alpha^2 \beta^{(\gamma-1)} = 1$, to retain self-similarity. We simply note that power-law spectra indicate the potential for self-similarity of fractal functions although this condition is not sufficient.

To strengthen our intuitive or physical feel for these functions, several bandlimited Weierstrass functions are plotted in Figures 6.8 and 6.9, and their power spectral density in Figure 6.10. In Figure 6.8 plots are displayed for discrete fractal dimensions ranging (top to bottom) from $D = 1.05$ (gently undulating) to $D = 1.95$ (wildly irregular and area-filling) for the case $s = 1$, $b = \sqrt{\pi}$ with $N = 6$ (left) and $N = 12$ (right) tones, all with random phases θ_n. With only a modest number of tones, noise-like signals typical of random fractals can be easily constructed. However, we note again that these band-limited fractal functions are only fractal over a limited regime of scale lengths so that as the resolution of these plots increases or decreases without bound, the curves no longer appear fractal. The addition of tones causes these graphs to become self-similar over a regime of increasing size. Increases in b cause these functions to be self-similar at increasingly wide, discrete, resolution lengths due to the spacing of the spectral tones of (6.9). However, as b approaches unity, these functions appear to be self-similar at all scales providing N is large enough.

A second display of the bandlimited Weierstrass function is given in Figure 6.9 where a perspective plot for the intermediate case of $N = 10$ displays the variation in roughness for a continuously increasing fractal dimension (back to front). As in the previous figure, $s = 1$ and $b = \sqrt{\pi}$, all with random phases

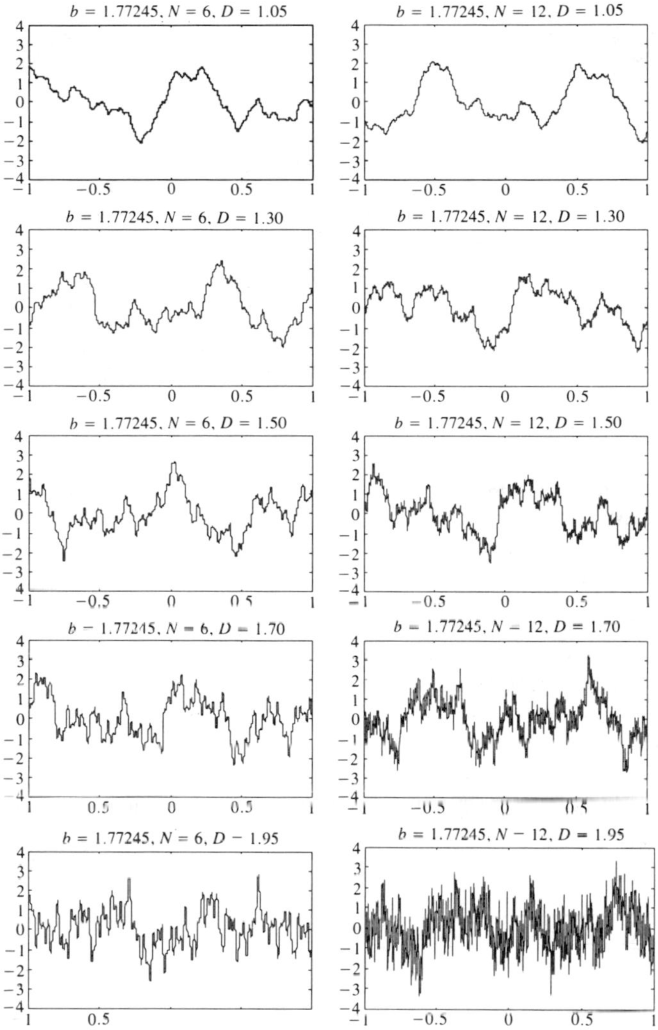

FIGURE 6.8. Bandlimited functions for increasing fractal dimension ($D = 1.05$ top to $D = 1.95$ bottom) and increasing number of tones ($N = 6$ left and $N = 12$ right). Here $b = \sqrt{\pi}$ and $s = 1$.

θ_n. This not only shows again that the fractal dimension D is a qualitative measure of roughness but it also provides some quantitative idea of the dependence of roughness on the fractal dimension.

In Figure 6.10 is shown the spatial frequency domain counterpart of Figure 6.9 as a function of spatial frequency ν and fractal dimension D. As in the previous figure, $N = 10$, $s = 1$ and $b = \sqrt{\pi}$, all with random phases θ_n. The plots of the power spectral density $S(\nu)$ as a function of scaled spatial frequency

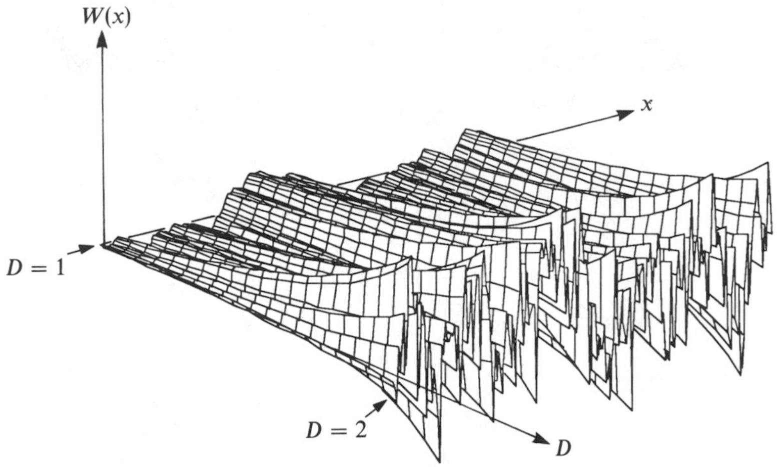

FIGURE 6.9. Bandlimited Weierstrass function $W(x)$ (height) as a function of coordinate x (left to right) for increasing fractal dimension ($D = 1.00$ back to $D = 2.00$ front) for $N = 10$. Note the gradually increasing roughness of $W(x)$ as D increases. Here $b = \sqrt{\pi}$ and $s = 1$.

log v and D visually demonstrates how the high-frequency components of the band-limited Weierstrass function $W(x)$ become the source of the roughness in these fractal functions The line spectra of Figure 6.10 for each fractal dimension D has the characteristic power-law decay and geometrical spacing needed to satisfy the requirements of self-similarity noted before. The continuous counterpart to this case occurs when b approaches unity and the line spectra merge.

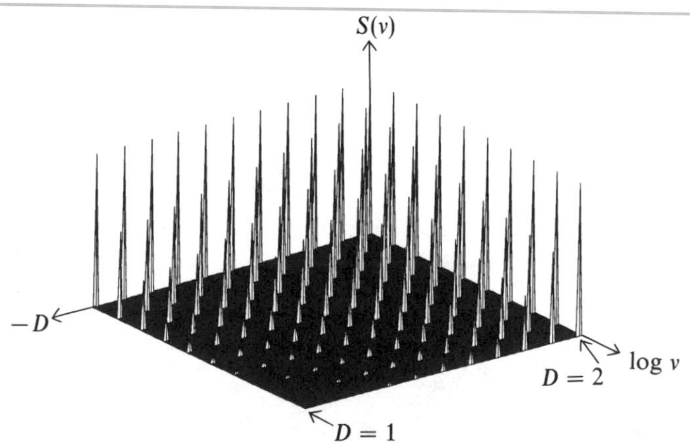

FIGURE 6.10. Amplitude of the bandlimited Weierstrass function power spectral density $S(v)$ as a function of scaled spatial frequency log v (increasing front to back) for $D = 1.00$ (left) to $D = 2.00$ (right) corresponding to the function of the previous figure. Here $N = 10$, $b = \sqrt{\pi}$ and $s = 1$.

6.3.3. *Electromagnetic Yardsticks*

The final connection between the characterization of fractal structures and fractal electrodynamics comes from the identification of an appropriate electromagnetic "yardstick." Just as a variable length physical yardstick can be used to measure the fractal characteristics of physical objects using the coastline method, it is desirable that an appropriate variable length electromagnetic yardstick be defined so that electromagnetic or physical properties of structures be determined as sketched in Figure 6.11. The natural electromagnetic yardstick is the wavelength. Therefore, we propose that an interrogating wave of variable wavelength be used to determine the fractal characteristics of electrical objects without *in situ* measurements. This is entirely reasonable since it is well known that waves interact most strongly or resonantly with objects containing scale sizes on the same order as the wavelength and has the added potential advantage that this interrogation by waves may yield this information remotely as the fractal characteristics of the wave become impressed on the scattered field.

We turn now to a series of canonical problems which demonstrate the use of fractal concepts, bandlimited fractals, and electromagnetic yardsticks in the interaction of waves with fractal structures. Since the details of many of these problems have been presented elsewhere, we will consider here the problem formulation, results, and salient features of these results. We also note that a different point of view is represented by the work of British researchers, circa 1980. At this time, Berry [11] coined the term "diffractal" to represent the interaction of electromagnetic waves with fractals and Berry and Blackwell considered the scattering of pulses from rough surfaces [12]. Shortly there-

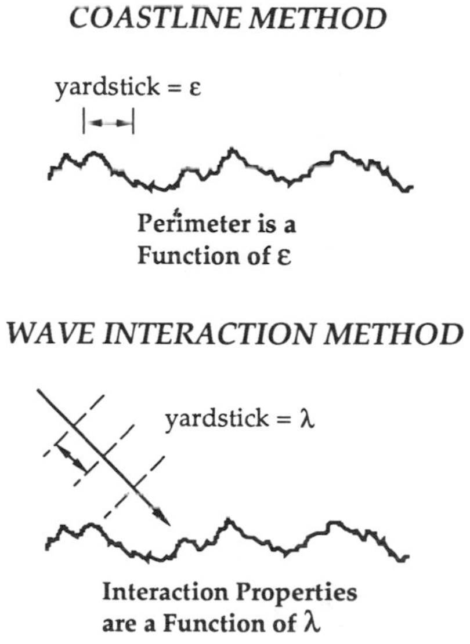

FIGURE 6.11. In the coastline method (top) a variable length yardstick ε is used to determine the perimeter and fractal properties of a rough curve. Likewise, in the wave interaction method (bottom) the variable wavelength λ of an interrogating wave can be used as a "yardstick" to remotely probe the characteristics of a rough structure.

after, Jakeman considered the scattering of electromagnetic waves from random surfaces with fractal slopes [13], [14]. More recent work in the optical regime and beyond involves light, x-ray, or neutron scattering from fractal clusters or aggregates as represented by the work of Teixeira [15] and Chen et al. [16].

As a general background to fractal electrodynamics, we note that in scattering from fractal structures, the differential scattering cross section $d\sigma/d\Omega$, under appropriate approximations, is found to vary characteristically as an inverse power of normalized wave number q for light, x-ray, and neutron illumination. Here q $[= 2k\sin(\theta/2)]$ is the magnitude of the difference between the scattered wave vector and the incident wave vector, θ is the scattering angle, and the incident illuminating wave with wavelength λ $(= 2\pi/k)$ has a spatial resolution of $2\pi/q$. In particular, for discrete fractal clusters with (volume) fractal dimension D, it is found that the scattered fields contain useful information regarding the fractal dimension. Since the scattered intensity is the product of the number of volume elements of size $1/q^3$ (which is q^D) and the mass or dipole moment in each volume element squared (which is q^{-2D}), we find immediately that

$$\frac{d\sigma}{d\Omega} \propto [q^D]\left[\frac{1}{q^{2D}}\right] \propto \frac{1}{q^D}. \tag{6.14}$$

Similar arguments for fractal surface scattering yield [15],

$$\frac{d\sigma}{d\Omega} \propto \frac{1}{q^{6-D}}, \tag{6.15}$$

where D is the (surface) fractal dimension. Since D has an upper bound of three, a variation of $q^{-\alpha}$ for $\alpha < 3$ indicates volume scattering while $\alpha > 3$ indicates surface scattering. This scattering and its variation according to the scattering mechanism have been confirmed by Schaefer et al. [17] for combined x-ray and light scattering form a colloidal aggregate of silica particles in which both variations of (6.14) and (6.15) are apparent. Using x-ray scattering from lignite coal, Bale and Schmidt [18] confirmed (6.15) for surface scattering.

These results for fractal aggregates and surfaces suggest that electromagnetic wave interactions with fractal structures indeed possess characteristic variations with wave number and so confirm the concept of using the wavelength as a variable yardstick and provide a basis for proceeding to investigate fractal electrodynamics. The remaining sections of this report examine several illustrative canonical examples which help us explore this topic. We start with the physically motivated problem of electromagnetic wave propagation through turbulence and progress through a series of problems where illuminating waves interact with a number of bandlimited fractal structures. This leads us to a progression of diffraction by one or more phase screens, scattering by fractal surfaces and finally scattering by a discrete fractally corrugated cylinders. In each case, the roughness of the structure as indicated by its fractal dimension, plays an important role in the form of the scattered field.

6.4. Turbulence and Diffraction by Fractal Phase Screens

6.4.1. *The Fractal Nature of Turbulence*

The inherently multiscale nature of turbulence makes electromagnetic wave propagation in turbulence an ideal physical example to consider in our exploration of fractal electrodynamics. Since the pioneering work of Tatarskii [19] and Chernov [20] in the early 1960s, wave propagation in the atmosphere and wave interactions in random media have been the subject of intense investigation. Much of this work is based on the power spectrum of atmospheric turbulence in which the celebrated $\frac{5}{3}$-law spectrum found by Kolmogorov in 1941 [21] was used. The $\frac{5}{3}$-law spectrum has been supported by an extensive set of experiments and is widely used today. In addition, however, experimental evidence reveals the intermittent characteristic of turbulence in which the small-scale structures of turbulence are found to be less space filling than their large scale counterparts.

Recently, Kim and Jaggard [22] proposed a model for atmospheric turbulence which makes use of the bandlimited Weierstrass function and replicates the $\frac{5}{3}$-law spectrum while allowing for intermittency. In this model, fluctuations in the refractive index are taken into account by the use of variably localized modulating functions. In the limit of no localization, the model is defined by the bandlimited Weierstrass function of eq. (6.8) in which the inner and outer scales naturally associated with turbulence are taken into account using a limited number of tones as discussed in the previous section. In this way, the inherent multiscale nature of turbulence is translated into the language of fractal geometry.

The validity of this model is tested by comparing the bandlimited spectrum and the spectrum within the inertial subrange. Although this test cannot guarantee uniqueness of the model, it suggests that the model is a viable candidate for use in the atmospheric problem since it produces the desired spectrum and is consistent with experimentally observed data. The virtue of this particular model is that it satisfies the general characteristics of turbulence (e.g., power-law spectrum, intermittency, randomness, multiscaling, existence of inner and outer scales) and is amenable to analytical and numerical analysis. We note that in 1926 Richardson suggested the Weierstrass function to model aspects of turbulence [23]. However, this suggestion was apparently not acted upon until the work summarized here was done. Likewise, Mandelbrot in his treatise on fractals [1] also notes possible applications to turbulence. Further work by Mandebrot [24], Frisch et. al. [25], and Siggia [26] are relevant to the discussion of this subsection.

A schematic of a multiscale model for turbulence is shown in Figure 6.12.

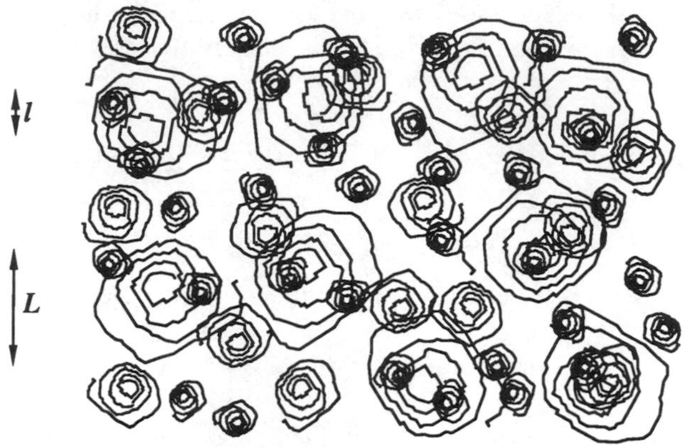

FIGURE 6.12. Schematic of multiscale turbulence with a structure ranging from the inner scale size l to the outer scale size L.

Here turbulent eddies of three sizes are displayed ranging from an inner scale length l to an outer scale length L. For the propagation in one dimension, we take a cut through these eddies and express the one-dimensional refractive index $n(x)$ of the atmosphere as

$$n(x) = \langle n(x) \rangle + \delta n(x), \tag{6.16}$$

where $\langle n(x) \rangle$ is the average background refractive index and $\delta n(x)$ is the spatial fluctuation. The latter, zero mean, function is taken to exist in Fourier space in the inertial subrange between the inner and outer scales. We express the fluctuation for the case where no localization is used (i.e., no intermittency) as

$$\delta n(x) = P \frac{\{2\langle n_f^2 \rangle [1 - b^{(2D-4)}]\}^{1/2}}{[1 - b^{(2D-4)(N+1)}]^{1/2}} \sum_{n=0}^{N} b^{n(D-2)} \cos\left(\frac{2\pi b^n x}{L} + \theta_n\right), \tag{6.17}$$

which is a modified version of the bandlimited Weierstrass function given in (6.8). Here P is the power ratio between the total fluctuation $\langle n_f^2 \rangle$ and the inertial subrange fluctuation $\langle \delta n(x) \rangle^2$ and L is the outer scale length. Since there are $N + 1$ scales present and b is the normalized fundamental frequency, the relation between the outer scale L, the inner scale l and b is determined by $b^N = L/l$. We take the phases θ_n to be independent and uniformly distributed over $[0, 2\pi]$.

When the discrete spatial frequencies are close ($b \to 1$), the power spectral density $V(k)$ of (6.17) can be averaged over wave number k ($= 2\pi v$) to produce the continuously approximated spectrum $\bar{V}(k)$[1] given as

$$\bar{V}(k) = 1.135\, P^2 \langle n_f^2 \rangle L^{-2/3} k^{-5/3}, \tag{6.18}$$

[1] For historical reasons, we use $V(k)$ in this section instead of $S(v)$ as used previously for the power spectral density.

which agrees in form with the Kolmogorov spectrum defined by [21]

$$V(k) = 0.24\langle n_f^2 \rangle L^{-2/3} k^{-5/3}. \tag{6.19}$$

A comparison of (6.18) with (6.19) yields the result that the one-dimensional Kolmogorov spectrum is proportional to k^{-D} with $D = \frac{5}{3}$. A similar calculation in three dimensions yields an analogous result in which $D = 11/3$ [22]. We note that the factor P in (6.18) represents the relative magnitude of the fluctuations and cannot be found from the scaling principles used here.

What is important is the frequency dependence of the power spectral density. These results indicate that a characteristic spectral dependence of k^{-D} exists for fluctuations using a bandlimited Weierstrass function and that the Kolmogorov spectrum can be replicated in one dimension with a Weierstrass

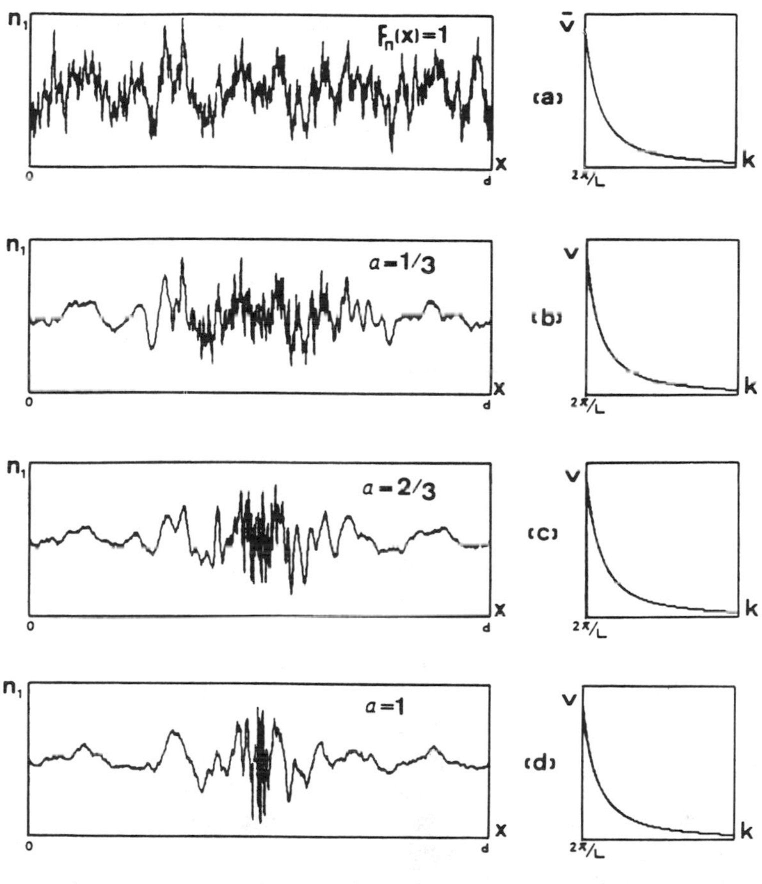

FIGURE 6.13. Plots of the bandlimited Weierstrass models for refractive index fluctuations (left) and their corresponding power spectral densities (right). At the top is an unlocalized spectrum (no intermittency) while the lower three plots indicate increasing localization. All spectra are graphically indistinguishable from the Kolmogorov spectrum (top). Adapted from [22].

fractal dimension of $D = \frac{5}{3}$. This is reminiscent of the differential scattering cross-section dependence noted in relation (6.14) and the spectral characteristics of self-similar functions.

A second result of this solution concerns the invariance of the spectrum of turbulence with respect to degree of intermittency or localization. Here the fluctuation indicated in (6.16) and (6.17) is modified by a wavepacket suitable for each tone so that each cosine function is modulated by a localization function which nulls each cosine outside of its wavepacket while leaving it undisturbed inside the wavepacket. This scheme is chosen in conformance with experimental evidence which suggests that a large fraction of the variance in the refractive index fluctuation comes from small portions of the space filled by turbulence, while as the scale size descreases it becomes less space filling.

While the mathematical details are given elsewhere [22], we display the results for intermittency in Figure 6.13. Here are shown plots of the bandlimited Weierstrass function for refractive index fluctuations $\delta n(x)$ (left) with increasing localization (top to bottom) and their corresponding power spectral densities (right). In each case, the power spectral densities $V(k)$ with intermittency present (lower three plots) are graphically indistinguishable from the nonlocalized result $\bar{V}(k)$ (top plot) given by (6.18).

For applications to wave propagation in the atmosphere, the bandlimited Weierstrass function and its localized modification provide models which are consistent with experimental data, physically motivated by the multiscale nature of turbulence, and yet are amenable to mathematical description and analysis. We turn now to one such application, the propagation of waves through one or more fractal phase screens. This in turn will lead to rough surface scattering problems involving both bounded and unbounded fractal structures.

6.4.2. *Diffraction by a Bandlimited Fractal Phase Screen*

Consider next the diffraction of electromagnetic waves by a fractal phase screen as a first approximation to propagation in random media and the reflection of electromagnetic waves from rough surfaces. In the former case, we can consider transverse variations in the fluctuation to be modeled by a bandlimited Weierstrass function which becomes imprinted on the phase planes of waves which traverse the medium while in the latter case we can consider a similar imprint on the phase of waves reflected from a rough reflective surface. This canonical problem is a straightforward application of fractal electrodynamics in which we build on previous results.

The scalar problem under consideration is shown in Figure 6.14 where a square fractal phase screen of side L in the x'–y' plane at $z = 0$ is illuminated by a normally incident time-harmonic plane wave. The aperture field at the $z = 0^-$ plane is given by $\psi(x', y')$. The transfer function of the screen is described by $t(x', y')$, and the time-harmonic diffracted field $\psi(x, y, z)$ or the intensity $I(x, y, z)$ $[= \psi(x, y, z)\, \psi^*(x, y, z)]$ is observed a distance z away from

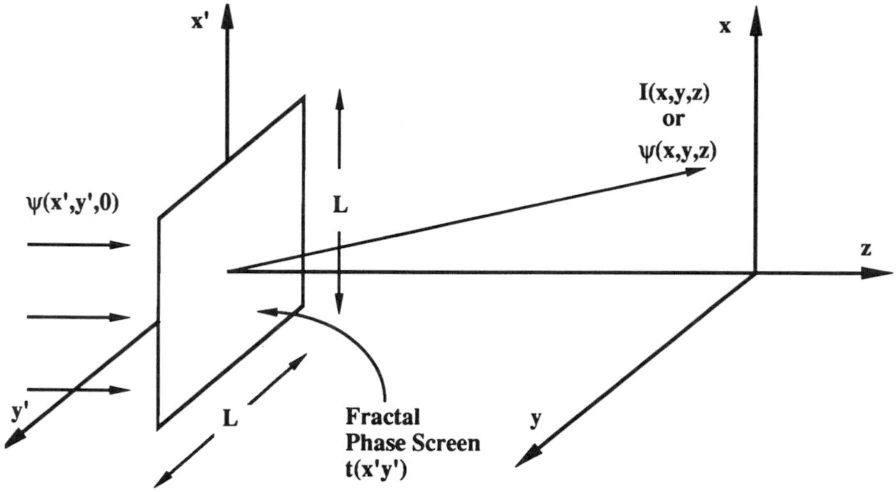

FIGURE 6.14. A square fractal phase screen of side L in the x'–y' plane at $z = 0$ illuminated by an incident plane wave $\psi(x', y')$. The transfer function of the phase screen is $t(x', y')$ and the diffracted field $\psi(x, y, z)$ or intensity $I(x, y, z)$ is observed at the point (x, y, z).

the screen. The phase screen is modeled by the phase transfer function

$$t(\mathbf{x}') = t(x', y') = \exp[iW(x')], \tag{6.20}$$

where $W(x)$ is the bandlimited Weierstrass function of eq. (6.8) with random phases θ_n uniformly distributed on $[0, 2\pi]$ and the pupil function $P(x', y')$ which defines the extent of the screen is given by

$$P(\mathbf{x}') = P(x', y') = \text{rect}\left(\frac{x'}{L}\right)\text{rect}\left(\frac{y'}{L}\right), \tag{6.21}$$

where as usual, $\text{rect}(x) \equiv 1$ for $|x| \leq \frac{1}{2}$ and is zero elsewhere and $\mathbf{x}' = (x', y') = (x', y', 0)$.

In the physical optics approximation for scalar waves, the time-harmonic diffracted field can be expressed as the integration of the aperture field over the x'–y' plane in the form

$$\psi(\mathbf{x}) = \psi(x, y, z) = (-2ik)\int_{-\infty}^{\infty}\int_{-\infty}^{\infty} \psi(\mathbf{x}')P(\mathbf{x}')t(\mathbf{x}')G(\mathbf{x}; x')\, d\mathbf{x}', \tag{6.22}$$

where the free-space Green's function is given by

$$G(\mathbf{x}; \mathbf{x}') = G(x, y; x', y') = \frac{\exp(ik|\mathbf{x} - \mathbf{x}'|)}{4\pi|\mathbf{x} - \mathbf{x}'|}, \tag{6.23}$$

k is the wave number of the incident field and $\mathbf{x} = (x, y, z)$.

Carrying out the integration of (6.22) in the undisturbed incident field approximation $[\psi(\mathbf{x}') = \psi(x', y') = 1]$ yields the result for the intensity averaged over the phase terms as [7]

$$\begin{aligned}\langle I(\mathbf{x})\rangle &= \langle I(x, y, z)\rangle \\ &= \frac{L^4}{\lambda^2 z^2} \sum_{q_1=-\infty}^{\infty} \sum_{q_2=-\infty}^{\infty} \cdots \sum_{q_N=-\infty}^{\infty} [J_{q_1}^2(C_{N_1}) J_{q_2}^2(C_{N_1+1}) \cdots J_{q_N}^2(C_{N_2})] \\ &\quad \times \left(\operatorname{sinc}^2\left\{L\left[\frac{x}{\lambda z} - aq_1 b^{N_1} - aq_2 b^{(N_1+1)} - \cdots - aq_N b^{N_2}\right]\right\}\right) \\ &\quad \times \left[\operatorname{sinc}^2\left(\frac{Ly}{\lambda z}\right)\right], \end{aligned} \tag{6.24}$$

where

$$C_n = \eta \frac{\sqrt{2}[1 - b^{(2D-4)} b^{(D-2)n}]^{1/2}}{[b^{(2D-4)N_1} - b^{(2D-4)(N_2+1)}]^{1/2}}, \tag{6.25}$$

and where $\operatorname{sinc}(x) \equiv \sin(\pi x)/(\pi x)$. This result displays a series of peaked spikes along the z-axis, each corresponding to diffraction by various combinations of the tones of the fractal phase screen. Alternatively, we see that each tone in the phase screen produces Raman–Nath diffraction as characterized by the Bessel function weighting.

Under conditions of small variance ($\eta^2 \ll 1$), (6.24) becomes

$$\begin{aligned}\langle I(x)\rangle &= \frac{L^4}{\lambda^2 z^2}\left\{(1-\eta^2)\operatorname{sinc}^2\left[L\left(\frac{x}{\lambda z}\right)\right] + \sum_{n=-\infty}^{\infty}\left(\frac{C_n^2}{4}\right)\operatorname{sinc}^2\left[L\left(\frac{x}{\lambda z} - sn^n\right)\right]\right\} \\ &\quad \times \left[\operatorname{sinc}^2\left(\frac{Ly}{\lambda z}\right)\right], \end{aligned} \tag{6.26}$$

which clearly demonstrates the physical diffraction processes. The largest term for the case of small variance is the nondiffracting component in (6.24) given by the condition $q_1 = q_2 = \cdots q_N = 0$ which becomes the first term in (6.26). The diffracted intensity is the second term of (6.26) which is proportional to the variance of the phase screen. The diffracted intensity peaks are positioned such that they satisfy the Bragg condition for each tone of the phase screen in the far zone. Thus, the fractal dimension D plays a role through C_n in the distribution of diffracted energy. For increasing fractal dimension, the diffracted field spread increases due to the increased strength of the high spatial frequencies in the phase screen.

Since the physics of the beam diffraction is now established in Fraunhofer regime, it is of interest to examine the evolution of the diffracted beam as it propagates from the screen to the far zone. Using the Fresnel approximation, the near-zone counterpart to (6.24) for the average intensity is given by

$$\begin{aligned}\langle I(x)\rangle &= \sum_{q_1=-\infty}^{\infty} \sum_{q_2=-\infty}^{\infty} \cdots \sum_{q_N=-\infty}^{\infty} [J_{q_1}^2(C_{N_1}) J_{q_2}^2(C_{N_1+1}) \cdots J_{q_N}^2(C_{N_2})] \\ &\quad \times \{[C(\xi_2) - C(\xi_1)]^2 + S(\xi_2) - S(\xi_1)]^2\} \\ &\quad \times \{[C(\eta_2) - C(\eta_1)]^2 + S(\eta_2) - S(\eta_1)]^2\}, \end{aligned} \tag{6.27}$$

where C and S denote the usual Fresnel integrals and their arguments are given by

$$\xi_1 = -\sqrt{\frac{k}{\pi z}}\left\{\frac{L}{2} + x - z\lambda s[q_1 b^{N_1} + q_2 b^{(N_1+1)} + \cdots + q_N b^{N_2}]\right\}, \tag{6.28}$$

$$\xi_2 = -\sqrt{\frac{k}{\pi z}}\left\{\frac{L}{2} - x + z\lambda s[q_1 b^{N_1} + q_2 b^{(N_1+1)} + \cdots + q_N b^{N_2}]\right\}, \tag{6.29}$$

$$\eta_1 = -\sqrt{\frac{k}{\pi z}}\left\{\frac{L}{2} + y\right\}, \tag{6.30}$$

$$\eta_2 = -\sqrt{\frac{k}{\pi z}}\left\{\frac{L}{2} - y\right\}. \tag{6.31}$$

In a similar manner as before, we can find the small variance approximation to (6.27) which is the Fresnel analogue to (6.26). This result is given by

$$\langle I(\mathbf{x})\rangle = \tfrac{1}{4}\Bigg((1-\eta^2)\{[C(\xi_4) - C(\xi_3)]^2 + [S(\xi_4) - S(\xi_3)]^2\}$$
$$+ \sum_{n=-\infty}^{\infty}\left(\frac{C_n^2}{4}\right)\{[C(\xi_6) - C(\xi_5)]^2 + [S(\xi_6) - S(\xi_5)]^2\}\Bigg)$$
$$\times \{[C(\eta_2) - C(\eta_1)]^2 + [S(\eta_2) - S(\eta_1)]^2\} \tag{6.32}$$

where the Fresnel integral arguments are expressed as

$$\xi_3 = -\sqrt{\frac{k}{\pi z}}\left\{\frac{L}{2} + x\right\}, \tag{6.33}$$

$$\xi_4 = +\sqrt{\frac{k}{\pi z}}\left\{\frac{L}{2} - x\right\}, \tag{6.34}$$

$$\xi_5 = -\sqrt{\frac{k}{\pi z}}\left\{\frac{L}{2} + x - z\lambda s b^n\right\}, \tag{6.35}$$

$$\xi_6 = +\sqrt{\frac{k}{\pi z}}\left\{\frac{L}{2} - x + z\lambda s b^n\right\}. \tag{6.36}$$

This result again demonstrates the division of intensity between diffracted and undiffracted components.

From both the Fraunhofer and Fresnel results, the ratio of total diffracted power to the total power is given by the simple expression

$$\frac{\iint \langle I(\mathbf{x})\rangle_{\text{diff}}\,dx\,dy}{\iint \langle I(\mathbf{x})\rangle_{\text{total}}\,dx\,dy} \approx 1 - \eta^2 \approx \exp(-\eta^2). \tag{6.37}$$

Although the left equality is valid for small variances, it can be shown from renormalization techniques that the right inequality is valid without constraint. This result implies that the variance η^2 controls the distribution of power between the diffracted and nondiffracted fields while the fractal dimension D, the number of tones N, and the screen size L determine the structure of the diffracted field.

Finally, in this section, we plot the evolution of the normalized diffracted average intensity $\langle I(\mathbf{x})\rangle/\langle I(0)\rangle$ and the instantaneous intensity $I(\mathbf{x})/I(0)$ for the fractal phase screen as a function of the normalized transverse coordinate x/L at various normalized longitudinal distances z/L. The aperture size is given by $\lambda/L = 10^{-5}$ and the values $s = 10^3$, $b = \sqrt{2\pi}$, $\eta^2 = 1$, $N_1 = 0$, $N = N_2 + 1 = 10$ and $D = 5/3$.[1] The average intensity variation (left) and its single-realization counterpart (right) are shown in Figure 6.15 at the distances $z/L = 10$, 100, 200, 400, 800, and 1600. While the top plots are deep in the Fresnel zone, the bottom plot approaches the far zone and at further distances evolve into a series of sinc^2 functions as given by (6.24) or (6.26).

Examining the evolution of the average beam intensity on the left of Figure 6.15 we note that very near the screen (top), this intensity simply mimics the screen distribution of energy. However, as the beam propagates away from the screen, energy sloughs off the shoulders of the distribution in a manner characteristic of Fresnel diffraction. Here, each tone of the Weierstrass function contributes a small spike (second plot from the top) which increases with distance until it separates (bottom plot). Eventually, the Fraunhofer diffraction patterns appear (not shown) in which a large number of sharply peaked sinc^2 distributions are in evidence, each centered at the middle of the Fresnel distributions seen in the bottom of Figure 6.15. As noted before, the fractal dimension controls the total spread of the diffracting field.

The single-realization results on the right of Figure 6.15 display the evolution of the beam in which order emerges from chaos. Deep in the near field (top), the single realization intensity takes on the chaotic character of the screen since the phase screen at this point has imprinted its variations on the incident field. However, as the field propagates away from the plane of the diffracting screen, the intensity increasingly becomes like its averaged counterpart. This is due to the disengagement of the diffracted orders due to each Weierstrass tone of the phase screen. Considering Bragg diffraction of the incident field by the screen and the conditions for the far-zone approximation, this disengagement is predicted to occur at distances z such that

$$z > \frac{L}{[\lambda s b N_1 (b-1)]} \qquad \text{and} \qquad a b^{N_1}(b-1) > \frac{2}{L}. \tag{6.38}$$

For the values used in Figure 6.15, this predicts that disengagement should occur for values of z/L which are larger than ~ 700. Clearly, Figure 6.15 confirms this since the left-hand and right-hand sides of the graph appear similar for $z/L = 800$ and beyond.

---

[1] This value for fractal dimension corresponds to the Kolmogorov spectrum as outlined in the previous section.

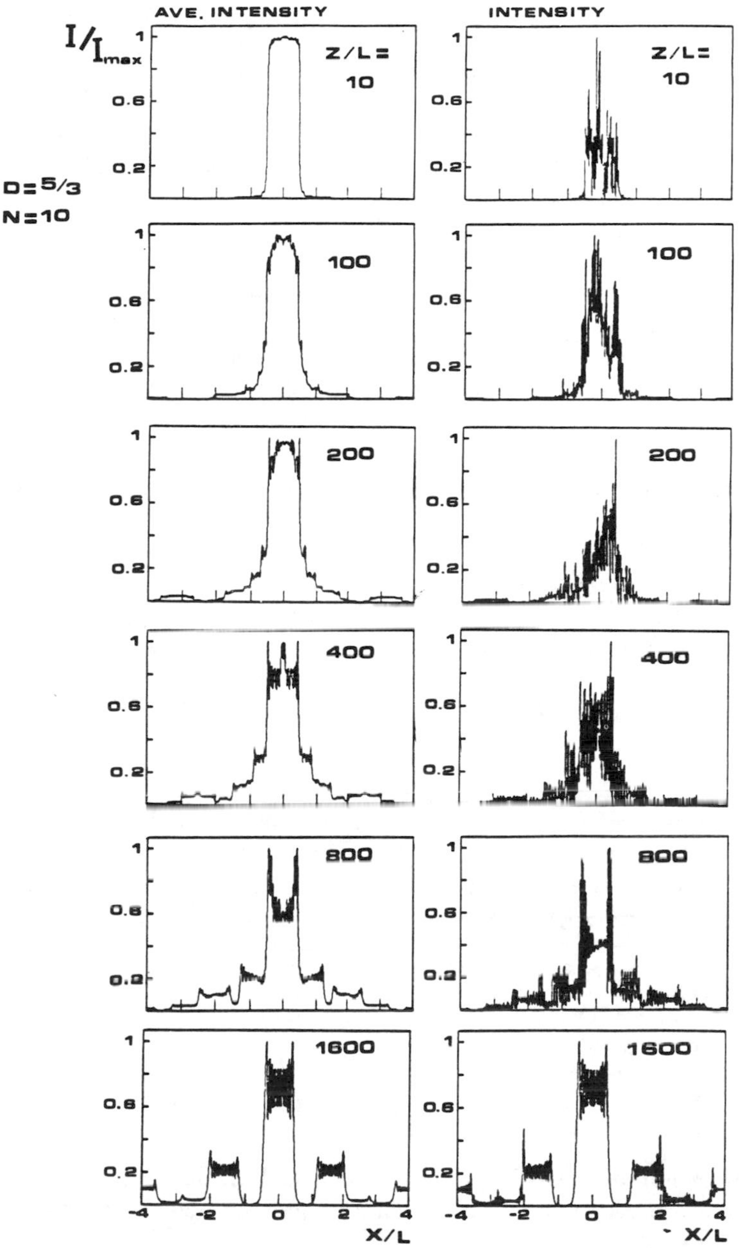

FIGURE 6.15. Evolution of the average intensity distribution (left) for a fractal phase screen of finite extent ($\lambda/L = 10^5$) as a function of the transverse normalized coordinate (x/L) for various distances from the screen (z/L = 10, 100, 200, 400, 800, and 1600) going from top to bottom. The analogous intensity distributions for a single realization are given on the right-hand side. For this plot, $D = \frac{5}{3}$ and all intensities are normalized to their maximum height. Figure adapted from [7].

6.4.3. *Diffraction by Multiple Fractal Phase Screens*

As demonstrated, turbulence can be modeled by use of the bandlimited Weierstrass function. Likewise, we have shown the effects of diffraction by a single fractal phase screen using the same Weierstrass function. Here we comment briefly on the approximation of continuous refractive variations by the use of multiple fractal phase screens.

The multiple-phase screen approach suitable for propagation through random media or a turbulent atmosphere is appropriate in the high-frequency regime when the refractive index fluctuations are small. However, this approach can be used over large distances in which the accumulated effects are significant due to its inherent conservation of energy. The reader is referred to the literature for the foundations of this approach [27], [28] and for a review of alternative techniques. The results for beam propagation through multiple fractal phase screens are useful in the study of laser propagation and high-frequency radio wave propagation in the atmosphere or ionosphere, to underwater acoustic propagation in imperfect ducts and to the characterization of photonic devices with defects.

The multiple-phase screen approach outlined here is based on the work of Kim and Jaggard [29] in which Bragg diffraction is used to produce the gross characteristics of beam evolution due to each screen and the Fresnel propagator or Green's function is used to account for free-space diffraction. The starting point for this method is the Helmholtz equation for the time-harmonic field $\psi(\mathbf{x})$ given as

$$\{\nabla^2 + k^2[1 + \delta n(\mathbf{x})]^2\}\psi(\mathbf{x}) = 0, \tag{6.39}$$

where k is the free space or background wave number and δn is the refractive index fluctuation of (6.17). Under the parabolic approximation and ignoring free-space diffraction, the solution to (6.39) for fields which propagated along the z-axis can be written

$$\psi(x, y, z + \Delta z) = \psi(x, y, z) \exp[ip(x, y, z)], \tag{6.40}$$

where the phase term $p(x, y, z)$ due to the refractive index fluctuation in the slice Δz is given by

$$p(x, y, z_m) = k \int_{z_m - \Delta z}^{z_m} \delta n(x, y, z')\, dz' \qquad (m = 1, 2, 3, \ldots). \tag{6.41}$$

This takes into account the propagation through the refractive index fluctuations and we use, as a model, a phase term composed of two bandlimited functions so that

$$p(x, y, z_m) = kW_m(x)V_m(y), \tag{6.42}$$

where $W_m(x)$ and $V_m(y)$ are of the form of relation (6.8).

For free-space diffraction between planes, we use the diffraction integral (6.22) with the Fresnel approximation to (6.18). The undisturbed aperture field $\psi(\mathbf{x}')$ is taken to be a Gaussian.

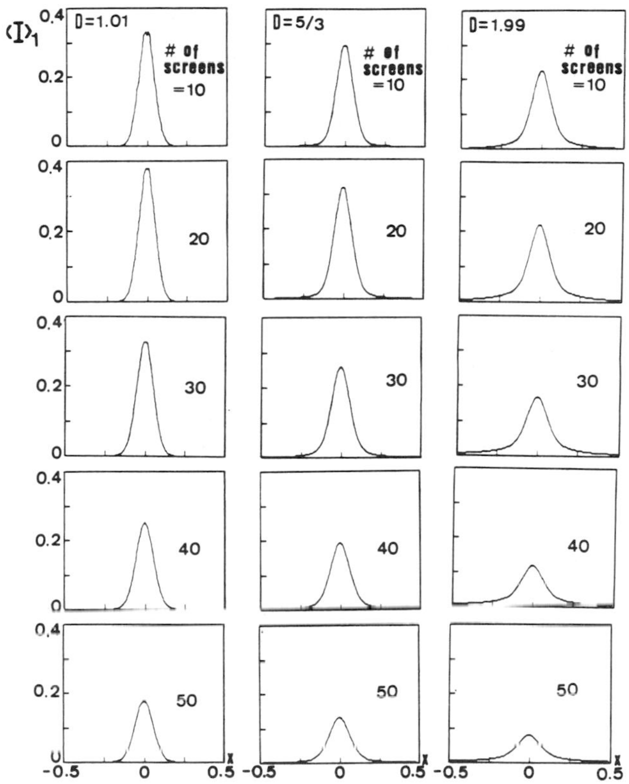

FIGURE 6.16. Evolution of the first-order diffracted beam $\langle I_1 \rangle$ for $D = 1.01$ (left), $\frac{5}{3}$ (center) and 1.99 (right) through increasing numbers of fractal phase screens (top to bottom). Figure adapted from [29].

The solution to the multiple-screen propagation problem is now completely specified. The incident field is diffracted by the first screen using (6.40)–(6.42) and propagated to the second screen using the Fresnel diffraction integral (6.22)–(6.23). The field is diffracted by the second screen using (6.40)–(6.42) and is propagated to the third screen using (6.22)–(6.23). This process is repeated until all phase screens have been traversed.

Figure 6.16 displays a representative example of the propagation of the first-order diffracted beam intensity $\langle I_1 \rangle$ through a series of fractal phase screens for the cases $D = 1.01$, $\frac{5}{3}$ (Kolmogorov spectrum), and 1.99. In each case, the first-order diffracted beam increases to approximately the twentieth phase screen and then monotonically decreases thereafter. Since energy is conserved, this indicates that the energy is being diffracted into higher-order beams past the twentieth screen as has been demonstrated (not shown). These higher-order diffracted terms in turn increase, peak, and decrease with additional phase screens. As expected from physical considerations and the discussion of diffraction by a single fractal phase screen, the larger fractal dimensions (right) tend to spread the beam more quickly than their lower dimension counterparts (left) due to the large high-spatial frequency components in the phase screen for large *D*.

6.5. Scattering by Fractally Rough Structures

6.5.1. *Reflection by Fractal Surfaces*

Turning from problems of transmission and diffraction, we examine several closely related canonical problems concerned with scattering and reflection. As before, these problems yield scattering characteristics which can be classified according to fractal dimension. This raises the possibility of using these characteristics to remotely identify structures according to roughness and other relevant parameters.

Scattering of optical, electromagnetic, and acoustical waves from rough surfaces has been of theoretical and practical interest for the past several decades to characterize wave interactions with microscopically rough interfaces, sea surfaces, ocean bottoms, and rough terrain, as well as in imaging and nondestructive evaluation of naturally occurring and man-made objects. In these past studies, both deterministic periodic functions and random functions have been used as mathematical models for rough surfaces. The introduction of fractal electrodynamics provides a new tool to describe naturally occurring rough structures. Surfaces formed through repetitive actions of nature yield a geometry which is a careful balance between periodic and aperiodic processes. As noted previously, this gives rise to the long-range order and short-range disorder characteristic of many fractals. Therefore, fractals can be used to describe both deterministic or random structures or an appropriate blend.

The function used here to model these surfaces is a modification of the bandlimited Weierstrass function given in (6.8) but is of similar form and shares many of the attributes of the Weierstrass function [30], [31]. Consider a zero-mean bandlimited fractal function, which we denote the *roughness fractal function* $f_r(x)$. This function has a finite band of spatial frequencies and exhibits self-similarity over a corresponding finite range of resolution. The fractal dimension *D* in the function gives a measure of the surface roughness ranging from $D = 1$ (smooth periodic curve) to $D = 2$ (rough, area-filling curve). This function is expressed as a weighted sum of periodic functions in the form,

$$f_r(x) = \sigma C \sum_{n=0}^{N-1} (D - 1)^n \sin(K_0 b^n x + \varphi_n), \tag{6.43}$$

where D $(1 < D < 2)$ is the roughness fractal dimension, K_0 is the fundamental spatial wave number, b (> 1) is the spatial frequency scaling parameter, φ_n are arbitrary phases, and N is the number of tones. The amplitude control factor

$$C = \sqrt{\frac{2[1-(D-1)^2]}{[1-(D-1)^{2N}]}} = \sqrt{\frac{2D(2-D)}{[1-(D-1)^{2N}]}} \tag{6.44}$$

is chosen so the function (6.43) has a standard deviation (r.m.s. height) σ while the value of b can be chosen such that the fractal function is almost-periodic. In eq. (6.43), the periodic functions of increasing frequency in the summation produce the fine structures. The dilation symmetry or self-similarity of (6.43) can be demonstrated by the relation

$$f_r(x) \approx \frac{1}{D-1} f_r(bx). \tag{6.45}$$

Clearly, other periodic functions could be used in function (6.43) to replace the sine function if desired.

As we can see from relation (6.43), the structural profile of the rough surface is determined by the parameters σ (r.m.s. height), D (fractal dimension), b (frequency scaling), K_0 (fundamental wave number), and N (number of tones). The traditional parameters used in random surface modeling are σ (r.m.s. height), Γ (correlation length), and σ_s (r.m.s. slope). It is of interest to find the relation between these two sets of parameters.

The r.m.s. height of the fractal surface is known to be σ. The r.m.s. slope σ_s of this fractal surface can be found by deriving the r.m.s. value of the first derivative of function (6.43). The result is

$$\sigma_s = K_0 \sigma \sqrt{\frac{[1-(D-1)^2]}{[1-(D-1)^{2N}]} \frac{[1-b^{2N}(D-1)^{2N}]}{[1-b^2(D-1)^2]}} \tag{6.46}$$

Note the special case $\sigma_s = K_0 \sigma$ when either $D = 1$ or $N = 1$.

Consider here the reflection of plane waves from a rough fractal conducting surface with variations along the x-axis as shown in Figure 6.17. The incident wave is characterized by wave vector $\mathbf{k_i}$ making an angle θ_i with respect to the z-axis while the scattered wave is characterized by the wave vector $\mathbf{k_s}$ making an angle θ_s with respect to the same axis. Both $\mathbf{k_i}$ and $\mathbf{k_s}$ are of magnitude k, the free-space eave number. We use here the Kirchhoff solution for scattering from rough surfaces which takes the exact roughness profile into account. This

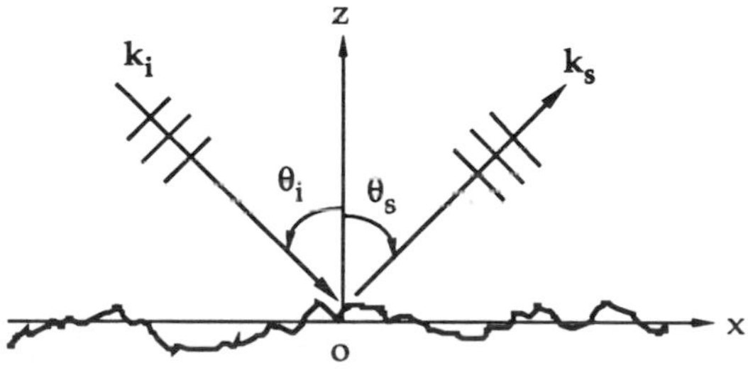

FIGURE 6.17. The geometry of electromagnetic wave scattering in a plane from a rough surface where the subscripts i and s indicate parameters associated with incident and scattered waves, respectively. Here $\mathbf{k}$ represents the wave vector and $\theta_i = 30°$.

Kirchhoff solution treats the rough surface as locally flat with the assumption that the wavelength of the incident wave is small relative to the radius of curvature of the surface irregularities. Angles close to normal incidence will be used to avoid the problem of shadowing.

If the incident plane wave of unit amplitude impinges on a square rough surface characterized by fractal function $f_r(x)$ extending from $x = -L$ to $x = L$, and $y = -L$ to $y = L$, as shown in Figure 6.17, the scattered field $\psi(\mathbf{x})$ at a distance $r = |\mathbf{x}|$ from the origin is given by the integral

$$\psi(\mathbf{x}) = \frac{ik \exp(ikr)}{4\pi r} \int_{-L}^{L} \int_{-L}^{L} (pf_r' - r) \exp[iv_x x' + iv_z f_r(x')] \, dx' \, dy' \quad (6.47)$$

with

$$p = (1 - R) \sin \theta_i + (1 + R) \sin \theta_s, \quad (6.48)$$

$$r = (1 + R) \cos \theta_s - (1 - R) \cos \theta_i, \quad (6.49)$$

$$v_x = +k(\sin \theta_i - \sin \theta_s), \quad (6.50)$$

$$v_z = -k(\cos \theta_i + \cos \theta_s), \quad (6.51)$$

where R is the reflection coefficient of the tangential plane at the point of interest, and θ_s is the direction of observer.

For computational simplicity, we consider here the case of scattering from a perfectly conducting rough surface as a suitable approximation to many naturally occurring situations, such as the problem of microwave scattering or imaging of ocean waves. The value of the Fresnel reflection coefficients become

$$R^{\perp} = 1, \qquad R^{\parallel} = -1, \quad (6.52)$$

where the superscripts $\perp$ and $\parallel$ indicate the wave polarization perpendicular and parallel to the incident plane, respectively.

For smooth perfectly conducting surfaces, we find, in the direction of specular reflection ($\theta_i = \theta_s$),

$$\psi(\mathbf{x}) = \frac{2ik \exp(ikr) L^2 \cos \theta_i}{\pi r}. \quad (6.53)$$

Introducing the scattering coefficient γ, we have from (6.47), after integrating by parts [30],

$$\begin{aligned} \gamma &= \frac{\psi(\mathbf{x})}{\psi_0(\mathbf{x})} \\ &= \frac{1}{4L \cos \theta_i} \left\{ \left(q + \frac{pv_x}{v_z} \right) \int_{-L}^{L} \exp(iv_x x + iv_z f_r(x)) \, dx - \right. \\ &\quad \left. \left[\frac{ip}{v_z} \exp(iv_x x + iv_z f_r(x)) \right]_{-L}^{L} \right\}. \end{aligned} \quad (6.54)$$

Inside the braces of eq. (6.54) the first term gives the major contribution to the scattering while the second term is an edge effect which is negligible when $L \gg \lambda$. After neglecting the edge effect, eq. (6.54) yields the closed form solution for the scattering coefficient

$$\gamma^{\perp}_{\parallel} = \pm \sec\theta_i \frac{1 + \cos(\theta_i + \theta_s)}{\cos\theta_i + \cos\theta_s}$$

$$\times \sum_{m_1, m_2, \ldots, m_{N-1} = -\infty}^{+\infty} \exp\left(i \sum_{n=0}^{N-1} m_n \varphi_n\right) \prod_{n=0}^{N-1} J_{m_n}[C(D-1)^n v_z \sigma]$$

$$\times \operatorname{sinc}\left[\left(v_x + K_0 \sum_{n=0}^{N-1} m_n b^n\right) L/\pi\right]. \tag{6.55}$$

where again the superscripts $\perp$ and $\parallel$ denote the perpendicular and parallel polarizations.

For a special case with $K_0 L \gg 1$, $k\sigma < 1$, and $\theta_i = \theta_s$, the scattering coefficient of the fractal surface in the specular direction yields

$$\gamma^{\perp}_{\parallel} = \pm \prod_{n=0}^{N-1} J_0[2C(D-1)^n k\sigma \cos\theta_i], \tag{6.56}$$

which is not a function of b or θ_n. A second-order expansion of eq. (6.56) for small $k\sigma$ yields

$$\gamma^{\perp}_{\parallel} \approx \pm[1 - 2(k\sigma)^2 \cos^2\theta_i], \tag{6.57}$$

which is not a function of either the fractal dimension D or the number of tones N. This equation shows that the decrease in the specular scattering is proportional to the variance of the rough surface. Equation (6.57) is consistent with the result derived for the average scattering coefficient of random surfaces and demonstrates that only the relative r.m.s. height of the surface determines the scattering intensity in the specular direction whether the surface is a fractal or a random.

We turn now to the numerical calculation of the scattering coefficient γ for different fractal dimensions using the Kirchhoff result (6.54). Plotted in logarithmic scale, Figure 6.18 shows examples of scattering patterns from fractal surfaces of several fractal dimensions with frequency scaling parameter $b = 2e/3 \approx 1.8122\ldots$ and six tones ($N = 6$). The illuminating plane wave is incident from 30° to the left ($\theta_i = 30°$). Here the r.m.s. height σ of the surface is 0.05λ and the patch size $2L$ is 40λ. Each plot in Figure 6.18 is the average result of ten members of the ensemble, each with a different set of randomly chosen phases θ_n for $f_r(x)$.

The scattering pattern for $D = 1.05$, where the roughness is minimal and the surface almost sinusoidal, shows a main beam at 30° to the right which is the specular scattering direction. The two large sidelobes, one on each side of the specular scattering lobe, are due to Bragg coupling by the dominant sinusoid. We call these lobes *coupling lobes.* With increasing roughness, for $D = 1.30$ and $D = 1.50$, more coupling lobes emerge and their intensities grow

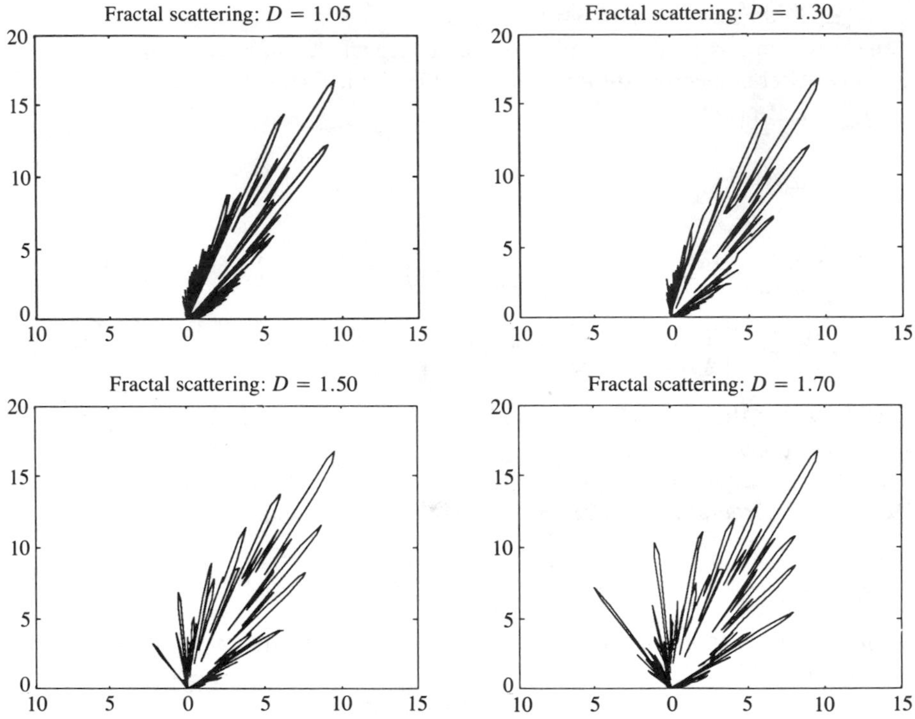

FIGURE 6.18. Patterns of the scattering coefficient magnitude for scattering from fractally corrugated surfaces with fractal dimensions $D = 1.05$, 1.30, 1.50, and 1.70, respectively, from top-left to bottom-right. Here the incident wave illuminates the surface from 30° to the left of normal. Adapted from [30]. Note that these are polar plots and the labels provide a dB scale for reference.

due to the additional coupling by significant sinusoids in the summation (6.43). The intensity of each coupling lobe approaches the same value as $D \rightarrow 2$. This effect can be observed in the plot for $D = 1.70$. The $D = 1.95$ case, where the intensities are more equal, is not included since its roughness is larger than that allowed by the Kirchhoff approximation. From these plots it is clear that the nulls occur in each member of the ensemble since nulls are evident in the ensemble average. The nulls are due to the nulls of the sinc functions of (6.55).

In order to study the relation of the fractal dimension to coupling lobe intensities, we plot again the $\theta_s = 30°\text{–}90°$ portion of Figure 6.18 as Figure 6.19. Plots in Figure 6.19 are the scattering coefficient versus $\sin[(\theta_s - 30°)/2]$, both in dB scales, for the same set of fractal dimensions as Figure 6.18. Two envelopes are indicated in the plots. The background envelope (slope ≈ -1) is due to the finite patch size and consists of the specularly reflected main beam and its sidelobes. The coupling lobe envelope (variable slope) is due to surface coupling and consists of the first-order coupling of the surface harmonics. The slope of the coupling lobe envelope is indicative of surface roughness. These

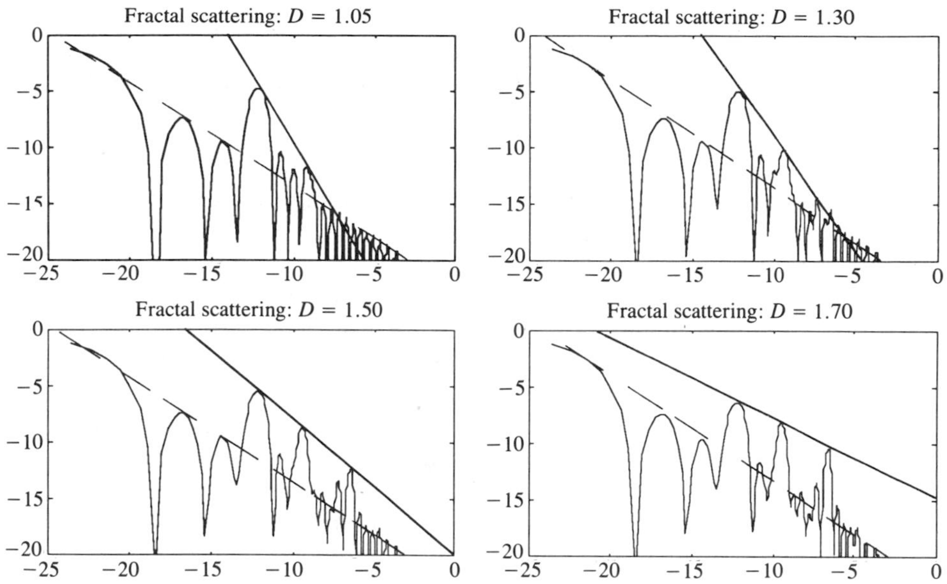

FIGURE 6.19. The scattering coefficient versus $\sin[(\theta_s - 30°)/2]$, both in dB scales, for fractal dimensions $D = 1.05$, 1.30, 1.50, and 1.70, respectively, from upper-left to lower-right. The envelope slopes of coupling sidelobes (solid lines) vary monotonically with the fractal dimension while the background slope (dashed lines) is constant for varying D.

slopes are found to be -2.50, -1.92, -1.22, and -0.72 for fractal dimensions $D = 1.05$, 1.30, 1.50, and 1.70, respectively. This implies that the magnitude of slope for our continuous bandlimited fractal surfaces varies approximately as $[2.7(2 - D)]$. This result is not unlike that of scattering from fractal aggregates as expressed by (6.15). Most importantly, the diffracted envelope slopes provide a remote means for quantifying surface roughness.

From the scattering calculations of this canonical problem, we make several observations regarding the fractal surfaces considered here. First, the scattering intensity in the specular direction depends only on the r.m.s. height of the rough surface. This result is in agreement with the average results of random surface models. Second, the fractal dimension D, which determines the surface roughness, controls the distribution of energy among scattering lobes. This result is typical of wave interactions with fractal structures. Third, the spatial frequency parameter b, which determines the separation of the spatial frequency of the harmonics, controls the angular separation of the coupling beams in accordance with conservation of momentum. Fourth, we note the slope of the coupling lobes provides a quantitative measure of surface roughness and the fractal dimension D predicted previously by heuristics. Finally, note that the formulation for this problem is almost identical to that for diffraction by a single fractal phase screen. In each case, an undisturbed incident field has impressed on its phase front the fractal information. The wave then propagates according to the laws of diffraction. An alternative approach to fractal surface scattering using a generalized Rayleigh approach is given elsewhere [32].

6.5.2. *Scattering by Fractal Fibers*

As the last canonical example we consider the scattering of electromagnetic waves from a fractally corrugated cylindrical surface. This structure can be imagined as the rough surface of the previous subsection which is wrapped around a right circular cylinder. The results of this problem have applications to light scattering by naturally occurring fibers or the scattering of light or microwaves from synthetically tailored surfaces. Studies of fractal spheres have been carried out using alternative means [33], [34].

The model used for these calculations is shown in Figure 6.20 where a normally incident plane wave impinges on an azimuthally symmetric dielectric cylinder of mean radius a and length l immersed in air. The incident and scattered wave vectors $\mathbf{k}_{inc}$ and $\mathbf{k}_{sc}$ form a plane with normal parallel to the cylinder axis. The longitudinal surface variations are described by the bandlimited Weierstrass function (6.8) and we examine the differential scattering cross section as a function of normalized size parameter for various fractal dimensions.

In cylindrical coordinates, the electric field $\mathbf{E}(\rho, \varphi, z)$ satisfies the Helmholtz equation for fiber refractive index n and vacuum wave number k

$$\{\nabla^2 + k^2[1 + (n^2 - 1)f(\rho, z)]\}\mathbf{E}(\rho, \varphi, z) = 0, \tag{6.58}$$

where the step function $f(\rho, z)$ contains the fractal information and is given by

$$f(\rho, z) = \begin{cases} 1, & \rho \leq a[1 + W(z)] \\ 0, & \rho > a[1 + W(z)], \end{cases} \tag{6.59}$$

and where $W(z)$ is the bandlimited Weierstrass function (6.8). Under the

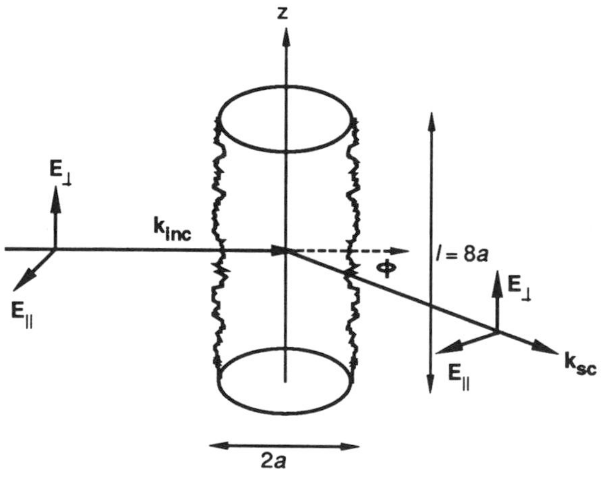

FIGURE 6.20. Scattering geometry for scattering of the incident plane wave by fractal fiber of length l and mean radius a. The scattering plane is defined by the incident and scattered wave vectors, $\mathbf{k}_{inc}$ and $\mathbf{k}_{sc}$ and field components are measured either parallel or perpendicular to this plane.

Rayleigh–Gans approximation, the differential scattering cross section $(d\sigma/d\Omega)_{\perp\to\perp}$ and $(d\sigma/d\Omega)_{\parallel\to\parallel}$ for the perpendicular and parallel polarization in the object plane of symmetry can be computed as a function of the azimuthal angle φ [35]. Assuming $(n-1) \ll 1$ and $(n-1)\,ka \ll 1$ we calculate

$$\left(\frac{d\sigma}{d\Omega}\right)_{\perp\to\perp} = \tfrac{1}{4}(n^2-1)^2 k^4 \left\{\int_{-l/2}^{l/2} dz \int_0^{a[1+W(z)]} J_0\left(2kb\sin\frac{\varphi}{2}\right) b\, db\right\}^2, \tag{6.60}$$

$$\left(\frac{d\sigma}{d\Omega}\right)_{\parallel\to\parallel} = \tfrac{1}{4}\cos^2\varphi(n^2-1)^2 k^4 \left\{\int_{-l/2}^{l/2} dz \int_0^{a[1+W(z)]} J_0\left(2kb\sin\frac{\varphi}{2}\right) b\, db\right\}^2. \tag{6.61}$$

Defining the dimensionless normalized size parameter $Q = 2ka\sin(\varphi/2)$ $(= qa)$ and representing the effective size of the fiber, eqs. (6.60) and (6.61) become

$$\left(\frac{d\sigma}{d\Omega}\right)_{\perp\to\perp} = \tfrac{1}{4}(n^2-1)^2(ka)^4 l^2 \left\{\int_{-1/2}^{1/2} [1+W(\alpha l)]\frac{1}{Q} J_1\{[1+W(\alpha l)]Q\, d\alpha\}\right\}^2 \tag{6.62}$$

$$\left(\frac{d\sigma}{d\Omega}\right)_{\parallel\to\parallel} = \cos^2\varphi\left(\frac{d\sigma}{d\Omega}\right)_{\perp\to\perp}. \tag{6.63}$$

For low frequency $ka \ll 1$ $(Q \ll 1)$, (6.62) yields

$$\left(\frac{d\sigma}{d\Omega}\right)_{\perp\to\perp} = \tfrac{1}{16}(n^2-1)^2(ka)^4 l^2(1+\eta^2)^2 \tag{6.64}$$

which is the volume scattering portion proportional to $k^4(\text{volume})^2$. Here η^2 is the variance of the Weierstrass function as defined in relation (6.8). For $\eta = 0$, the formulation (6.62) of the differential scattering cross section reduces to that of smooth cylinder case

$$\left(\frac{d\sigma}{d\Omega}\right)_{\perp\to\perp} = \tfrac{1}{4}(n^2-1)^2(ka)^4 l^2\left[\frac{1}{Q}J_1(Q)\right]^2. \tag{6.65}$$

Comparing the corrugated surface formulation and the smooth surface formulation, we see that the surface corrugation function $W(z)$ which contains fractal information is the varying portion of the weight function $[1 + W(z)]$ in the weighted average. Equation (6.24) implies that the effective volume of this finite fiber has been increased by a factor of $(1+\eta^2)$ due to the surface corrugation. This factor can also be obtained by calculating the average volume of the corrugated fiber directly. Note the positions of nulls in both the smooth and the fractal cases are given by the zeros of $J_1(Q)$ regardless of the phases θ_m given in relation (6.8). This is reminiscent of the case of scattering from a fractally rough surface given in the previous section.

We have simulated a large number of samples for the differential scattering cross section of the fractal fibers using eq. (6.62) with a fiber refractive index

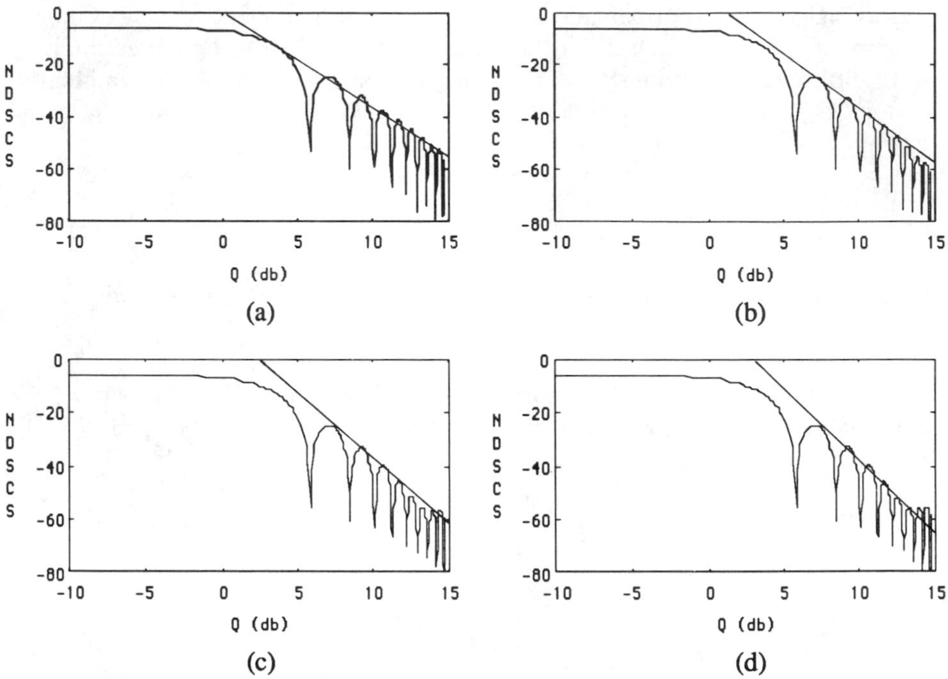

FIGURE 6.21. Simulation results for the perpendicular normalized differential scattering cross section (NDSCS) $[= 4(n^2 - 1)^{-2}(ka)^{-4}l^{-2}(d\sigma/d\Omega)_{\perp\to\perp}]$ are plotted against the normalized size parameter $Q = 2ka\sin(\varphi/2)$ in logarithmic scale. For the same variance $\eta = 0.1$, the slopes of the envelopes are measured for different fractal dimension D. (a) $D = 1.05$, slope $= -3.88$; (b) $D = 1.50$, slope $= -4.00$; (c) $D = 1.75$, slope $= -4.54$; and (d) $D = 1.95$, slope $= -4.82$. Adapted from [35].

$n = 1.015$. A spatial frequency parameter $b = \sqrt{\pi} = 1.77245\ldots$ and nine tones ($N = 9$) are consistently used in the bandlimited Weierstrass function (6.8). Plots of the relative differential scattering cross section versus the normalized size parameter $Q = 2ka\sin(\varphi/2)$ for varying fractal dimension D and a specified value of η^2 are shown in Figure 6.21. Each plot shown here is the average result over ten samples each with a different set of random phases in the Weierstrass corrugation function.

Note that the intermediate frequency regime shown in the plots of Figure 6.21 represent fractal scattering while the low- and high-frequency regimes and nonfractal. Here the fractal and nonfractal regimes of the rough fiber are easily identified through electromagnetic interrogation. The low-frequency ($Q \lesssim 1$) regime represents Rayleigh scattering and so depends on the total volume of the fiber segment which is not a function of the fractal parameters. High-frequency scattering ($Q \lesssim 50$) is due to the wave interaction with the details of the scatterer. The fractal information can be "seen" by wavelengths comparable to the size of the fiber. The Rayleigh–Gans approximation we use here is valid under the condition $(n - 1)ka \ll 1$, which includes low- and intermediate-frequency regimes ($Q \lesssim 15$) since the tenuous refraction index used here is $n = 1.015$.

Figure 6.21 demonstrates the monotonic dependence of this slope value on the fractal dimension D of the surface corrugation for a fixed variance $\eta = 1.0$. With increasing fractal dimension D, (a) $D = 1.05$ (relatively smooth corrugation), (b) $D = 1.50$, (c) $D = 1.75$, and (d) $D = 1.95$ (relatively rough corrugation), the increasing slope values for the lobe envelopes are found to be -3.88, -4.00, -4.54, and -4.82, respectively. It is apparent that increasing roughness (increasing D) leads to a faster fall off of the differential scattering cross section with frequency or normalized size parameter.

These scattering results can be interpreted through the concept of the quality factor of the fiber. Resonant scattering will increase for fibers with a high-quality factor and will decrease for low-quality factor fibers. Here the quality factor decreases for increasing roughness (increasing D) or for increasing frequency Q (the normalized size parameter). Therefore, for rough fibers at high frequencies, the scattering envelope drops off more quickly than for its smooth counterpart as observed here.

From the four samples in Figure 6.21, we note that the positions of the nulls and the peaks are fixed. The amplitude of each lobe varies with the defining characteristics of the surface corrugation. From the results of similar simulations (not shown), we also conclude that the scattering field is insensitive to the number of tones N used in the Weierstrass corrugation function if N is sufficiently large. Since the small-amplitude extremely fine structures cannot be probed by the relatively long-wavelength electromagnetic waves we use here, this result is expected from physical arguments. The fiber length l, we can see from the formulations, play a significant rôle only in the overall scaling of the scattering power. It can be shown that the choice of the fundamental frequency sb^{N_1} has no impact on the scattered field variation except for minor edge effects.

The slope of the envelope of the differential scattering cross-section pattern, as expected from the results of other fractal volume scattering work, is monotonically dependent on the fractal dimension D. The increasing steepness of the slope of the scattering cross section with fractal dimension in the fractal regime again provides a remote means for characterizing these rough fibers by their scattering characteristics. These results also indicate possibilities for the tailoring of surface geometry to achieve desired scattering properties. Note that as a result of the fractal corrugation, the forward scattering is enhanced at the expense of backscatter.

The scattering of electromagnetic waves from cylinders which are fractally fluted in the azimuthal direction have also been considered [36].

6.6. Epilogue

Here the framework for fractal electrodynamics has been laid through the integration of nineteenth-century electromagnetics with twentieth-century geometry. Two important concepts are used to make this connection. First, the introduction of bandlimited fractals allows us to overcome many of the calculational obstacles inherent in applying differential operators to mathematical fractals. An auxiliary benefit is the ability to easily specify complicated boundaries with functions of variable roughness and other desired attributes. Second, the concept of using a variable wavelength as an interrogating wave provides the electromagnetician with an appropriate yardstick suitable for remote measurements and characterization.

The canonical problems examined here suggest that wave interactions with fractal objects have a commonality which allows us to characterize these complex interactions. These concepts may be applied to electromagnetic wave interactions with naturally occurring structures, to the characterization of imperfect devices, or to the tailoring of boundaries for specified scattering characteristics. A host of related problems not covered here include the use of fractal electrodynamics for the design of robust antenna arrays, for investigations of geophysical and acoustical remote sensing, and for the calculation of electromagnetic radiation from lightning.

Acknowledgments

There are many individuals, colleagues, and students, who are responsible for my past and current interest in the relation of geometry and electromagnetics. Most of all, I appreciate the inspiration of Professor C.H. Papas (California Institute of Technology) who introduced me to simplicity and physical insight when confronting problems in electromagnetic theory. It was his quest to understand the "baby problem" behind all other problems and to understand the physics behind the mathematics that led to my interest in connecting the complexities of electromagnetics to the simplicity of geometry.

To my colleagues at Penn, especially Professors Kritikos and Engheta, who are a constant source of new ideas and encouragement, I am indebted. Finally, I note that students, both former and present, who are now my colleagues, have helped forge the area of fractal electrodynamics during the past five years. These include Dr. Y. Kim (Jet Propulsion Laboratory) and Dr. X. Sun (University of Pennsylvania) who collaborated with me in many of the canonical problems summarized here and who were not afraid to tackle a new research area before it became established.

References

[1] B.B. Mandelbrot (1983), *The Fractal Geometry of Nature*, W.H. Freeman, San Fransisco, based on earlier versions of his work such as *Fractals*: *Form*, *Chance and Dimension*, W.H. Freeman, San Fransisco, 1977.

[2] H.E. Stanley (1986), From: An introduction to self-similarity and fractal behavior, in *On Growth and Form*, edited by H.E. Stanley and N. Ostrowsky, Martinus Nijhoff, Dordrecht and Boston.

[3] This discussion is based, in part, on a lecture of Leo P. Kadanoff, Measuring the properties of fractals, presented at AT&T Bell Laboratories, Holmdel, NJ

(Feb. 14, 1986); also see M. Barnsley, *Fractals Everywhere*, Academic Press, Boston, 1988.

[4] D.L. Jaggard, S.D. Bedrosian, and J. Dayanim (1987), A fractal-graph approach to networks, *1986 Circuits and Systems Symposium*, Philadelphia, PA (May 4–7, 1987); S.D. Bedrosian and D.L. Jaggard, A fractal-graph approach to large networks, *Proc. IEE-E*, **75**, 966–968.

[5] Y. Kim and D.L. Jaggard (1986), Fractal random arrays, *Proc. IEE-E*, **74**, 1278–1280.

[6] M.V. Berry and Z.V. Lewis (1980), On the Weierstrass–Mandelbrot fractal function, *Proc. Roy. Soc. London*, **ser. A370**, 459–484.

[7] D.L. Jaggard and Y. Kim (1987), Diffraction by bandlimited fractal screens, *J. Opt. Soc. Amer.*, **A4**, 1055–1062.

[8] K.J. Falconer (1985), *The Geometry of Fractal Sets*, Cambridge University Press, Cambridge.

[9] A.S. Besicovitch (1932), *Almost Periodic Functions*, Cambridge University Press, London.

[10] Y. Kim (1987), Wave propagation in bandlimited fractal media, Ph.D. thesis, Department of Electrical Engineering, University of Pennsylvania.

[11] M.V. Berry (1979), Diffractals, *J. Phys.*, **A12**, 781–797.

[12] M.V. Berry and T.M. Blackwell (1981), Diffractal echoes, *J. Phys.*, **A14**, 3101–3110.

[13] E. Jakeman (1982), Scattering by a corrugated random surface with fractal slope, *J. Phys.*, **A15**, L55–L59; and Fresnel scattering by a corrugated random surface with fractal slope, *J. Opt. Soc. Amer.*, **72**, 1034–1041.

[14] E. Jakeman (1983), Fraunhofer scattering by a sub-fractal diffuser, *Optica Acta*, **30**, 1207–1212.

[15] J. Teixeira (1986), Experimental methods for studying fractal aggregates, in *On Growth and Form*, edited by H.E. Stanley and N. Ostrowsky, Martinus Nijhoff, Dordrecht and Boston, pp. 145–162.

[16] Z. Chen, P. Sheng, D.A. Weitz, H.M. Lindsay, M.Y. Lin, and P. Meakin (1988), Optical properties of aggregate clusters, *Phys. Rev.*, **B37**, 5232–5235.

[17] D.W. Schaefer (1984), J.E. Martin, P. Wiltzius, and D.S. Cannell, *Phys. Rev. Lett.*, **52**, 2371.

[18] H.D. Bale and P.W. Schmidt (1984), Small-angle X-ray-scattering investigation of submicroscopic porosity with fractal properties, *Phys. Rev. Lett.*, **53**, 596–599.

[19] V.I. Tatarskii (1961), *Wave Propagation in a Turbulent Medium*, McGraw-Hill, New York.

[20] L.A. Chernov (1960), *Wave Propagation in a Random Medium*, McGraw-Hill, New York.

[21] A.N. Kolmogorov (1961), The local structure of turbulence in incompressible viscous fluid for very large reynolds' number, and Dissipation of energy in the locally isotropic turbulence, in *Turbulence, Classical Papers on Statistical Theory*, edited by S.K. Friedlander and L. Topper, Interscience, New York.

[22] Y. Kim and D.L. Jaggard (1988), A bandlimited fractal model of atmospheric refractivity fluctuation, *J. Opt. Soc. Amer.*, **A5**, 475–480.

[23] L.F. Richardson (1926), Atmospheric diffusion shown on a distance-neighbor graph, *Proc. Roy. Soc. London*, **Ser. A110**, 709–737.

[24] B.B. Mandelbrot (1974), Intermittent turbulence in self-similar cascades; divergence of high moments and dimension of the carrier, *J. Fluid Mech.*, **62**, 331–358.

[25] U. Frisch, P. Sulem, and M. Nelkin (1978), A simple dynamical model of intermittent fully developed turbulence, *J. Fluid Mech.*, **87**, 719–736.

[26] E.D. Siggia (1978), Model of intermittency in three-dimensional turbulence, *Phys. Rev.*, **A17**, 1166–1176.

[27] R.W. Lee and J.C. Harp (1969), Weak scattering in random media, with applications to remote probing, *Proc. IEEE*, **57**, 375–406.

[28] J. Van Roey, J. Van der Donk, and P.E. Lagasse (1981), Beam propagation method: Analysis and assessment, *J. Opt. Soc. Amer.*, **71**, 803–810.

[29] Y. Kim and D.L. Jaggard (1988), Optical beam propagation in a bandlimited fractal medium, *J. Opt. Soc. Amer.*, **A5**, 1419–1426.

[30] D.L. Jaggard and X. Sun (1990), Scattering by fractally corrugated surfaces, to appear in *J. Opt. Soc. Amer.* **A7** (Spring 1990).

[31] X. Sun (1989), Electromagnetic Wave Scattering from Fractal Structures, Ph.D. thesis, Department of Electrical Engineering, University of Pennsylvania.

[32] See X. Sun and D.L. Jaggard (1990), Scattering from non-random fractal surfaces, to appear in *Optics Comm.* (1990); or, D.L. Jaggard and X. Sun (1990), Rough surface scattering: A generalized Rayleigh solution, submitted for publication.

[33] C. Bourrely, P. Chiappetta, and B. Torresani (1986), Light scattering by particles of arbitrary shape: A fractal approach, *J. Opt. Soc. Amer.*, **A3**, 250–255.

[34] C. Bourrely and B. Torresani (1986), Scattering of an electromagnetic wave by an irregularly shaped object, *Optics Comm.*, **58**, 365–368.

[35] D.L. Jaggard and X. Sun (1989), Backscatter cross-section of bandlimited fractal fibers, *IEEE Trans. Antennas and Propagation*, **AP-37**, 1591–1597.

[36] X. Sun and D.L. Jaggard (1990), Scattering from fractally fluted cylinders, to appear in *J. Electromagnetic Wave Appl.*

7
Fringing Field Effects in VLSI Structures

CHING-LIN JIANG*

7.1. Introduction

Stray capacitances in transistor and interconnect structures play an important role in Very Large Scale Integration (VLSI) circuit designs. The application of electrostatics to the estimation and understanding of various fringing field effects is essential in the optimization of device performances [1], [2].

Figure 7.1 depicts a simplified cross section of three circuit arrangements on a silicon substrate. In Figure 7.1(a) a metal-oxide-semiconductor (MOS) transistor is shown. The gate (G) which acts as the control valve of the device is made of conductive polysilicon 3000–4000 Å thick (X_p). The interface area underneath the gate forms the channel through which electrons or holes flow. The source (S) and drain (D) are highly conducting circuit nodes of heavily doped regions 2000–4000 Å deep (X_j). These heavily doped regions are also referred to as "active" regions. The oxide between the gate and the silicon is about 200–400 Å (t_{ox}). The length (L) of the gate can be as short as 1 μm or less. The overlap capacitances between the gate and source–drain regions can greatly affect the switching speed of the device because they contribute significantly to the time constant of the charging and discharging operations. Here the fringing field effect becomes more pronounced as the process technology advances to the point that the dimensions of the gate, source, and drain (X_p, X_j) are comparable to the gate length L.

In Figure 7.1(b), a metal interconnect line of thickness T and width W is shown to pass over a conducting active region. The oxide separation H varies from 3500 Å to 10,000 Å. The fringing fields modify greatly the line to active region capacitance, especially when the linewidth gets down to 1–2 μm. Another commonly seen situation is depicted in Figure 7.1(c), where two metal or polysilicon lines of thickness T, width W, and separation S are running in parallel. The line-to-line capacitance consists mostly of the fringing field contributions. The noise coupling associated with these line capacitances could result in malfunctions, especially in analog or dynamic digital designs.

* Dallas Semiconductor Corporation, Dallas, TX 75244-3219, USA.

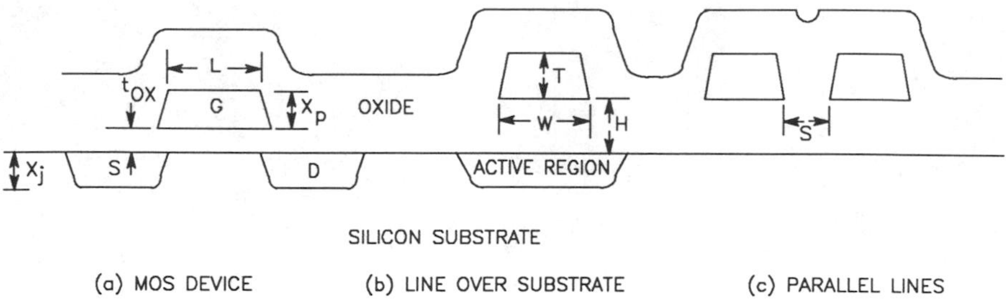

FIGURE 7.1. Simplified cross section of VLSI structures (not to scale).

Therefore, attention must be paid to their estimation and minimization in order to avoid severe circuit degradations.

To calculate the capacitances, we start with the elementary equation

$$C = \frac{Q}{V}, \tag{7.1}$$

where Q is the charge on the structure and V is the potential difference between the electrodes. In terms of fields, eq. (7.1) can be rewritten as

$$C = \frac{\int_S \mathbf{D} \cdot \mathbf{dS}}{\int_l \mathbf{E} \cdot \mathbf{dl}}, \tag{7.2}$$

where **D** is the electric flux density, integrated over the surface of the electrode, and **E** is the electric field intensity, integrated along a path between electrodes. The fields **D** and **E** can be determined from the solution of Laplace's equation $\nabla^2 \varphi = 0$ with volume charge density $\rho = 0$, and subject to the proper boundary conditions at the electrode surfaces and the different dielectric interfaces. Then the capacitance value can be calculated from eq. (7.2). Typically, numerical techniques have to be employed in many cases of practical interest.

However, approximate analytical expressions for stray capacitance are very desirable to provide insights and guidelines for VLSI circuit designs, simulations, and process improvements. It is the purpose of this chapter to review some of the simplified models and techniques used in analyzing the fringing field effects of VLSI devices. The three circuit structures shown in Figure 7.1(a–c) will be considered.

7.2. Overlap Capacitance of a VLSI MOS Device

The MOS device in Figure 7.1(a) can be modeled as shown in Figure 7.2 to estimate its overlap capacitance. The following simplifying assumptions are made [1]:

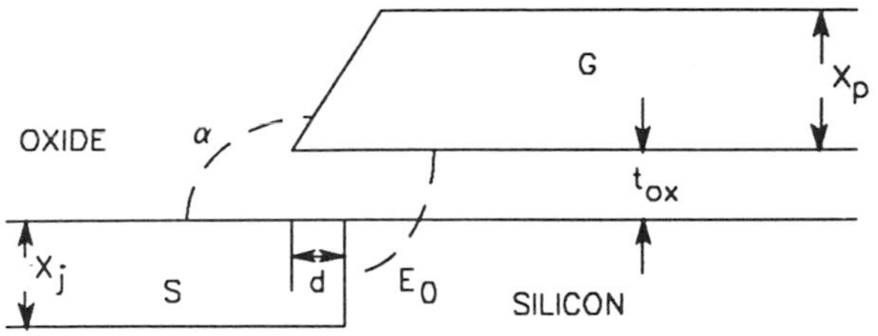

FIGURE 7.2. Approximate MOS overlap model.

(a) Since $t_{ox} \ll X_p$, $t_{ox} \ll X_j$, the regions away from the oxide gap becomes insignificant, thus the junction sidewall is approximated as a straight surface.
(b) The electric field lines across the silicon–oxide interface are assumed to be normal.
(c) Interface charge is negligible.
(d) The fields in the silicon substrate consist of the component in the φ direction only and its magnitude depends on the radial distance only.

From the above, we obtain

$$\varepsilon_{ox} E_{ox}|_{Si-ox} = \varepsilon_{Si} E_{Si}|_{Si-ox}. \tag{7.3}$$

At this point, a transformation can be made to convert the structure in Figure 7.2 to a uniform dielectric model in Figure 7.3, where the silicon substrate is replaced with oxide. It is noted that the source sidewall is now at an angle β ($\neq \pi/2$) to the gate surface. To calculate β, we notice that the potential difference $\int_l \mathbf{E} \cdot \mathbf{dl}$ between the Si–ox interface and the source region is invariant under the transformation. Thus

$$\beta E_{\varphi,ox} = \frac{\pi}{2} E_{\varphi,Si}, \tag{7.4}$$

where $E_{\varphi,ox}$ represents the field in the φ direction with silicon replaced by oxide as shown in Figure 7.3, and $E_{\varphi,Si}$ the corresponding field in the original configuration (Figure 7.2). Since E_φ's depend only on radial distance and at

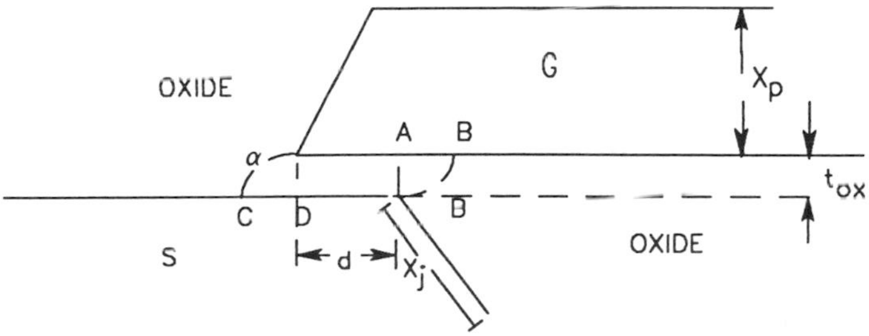

FIGURE 7.3. Overlap structure after transformation.

the original interface

$$E_{\varphi,\mathrm{ox}} = E_{\mathrm{ox}}|_{\mathrm{Si-ox}},$$

$$E_{\varphi,\mathrm{Si}} = E_{\mathrm{Si}}|_{\mathrm{Si-ox}},$$

therefore, from eqs. (7.3) and (7.4) we obtain

$$\beta = \frac{\pi}{2} \cdot \frac{\varepsilon_{\mathrm{ox}}}{\varepsilon_{\mathrm{Si}}}. \tag{7.5}$$

Now the total overlap capacitance per unit width for the two-dimensional device structure (Figure 7.3) can be obtained by combining the following three components:

(1) capacitance from plates at angle α and length $X_p/\sin\alpha$;
(2) parallel plate capacitance with length $d + \Delta$; and
(3) capacitance from plates at angle β and length X_j.

The unit capacitance between two conducting planes at an angle θ is given by [3]

$$C = \frac{\varepsilon \ln(d_2/d_1)}{\theta}, \tag{7.6}$$

where d_2, d_1, and θ are shown in Figure 7.4. It should be noted that the parallel plate component includes a correction term Δ which is due to the contribution from segments AB and CD. This correction is of higher order and can be approximated by the quasi-empirical formula $\Delta = AB + CD$. Hence

$$\Delta = \left(\frac{1 - \cos\alpha}{\sin\alpha} + \frac{1 - \cos\beta}{\sin\beta}\right). \tag{7.7}$$

The three unit capacitance components are thus given by

$$C_1 = \frac{\varepsilon_{\mathrm{ox}}}{\alpha} \ln\left(1 + \frac{X_p}{t_{\mathrm{ox}}}\right), \tag{7.8}$$

$$C_2 = \varepsilon_{\mathrm{ox}} \frac{d + \Delta}{t_{\mathrm{ox}}}, \tag{7.9}$$

$$C_3 = \frac{\varepsilon_{\mathrm{ox}}}{\beta} \ln\left(1 + \frac{X_j \sin\beta}{t_{\mathrm{ox}}}\right). \tag{7.10}$$

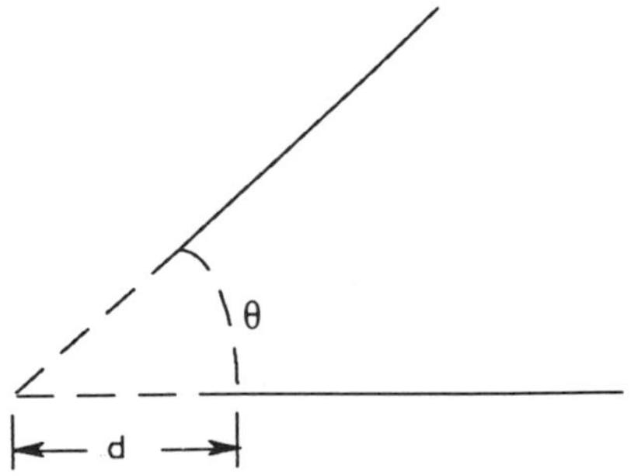

FIGURE 7.4. Two conducting planes at an angle.

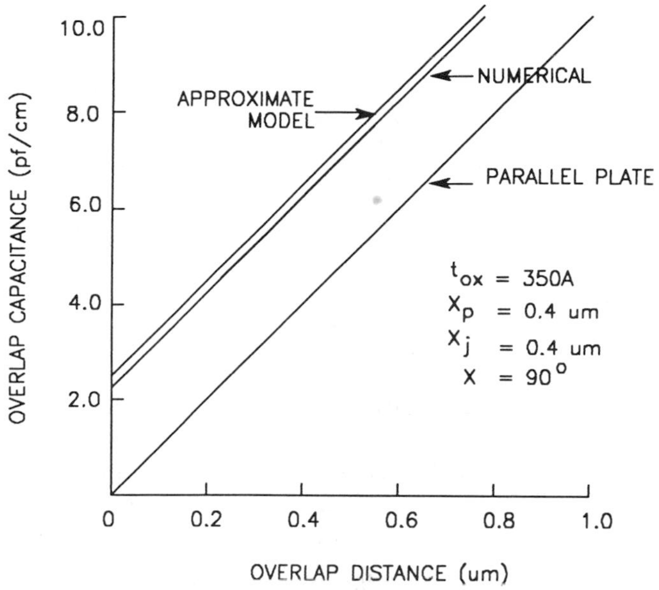

FIGURE 7.5. Overlap capacitance in a MOS device.

Substituting β from (7.5), the total overlap capacitance per unit width of the MOS device is

$$C_{ov} = \frac{\varepsilon_{ox}}{\alpha} \ln\left(1 + \frac{X_p}{t_{ox}}\right) + \varepsilon_{ox}\frac{d + \Delta}{t_{ox}} + 2\frac{\varepsilon_{Si}}{\pi}\ln\left[1 + \frac{X_j}{t_{ox}}\sin\left(\frac{\pi}{2}\frac{\varepsilon_{ox}}{\varepsilon_{Si}}\right)\right]. \quad (7.11)$$

A plot of the overlap capacitance C_{ov} versus the overlap distance d is shown in Figure 7.5. Also shown are a plot of the simple parallel-plate component without fringing fields and a plot of the numerical results computed with a two-dimensional program using the Gauss–Seidel iteration scheme [1]. For a typical case, $d = 0.2\ \mu$m, the deviation of the model from the numerical result is less than 5% whereas the parallel plate capacitance yields an error greater than 50%. It can be readily seen that as the overlap is reduced from $d = 0.2\ \mu$m to zero, a good 50% of the stray capacitance is still present due to the pure fringing field components. This is important in assessing the performance improvement from advanced zero overlap VLSI processes. It may also be noted that C_{ov} is a linear function of d indicating that the fringing components are independent of the overlap in a typical structure. Additionally, it is interesting to see that, from eq. (7.11), the fringing component on the channel side (C_3, eq. (7.10)) is much larger than that on the outer side (C_1, eq. (7.18)) because $\varepsilon_{Si} \sim 3\varepsilon_{ox}$ and α is usually greater than $\pi/2$.

A comparison also has been made between the calculated value from eq. (7.11) and measured data and good agreement (typically within 5%) was indeed confirmed [1].

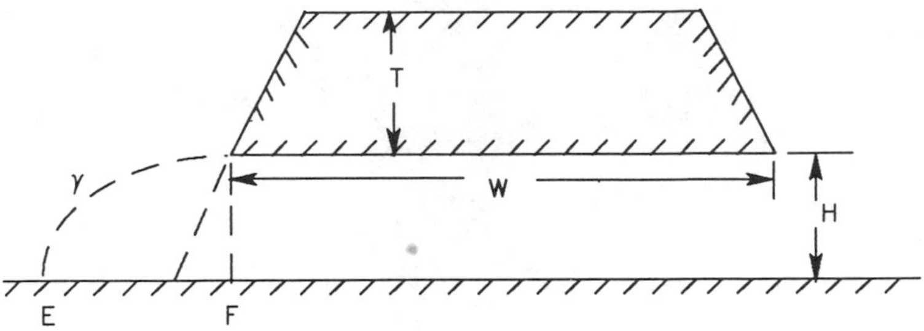

FIGURE 7.6. Line-to-substrate capacitance model.

7.3. Line-to-Substrate Capacitance

For the case in Figure 7.1(b), the simple model of a metal line above a conducting plane will be used (see Figure 7.6). While the top surface contribution is assumed negligible, the following two components will be considered:

(1) capacitances from the plates at an angle γ and length $T/\sin\gamma$;
(2) parallel-plate capacitance with length $W + \Delta_m$.

Here again a quasi-empirical approach will be used to estimate Δ_m:

$$\Delta_m = 2 \cdot EF = 2\frac{1 - \text{Cos}\,\gamma}{\sin\gamma} \cdot H = 2H\tan\left(\frac{\gamma}{2}\right). \tag{7.12}$$

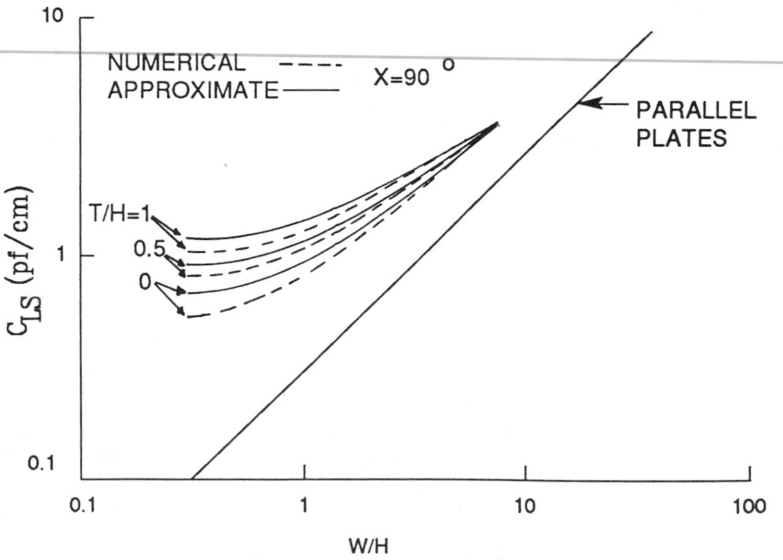

FIGURE 7.7. Line-to-substrate capacitance.

Since the full segments $2 \cdot EF$ are used, this term will overestimate the capacitance, especially when $W < H$ or $T < H$. The line-to-substrate stray capacitance is then given by

$$C_{LS} = 2\frac{\varepsilon_{ox}}{\gamma}\ln\left(1 + \frac{T}{H}\right) + \varepsilon_{ox}\frac{W}{H} + 2\varepsilon_{ox}\tan\left(\frac{\gamma}{2}\right). \tag{7.13}$$

A plot of C_{LS} versus W/H for various values of T/H is shown in Figure 7.7. A rectangular conductor is considered, that is, $\gamma = 90°$. Also shown in Figure 7.7 are the corresponding curves derived from a numerical model [2]. It can be seen that the approximate results are in good agreement with the numerical computations. The fringing field effects are very pronounced for $W/H \lesssim 4$. The usual parallel plate formula gives values off by a few hundred percent. It is also noted that as the conductor becomes narrower from $W/H = 1$ to $W/H = 0.3$, a reduction of only 25% in C_{LS} can be obtained. Similarly, as the conductor thickness changes from $T/H = 1$ to $T/H = 0.5$, a decrease of only 15% results.

7.4. Line-to-Line Coupling Capacitance

Figure 7.8 is a model for the line-to-line interaction shown in Figure 7.1(c), where the substrate is assumed to be a conducting ground plane. By symmetry, the mutual coupling capacitance between the two lines is the sum of the following three components:

(1) Side-to-side capacitance

$$C_{ss} = \frac{\varepsilon_{ox}}{2\lambda - \pi}\ln\left(1 - 2\frac{T}{S}\cot\lambda\right). \tag{7.14}$$

(2) Top-to-top capacitance

$$C_{tt} = \frac{\varepsilon_{ox}}{\pi}\ln\left(1 + 2\frac{W + T\cot\lambda}{S - 2T\cot\lambda}\right). \tag{7.15}$$

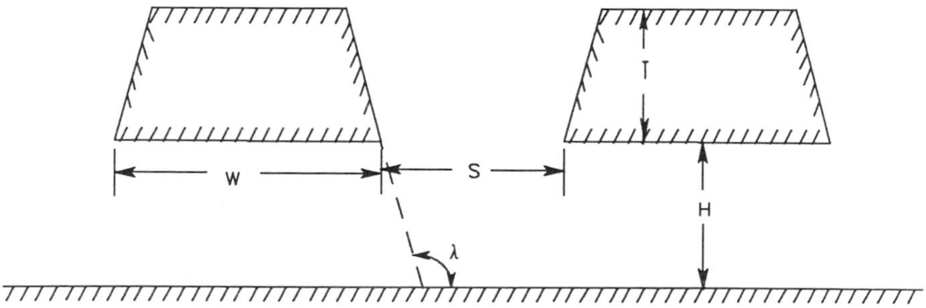

FIGURE 7.8. Line-to-line capacitance model.

(3) Bottom-to-bottom capacitance. The ground plane effect needs to be considered. It is noted that when $S \ll H$, full bottom side coupling should be included, while for $S \gg H$, the contribution from the bottoms is small. A quasi-empirical factor $1/(1 + S/H)$ will be used

$$C_{bb} = \frac{\varepsilon_{ox}}{\pi} \cdot \frac{1}{1 + S/H} \ln\left(1 + 2\frac{W}{S}\right). \tag{7.16}$$

Thus the line-to-line coupling capacitance is given by

$$C_{LL} = C_{ss} + C_{tt} + C_{bb}. \tag{7.17}$$

For rectangular lines, $\lambda = \pi/2$ and eq. (7.17) can be reduced to

$$C_{LL} = \varepsilon_{ox}\frac{T}{S} + \frac{\varepsilon_{ox}}{\pi}\left(1 + \frac{1}{1 + S/H}\right) \ln\left(1 + 2\frac{W}{S}\right). \tag{7.18}$$

Figure 7.9 depicts C_{LL} as a function of W/H with different sets of T/H and S/W values. The dashed curves are derived from a numerical model [2]. The parallel plate capacitance and the line-to-substrate C_{LS} at $T/H = 1$ are also shown. It should be noted that all quantities in Figure 7.9 are magnified by a factor 2. Here again excellent agreement is obtained between the approximate and the numerical models. The relatively large deviations for $W/H > 2$ are due to the overestimation of bottom-to-bottom coupling C_{bb}. In this region, C_{LL} is of little practical significance because $C_{LL} \ll C_{LS}$. It is also noted that for $W/H \lesssim 1$, C_{LL} increases to more then 40% of C_{LS} ($T/H = 1, S/W = 1$). This could result in more than 25% mutual coupling between two very long parallel

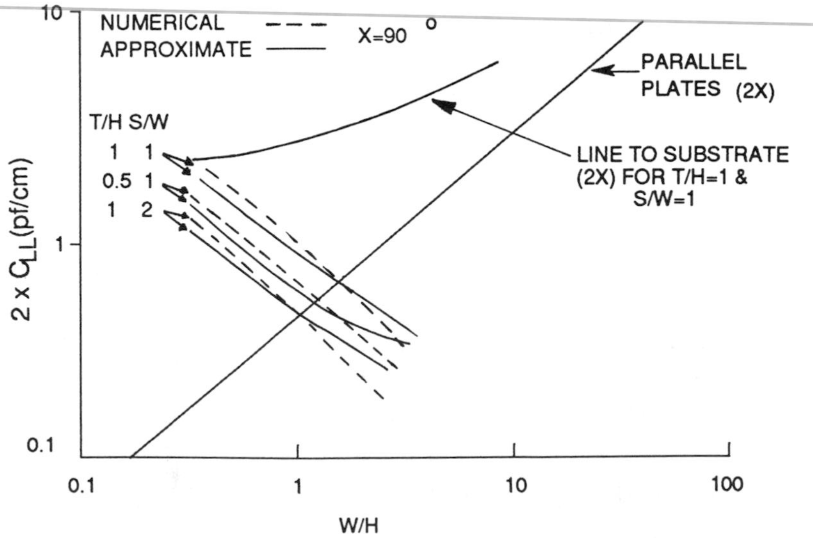

FIGURE 7.9. Line-to-line capacitance (2 ×).

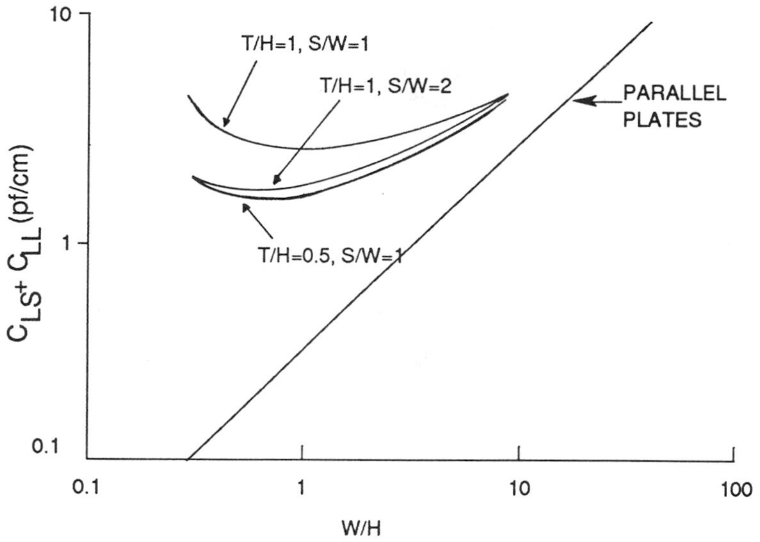

FIGURE 7.10. $C_{LS} + C_{LL}$.

interconnect lines. Therefore, in certain cases isolation lines must be added, or larger spacing must be used to minimize interference between neighboring signals.

In Figure 7.10, C_{LS} and C_{LL} are combined for various T/H and S/W. The total capacitances go through a region of minimum values as W/H varies from small to large. It is clear that when W/H is small, the C_{LL} contribution is significant. As W/H becomes larger, C_{LL} decreases sharply while C_{LS} starts to increase, resulting in concave curves as shown in Figure 7.10. It can be observed that decreasing the line thickness T and/or increasing the line spacing S reduces the total line capacitance and at the same time shifts the minimum point to a smaller line width. However, the thinner T will yield a higher line resistance while a larger S will lower the packing density. These trade-offs must be carefully considered as VLSI technology drives dimensions toward $W/H \lesssim 1$.

7.5. Conclusions

Simple, approximate models have been employed to derive expressions for stray capacitances in various VLSI structures. Good agreements with numerical solutions and experimental results have also been shown. Since stray capacitance plays a major role in determining circuit speed and signal coupling, the approach presented in this article should provide valuable insights and guidelines for the design and optimization of future VLSI circuits and device structures.

Acknowledgments

The author wishes to thank Tom Harrington for many helpful discussions. He also wants to thank Nancy Lee for typing the manuscript, and Kim Memar for her help.

References

[1] R. Shrivastava and K. Fitzpatrick (1982), A simple model for the overlap capacitance of a VLSI MOS device, *IEEE Trans. Electron Devices*, **ED-29**, 12, 1870–1875.

[2] R.L.M. Dang and N. Shigyo (1981), Coupling capacitances for two-dimensional wires, *IEEE Electron Device Letters*, **EDL-2**, 8, 196–197.

[3] M. Zahn (1979), *Electromagnetic Field Theory: A Problem Solving Approach*, Wiley, New York, pp. 272–273.

8
Some Methods of Reducing the Sidelobe Radiation of Existing Reflector and Horn Antennas

VASSILIS KERDEMELIDIS,* GRAEME LESLIE JAMES,† and
BRIAN ALEXANDER SMITH‡

8.1. Introduction

Reflector and horn antennas find extensive use over a wide range of applications such as communications, radioastronomy, remote sensing, medicine, and industry. The ever-increasing use of the electromagnetic spectrum has for some time now required methods of adapting or designing antennas to reduce the possible interference between the wanted and undesired signals. Over the years many techniques have been proposed and a number of methods have received acceptance and appear in general use. They fall into two broad categories:

(a) general reduction of sidelobes; and
(b) gain reduction in specific narrow directions, in effect, null placement in the radiation patterns.

If we begin with a new antenna, then for general reduction of sidelobes in a reflector antenna design a number of options are available, as summarized in the article by Schrank [1]. For horn antennas an obvious choice for low-sidelobe performance is the corrugated horn (see Clarricoats and Olver [2] for a general reference). In many cases, however, the antenna engineer is confronted with finding means of modifying the radiation patterns of existing or off-the-shelf antennas, to reduce the crosstalk between systems and thereby allowing greater utilization of the available frequency bands. This adaptation of existing antennas is especially the case with null placement requirements, as these often change over time.

It is the purpose of this chapter, therefore, to review some of the methods that have been investigated over the past twenty years, both at the authors' laboratory (University of Canterbury, New Zealand) and elsewhere, in

* Department of Electrical and Electronic Engineering, University of Canterbury, Christchurch, New Zealand.

† Division of Radiophysics, CSIRO, Sydney, Australia.

‡ Telecom Corporation of New Zealand, Christchurch, New Zealand.

modifying the radiation patterns of *existing* or off-the-shelf reflector and horn antennas rather than in the design of new ones.

8.2. General Overview of Methods

The techniques applicable to the reduction of sidelobe levels in the forward direction of an antenna differ, in general, from those used exclusively to reduce the back radiation. For instance, the main beam width of the antenna may be reduced so as to lessen the interference pickup from directions close to the main beam. Among the techniques used for the purpose is the placement of low-loss dielectric on the antenna reflector surface. The price paid is an increased sidelobe level elsewhere (Hamid and Mohsen [3]).

Sidelobe levels in the general forward direction may be reduced overall by the placement on the reflector of mylar strips coated with graphite or carbon-impregnated sponge material (Lader and Winderman [4]).

Wider angle sidelobes in the forward direction may be reduced further by attaching "choke blinders" to the exciter feed (Johnson [5], Cain and Johnson [6]).

The radiation behind the reflector or horn antenna is attributed to the field diffracted by the edges. For this reason, most of the techniques aimed at reducing the back radiation concentrate on the edge geometry (Cornbleet [7], Lewin [8], James [9], Koch [10], Middleton [11], James and Kerdemelidis [12], Smith [13]).

In addition to the modifications of the edge geometry, radiation through (holes in) the reflector has been suggested as a possible means of controlling back lobes in predetermined directions (Sletten and Blacksmith [14], Smith [13], Hamilton and Kerdemelidis [15]).

In the remaining sections we consider a number of these approaches in more detail.

8.3. Effect of Lossy Films on Sidelobes from Reflector Antennas

Lader and Winderman [4] have investigated the effects of treating the surface of a reflector antenna with radio-frequency absorbing material on its radiation pattern. They analyzed the influence of amplitude and phase tapering of the aperture field on the resultant pattern of an antenna and then tried to reproduce the results by the use of various materials. Linear, parabolic, and Gaussian tapers were considered analytically. Experiments were then performed for absorptive tapered conical annuli. These annuli were applied to the outer, nearly linear, portions of the reflector. Mylar and carbon-impregnated sponge were used as absorptive material. The results obtained showed considerable pattern modification—a general lowering of forward sidelobes. No results for back radiation are given.

Middleton [11] has further studied the possibility of reducing radiation in predetermined directions for parabolic and plane reflector antennas. His results allow the following observations:

(i) A greater absorption and, consequently, greater effect on sidelobe level is produced by substituting the lossy thin film by a resonant thickness of low conductivity nonmagnetic material.
(ii) Ferrite material, attached to the center or an edge, causes a dramatic drop in the main lobe magnitude. This indicates the areas of the reflector which contribute to the main lobe, i.e., the specularly reflecting region and the edges. In particular, with ferrite material placed at the center, both the main lobe and the near sidelobes are all reduced. Some of the back lobes are also attenuated. With a ferrite strip placed only at the edge of the parabolic reflector, most of the forward sidelobes are reduced. This implies that the edge diffraction is the main cause of the forward sidelobes.

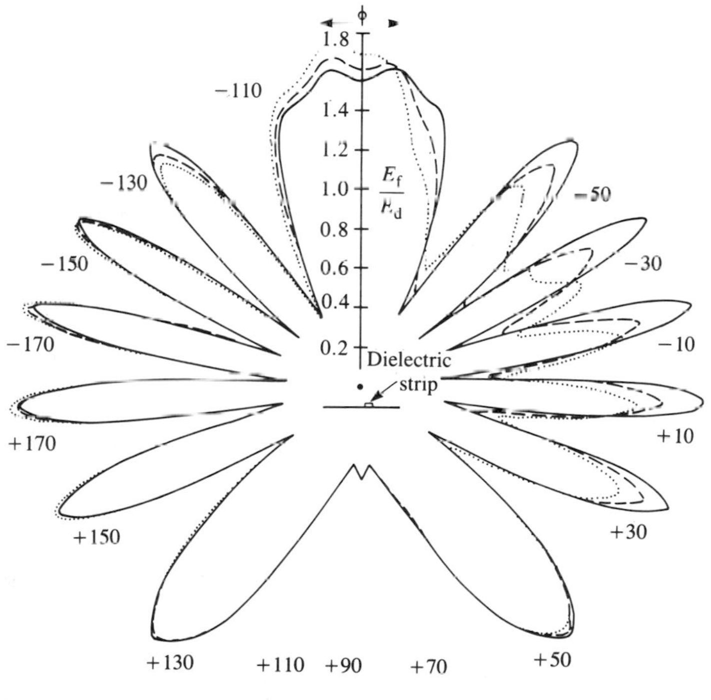

FIGURE 8.1. Far-field pattern of flat reflector antenna with dielectric strip excited with a line source (dot). ——, pattern without dielectric strip; ----, pattern with ferrite strip; ·····, low-conductivity nonmagnetic material. (Reproduced by permission, H.T. Middleton [11].)

(iii) For sidelobe reduction, there exists an optimum width of lossy strip. Lossy strips wider than this optimum tend to reduce antenna gain.
(iv) With the particular combinations of strip dimensions, Middleton noted no great reduction in the sidelobe amplitudes.

Thus, although changes were observed, the results obtained with the use of dielectric or ferrite strips were marginal (see Figure 8.1).

8.4. Effect of Holes or Slits in the Reflector

Sletten and Blacksmith [14] suggested that if radiation is allowed to leak through (holes in) the reflector near the rim, and is of appropriate amplitude and phase, it may cancel the radiation diffracted from the rim. Smith [13] has studied this possibility by placing slits in parabolic reflector antennas.

To effectively control the resultant field from two sources we need to be able to vary their relative phases and amplitudes. The difficulty in this procedure is in independently controlling the phase and amplitude of the radiation through the slit, and is further compounded by the multiplicity of effective diffracting sources on the rim.

The radiation pattern or field from a slit depends on:

(i) the slit size (Smith [13]);
(ii) the angle of incidence of the primary or exciter field on the slit (Smith [13]);
(iii) the relative inclination of the two effective "screens" that make up the slit (Hamilton and Kerdemelidis [15]).

To appreciate some of the problems involved we can observe the diffracting fields for the case of slits in a plane screen (Figure 8.2). The effect of the aperture width and of the angle of incidence of the primary fields are shown in Figures 8.3 and 8.4, where:

Φ_0 is the angle that the incident wave makes with the normal to the aperture; and
Φ is the angle of the diffracted field as shown in Figure 8.2.

Fields in Figures 8.3 and 8.4 were determined by means of Keller's Geometrical Theory of Diffraction (GTD) and taking into account multiply scattered rays. We can see from Figures 8.3 and 8.4 that as the slit width decreases, the diffracted field amplitude decreases and the main diffracted lobe tends to move towards the normal of the slit plane. Note that, as the result of the formulation using plane wave diffraction coefficients, the computed diffracted field diverges along the slit edges. Using cylindrical diffraction coefficients this divergence may be avoided (Smith [13]).

Another complication in using slits to produce nulls in desired directions

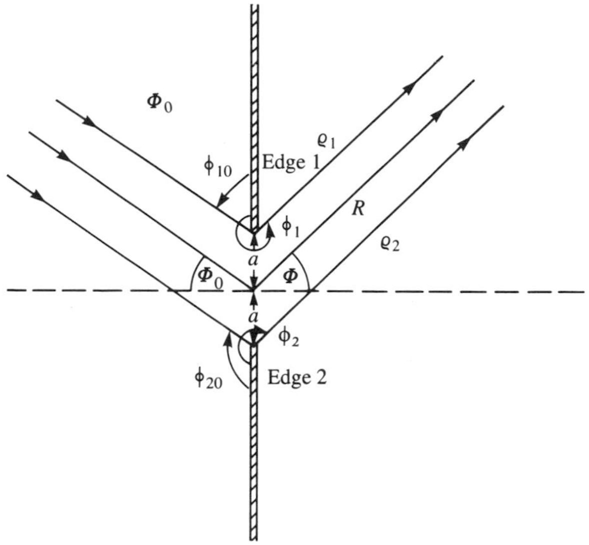

FIGURE 8.2. Geometry of diffraction through slit of width $2a$ in a plane screen. Φ_0 angle of incidence; Φ angle of diffraction. (By permission, B.A. Smith [13].)

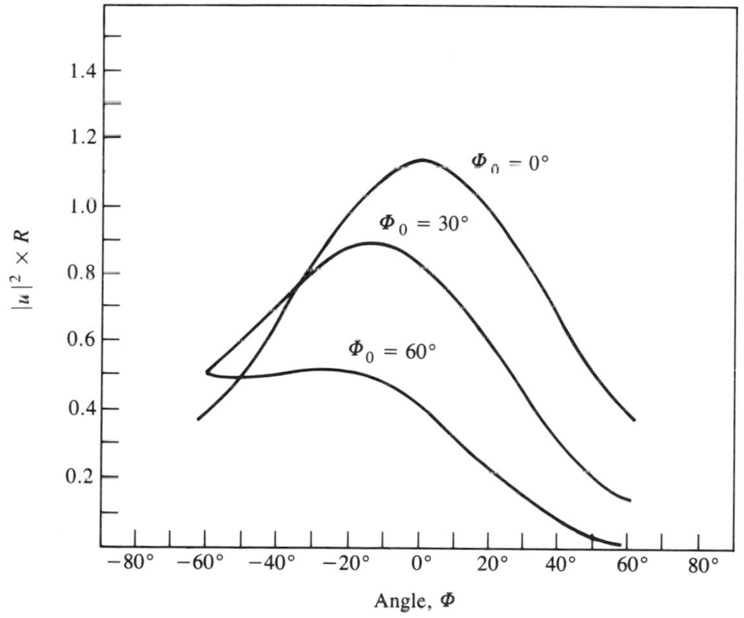

FIGURE 8.3. Angular distribution of scattered intensity for diffraction by a slit of width of half-wavelength using GTD with multiply diffracted rays. (By permission, B.A. Smith [13].)

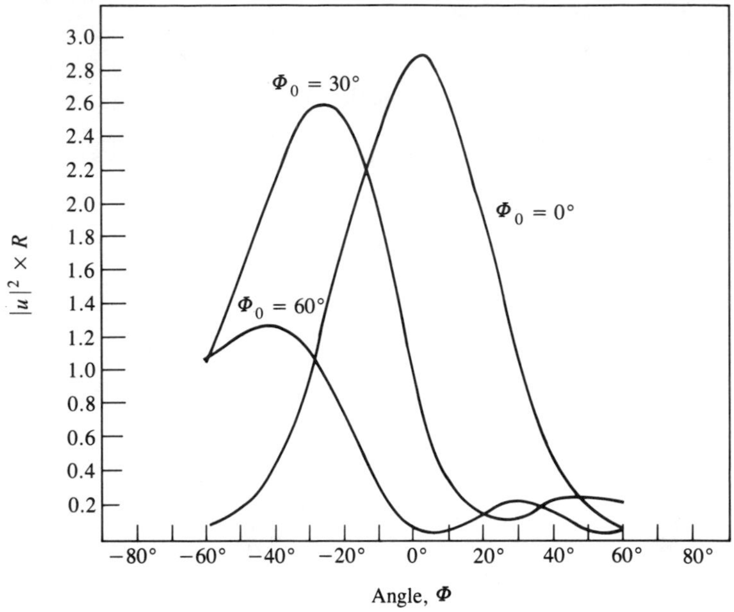

FIGURE 8.4. As for Figure 8.3 but with slit width one wavelength. (By permission, B.A. Smith [13].)

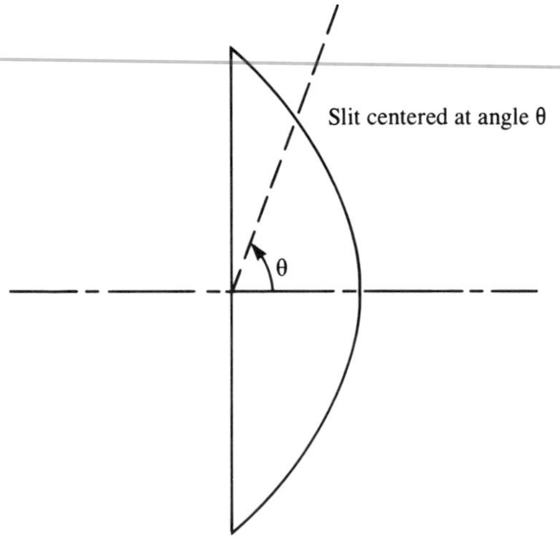

FIGURE 8.5. Parabolic reflector geometry for back radiation computations. (By permission, B.A. Smith [13].)

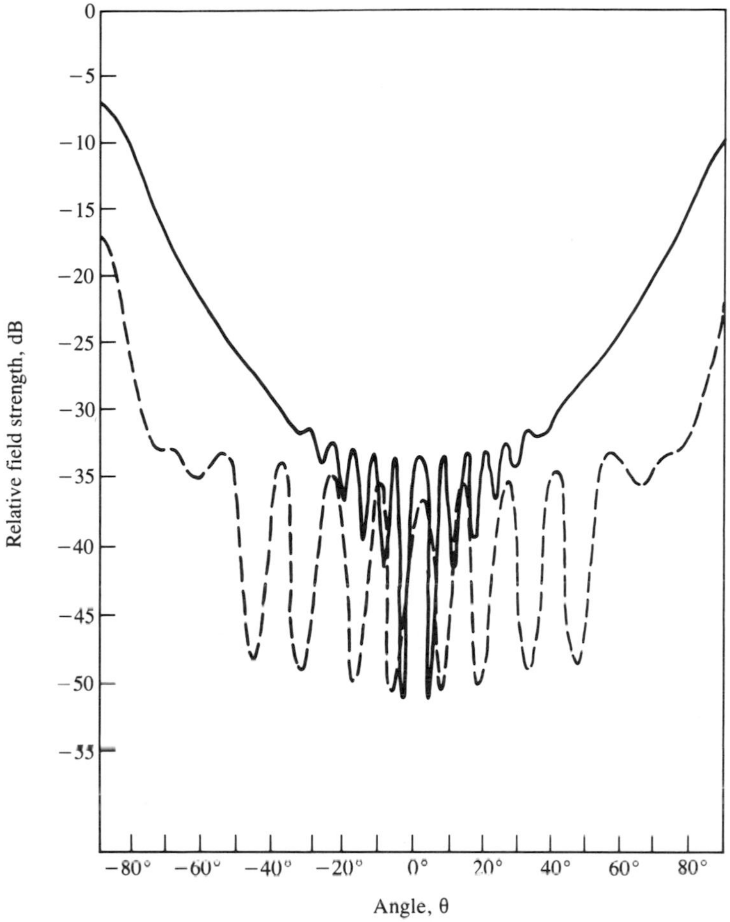

FIGURE 8.6. Back radiation pattern for unperturbed parabolic cyclinder reflector excited by an electric dipole parallel and in focal axis of the parabola. ——, edge diffraction theory; , measured. (By permission, B.A. Smith [13].)

is that a large contribution to the scattered backfield, along the direction θ (Figure 8.5), originates around the area of the reflector, defined by the intersection of the radial at θ and the reflector surface. Thus locating nulls in desired directions by means of slits is a rather complex undertaking.

Figure 8.6 shows the diffraction field for the unperturbed parabolic cylinder antenna, where the aperture $D = 10\lambda$ and the focal length $f = 0.25D$. Figures 8.7, 8.8, and 8.9 show the computed and measured radiation patterns for slits of different widths placed at $\theta = -70°$.

Both the theoretical and the experimental results indicate that as the *slit width increases*:

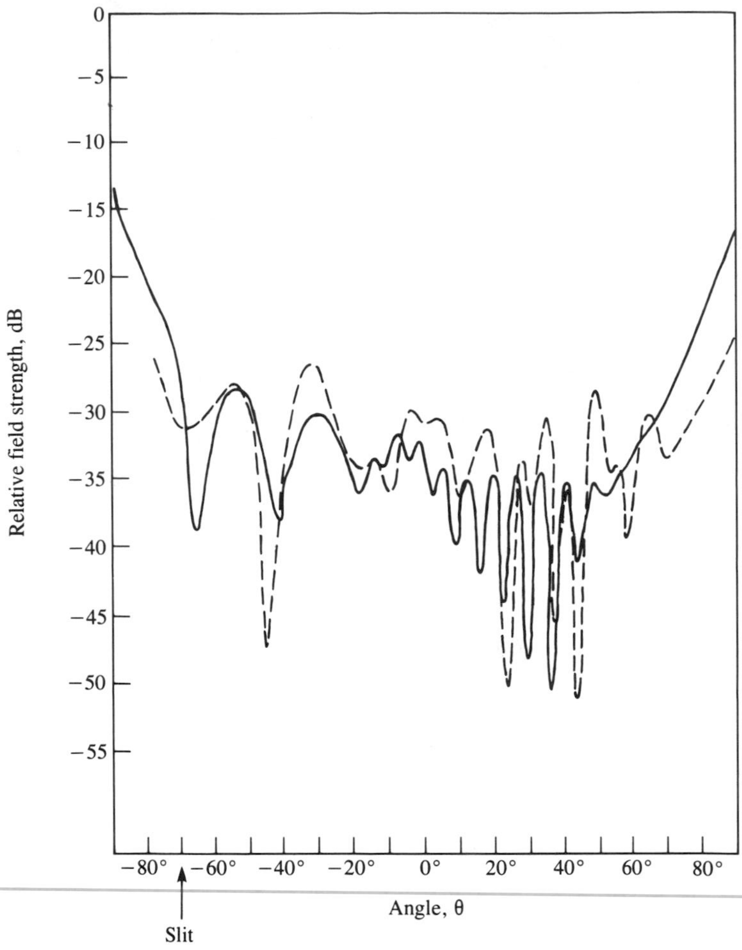

FIGURE 8.7. Back radiation pattern for parabolic cylinder reflected with a slit of width $\lambda/4$ at $\theta = -70°$. ——, (solid line) edge diffraction theory; ----, measured. (By permission, B.A. Smith [13].)

(a) the general sidelobe level increases; and
(b) the observed null in the pattern approaches the line through the slit.

It should be noted that:

(i) No phase control of the slit diffracted field was attempted.
(ii) No fully perforated reflectors were investigated. It may be that a random distribution of slits or holes would produce a general reduction in the back lobe levels.
(iii) The fields were computed using a modified GTD method and thus tend to diverge along the caustics.

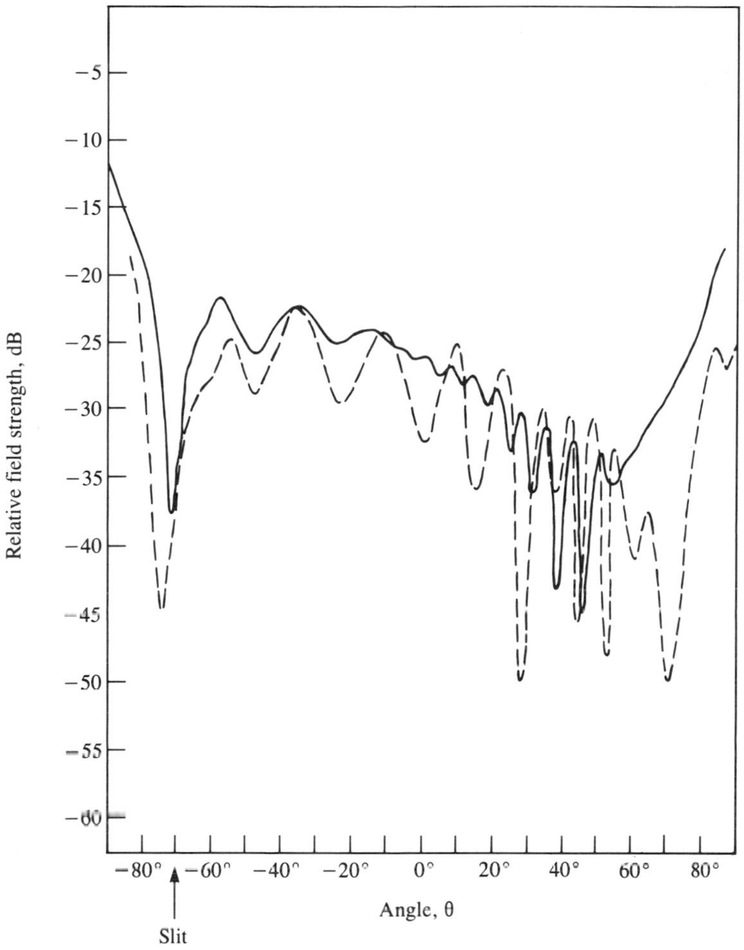

FIGURE 8.8. Back radiation pattern for parabolic cylinder reflector, with a $\lambda/2$ slit at $\theta = -70^\circ$. ——, edge diffraction theory; ----, measured. (By permission, B.A. Smith [13].)

The phase change in transmission through slits is given in Hamilton and Kerdemelidis [15].

8.5. Effect of Edge Geometry on Back Radiation

Aperture field methods give adequate results for radiation pattern near the main beam. To determine the wide angle and back sidelobes, the currents flowing near the aperture edges must be taken into account.

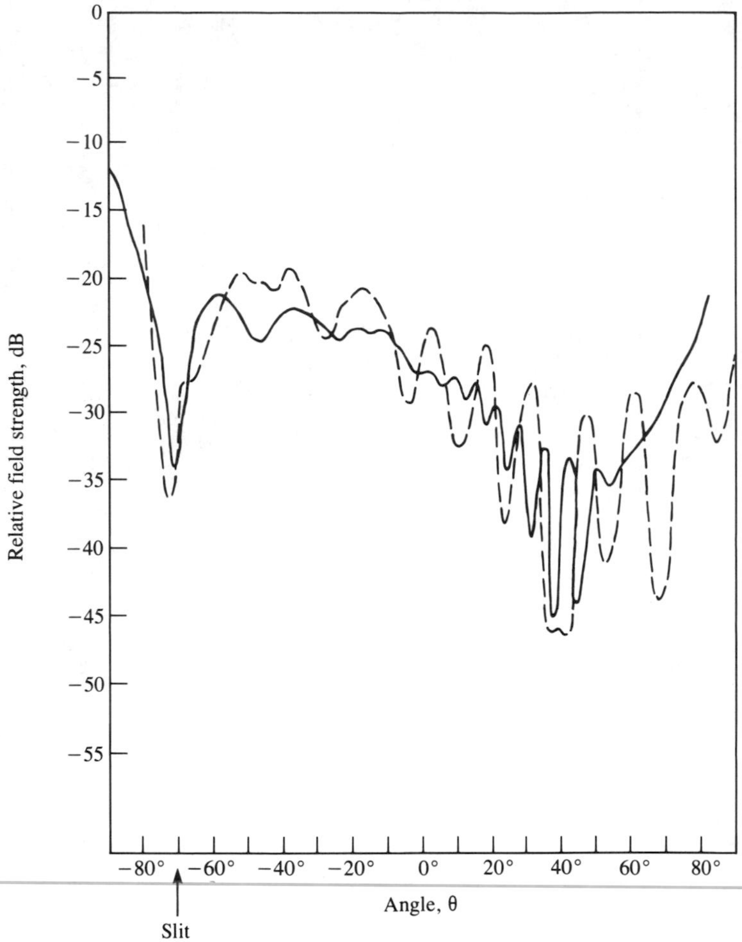

FIGURE 8.9. Back radiation pattern for parabolic cylinder reflector, with $2/3\lambda$ slit at $\theta = 0.70$. ——, edge diffraction theory; ----, measured. (By permission, B.A. Smith [13].)

For antennas such as horns and reflector types, the aperture ends in an edge and it is possible to apply edge diffraction methods. Thus, by considering each portion of the antenna rim as an isolated edge we can use edge diffraction techniques.

In the sections that follow we use edge diffraction concepts to control the radiation pattern of an antenna. The sections cover:

(i) null placement by stepped edge;
(ii) flanged aperture antennas; and
(iii) other methods of sidelobe reduction in horn and reflector antennas.

8.5.1. *Null Placement by a Stepped Edge*

A simple method of obtaining null placement for antennas involving edges is to shape the antenna edge in the form of a step at the critical point of radiation (James and Kerdemelidis [16]). This can be seen if we first consider the modified half-plane in Figure 8.10, with an incident wave at an angle φ_0 to the screen. When $\sigma = 0$ we have the classical half-plane problem where the diffracted far field is proportional to $D(\varphi, \varphi_0)e^{-jkr}r^{-1/2}$ where $D(\varphi, \varphi_0)$ is the diffraction coefficient. The edge is seen to behave as a line source for the diffracted field. If $\sigma > 0$ then we have two edges at $z = 0$, σ, and in the x–y-plane these behave as two lines sources. Summing these sources in the far field gives nulls in the x–y-plane when

$$\varphi = \cos^{-1}\left\{\cos\varphi_0 - \frac{n\lambda}{2\sigma}\right\}; \qquad n \text{ odd.} \tag{8.1}$$

A simple application of this result is null placement in the E-plane radiation pattern of an H-plane sector horn. By considering the diffraction mechanism

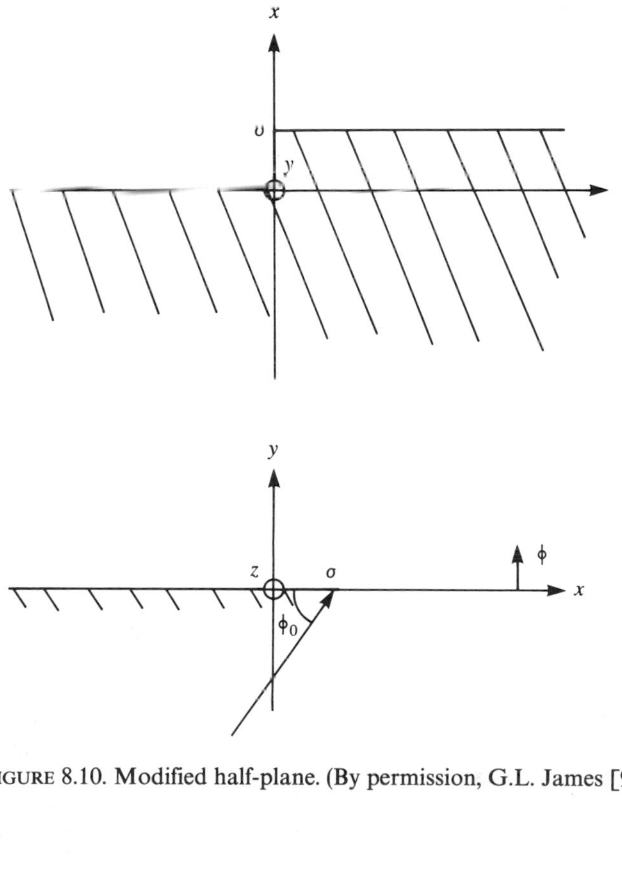

FIGURE 8.10. Modified half-plane. (By permission, G.L. James [9].)

at the horn aperture edges we can readily show that the major contribution of the diffracted field into the E-plane comes from the midpoint of the E-plane edges. We call these the critical points of radiation for the E-plane. At other points the field is diffracted away from the E-plane. If we now place a step at these critical points we can place nulls in the E-plane at angles given by (8.1). (Although there are two edges involved, (8.1) holds because the E-plane walls are parallel to each other.) When this is not the case—as for a pyramidal horn or reflector antenna—it does not apply in those regions where both edges contribute to the field (see James and Kerdemelidis [16], [12]).

Typical results are given in Figure 8.11. It is seen that a 15–20 dB reduction in field strength is readily achieved at the angles predicted by (8.1) where $\varphi_0 = 0$. Note that the null placement is independent of the horn length provided the latter is not excessively small, i.e., less than a wavelength. Measurements indicated that the on-axis gain was unaffected for values of σ up to

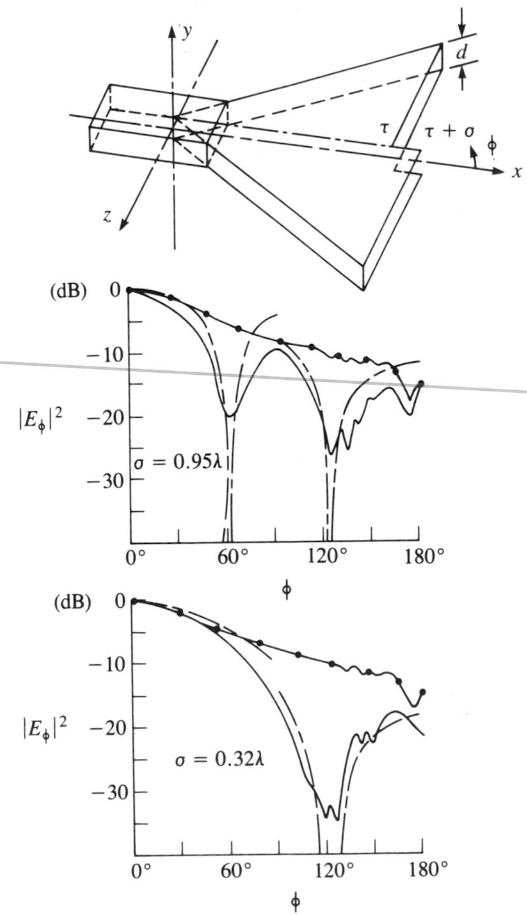

FIGURE 8.11. E-plane radiation pattern of H-plane sectoral horn: $d = 0.3\lambda$. —•—•—, unperturbed horn pattern (measured); ———, horn with perturbation (measured); — - — - —, GTD. (By permission G.L. James [9].)

3λ and that shaping the E-plane edges had little effect in the other principal plane.

The same procedure was carried out on other antennas, namely a pyramidal horn, paraboloidal reflector, and an E-plane sectoral horn giving similar results with the exception of the latter antenna where a loss of one-axis gain is obtained. In fact, it is possible to place a null on-axis for this antenna. For a circularly symmetrical reflector antenna the axis is a caustic of the edge diffracted rays. In other words, the entire rim contributes to the edge diffracted field along the axis. Thus to achieve a null on the back axis it is necessary for the entire rim to be stepped in some way. This is achieved by castellating the edge as in James and Kerdemelidis [12]. The depth of the castellations is still given by (8.1).

8.5.2. *Flanged Aperture Antennas*

The effect on the radiation pattern of metal flanges attached to antenna aperture edges was originally investigated by Owen and Reynolds [17] for the E-plane of a small H-plane sector horn. Experimental results were obtained with flanges attached to the long sides of the horn. They suggested that the effect of the flanges on the radiation pattern is similar to having a line source at the edge of each flange. A limited study was made of this in a later paper by Butson and Thompson [18], who also considered a flanged aperture waveguide and E-plane sectoral horn. Recently, further work has been carried out on both E- and H-plane sectoral horns, where the main contribution has been to investigate the effect of asymmetrical flanges, the position of the flanges from the open end of the horn, and the effect of flanges on the on-axial gain (Nair and Srivastava [19], Koshy et al., [20]).

All of the above work has been confined mainly to an experimental approach to shaping the forward sector of the radiation pattern. This problem, however, readily lends itself to a theoretical treatment using the GTD. By considering each edge as an isolated semi-infinite wedge we can readily obtain an approximate expression for the radiation field. When the metal flanges are less than a wavelength in extent it becomes necessary to account for multiple coupling between edges. This somewhat complicates the analysis but for flanges of practical length, i.e., $W > 0.3\lambda$, good results can then be achieved by considering first-order coupling only. Figure 8.12 gives three examples to illustrate the comparison between theoretical and measured results. It is seen that good agreement is achieved except for the example when $\beta > \pi/2$ where considerable error exists in some forward directions. This may be attributed to the neglected coupling between the flanges—especially the reflection of the edge sources at the opposite flange—not present when $\beta < \pi/2$. Coupling can be accounted for by the system of images between the flanges and higher-order diffractions between the edges. This considerably complicates the analysis.

The effect of flange length W and the angle β has been thoroughly investigated by previous authors for forward directions of the radiation pattern.

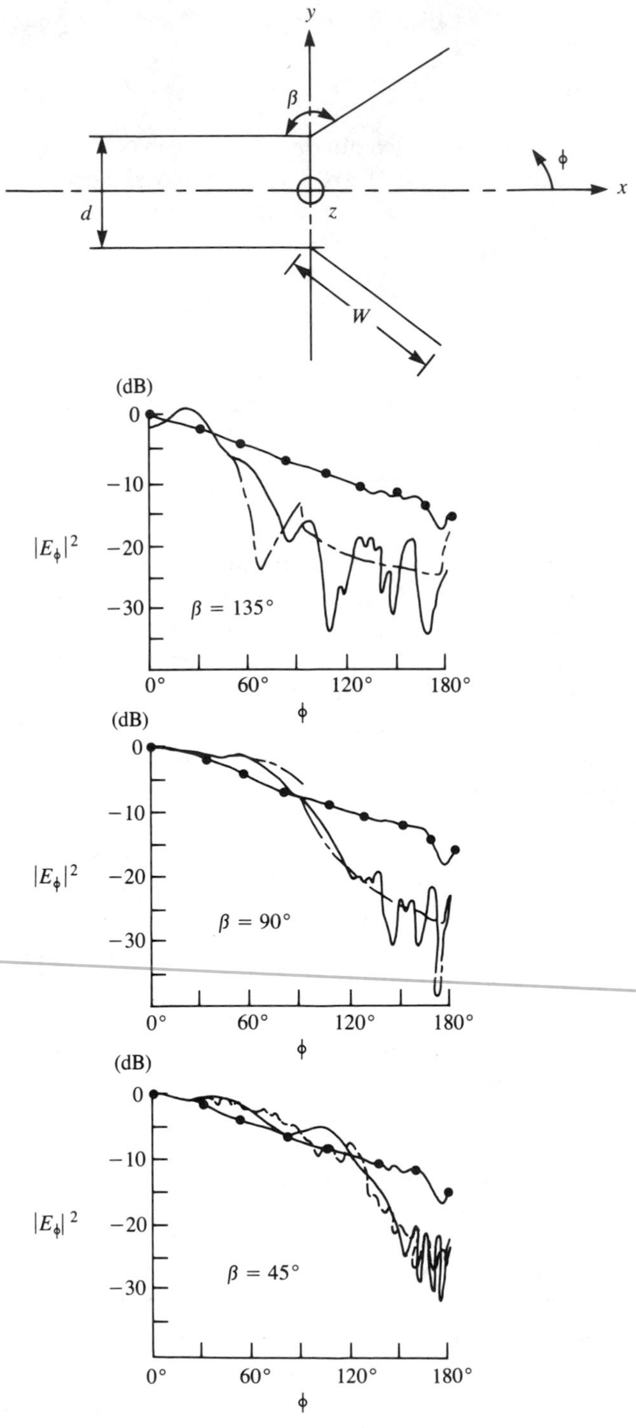

FIGURE 8.12. E-plane of flanged aperture H-plane sectoral horn: $W = 2\lambda$, $d = 0.3\lambda$. —•—•—•—, unperturbed horn pattern (measured); ———, horn with perturbation (measured); —-—-—, GTD. (By permission, G.L. James [9].)

For backward directions a reduction in radiation level is achieved. The angle β determines the point where reduction begins, as we see in Figure 8.12, and the flange length, W, the amount of reduction. For $W > \lambda$ a reduction of approximately 3 dB is achieved in the sidelobe level affected by the flange every time W is doubled.

Although greater sidelobe reduction is usually achieved with a stepped edge, the flanged aperture can give reduction over a wider sector of the radiation pattern. It also has the advantage of being relatively insensitive to frequency.

8.5.3. *Other Methods of Sidelobe Reduction in Horn and Reflector Antennas*

An attempt to control the radiation pattern of the E-plane of a pyramidal horn was made with one or two narrow metal strips placed inside the horn and perpendicular to the E-field. Our intention was to obtain control in the forward direction of the pyramidal horn as the flanged aperture and stepped edge techniques had little effect in this region. Theoretical and experimental investigation, however, did not give encouraging results. Some success has been reported using sectoral horns. Narrowing of the E-plane pattern of an E-plane sectoral horn has been achieved with a strip placed inside the horn parallel to the E-field (Silver [21]). The strip was of considerable width and gave a troublesome mismatch. More recently (Nair et al. [22]), a successful method has been found by placing two metal strips of half-wavelength in width at the aperture of the E-plane sectoral horn and perpendicular to the E-field. This gave an improvement in both gain and beam width, and slightly improved the matching.

Reduction of the sidelobe levels in the E plane of a pyramidal horn has been obtained by choke slots or a corrugated surface cut into the E-plane walls to reduce the illumination of the E-plane edges (Lawrie and Peters [23]). This principle has also been applied to conical horns (Clarricoats and Saha [24]). Corrugated horns have the advantage of a nearly circularly symmetrical radiation pattern over a bandwidth of the order of 1.5:1 with low sidelobes. A further reduction of the sidelobe level in the E-plane of a pyramidal horn is achieved if higher-order modes are allowed to exist in the aperture (Bahret and Peters [25]). A similar technique was used earlier by Potter [26] for conical horns.

Dielectrics are often inserted in electromagnetic horns to alter the phase distribution at the aperture of the horn and hence control the radiation pattern (e.g., Quddus and German [27], Hamid et al. [28], Oh et al. [29]). There are several disadvantages however in the use of dielectrics. One which can directly interfere with the radiation pattern is the excitation of unwanted modes at the interface between the dielectric and the air in the horn. A phase-correcting method without resorting to dielectrics has been obtained for E-plane sectoral horns by varying the guide wavelength within the horn (Craddock [30]). The technique gives a slight taper in the H-plane.

Finally, Smith [13] has considered the effect of rounding of the edge of a parabolic reflector. The general radiation level behind the reflector was reduced by 10 dB, relative to the unperturbed reflector. The edge was made in the form of a cylinder with the diameter equal to one wavelength with its axis parallel to the parabolic reflector axis. Larger diameter edges gave greater sidelobe reduction.

References

[1] H.E. Schrank (1985), Low sidelobe reflector antennas, *AP Society Newsletter*, **27**, 2, 5–16.

[2] P.J.B. Clarricoats and A.D. Olver (1984), *Corrugated Horns for Microwave Antennas*, Peter Peregrinus, London.

[3] M.A.K. Hamid and A. Mohsen (1969), Diffraction by dielectric-loaded horns and corner reflectors, *IEEE Trans. Antennas and Propagation*, **AP-17**, 5, 660–662.

[4] L.J. Lader and Winderman (1966), Method of reducing side lobes of directive antennas, *Canad. J. Phys.* **44**, 11, 2765–2773.

[5] R.C. Johnson (1966), Reducing wide-angle sidelobes in radar antennas, *Microwave J.*, **9**, 8, 48.

[6] F.L. Cain and R.C. Johnson (1967), Investigation of choke blinders for feed horns of radar antennas, *IEEE Trans.*, **EMC-9**, 2, 65–72.

[7] S. Cornbleet (1967), Progress in microwave communication antennas in the U.K., *Microwave J.*, **10**, 12, 84.

[8] L. Lewin (1972), Main reflector rim diffraction in back direction, *Proc. IEEE*, **119**, 1100–1102.

[9] G.L. James (1973), Electromagnetic effect of edges, Ph.D. thesis, Department of Electrical Engineering, University of Canterbury, Christchurch, New Zealand.

[10] G.F. Koch (1966), Paraboloidal antennas with a low noise temperature, *Nachrichtentech, Z-CJ*, **19**, 125.

[11] H.T. Middleton (1970), Control of radiation from reflector antennas by means of lossy films, M.E. Report, Department of Electrical Engineering, University of Canterbury, Christchurch, New Zealand.

[12] G.L. James and V. Kerdemelidis (1973), Selective reduction in back radiation from paraboloidal reflector antennas, *IEEE Trans. Antennas and Propagation*, **AP-21**, 886.

[13] B.A. Smith (1974), Some methods of selectively reducing back radiation in parabolic reflector antennas, M.E. Report, Department of Electrical Engineering, University of Canterbury.

[14] C.J. Sletten and P. Blacksmith (1965), The paraboloidal mirror, *J. Appl. Optics*, **4**, 1239.

[15] E.J. Hamilton and V. Kerdemelidis (1982), Transmission through slits formed by inclined planes, *IEEE Trans. Antennas and Propagation*, **AP-30**, 2, 199–204.

[16] G.L. James and V. Kerdemelidis (1972), Null placing in radiation patterns by shaping antenna edges, *Electron. Lett.*, **8**, 439.

[17] A.R.G. Owen and L.G. Reynolds (1946), The effect of flanges on the radiation patterns of small horns, *J. Inst. Elect. Engrs.*, **93**, IIIA, 1528.

[18] P.C. Butson and G.T. Thompson (1959), The effect of flanges on the radiation patterns of waveguide and sectoral horns, *Proc. IEE*, **106**, 422.

[19] K.G. Nair and G.P. Srivastava (1967), Beam shaping of H-plane sectoral horns by metal flanges, *J. Telecom. Engrs.* (India), **13**, 76.

[20] V.K. Koshy, K.G. Nair, and G.P. Srivastava (1968), An experimental investigation of the effect of conducting flanges on the H-plane radiation patterns of E-plane radiation patterns of E-plane sectoral horns, *J. Telecom. Engrs.* (India), **14**, 519.

[21] S. Silver (1949), *Microwave Antenna Theory and Design*, McGraw-Hill, New York, p. 383, p. 344.

[22] K.G. Nair, G.P. Srivastava, and S. Hariharan (1969), Sharpening of E-plane radiation patterns of E-plane sectoral horns by metallic grills, *IEEE Trans. Antennas and Propagation*, **AP-17**, 91.

[23] R.E. Lawrie and L. Peters (1966), Modifications of horn antennas for low sidelobe levels, *IEEE Trans. Antennas and Propagation*, **AP-14**, 605.

[24] P.J.B. Clarricoats and P.K. Saha (1971), Propagation and radiation behaviour of corrugated feeds—Part 2, *Proc. IEE*, **118**, 9, 1177.

[25] W.F. Bahret and L. Peters (1968), Small-aperture small-flare-angle corrugated horns, *IEEE Trans. Antennas and Propagation*, **AP-16**, 494.

[26] P.D. Potter (1963), A new horn with suppressed sidelobes and equal beamwidths, *Microwave J.*, **6**, 71.

[27] M.A. Quddus and J.P. German (1961), Phase correction by dielectric slabs in sectoral horn antennas, *IRE Trans. Antennas and Propagation*, **AP-9**, 413.

[28] M.A.K. Hamid, R.J. Boulanger, N.J. Mostowy, and A. Mohsen (1970), Radiation characteristics of dielectric-loaded horn antennas, *Electron. Lett.*, **6**, 20.

[29] L.L. Oh, S.Y. Peng, and C.D. Lunden (1970), Effects of dielectrics on the radiation patterns of an electromagnetic horn, *IEEE Trans. Antennas and Propagation*, **AP-18**, 553.

[30] I.D. Craddock (1964), Phase-corrected E-plane horns *Proc. IEE*, **III**, 1379.

9
Group Symmetries of Antenna Arrays

H.N. KRITIKOS*

9.1. Introduction

The analysis of the radiation patterns of three-dimensional arrays has, up to the present time, been limited to a few well-known structures which were introduced as natural extensions of simple geometric shapes, such as spheres or cylinders, etc. At the present time no systematic procedure exists for choosing the proper three-dimensional arrangement of radiators in space. New geometrical concepts are needed to enrich the presently available techniques of analysis of antenna theory. An area where similar geometrical concepts have been developed with great succes is crystallography. In this area, using group theory as the underlined mathematical structure, a large class of crystal lattices in space were studied and a very elegant and systematic formalism was developed for the classification of crystal structures.

In this chapter a new class of three-dimensional lattices will be studied and a number of their radiation properties will be examined. The basic formalism of treating these problems has been transplanted from crystallography and solid state theory. In examing the properties of the various crystal lattices, group theory is used as a powerful tool for the systematic analysis of the various three-dimensional structures and their associated wave functions. In an analogous manner, group theory will be used for the systematic classification of array lattices and their associated radiation patterns.

The groups which are of interest in this work are those whose elements correspond to certain operations in space. These operations are translation, reflection, rotation, and inversion. An example of these groups is the full rotation group O_3, which includes infinitesimal rotations plus coordinate inversions. A number of investigators have used the properties of the O_3 group to examine the symmetries of electromagnetic fields. Liubarskii [1], Tinkham [2], Gelfand [3], and Naimark [4] have studied the rotational properties of

* Department of Electrical Engineering, University of Pennsylvania, Philadelphia, PA 19104-6390, USA.

Maxwells' equations. Rose [5] has investigated the rotational properties of radiation fields due to multiple sources. Kritikos [6], [7] has reported on the symmetries of radiation arrays and the Poynting vector. Richie [8] has investigated the symmetries of radiation patterns.

The groups which are pertinent to the development of the present work are the finite rotational groups which are known as the crystallographic point groups. The point groups are a subgroup of the rotation group O_3 and were successfully employed in the classification of crystal lattices. This formalism can be used very effectively for the systematic classification of array lattices. Alternative geometries can be examined and the global properties of the radiation patterns can be analyzed without having to carry out extensive integrations of the radiation integrals. It is also a powerful tool that can be used to design new types of antennas whose radiation fields are orthogonal, thus introducing a new synthesis procedure.

9.2. Group Theory Fundamentals

For the sake of completeness a number of basic properties of groups will be introduced here to lay the groundwork for the following discussion.

Group. A group is a set of elements $g_i \in G$ for which an operation is defined with the properties:

(a) $g_i g_i = g_k$, $g_k \in G$.
(b) Associativity, $g_i(g_j g_k) = (g_i g_j) g_k$.
(c) There exists an identity element E; $E g_j = g_j = g_j E$.
(d) For every element g_i there exists an inverse element g_i^{-1} such that $g_i^{-1} g_i = E$.

Representation of a Group. For each element g_i we associate a matrix $\Gamma_{\mu\nu}^{(i)}(g_i)$ such that it has the same operational properties under matrix multiplication as the elements of the group

$$g_i g_j = g_k \quad \Rightarrow \quad \Gamma_{\mu\kappa}^{(i)}(g_i)\Gamma_{\kappa\nu}^{(j)}(g_j) = \Gamma_{\mu\nu}^{(k)}(g_k). \tag{9.1}$$

The Irreducible Representation. For a given element of a group there exists a multiplicity of matrix representations obtained by similarity transformations. The representation of the lowest dimensionality in a block form is called irreducible.

Character of a Representation. It is defined as $\chi(g_i) = \mathrm{Tr}\ \Gamma_{\mu\mu}^{(i)}(g_i)$. The character is a constant and is the same for all representations within the same class.

The basis Functions. A set of functions φ_ν exists such that for every operation they transform as

$$\varphi_\mu^{(i)} = \Gamma_{\mu\nu}^{(i)}(g_i)\varphi_\nu^{(i)}. \tag{9.2}$$

Orthogonality of the Basis Functions. Functions that belong to two different irreducible representations or to two different rows of the same unitary representation are orthogonal

$$\langle \varphi_k^{(i)}, \varphi_\mu^{(j)} \rangle = N_\kappa \delta_{ij} \delta_{\kappa\mu}. \tag{9.3}$$

Dimensionality Theorem. *Let h be the number of elements and let l_i be the dimensionality of the irreducible block of the matrix $\Gamma_{\mu\nu}^{(i)}(g_i)$, then*

$$h = \sum_i l_i^2. \tag{9.4}$$

Expansion Theorem. *Every function F on the topological space of the group (having the operation $P(R)$, $g_i = R$) can be expanded in the form $F = \sum_i \alpha_i f_i$, where f_i transforms according to one of the classes of the irreducible representations. Quantity α_i is a constant to be determined from the projection of F in the function space.*

The Class Projection Operator $\mathscr{P}^{(i)}$. This operator can be used to generate functions belonging to the same class only. It is defined as

$$\mathscr{P}^{(i)} = \frac{l_i}{h} \sum_R \chi(R) P(R), \tag{9.5}$$

where $\chi(R)$ is the character of the irreducible representation and $P(R)$ is the geometrical symmetry operator. It is used to find the projection of the function F onto a class

$$f_i = \mathscr{P}^{(i)} F,$$

where F is any function on the space operated by $P(R)$.

The Companion Function Projection Operator $\mathscr{P}_{\lambda\kappa}^{(i)}$. This operator generates the companion functions within the same class. It is defined as

$$\mathscr{P}_{\lambda\kappa}^{(i)} = \frac{l_i}{h} \sum_R \Gamma_{\lambda\kappa}^{(i)} P(R). \tag{9.6}$$

It is used to find the companion functions of $f_\kappa^{(i)}$. We have

$$f_\lambda^{(i)} = \mathscr{P}_{\lambda\kappa}^{(i)} f_\kappa^{(i)}. \tag{9.7}$$

9.3. Symmetry Operations

The covering geometrical operations which retain the symmetry and leave lattices invariant are the elements of groups. Using the Shoenflies notation we recognize the following operations:

E, identity;
C_n, rotation about an axis through the origin through an angle of $2\pi/n$;

C'_n, rotation about an axis through the origin perpendicular to the axis of highest symmetry through an angle of $2\pi/n$;
σ_h, reflection about a plane through the origin perpendicular to the axis of highest symmetry;
σ_v, reflection about a plane through the origin passing through the axis of highest symmetry;
i, inversion; and
S_n, improper rotation, i.e., a rotation C_n and then an inversion.

9.4. Elementary Radiating Elements

The symmetries of the elementary radiators are of importance because they are the simplest elements with which more complex arrays can be constructed. The elementary elements are the dipole and the loop. Consider the elementary dipole f at the origin aligned along the z-axis (see Figure 9.1).

The axis of highest symmetry is z and horizontal plane is xy. The geometric symmetry operations are symbolically designated as follows:

$$P(R)f - \lambda_R f, \tag{9.8}$$

where λ_R is the eigenvalue of the operator R. For example, a rotation C'_2 through the x- or y-axis perpendicular to the highest axis of symmetry is symbolically represented as

$$P(C'_2)f = (-1)f. \tag{9.9}$$

Similarly, for a loop at the origin whose dipole moment g is oriented along the z-axis we observe

$$P(C'_2)g = (-1)g. \tag{9.10}$$

The collective behavior of both the magnetic and electric dipole is shown in Table 9.1.

It is instructive to note that the reflection operations σ_v, σ_h, i, S_2, are the only ones that are capable of distinguishing the loop from the dipole. We, therefore, can differentiate between magnetic and electric sources by observing the resulting configurations and associated fields in reflection operations.

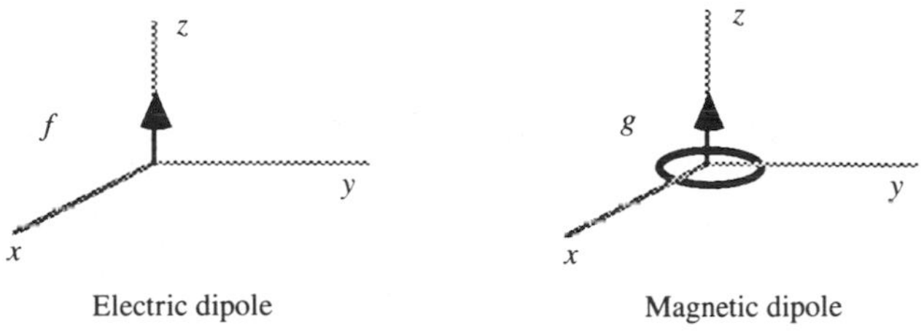

FIGURE 9.1. Elementary radiation elements.

TABLE 9.1. Eigenvalues of the elementary operations.

	E	C_2	C_2'	σ_v	σ_h	i	S_2
Dipole	1	1	−1	1	−1	−1	−1
Loop	1	1	−1	−1	1	1	1

9.5. Array Lattices and Their Symmetries

Once the symmetries of single radiation elements have been established, the next task is to arrange them in three-dimensional lattices in space and investigate the symmetries of the composite structure. In general, the excitation sources can be any radiator of any degree of complexity for simplicity, however, in this chapter only the elementary electric and magnetic dipoles will be used. The symmetries of lattices and their associate representations have been extensively studied in crystallography [2] and reported in the literature. Koster [9] has reported on the space groups and Tinkham [2] has examined the point groups. In this chapter only the point groups, a class of 32 only, will be considered because they represent a simple and convenient way to introduce group theory into antenna synthesis. The other space groups can be potentially utilized and will be the topic of future work.

As an example consider the group C_2 which is associated with a two-point lattice (see Figure 9.2). The covering operations of the group are the identity E and a rotation C_2 through an axis perpendicular to the plane of the paper. Table 9.2 shows the character table of the group.

In the character table we observe that there are two classes of elements. From this observation we imply that there are two orthogonal excitations of the two-point lattice. The two excitations are found by using the projection theorem and any convenient elementary source. Let us consider an elementary dipole source situated at B. The symmetry operations give

$$P(E)f_1 = f_1, \tag{9.11}$$

$$P(C_2)f_1 = f_2. \tag{9.12}$$

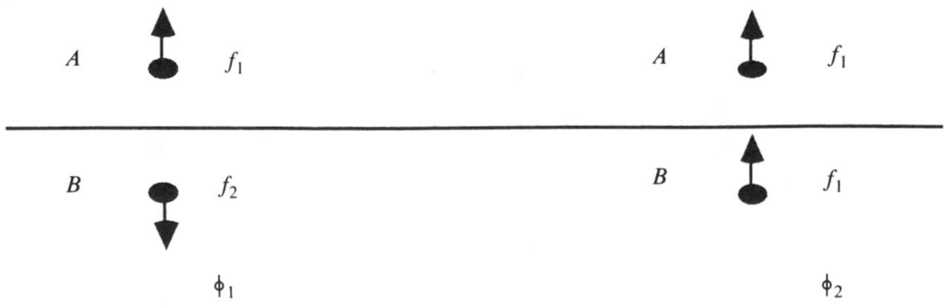

FIGURE 9.2. The two-point lattice. Group C_2.

TABLE 9.2. Character table for point group C_2.

	E	C_2
f_1	1	1
f_2	1	-1
$h = 2$	$l_1 = 1$	$l_2 = 1$

Starting with f_1 and using the projection operator we can generate the orthogonal basis functions corresponding to the classes of the group. There are two classes here and we have the basis functions

$$\varphi_1 = \tfrac{1}{2}(f_1 + f_2), \tag{9.13}$$

$$\varphi_2 = \tfrac{1}{2}(f_1 - f_2). \tag{9.14}$$

These two configurations are shown in Figure 9.2(a) and (b). The two excitations are orthogonal because of Theorem 7. We have therefore

$$\langle \varphi_1, \varphi_2 \rangle = 0, \tag{9.15}$$

or

$$\int_{-\infty}^{\infty} \int_{-\infty}^{\infty} \int_{-\infty}^{\infty} (\varphi_1, \varphi_2)\, dx\, dy\, dz = 0. \tag{9.16}$$

This an elementary result to verify, because while φ_1 is an even function, φ_2 is an odd one. While this is an obvious result it is indicative of the process which we have to follow to generate the orthogonal excitations.

9.6. The Point-Group Lattices

In general, any lattice in space can be symmetrized and any set of elementary excitation sources can be utilized to generate orthogonal patterns. To achieve this we can proceed in the following manner. Given any lattice in space we initiate the following process:

(a) Determine the covering symmetry operations.
(b) Determine the irreducible group corresponding to the symmetry operations.
(c) Determine the character table of the group.
(d) Starting with an arbitrary elementary source f_1 generate its companion sources $f_2, f_3, \ldots, f_n$ by using the projection operator $P(R)$.
(e) Generate the orthogonal excitation φ_n with the help of the projection operator.

For the case of lattices which can be derived from the point groups, items *a*, *b*, and *c* can be easily obtained from results available in the open literature. We can then proceed immediately to study the radiation properties of the antenna arrays and generate their corresponding orthogonal excitations.

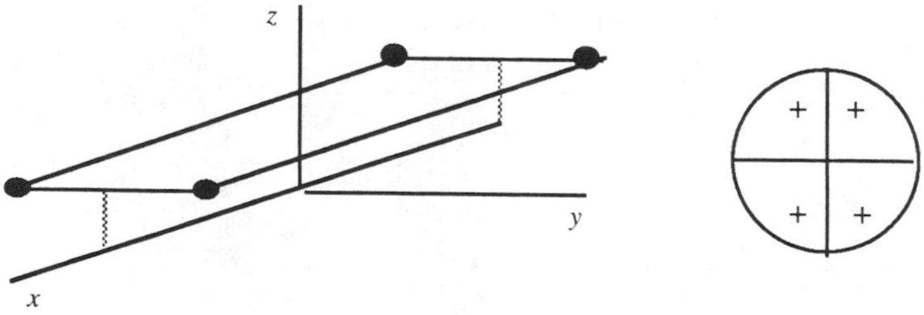

FIGURE 9.3. The four-point lattice. Group C_{2v} and its stereographic projection.

The point groups fall into two general categories: the simple rotation groups, i.e., C_n, C_{nh}, C_{nd}, S_n, D_n, D_n, D_{nh}, D_{nd} and those of higher symmetry T, O. The lattices that correspond to these groups can be vizualized by their stereographic projections. In this presentation scheme the points of the lattice are projected into the x–y-plane. Points lying above the plane are shown as crosses and those lying below as circles.

A number of typical examples are examined below.

Group C_{2v}

The covering operations of the group are (see Figure 9.3): the identity E and C_2 around the z-axis and σ_v about the x–z-plane and σ'_v about the y–z-plane.

Starting with the dipole elementary radiator f_1 the companion sources are generated below:

$$P(E)f_1 = f_1, \qquad P(C_2)f_1 = f_3, \qquad P(\sigma_v)f_1 = f_2, \qquad P(\sigma'_v)f_1 = f_4. \tag{9.17}$$

The character table of the group is shown in Table 9.3.

The projection theorem gives

$$\begin{aligned}
\varphi_1 &= \tfrac{1}{4}(f_1 + f_2 + f_3 + f_4),\\
\varphi_2 &= \tfrac{1}{4}(f_1 - f_2 + f_3 - f_4),\\
\varphi_3 &= \tfrac{1}{4}(f_1 + f_2 - f_3 - f_4),\\
\varphi_4 &= \tfrac{1}{4}(f_1 - f_2 - f_3 + f_4).
\end{aligned} \tag{9.18}$$

Figure 9.4 shows the orthogonal excitations of the group C_{2v}.

TABLE 9.3. Character table of point group C_{2v}.

	E	C_{2v}	σ_v	σ'_v
φ_1	1	1	1	1
φ_2	1	1	-1	-1
φ_3	1	-1	1	-1
φ_4	-1	-1	1	1
$h = 4$	$l_1 = 1,$	$l_2 = 1,$	$l_3 = 1,$	$l_4 = 1$

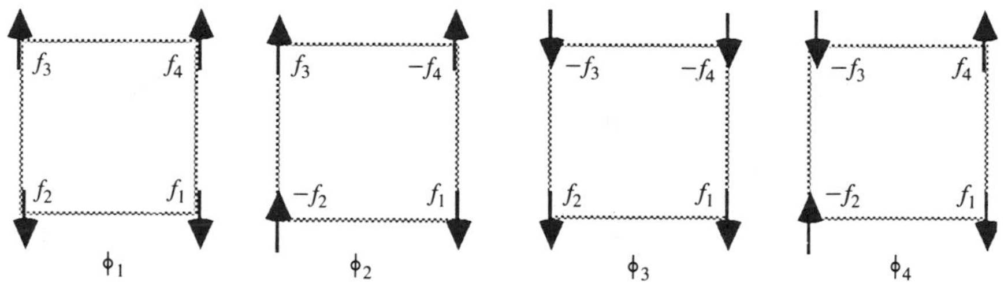

FIGURE 9.4. The orthogonal excitations of group C_{2v}.

Group D_2

The covering operations of this group are the identity E and three C_2 rotations about the x-, y-, z-axes. Starting with the elementary source f_1 as being a electric dipole shown in Figure 9.5, the companion sources are

$$P(E)f_1 = f_1, \qquad P(C_2, x)f_1 = f_2, \qquad P(C_2, z)f_1 = f_3, \qquad P(C_2, y)f_1 = f_4. \tag{9.19}$$

The character table is shown in Table 9.4.

The orthogonal basis functions are

$$\begin{aligned} \varphi_1 &= \tfrac{1}{4}(f_1 + f_2 + f_3 + f_4), \\ \varphi_2 &= \tfrac{1}{4}(f_1 + f_2 - f_3 - f_4), \\ \varphi_3 &= \tfrac{1}{4}(f_1 - f_2 + f_3 - f_4), \\ \varphi_4 &= \tfrac{1}{4}(f_1 - f_2 - f_3 + f_4). \end{aligned} \tag{9.20}$$

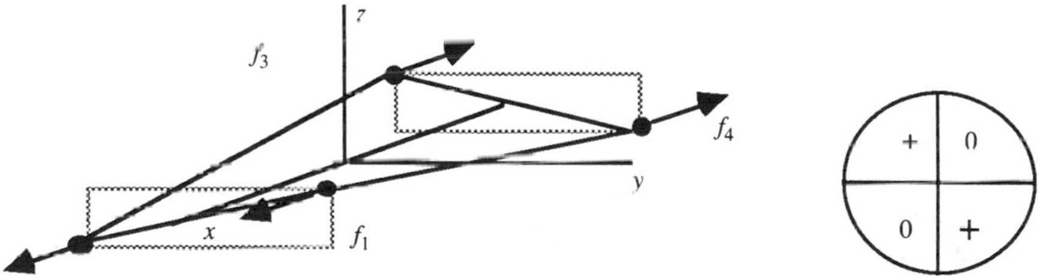

FIGURE 9.5. The four-point lattice. Group D_2 and its stereographic projection.

TABLE 9.4. Character table of point group D_2.

	E	C_2, z	C_2, y	C_2, x
φ_1	1	1	1	1
φ_2	1	1	-1	-1
φ_3	1	-1	1	-1
φ_4	1	-1	-1	1
$h = 4$	$l_1 = 1$	$l_2 = 1$	$l_3 = 1$	$l_4 = 1$

Group C_{3v}

This covers the symmetries of a triangle-like lattice shown with its stereographic projection in Figure 9.6.

The covering operations are the identity E, two C_3 rotations around the z-axis which is perpendicular to the plane of the paper, three σ_v reflections around the diagonal planes passing through the y-axis, and diagonals L_1, L_2, L_3. The companion sources are

$$P(E)f_1 = f_1, \qquad P\begin{pmatrix} C_3 \\ \bar{C}_3 \end{pmatrix} f_1 = \begin{pmatrix} f_2 \\ f_3 \end{pmatrix}, \qquad P\begin{pmatrix} \sigma_v \cdot L_1 \\ \sigma_v \cdot L_2 \\ \sigma_v \cdot L_3 \end{pmatrix} f_1 = \begin{pmatrix} f_1 \\ f_3 \\ f_2 \end{pmatrix}. \tag{9.21}$$

The normal procedure here of using the class projection operator $\mathscr{P}^{(i)}$ does not produce interesting results. The application, however, of the companion function projection operator $\mathscr{P}^{(i)}_{\lambda\kappa}$ produces an interesting structure. For this applicaton it is necessary that the irreducible matrix representation $\Gamma^{(i)}_{\mu\nu}(g_i)$ is known. For the group C_{3v} it is readily available in the literature [2]. From the character table which is shown in Table 9.5 we recognize that group C_{3v} has three classes, two of which are one dimensional and one is two dimensional.

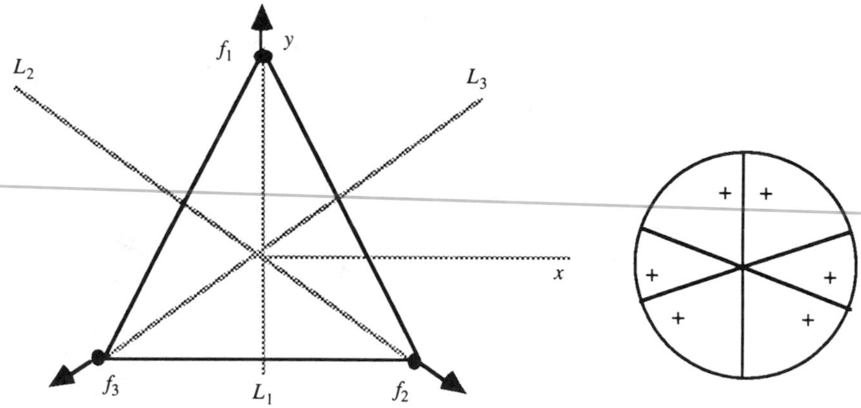

FIGURE 9.6. The three-point lattice. Group C_{3v} and its stereographic projection.

TABLE 9.5. Character table of point group C_{3v}.

	E	$2C_3$	$3\sigma_v$
$\varphi^{(1)}$	1	1	-1
$\varphi^{(2)}$	1	1	-1
$\varphi^{(3)}_{\mu\nu}$	2	-1	0
$h = 6$	$l_1 = 1$	$l_2 = 1$	$l_3 = 2$

The irreducible matrix representation is

	E	C	$\bar{C}$	σ_1	σ_2	σ_3
$\Gamma^{(1)}$	1	1	1	1	1	1
$\Gamma^{(2)}$	1	1	1	-1	-1	-1
$\Gamma^{(3)}$	$\begin{pmatrix} 1 & 0 \\ 0 & 1 \end{pmatrix}$	$\begin{pmatrix} -1/2 & \sqrt{3}/2 \\ -\sqrt{3}/2 & -1/2 \end{pmatrix}$	$\begin{pmatrix} -1/2 & -\sqrt{3}/2 \\ \sqrt{3}/2 & -1/2 \end{pmatrix}$	$\begin{pmatrix} -1 & 0 \\ 0 & 1 \end{pmatrix}$	$\begin{pmatrix} 1/2 & -\sqrt{3}/2 \\ -\sqrt{3}/2 & -1/2 \end{pmatrix}$	$\begin{pmatrix} 1/2 & \sqrt{3}/2 \\ \sqrt{3}/2 & -1/2 \end{pmatrix}$

The companion function projection operator gives the following:

$$\begin{aligned} \varphi^{(1)} &= \mathscr{P}^{(1)} f_1 = \tfrac{1}{3}(f_1 + f_2 + f_3), \\ \varphi^{(2)} &= \mathscr{P}^{(2)} f_1 = 0, \\ \varphi^{(3)}_{21} &= \mathscr{P}^{(3)}_{21} f_1 = \frac{\sqrt{3}}{6}(f_2 - f_3), \\ \varphi^{(3)}_{22} &= \mathscr{P}^{(3)}_{22} f_1 = \tfrac{1}{6}(2f_1 - f_2 - f_3), \\ \varphi^{(3)}_{11} &= \mathscr{P}^{(3)}_{11} f_1 = 0, \qquad \varphi^{(3)}_{12} = \mathscr{P}^{(3)}_{12} f_1 = 0. \end{aligned} \tag{9.22}$$

The orthogonal excitations (Figure 9.7) may take many different forms depending on the form of the seed function f_1 which is arbitrary. For example, for a seed scalar function $\psi_1 = \delta(r - r_1)$ and its companions $\psi_2 = \delta(r - r_2)$ and $\psi_3 = \delta(r - r_3)$ the orthogonal excitations are

$$\begin{aligned} \varphi^{(1)} &= \tfrac{1}{3}(\psi_1 + \psi_2 + \psi_3), \\ \varphi^{(2)} &= 0, \\ \varphi^{(3)}_{21} &= \frac{\sqrt{3}}{6}(\psi_2 - \psi_3), \\ \varphi^{(3)}_{22} &= \tfrac{1}{6}(2\psi_1 - \psi_2 - \psi_3), \\ \varphi^{(3)}_{22} &= 0, \qquad \varphi^{(3)}_{22} = 0, \end{aligned} \tag{9.23}$$

and so on.

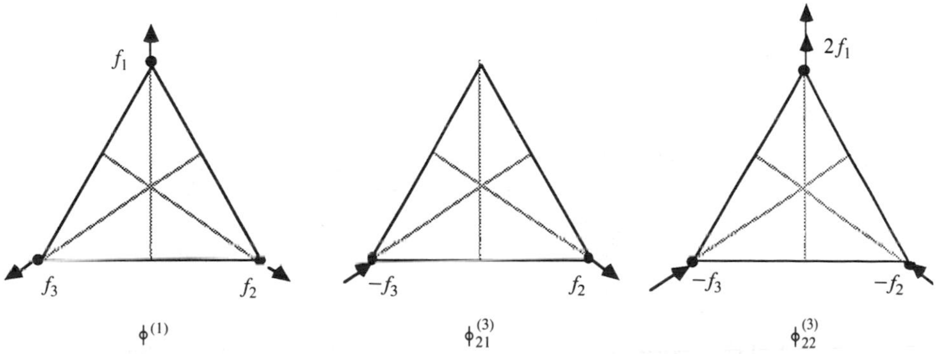

FIGURE 9.7. The orthogonal excitations of group C_{3v}.

9.7. Applications

Group theory does not provide the actual functional forms of the excitation patterns, neither does it provides the form of the radiation patterns. Its power lies in the classification of the array excitations and the resulting radiation patterns. It also offers a systematic way by which it is possible to explore the properties of array lattices in three dimensions. The most important property, however, is to be found in the synthesis of orthogonal patterns which only interact with targets having the same symmetries. As an example, consider an incident signal E interacting with an array defined by a current excitation function J. The received power P is given by the inner product

$$P = \langle J \cdot E \rangle = \int_{-\infty}^{\infty} \int_{-\infty}^{\infty} \int_{-\infty}^{\infty} J \cdot E \, dx \, dy \, dz.$$

The inner product only exists if J and E belong to the same symmetry classes.

With some care the same argument also can be extended to include antenna arrays excited by a current excitation $J(x, y, z = 0)$ and far-field source distributions $g(\xi, \eta, r)$ (ξ, η are the directional cosines). We shall first consider scalar sources. For symmetries involving the x–y-plane only and for scalar sources the far-field pattern $\psi(\xi, \eta, r)$ has the same symmetries as the scalar excitation sources because the Greens' function $G(r, 0)$ is spherically symmetric. We can easily demonstrate this by showing that the Greens' function commutes with the symmetry operators mentioned in Section 9.3. We conclude, therefore, that the far-field radiation pattern $\psi(\xi, \eta, r)$ will only interact with source distributions $g(\xi, \eta, r)$ having the same symmetries. This can be demonstrated by the group C_{3v}. From the Character Table 9.5 we observe that only two classes, the first and third, contribute to the basis functions. The first is the identity and the third deals only with operations on the x–y-plane. We can therefore conclude that source distributions $g(\xi, \eta, r)$ belonging to different classes from those of the radiatiation pattern will not be received by the array. Typical orthogonal source distributions derived from eq. (9.23) are shown in Figure 9.8.

For vector sources such as electric current distributions J we must exercise some care in applying the above analysis because the dyadic Greens' function

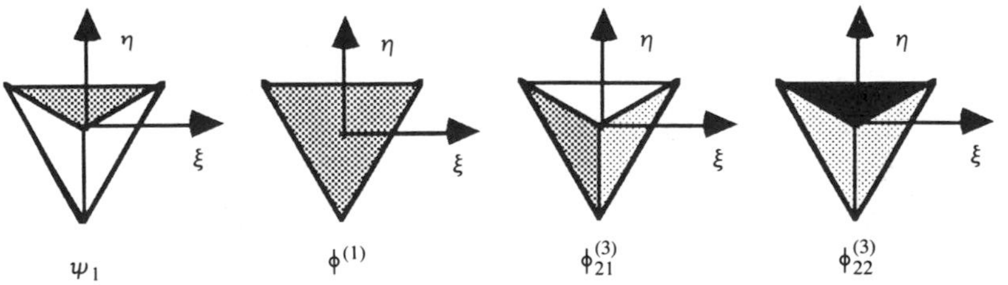

FIGURE 9.8. Typical orthogonal source distributions with ψ_1 as the seed function.

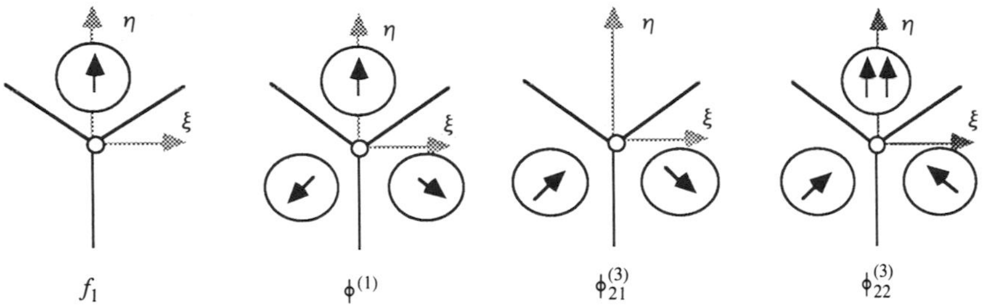

FIGURE 9.9. Typical vector orthogonal distributions with f_1 as the seed vector.

is not spherically symmetric. For small ξ and η, however, the polarization of the far-field pattern is the same as that of the source current vectors. The orthogonal excitations of the previous section can be applied and we recognize the typical polarization clusters shown in Figure 9.9.

Depending on the seed vector a rich variety of primitive clusters or basis functions can be generated. The same procedure can be used for the other groups mentioned in this chapter, and in fact all of the 32 point groups can be utilized to provide a rich reservoir of lattice structures and their corresponding orthogonal excitations.

References

[1] G. Liubarskii, *The Application of Group Theory in Physics*, Pergamon Press, Oxford.

[2] M. Tinkham (1964), *Group Theory and Quantum Mechanics*, McGraw-Hill, New York.

[3] N. Gelfand et al. (1963), *Representation of the Rotation and Lorentz Group and Their Applications*. Macmillan, New York.

[4] M. Naimark (1964), *Linear Representation of the Lorentz Group*, Pergamon Press, Oxford.

[5] M. Rose (1955), *Multipole Fields*, Wiley, New York.

[6] H. Kritikos (1973), Radiation symmetries of antenna arrays *J. Franklin Inst.*, **295**, 4, 283.

[7] H.N. Kritikos (1983), Rotational symmetries of the electromagnetic radiation fields. *IEEE Trans. Antennas and Propagation*, **AP-31**, 2.

[8] Richie an H.N. Kritikos, *An Application of Group Theory to Arrays*, to be published.

[9] G.F. Koster (1957), *Space Groups and Their Representations*, Academic Press, New York.

10
Many-Body Problems in Electromagnetic Theory

JOHN LAM*

La lune blanche
Luit dans les bois;
De chaque branche
Part une voix
Sous la ramée . . .

O bien-aimée.

Paul Verlaine

10.1. Prolegomena

A systematic method for formulating and solving many-body problems in electromagnetic theory is illustrated with the many-sphere problem in electrostatics. The general problem of an arbitrary number of spheres of arbitrary radii and arbitrary uniform compositions in an arbitrary configuration and under the excitation of a uniform electric field is formulated in terms of a pair of coupled, linear, inhomogeneous matrix equations. The equations can be solved analytically by iteration. The method is applied to the problem of an infinite number of identical spheres arranged in a simple cubic lattice. A formula for the effective dielectric constant of the simple cubic lattice of spheres is derived.

For those undaunted souls who are chosen by the Fates to tame boundary-value problems in electromagnetic theory, the one-body problem is very often already one body too many. Two bodies are definitely a crowd. A problem involving three or more bodies seldom fails to overwhelm the enquiring mind even in the present heroic age. Our intention in this chapter is to present a powerful procedure developed by Rayleigh [1] to solve many-body problems in electromagnetic theory, or at least a beautiful subset of them, rich in the possibilities of application to everyday life. We want to declare from the outset that by solution we really mean exact, analytic solution. That is to say, it must either be obtainable in closed, analytic form, or reducible to a system of algebraic equations with explicit coefficients, which can be solved either numerically to an arbitrary degree of accuracy, or by means of power-series

* Northrop Corporation, B-2 Division, 8900 Washington Boulevard, Pico Rivera, CA 90660, USA.

expansion in a small parameter, whose coefficients can be worked out systematically one by one, if we only have patience enough and time. Purely numerical methods, such as the finite-element method, will not be given consideration here.

Historically, the one-body boundary-value problem in electromagnetic theory was solved systematically by the method of separation of variables. This method works only in coordinate systems in which Maxwell's equations in one form or another are separable. The surface of the body must also coincide with an entire coordinate surface. In this fashion the one-cylinder and one-sphere problems are exactly soluble. The one-halfplane problem does not meet the separability criterion, but is fortunately also exactly soluble by the Wiener–Hopf method.

The two-cylinder and two-sphere problems are likewise soluble by the method of separation of variables [2], [3]. The characteristic feature of the two-body problem in particular, and the many-body problem in general, is the mutual interaction of the bodies. This leads to a self-consistency problem: The total electromagnetic excitation exerted on one body is the sum of that exerted by the external applied field and that emanating from the excitation induced on the other body. These excitations on the bodies must be calculated self-consistently. To perform the calculation, we have to translate the excitation on one body into incident fields converging on the other. In this endeavor, we need mathematical relations which generally come under the name of addition theorems for eigenfunctions in the coordinate system in point. The two-halfplane problem, on the other hand, is soluble by the Wiener–Hopf method [4–6].

The same self-consistency problem and the same need of addition theorems carry over into the problems of three or more bodies. The complication of a many-body problem over that of one body is essentially one of bookkeeping. We must keep track of the plethora of interaction terms. The situation simplifies dramatically, however, in problems of infinitely-many identical bodies arranged in a periodic lattice. In as early as 1892, Rayleigh [1] initiated the solution of the electrostatic and magnetostatic problems of infinitely-many identical cylinders and infinitely-many identical spheres in rectangular lattices. Not surprisingly, the periodic, infinitely-many-halfplane problem is also soluble by the Wiener–Hopf method [7], [8].

In this chapter, we complete and extend Rayleigh's solution of the electrostatic many-sphere problem. Specifically, we first consider the general electrostatic problem of an arbitrary number of spheres of arbitrary radii and arbitrary uniform compositions in an arbitrary configuration. We reduce this problem to a pair of coupled matrix equations. Then we focus our treatment on the infinitely-many-identical-sphere problem in a simple cubic lattice as originally investigated by Rayleigh. We complete his formulation of the problem by deriving a matrix equation. We obtain an explicit analytic solution of the matrix equation by iteration. A formula for the effective dielectric constant of the simple cubic lattice of identical spheres is derived. This solution serves as a paradigm for the study of other many-body problems in electromagnetic theory.

10.2. Boundary Conditions

Consider an arbitrary configuration of N spheres embedded in an infinite, homogeneous dielectric medium. The spheres may have arbitrary different sizes and arbitrary different homogeneous compositions. Let the spheres be labeled by an index ν, with $\nu = 1, 2, 3, \ldots N$. Denote the radius of the νth sphere by a_ν, and its dielectric constant by ε_ν. Denote the dielectric constant of the ambient medium by $\varepsilon_a \cdot \varepsilon_\nu$ and ε_a are in general complex. Let a uniform external electric field $\mathbf{E}_0$ be applied to the system of spheres. The spheres are excited by the field and interact with one another. Our objective is to calculate the resultant pattern of excitation induced on each sphere.

We formulate the problem by following Rayleigh's procedure in [1]. Rayleigh considered an infinite system of identical spheres in a simple cubic lattice. His procedure can be extended to a finite system of arbitrary spheres in an arbitrary configuration. We assume that the time variation of the applied electric field $\mathbf{E}_0$ is slow, so that the typical wavelength is much greater than the typical sphere separation. Then the problem is one of electrostatics. It amounts to determining the electrostatic potential V which obeys the Laplace equation

$$\nabla^2 V = 0. \tag{10.1}$$

The electric field $\mathbf{E}$ is obtained by taking the negative gradient of V

$$\mathbf{E} = -\nabla V. \tag{10.2}$$

We set up a rectangular coordinate system with origin at the center of, say, the νth sphere, as shown in Figure 10.1. The z-axis is aligned along the direction of the applied field. We consider the total potential V inside and around the νth sphere. In the immediate exterior of the νth sphere, at a point $\mathbf{r}_\nu$, we can expand V in a series of spherical harmonics

$$\begin{aligned} V = A_{00}^\nu &+ \sum_{n=1}^{\infty} \sum_{m=0}^{n} \left[A_{nm}^\nu \left(\frac{r_\nu}{a_\nu}\right)^n + B_{nm}^\nu \left(\frac{a_\nu}{r_\nu}\right)^{n+1} \right] Y_{nm}^c(\theta_\nu, \varphi_\nu) \\ &+ \sum_{n=1}^{\infty} \sum_{m=0}^{n} \left[C_{nm}^\nu \left(\frac{r_\nu}{a_\nu}\right)^n + D_{nm}^\nu \left(\frac{a_\nu}{r_\nu}\right)^{n+1} \right] Y_{nm}^s(\theta_\nu, \varphi_\nu), \end{aligned} \tag{10.3}$$

where Y_{nm}^c and Y_{nm}^s are the surface harmonics

$$\begin{aligned} Y_{nm}^c(\theta, \varphi) &= P_n^m(\cos\theta)\cos(m\varphi), \\ Y_{nm}^s(\theta, \varphi) &= P_n^m(\cos\theta)\sin(m\varphi), \end{aligned} \tag{10.4}$$

P_n^m denotes an associated Legendre function; r_ν, θ_ν, and φ_ν are the standard

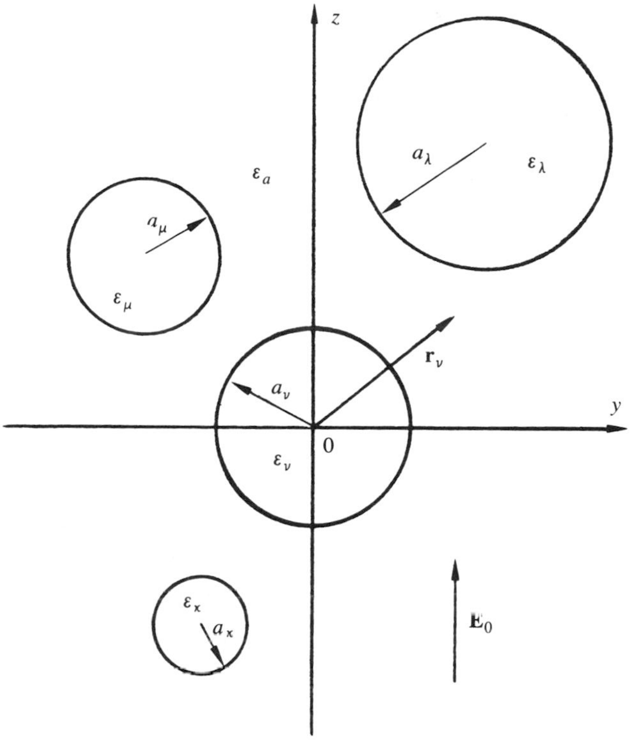

FIGURE 10.1. An arbitrary configuration of spheres embedded in a dielectric medium and excited by an applied electric field $\mathbf{E}_0$.

spherical polar coordinates of $\mathbf{r}_\nu$ referred to the center of the νth sphere as the origin. Inside the νth sphere, we can similarly expand V as

$$V = F_{00}^\nu + \sum_{n=1}^{\infty} \sum_{m=0}^{n} \left(\frac{r_\nu}{a_\nu}\right)^n [F_{nm}^\nu Y_{nm}^c(\theta_\nu, \varphi_\nu) + G_{nm}^\nu Y_{nm}^s(\theta_\nu, \varphi_\nu)]. \quad (10.5)$$

At the surface of the sphere $r_\nu = a_\nu$, the external and internal values of V must satisfy the boundary conditions that V and $\varepsilon(\partial V/\partial r_\nu)$ be continuous. By the linear independence of the surface harmonics, these boundary conditions lead to the relations

$$\begin{aligned} A_{00}^\nu &= F_{00}^\nu, \\ A_{nm}^\nu + B_{nm}^\nu &= F_{nm}^\nu, \\ C_{nm}^\nu + D_{nm}^\nu &= G_{nm}^\nu, \\ \varepsilon_a[nA_{nm}^\nu - (n+1)B_{nm}^\nu] &= \varepsilon_\nu n F_{nm}^\nu, \\ \varepsilon_a[nC_{nm}^\nu - (n+1)D_{nm}^\nu] &= \varepsilon_\nu n G_{nm}^\nu. \end{aligned} \quad (10.6)$$

Eliminating F^{ν}_{nm} and G^{ν}_{nm}, we obtain a pair of simple relations

$$\begin{aligned} B^{\nu}_{nm} &= R^{\nu}_{n} A^{\nu}_{nm}, \\ D^{\nu}_{nm} &= R^{\nu}_{n} C^{\nu}_{nm}, \end{aligned} \tag{10.7}$$

where

$$R^{\nu}_{n} = \frac{n(\varepsilon_a - \varepsilon_\nu)}{(n+1)\varepsilon_a + n\varepsilon_\nu}. \tag{10.8}$$

Equations (10.7) hold for every sphere in the assembly.

10.3. Self-Consistency Relation

To determine the expansion coefficients A^{ν}_{nm}, B^{ν}_{nm}, C^{ν}_{nm}, and D^{ν}_{nm} of the electrostatic potential V around the νth sphere, we need to supplement (10.7) with additional relations. Equations (10.7) are derived on information pertaining to the radius and material constitution of the individual spheres only. They take no account yet of their interaction. Following Rayleigh's approach in [1], we now construct a self-consistency relation describing the mutual interaction of the spheres.

In the expansion (10.3), the terms in negative powers of r_ν are the contributions to the total potential in the immediate exterior of the νth sphere from the electrostatic multipole moments induced on the νth sphere. The terms in positive powers of r_ν, on the other hand, are contributions from the applied field and the multipole moments induced on all the other spheres. The potential of the applied field $\mathbf{E}_0$ is given by

$$V_0 = U_\nu - E_0 r_\nu \cos\theta_\nu, \tag{10.9}$$

where U_ν is a constant dependent on the location of the νth sphere. Then the above observation concerning the terms of (10.3) can be expressed mathematically by the following equation

$$\begin{aligned} A^{\nu}_{00} &+ \sum_{nm} \left(\frac{r_\nu}{a_\nu}\right)^{n} [A^{\nu}_{nm} Y^{c}_{nm}(\theta_\nu, \varphi_\nu) + C^{\nu}_{nm} Y^{s}_{nm}(\theta_\nu, \varphi_\nu)] \\ &= U_\nu - E_0 r_\nu \cos\theta_\nu + \sum_{\mu \neq \nu}^{N} \sum_{nm} \left(\frac{a_\mu}{r_\mu}\right)^{n+1} [B^{\mu}_{nm} Y^{c}_{nm}(\theta_\mu, \varphi_\mu) + D^{\mu}_{nm} Y^{s}_{nm}(\theta_\mu, \varphi_\mu)]. \end{aligned} \tag{10.10}$$

Equation (10.10) is a self-consistency relation. It connects the induced multipole moments on all the spheres. Whereas (10.7) results from the boundary conditions on the surfaces of the spheres, (10.10) describes their interaction. Together, these two sets of equations contain the complete formulation of the problem, and determine the expansion coefficients of the total potential. But before we can carry out this determination, we must first reduce (10.10) to a usable form.

Equation (10.10) is an identity. It is valid for every point $\mathbf{r}_\nu$ in the immediate exterior of the νth sphere. We note that if we can express the right-hand side of (10.10) in terms of spherical harmonics in $\mathbf{r}_\nu$, we can then equate the coefficients term by term, and (10.10) will be seen to be equivalent to an infinite system of algebraic equations. But the right-hand side contains spherical harmonics in $\mathbf{r}_\mu$ with $\mu \neq \nu$. $\mathbf{r}_\nu$ is the position vector of a point referred to the center of the νth sphere as origin. $\mathbf{r}_\mu$ is the position vector of the same point referred to the center of a different sphere, the μth sphere, as origin. We must first express the spherical harmonics in $\mathbf{r}_\mu$ in terms of those in $\mathbf{r}_\nu$. In other words, we need mathematical identities which are generally known as addition theorems. This type of addition theorem for the spherical harmonics does not seem to exist in the literature. We shall derive them in the following section.

10.4. Addition Theorems

The addition theorems we need may be stated as the explicit expansions of the spherical harmonics defined with respect to the center of the μth sphere in terms of those defined with respect to the center of the νth sphere

$$\frac{1}{r_\mu^{n+1}} Y_{nm}^c(\theta_\mu, \varphi_\mu) = \sum_{n'=0}^{\infty} \sum_{m'=0}^{n'} \Gamma_{nm;n'm'}^{cc}(\mathbf{R}_{\mu\nu}) r_\nu^{n'} Y_{n'm'}^c(\theta_\nu, \varphi_\nu) + \sum_{n'=1}^{\infty} \sum_{m'=1}^{n'} \Gamma_{nm;n'm'}^{cs}(\mathbf{R}_{\mu\nu}) r_\nu^{n'} Y_{n'm'}^s(\theta_\nu, \varphi_\nu), \tag{10.11}$$

$$\frac{1}{r_\mu^{n+1}} Y_{nm}^s(\theta_\mu, \varphi_\mu) = \sum_{n'=0}^{\infty} \sum_{m'=0}^{n'} \Gamma_{nm;n'm'}^{sc}(\mathbf{R}_{\mu\nu}) r_\nu^{n'} Y_{n'm'}^c(\theta_\nu, \varphi_\nu) + \sum_{n'=1}^{\infty} \sum_{m'=1}^{n'} \Gamma_{nm;n'm'}^{ss}(\mathbf{R}_{\mu\nu}) r_\nu^{n'} Y_{n'm'}^s(\theta_\nu, \varphi_\nu), \tag{10.12}$$

where $\mathbf{R}_{\mu\nu}$ denotes the position vector extending from the center of the μth sphere to that of the νth sphere, as shown in Figure 10.2. We know such expansions exist since the left-hand sides are solutions of the Laplace equation. They are expandable in terms of a complete set of eigenfunctions. We establish the addition theorems by working out the expansion coefficients.

Let us first work out the coefficient $\Gamma_{nm;n'm'}^{cc}(\mathbf{R}_{\mu\nu})$. Multiplying (10.11) by $Y_{n'm'}^c(\theta_\nu, \varphi_\nu)$ and integrating θ_ν and φ_ν over the entire solid angle of 4π around the center of the νth sphere, we get

$$\Gamma_{nm;n'm'}^{cc}(\mathbf{R}_{\mu\nu}) = \frac{1}{K_{n'm'} r_\nu^{n'}} \int_0^{2\pi} d\varphi_\nu \int_0^{\pi} d\theta_\nu \sin\theta_\nu \frac{1}{r_\mu^{n+1}} Y_{nm}^c(\theta_\mu, \varphi_\mu) Y_{n'm'}^c(\theta_\nu, \varphi_\nu), \tag{10.13}$$

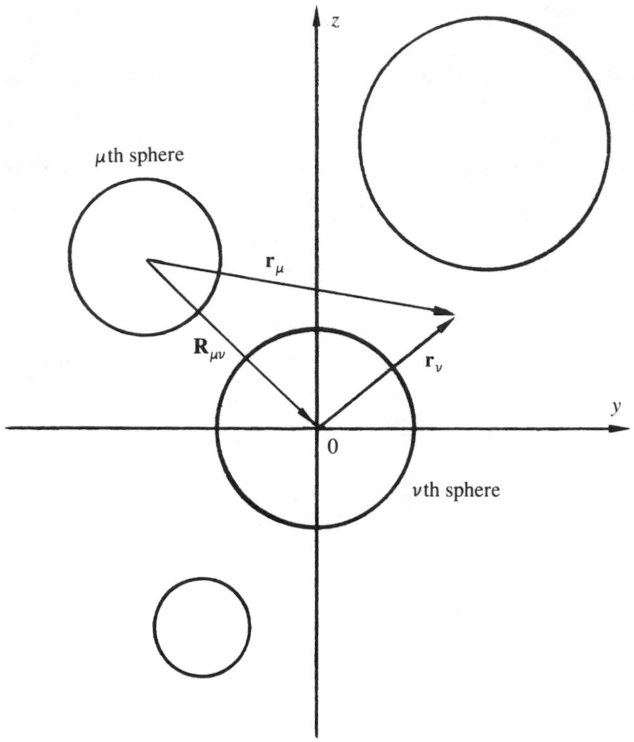

FIGURE 10.2. Geometry for the addition theorems.

where

$$K_{nm} = \frac{(n+m)!}{(n-m)!}\frac{2\pi}{2n+1}(1+\delta_{m0}). \tag{10.14}$$

We have used the orthogonality relation of the associated Legendre functions

$$\int_0^\pi d\theta \sin\theta P_n^m(\cos\theta)P_{n'}^m(\cos\theta) = \frac{(n+m)!}{(n-m)!}\frac{2}{2n+1}\delta_{nn'}, \tag{10.15}$$

so that

$$\int_0^{2\pi} d\varphi \int_0^\pi d\theta \sin\theta Y_{nm}^c(\theta,\varphi)Y_{n'm'}^c(\theta,\varphi) = K_{nm}\delta_{nn'}\delta_{mm'}, \tag{10.16}$$

$$\int_0^{2\pi} d\varphi \int_0^\pi d\theta \sin\theta Y_{nm}^s(\theta,\varphi)Y_{n'm'}^s(\theta,\varphi) = K_{nm}\delta_{nn'}\delta_{mm'}, \tag{10.17}$$

and

$$\int_0^{2\pi} d\varphi \int_0^\pi d\theta \sin\theta Y_{nm}^c(\theta,\varphi)Y_{n'm'}^s(\theta,\varphi) = 0. \tag{10.18}$$

We are able to evaluate the double integral in (10.13) analytically by using the following integral representations of the spherical harmonics [9]

$$\begin{aligned} r^n Y^c_{nm}(\theta, \varphi) &= \frac{(n+m)!}{2\pi i^m n!} \int_0^{2\pi} W^n(\mathbf{r}, u) \cos(mu)\, du, \\ r^n Y^s_{nm}(\theta, \varphi) &= \frac{(n+m)!}{2\pi i^m n!} \int_0^{2\pi} W^n(\mathbf{r}, u) \sin(mu)\, du, \\ \frac{1}{r^{n+1}} Y^c_{nm}(\theta, \varphi) &= \frac{i^m n!}{2\pi (n-m)!} \int_0^{2\pi} W^{-n-1}(\mathbf{r}, u) \cos(mu)\, du, \\ \frac{1}{r^{n+1}} Y^s_{nm}(\theta, \varphi) &= \frac{i^m n!}{2\pi (n-m)!} \int_0^{2\pi} W^{-n-1}(\mathbf{r}, u) \sin(mu)\, du. \end{aligned} \tag{10.19}$$

The function W is defined by

$$W(\mathbf{r}, u) = z + ix \cos u + iy \sin u \tag{10.20}$$

with x, y, and z denoting the rectangular coordinates of $\mathbf{r}$, corresponding to the spherical polar coordinates r, θ, and φ

$$x = r \sin\theta \cos\varphi, \qquad y = r \sin\theta \sin\varphi, \qquad z = r\cos\theta. \tag{10.21}$$

Now, from Figure 10.2, we have

$$\mathbf{r}_\mu = \mathbf{r}_\nu + \mathbf{R}_{\mu\nu}. \tag{10.22}$$

We take that the rectangular coordinate systems with origins at the centers of the μth and νth spheres have parallel axes. Using (10.19), (10.20), and (10.22), we write

$$\frac{1}{r_\mu^{n+1}} Y^c_{nm}(\theta_\mu, \varphi_\mu) = \frac{i^m n!}{2\pi(n-m)!} \int_0^{2\pi} \frac{\cos(mu)\, du}{[W(\mathbf{r}_\nu, u) + W(\mathbf{R}_{\mu\nu}, u)]^{n+1}}. \tag{10.23}$$

Since $R_{\mu\nu} > r_\nu$, we use the binomial theorem to obtain

$$\frac{1}{[W(\mathbf{r}_\nu, u) + W(\mathbf{R}_{\mu\nu}, u)]^{n+1}} = \frac{1}{W^{n+1}(\mathbf{R}_{\mu\nu}, u)} \sum_{M=0}^{\infty} (-1)^M \frac{(n+M)!}{n!M!} \frac{W^M(\mathbf{r}_\nu, u)}{W^M(\mathbf{R}_{\mu\nu}, u)}. \tag{10.24}$$

We substitute (10.24) into (10.23), and (10.23) into (10.13). Now the double integral of (10.13) can be evaluated by first integrating over φ_ν using (10.19), and then over θ_ν using (10.15). The result is as follows:

$$\Gamma^{cc}_{nm;n'm'}(\mathbf{R}_{\mu\nu}) = \frac{(-1)^{n'}}{K_{n'm'}} \frac{i^{m+m'}}{2n'+1} \frac{(n+n')!}{(n-m)!(n'-m')!} \int_0^{2\pi} \frac{2\cos(mu)\cos(m'u)}{W^{n+n'+1}(\mathbf{R}_{\mu\nu}, u)}\, du. \tag{10.25}$$

Next, we resolve the product of two cosines in the numerator of the integrand into a sum of two cosines. Then, by (10.19), the integral can be reduced to spherical harmonics in $\mathbf{R}_{\mu\nu}$. We finally obtain

$$\Gamma^{cc}_{nm;n'm'}(\mathbf{R}_{\mu\nu}) = \frac{(-1)^{n'}}{K_{n'm'}} \frac{2\pi}{2n'+1} \left[\frac{(n+n'-m-m')!}{(n-m)!(n'-m')!} \frac{1}{R_{\mu\nu}^{n+n'+1}} Y^c_{(n+n')(m+m')}(\Theta_{\mu\nu}, \Phi_{\mu\nu}) + (-1)^{m'} \frac{(n+n'-m+m')!}{(n-m)!(n'-m')!} \frac{1}{R_{\mu\nu}^{n+n'+1}} Y^c_{(n+n')(m-m')}(\Theta_{\mu\nu}, \Phi_{\mu\nu}) \right], \quad (10.26)$$

where $R_{\mu\nu}$, $\Theta_{\mu\nu}$, and $\Phi_{\mu\nu}$ denote the spherical polar coordinates of $\mathbf{R}_{\mu\nu}$.

The same calculation can be carried out for the other three types of expansion coefficients. For the expansion coefficient $\Gamma^{cs}_{nm;n'm'}(\mathbf{R}_{\mu\nu})$, we also arrive at (10.25), except that the numerator of the integrand is replaced by $2\cos(mu)\sin(m'u)$. For $\Gamma^{sc}_{nm;n'm'}(\mathbf{R}_{\mu\nu})$, it is replaced by $2\sin(mu)\cos(m'u)$. And for $\Gamma^{ss}_{nm;n'm'}(\mathbf{R}_{\mu\nu})$, it is replaced by $2\sin(mu)\sin(m'u)$. Taking these differences into account, we obtain

$$\Gamma^{cs}_{nm;n'm'}(\mathbf{R}_{\mu\nu}) = \frac{(-1)^{n'}}{K_{n'm'}} \frac{2\pi}{2n'+1} \left[\frac{(n+n'-m-m')!}{(n-m)!(n'-m')!} \frac{1}{R_{\mu\nu}^{n+n'+1}} Y^s_{(n+n')(m+m')}(\Theta_{\mu\nu}, \Phi_{\mu\nu}) - (-1)^{m'} \frac{(n+n'-m+m')!}{(n-m)!(n'-m')!} \frac{1}{R_{\mu\nu}^{n+n'+1}} Y^s_{(n+n')(m-m')}(\Theta_{\mu\nu}, \Phi_{\mu\nu}) \right], \quad (10.27)$$

$$\Gamma^{sc}_{nm;n'm'}(\mathbf{R}_{\mu\nu}) = \frac{(-1)^{n'}}{K_{n'm'}} \frac{2\pi}{2n'+1} \left[\frac{(n+n'-m-m')!}{(n-m)!(n'-m')!} \frac{1}{R_{\mu\nu}^{n+n'+1}} Y^s_{(n+n')(m+m')}(\Theta_{\mu\nu}, \Phi_{\mu\nu}) + (-1)^{m'} \frac{(n+n'-m+m')!}{(n-m)!(n'-m')!} \frac{1}{R_{\mu\nu}^{n+n'+1}} Y^s_{(n+n')(m-m')}(\Theta_{\mu\nu}, \Phi_{\mu\nu}) \right], \quad (10.28)$$

$$\Gamma^{ss}_{nm;n'm'}(\mathbf{R}_{\mu\nu}) = \frac{(-1)^{n'}}{K_{n'm'}} \frac{2\pi}{2n'+1} \left[-\frac{(n+n'-m-m')!}{(n-m)!(n'-m')!} \frac{1}{R_{\mu\nu}^{n+n'+1}} Y^c_{(n+n')(m+m')}(\Theta_{\mu\nu}, \Phi_{\mu\nu}) + (-1)^{m'} \frac{(n+n'-m+m')!}{(n-m)!(n'-m')!} \frac{1}{R_{\mu\nu}^{n+n'+1}} Y^c_{(n+n')(m-m')}(\Theta_{\mu\nu}, \Phi_{\mu\nu}) \right], \quad (10.29)$$

10.5. Reduction to Matrix Equations

Having established the addition theorems, we return to the self-consistency relation (10.10). Multiplying (10.10) by $Y^c_{nm}(\theta_\nu, \varphi_\nu)$, integrating over θ_ν and φ_ν, and using (10.11) and (10.12), we obtain

$$K_{nm} a_\nu^{-n} A_{nm}^\nu = -E_0 K_{10} \delta_{n1} \delta_{m0} + \sum_{\mu \neq \nu}^{N} \sum_{n'=1}^{\infty} \sum_{m'=0}^{n} [K_{nm} \Gamma^{cc}_{n'm';nm}(\mathbf{R}_{\mu\nu}) a_\mu^{n'+1} B^\mu_{n'm'} + K_{nm} \Gamma^{sc}_{n'm';nm}(\mathbf{R}_{\mu\nu}) a_\mu^{n'+1} D^\mu_{n'm'}]. \tag{10.30}$$

Again, multiplying (10.10) by $Y^s_{nm}(\theta_\nu, \varphi_\nu)$ instead, we obtain

$$K_{nm} a_\nu^{-n} C_{nm}^\nu = \sum_{\mu \neq \nu}^{N} \sum_{n'=1}^{\infty} \sum_{m'=0}^{n'} [K_{nm} \Gamma^{cs}_{n'm';nm}(\mathbf{R}_{\mu\nu}) a_\mu^{n'+1} B^\mu_{n'm'} + K_{nm} \Gamma^{ss}_{n'm';nm}(\mathbf{R}_{\mu\nu}) a_\mu^{n'+1} D^\mu_{n'm'}]. \tag{10.31}$$

These two equations, deriving from the self-consistency relation (10.10), express the interaction of the spheres. We recall from Section 10.2 that the boundary conditions on each sphere lead to another pair of equations

$$\begin{aligned} B^\nu_{nm} &= R^\nu_n A^\nu_{nm}, \\ D^\nu_{nm} &= R^\nu_n C^\nu_{nm}. \end{aligned} \tag{10.32}$$

Eliminating A^ν_{nm} and C^ν_{nm} from (10.30) and (10.31) with (10.32), we obtain a pair of coupled, linear, inhomogeneous matrix equations in B^ν_{nm} and D^ν_{nm}. They can be solved analytically by iteration, or numerically on the computer. This completes our reduction of the general N-sphere problem to matrix equations

10.6. Identical Spheres in a Simple Cubic Lattice

As an illustration, we apply the general formalism developed above to a specific many-body problem. We consider an infinite number of identical, homogeneous spheres arranged in a simple cubic lattice, as shown in Figure 10.3. Let the radius of the spheres be denoted by a, and the lattice constant by b. Also, let the dielectric constant of the spheres be denoted by ε_2, and that of the ambient medium by $\varepsilon_1 \cdot \varepsilon_1$ and ε_2 are generally complex.

We introduce a rectangular coordinate system with origin at the center of an arbitrarily chosen sphere, and with the axes parallel to the edges of the primitive cubic cell. The rectangulr coordinates of the center of any sphere are then integral multiples of b. Let a uniform electric field $\mathbf{E}_0$ be applied to the lattice of spheres. Without loss of generality, we can choose $\mathbf{E}_0$ to point in the positive z direction, as indicated in Figure 10.3. An applied electric field pointing in an arbitrary direction can always be resolved into components along the three coordinate axes, and their effects added up vectorially. We want to calculate the electric multipole moments induced on each sphere as a result of the excitation by the applied field and the mutual interaction of the spheres.

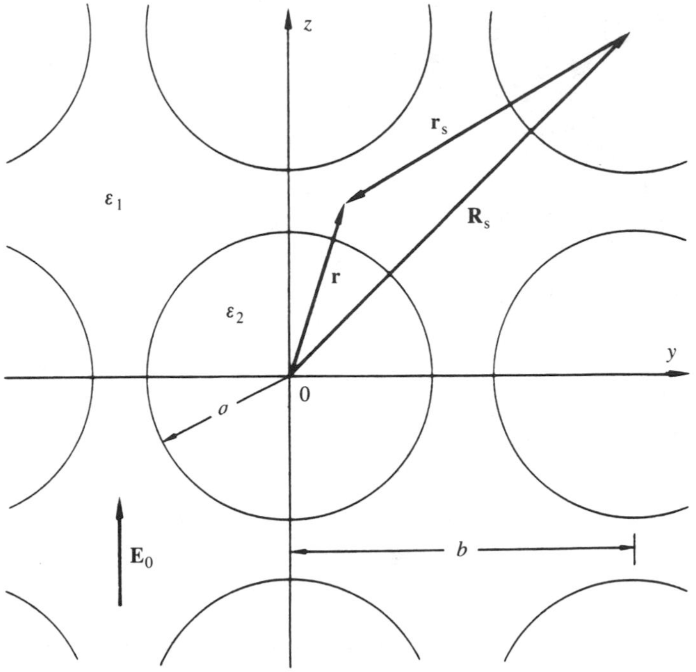

FIGURE 10.3. A simple cubic lattice of identical spheres embedded in a dielectric medium.

Our many-body problem here has a high degree of symmetry. Rather than treating it as a special limit of the general many-body problem studied in the preceding sections, we will take advantage of the symmetry and start anew from the very beginning, even though we shall proceed exactly as before. We consider the total electrostatic potential V at the sphere centered at the origin. In the immediate exterior of the sphere, we have

$$V = A_{00} + \sum_{nm} \left[A_{nm} \left(\frac{r}{a} \right)^n + B_{nm} \left(\frac{a}{r} \right)^{n+1} \right] Y_{nm}^c(\theta, \varphi). \qquad (10.33)$$

This expression is considerably simpler than (10.3). First of all, because of the simple cubic symmetry, Y_{nm}^s does not appear. Moreover, the summation index n is restricted to odd integers ($n = 1, 3, 5, \ldots$) and m to integral multiples of 4 ($m = 0, 4, 8, \ldots$). And because the spheres are all identical and equivalent, the sphere index v has been dropped from the expansion coefficients A_{nm} and B_{nm}. r, θ, and φ are the spherical polar coordinates of a point $\mathbf{r}$ referred to the coordinate origin. Similarly, in the interior of the sphere, we write

$$V = F_{00} + \sum_{nm} F_{nm} \left(\frac{r}{a} \right)^n Y_{nm}^c(\theta, \varphi) \qquad (10.34)$$

in place of (10.5). The ranges of summation of n and m are as in (10.33). Applying the boundary conditions at the surface of the sphere $r = a$ and eliminating F_{nm}, we obtain a relation between A_{nm} and B_{nm}

$$B_{nm} = R_n A_{nm} \tag{10.35}$$

with

$$R_n = \frac{n(\varepsilon_1 - \varepsilon_2)}{(n+1)\varepsilon_1 + n\varepsilon_2}. \tag{10.36}$$

Next we write down the self-consistency relation describing the interaction of the spheres

$$A_{00} + \sum_{nm} A_{nm}\left(\frac{r}{a}\right)^n Y_{nm}^c(\theta, \varphi) = -E_0 r \cos\theta + \sum_s \sum_{nm} B_{nm}\left(\frac{a}{r_s}\right)^{n+1} Y_{nm}^c(\theta_s, \varphi_s). \tag{10.37}$$

The first term on the right-hand side is the potential of the applied field

$$V_0 = -E_0 r \cos\theta. \tag{10.38}$$

The index s labels a sphere other than the one centered at the origin. The summation over s goes over all the spheres in the lattice except the one at the origin. r_s, θ_s, and φ_s are the spherical polar coordinates of the point $\mathbf{r}$ in a coordinate system centered at the center of the sth sphere, as shown in Figure 10.3. That is to say

$$\mathbf{r}_s = \mathbf{r} - \mathbf{R}_s, \tag{10.39}$$

where $\mathbf{R}_s$ is the position vector of the center of the sth sphere. $\mathbf{R}_s$ is the counterpart of $-\mathbf{R}_{\mu\nu}$ in Figure 10.2.

Equation (10.37) is an identity. We project out algebraic equations by multiplying it by $Y_{nm}^c(\theta, \varphi)$, integrating over θ and φ, and applying the addition theorem in the form (10.13). The result is

$$K_{nm} a^{-n} A_{nm} = -E_0 K_{10} \delta_{n1} \delta_{m0} + \sum_s \sum_{n'm'} K_{nm} \Gamma_{n'm';nm}^{cc}(-\mathbf{R}_s) a^{n'+1} B_{n'm'}. \tag{10.40}$$

Eliminating A_{nm} from (10.40) with (10.35) and using (10.26), we obtain a matrix equation in B_{nm}

$$\frac{K_{nm}}{R_n} B_{nm} = -E_0 a K_{10} \delta_{n1} \delta_{m0} + \sum_{n'm'} \lambda^{n+n'+1} E_{nm;n'm'} B_{n'm'}, \tag{10.41}$$

where

$$\lambda = \frac{a}{b} \tag{10.42}$$

and

$$E_{nm;n'm'} = (-1)^n \frac{2\pi}{2n+1}$$

$$\times \left(\frac{(n+n'-m-m')!}{(n-m)!(n'-m')!} S_{n+n'}^{m+m'} + i^{m+m'-|m-m'|} \frac{(n+n'-|m-m'|)!}{(n-m)!(n'-m')!} S_{n+n'}^{|m-m'|} \right). \tag{10.43}$$

S_n^m is a dimensionless lattice sum

$$S_n^m = \sum_s \left(\frac{b}{R_s}\right)^n Y_{nm}^c(\Theta_s, \Phi_s) \tag{10.44}$$

with R_s, Θ_s, and Φ_s denoting the spherical polar coordinates of $\mathbf{R}_s$. In deriving (10.43), we have recast (10.26) slightly by introducing $|m - m'|$ to make the expression more convenient for computation.

Equation (10.41) is the complete formulation of the many-sphere problem in a simple cubic lattice. The study of this problem was pioneered by Rayleigh in [1], using the self-consistency relation (10.37). But he did not have the addition theorems, and was compelled to resort to partial differentiation of (10.37) to obtain algebraic equations one by one. In this way, he derived the first few terms of the first few equations of the infinite set (10.41). Other authors continued Rayleigh's work. They braved the tedium of interminable partial differentiation, and added more and longer algebraic equations to his set [10], [11]. By using the addition theorems, we were able to derive the complete set in one fell swoop [12].

10.7. Effective Dielectric Constant

The simple cubic lattice of spheres is a model of a so-called artificial dielectric. The embedding of spheres in a homogeneous dielectric medium alters the dielectric constant of the medium in a controllable fashion. The effective dielectric constant of the sphere-loaded medium depends on the dielectric constant of the spheres as well as on their volume fraction in the medium. We solve (10.41) analytically and derive a formula for the effective dielectric constant.

The effective dielectric constant ε of the lattice of spheres is determined by the polarization P of the lattice in the applied electric field $\mathbf{E}_0$

$$\varepsilon = \varepsilon_1 + \frac{P}{E_0}. \tag{10.45}$$

P is given by the induced electric dipole moment per unit volume. The dipole moment induced on a sphere is proportional to the coefficient B_{10}. It can be shown [12] that

$$\frac{\varepsilon}{\varepsilon_1} = 1 + \frac{4\pi a^2 B_{10}}{b^3 E_0}. \tag{10.46}$$

The concept of an effective dielectric constant is meaningful so long as the wavelength of the applied field is much greater than the lattice constant b.

Next we write down (10.41) for $n = 1$ and rearrange

$$\frac{E_0 a}{B_{10}} = -\frac{1}{R_1} + \lambda^3 \frac{E_{10;10}}{K_{10}} + \sum_{n=3}^{\infty} \sum_{m=0}^{n} \lambda^{n+2} \frac{E_{10;nm}}{K_{10}} \frac{B_{nm}}{B_{10}}. \tag{10.47}$$

The summation indices are of course restricted by symmetry as in (10.33). The left-hand side is precisely what we need in (10.46). We calculate the right-hand side of (10.47) by writing down and rearranging (10.41) for $n \geq 3$ as follows:

$$\frac{B_{nm}}{B_{10}} = \lambda^{n+2} Q_{nm;\,10} + \sum_{n'=3}^{\infty} \sum_{m'=0}^{n'} \lambda^{n+n'+1} Q_{nm;\,n'm'} \frac{B_{n'm'}}{B_{10}}, \qquad n \geq 3, \tag{10.48}$$

with

$$Q_{nm;\,n'm'} = \frac{R_n}{K_{nm}} E_{nm;\,n'm'}. \tag{10.49}$$

Equation (10.48) can be solved systematically for B_{nm}/B_{10} by iteration. The solution can be arranged in a power series in λ. Substituting it into (10.47), we obtain

$$\begin{aligned}\frac{E_0 a}{B_{10}} = {} & -\frac{1}{R_1} - 2S_2^0\lambda^3 + 16(S_4^0)^2 R_3 \lambda^{10} + \left[36(S_6^0)^2 + \frac{8}{9!}(S_6^4)^2\right] R_5 \lambda^{14} \\ & - 320(S_4^0)^2 S_6^0 R_3^2 \lambda^{17} + \left[64(S_8^0)^2 + \frac{192}{11!}(S_8^4)^2\right] R_7 \lambda^{18} + \cdots. \end{aligned} \tag{10.50}$$

Recall from (10.42) that the expansion parameter λ lies between 0 and $\frac{1}{2}$. The structure of the lattice enters through the lattice sums S_n^m. The material composition of the spheres enters through the factors R_n. The loading density of the spheres enters through the parameter λ.

The lattice sums S_n^m have been published in [10]:

$$\begin{aligned} S_2^0 &= 2\pi/3, & S_4^0 &= 3.1082, \\ S_6^0 &= 0.57333, & S_6^4 &= -1444.79, \\ S_8^0 &= 3.2593, & S_8^4 &= 5475.61. \end{aligned} \tag{10.51}$$

We want to point out that the lattice sum S_2^0 is only conditionally convergent. Its value is not unique,but depends on the shape of the volume of summation at intermediate stages before the limit of the infinite lattice is taken. This mathematical shape dependence is related to the physical occurrence of depolarization fields arising from exposed electric charges on the surface of the volume of summation. The value of $2\pi/3$ is obtained by choosing the intermediate volume of summation to be a long rectangular cylinder aligned along the z direction, as described by Rayleigh in [1]. For the long cylinder, the depolarization field is zero. It is the right choice since the depolarization field should not figure in the definition of the effective dielectric constant.

We introduce the volume fraction p occupied by the spheres

$$p = \frac{4\pi a^3}{3b^3} = \frac{4\pi}{3}\lambda^3 \tag{10.52}$$

and define a dimensionless function

$$\Lambda(p) = \frac{E_0 a}{B_{10}}. \tag{10.53}$$

Substituting (10.51) into (10.50), we obtain

$$\Lambda(p) = -\frac{1}{R_1} - p + 1.3047R_3p^{10/3} + 0.0723R_5p^{14/3} - 0.5289R_3^2p^{17/3} + 0.1526R_7p^6 + \cdots. \tag{10.54}$$

By (10.46), the effective dielectric constant of the sphere-loaded medium is given by

$$\frac{\varepsilon}{\varepsilon_1} = 1 + \frac{3p}{\Lambda(p)}. \tag{10.55}$$

Note that since $\Lambda(p)$ occurs in the denominator, a truncation of $\Lambda(p)$ at a finite power of p still leads to an infinite series for ε. A truncation approximation for $\Lambda(p)$ corresponds to the summation of subsets of perturbation terms for ε to infinite order.

We use formula (10.55) to calculate the effective dielectric constant of a simple cubic lattice of perfectly conducting spheres. In the limit of perfect conductivity, (10.36) reduces to

$$R_n = -1, \qquad \varepsilon_2 \to i\infty. \tag{10.56}$$

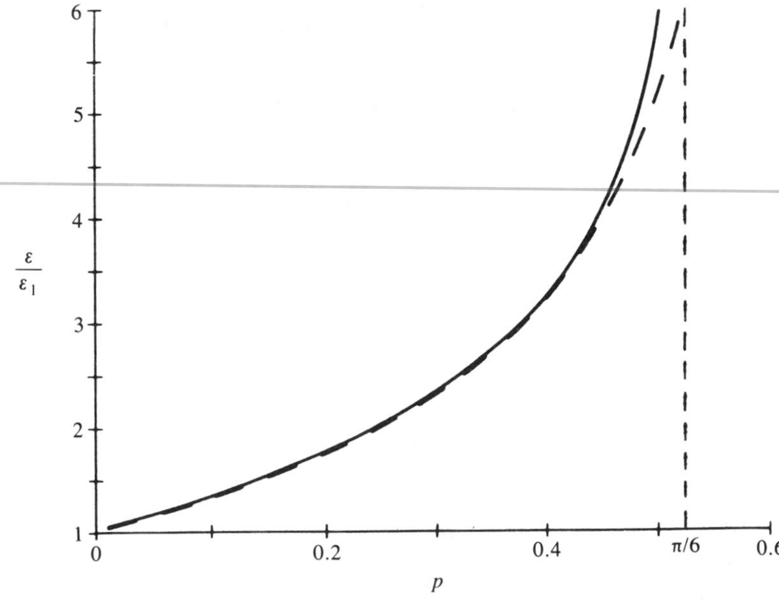

FIGURE 10.4. Effective dielectric constant ε of a simple cubic lattice of perfectly conducting spheres versus volume fraction p of spheres. The solid line is from the computer calculations of [10] and [11]. The broken line is from eqs. (10.54), (10.55), and (10.56) of this work.

Substituting (10.56) into (10.54) and neglecting terms in $\Lambda(p)$ beyond order p^6, we compute ε as a function of p, and plot the result in Figure 10.4 as the broken line. The solid line is taken from the large-scale computer calculations of [10] and [11]. The two curves agree very well except near the maximum volume fraction $p_{\max}$ of the simple cubic lattice

$$p_{\max} = \frac{\pi}{6} = 0.523\ldots \tag{10.57}$$

at which the spheres just touch. This deviation near $p_{\max}$ is due to the fact that, for perfectly conducting spheres, ε has a logarithmic singularity at $p = p_{\max}$ [13]. Therefore the truncated power-series expansion of $\Lambda(p)$ becomes inadequate near $p_{\max}$. We should develop a new solution of (10.48) useful for $p \cong p_{\max}$.

It has been found [10], [11] that the computer solution for perfectly conducting spheres can be fitted extremely well over the entire range of p by the interpolation formula

$$\frac{\varepsilon}{\varepsilon_1} = 1 - \frac{\pi}{2} \ln\left(1 - \frac{6p}{\pi}\right). \tag{10.58}$$

The logarithmic singularity shows that we can achieve arbitrarily large values of the effective dielectric constant of the artificial dielectric by using perfectly conducting spheres. The theoretical calculation is in excellent agreement with measurement [10], [11].

10.8. Epilegomena

By establishing the addition theorems for spherical harmonics, we are able to complete the formulation of the electrostatic many-body problem of a simple cubic lattice of identical spheres, originally investigated by Rayleigh in 1892. The formulation is succinctly contained in the matrix eq. (10.41). The equation can be solved numerically, as well as analytically by iteration. The advantage of an analytic solution is that, unlike for a numerical solution, it shows the explicit analytic dependence of the solution on the parameters of the problem.

We have also extended the calculation to solve the magnetostatic many-body problem of a simple cubic lattice of conducting magnetic spheres [12]. If the magnetic permeability and the electrical conductivity of the spheres are denoted by μ_2 and σ, respectively, and the magnetic permeability of the ambient medium, which is assumed to be nonconducting, is denoted by μ_1, then the effective magnetic permeability μ of the sphere-loaded medium is given by an analogue of (10.55)

$$\frac{\mu}{\mu_1} = 1 + \frac{3p}{\Lambda(p)}. \tag{10.59}$$

$\Lambda(p)$ is given as before by (10.54), except that R_n now assumes the new form

$$R_n = \frac{n\mu_1[aj_n(ka)]' - n(n+1)\mu_2 j_n(ka)}{(n+1)\mu_1[aj_n(ka)]' + n(n+1)\mu_2 j_n(ka)}, \tag{10.60}$$

where

$$k = \sqrt{i\omega\mu_2\sigma} \tag{10.61}$$

with ω denoting the angular frequency of the external applied magnetic field. j_n is a spherical Bessel function, and the prime denotes partial differentiation with respect to a. Besides paramagnetic effects, (10.60) contains the diamagnetic, lossy effects of eddy currents in the spheres. The details of the calculation are presented in [12].

Having gone over the details of the solution of the many-identical-sphere problem in a simple cubic lattice, we should experience no greater conceptual difficulty in deriving solution of the general many-sphere problem formulated in (10.30), (10.31), and (10.32). Indeed, we should be able to extend the formulation and solution procedures—namely, along the well-trodden pathway of boundary conditions, self-consistency relation, addition theorems, matrix equations and iteration—to other lattices, bodies of other shapes, as well as to dynamic problems. Our success will depend on our ability to establish the appropriate addition theorems to project out algebraic equations from the self-consistency relation, especially in the generally vectorial dynamic problems.

Acknowledgment

The author considers it his great fortune to be a student of Professor Papas. And on this felicitous occasion, when so much blossoms and so much comes to fruition, the author would like to thank him for his guidance in the perfect, fertile, esoteric world of electromagnetism which we have all come to love to explore.

References

[1] Lord Rayleigh (1892), On the influence of obstacles arranged in rectangular order upon the properties of a medium, *Philos. Mag.*, **34**, 481.

[2] G.O. Olaofe (1970), Scattering by two cylinders, *Radio Sci.*, **5**, 1351.

[3] C. Liang and Y.T. Lo (1967), Scattering by two spheres, *Radio Sci.*, **2**, 1481.

[4] A.E. Heins (1948), The radiation and transmission properties of a pair of semi-infinite parallel plates, *Quart. Appl. Math.*, **6**, 157 and 215.

[5] L.A. Vainshtein (1948), Strogoe reshenie zadachi o ploskom volnovode s otkrytym kontsom, *Izvestia Akad. Nauk SSSR*, **12**, 144.

[6] L.A. Vainshtein (1948), K teorii diffraktsii na dvukh parallel'nykh poluploskostyakh, *Izvestia Akad. Nauk SSSR*, **12**, 166.

[7] J.F. Carlson and A.E. Heins (1948), The reflection of an electromagnetic plane wave by an infinite set of plates, *Quart. Appl. Math.*, **4**, 313 and **5**, 82.

[8] J. Lam (1967), Radiation of a point charge moving uniformly over an infinite array of conducting half-planes, *J. Math. Phys.*, **8**, 1053.

[9] P.M. Morse and H. Feshbach (1953), *Methods of Theoretical Physics*, Part II, McGraw-Hill, New York, p. 1270.

[10] R.C. McPhedran and D.R. McKenzie (1978), The conductivity of lattices of spheres I. The simple cubic lattice, *Proc. Roy. Soc. London*, Ser A. **359**, 45.

[11] W.T. Doyle (1978), The Clausius–Mossotti problem for cubic arrays of spheres, *J. Appl. Phys.*, **49**, 795.

[12] J. Lam (1986), Magnetic permeability of a simple cubic lattice of conducting magnetic spheres, *J. Appl. Phys.*, **60**, 4230.

[13] J.B. Keller (1963), Conductivity of a medium containing a dense array of perfectly conducting spheres or cylinders or nonconducting cylinders, *J. Appl. Phys.*, **34**, 991.

11
Electromagnetic Shielding

K.S.H. Lee*

11.1. Introduction

The most common method of protecting a system from undesirable electromagnetic interferences is by placing it inside a shielded enclosure of highly conducting walls. In practice, an enclosure is never made perfect in the sense that it has cracks and seams to allow for energy penetration through apertures, and the conductivity of its walls, albeit high, is not infinite, thus permitting low-frequency magnetic field penetration through diffusion. The diffusive penetration mechanism is particularly effective at low frequencies where the skin depth of the enclosure's wall is greater than the wall's thickness. At these frequencies the familiar attenuation and scattering losses of the wall are negligible and the only means to shield against the magnetic field penetration is by way of the induced currents in the shield. Hence the geometrical shape of the enclosure becomes most critical, since it determines the induced current distribution.

We start with the general theory of magnetic diffusion in Section 11.2, where we will see how the problem of solving Maxwell's equations without the displacement-current term is reduced to the problem of solving a Laplace equation subject to a set of boundary conditions that duplicate the shielding effect of the enclosure's wall. From these boundary conditions we will derive simple perturbation procedures for calculating the penetrant magnetic field into enclosures with electrically thick or electrically thin shield walls.

We apply the equivalent scalar theory in Section 11.2 to several canonical enclosure shapes in Section 11.3 and obtain simple engineering formulas for the penetrant field. We then show how these simple formulas may be generalized to arbitrary-shaped enclosures.

Enclosures of multilayered shields can be treated straightforwardly with the equivalent scalar theory, although more algebraic manipulations would be involved. The shielding of N-layered shields is always less than the sum (in

---

* Kaman Sciences Corporation, Dikewood Division, 2800 28th Street, Suite 370, Santa Monica, CA 90405, USA.

the logarithmic sense) of the N individual shields due to the presence of interactions among the shields. These aspects and others will be taken up in Section 11.4.

The last section, Section 11.5, is a revisit to the exact solution of the spherical shield problem. We will show how the exact solution, under certain specific approximations, is reduced to that obtained from the equivalent scalar theory. The penetrant electric field is also calculated along with the penetrant magnetic field in order to show why the electric field penetration problem is of no concern in comparison to the corresponding problem of magnetic field diffusion.

11.2. General Theory of Magnetic Diffusion

In the magnetic diffusion problem displacement currents can be neglected everywhere, and the basic equations in the Laplace domain(s) are [1]

$$\nabla \times \mathbf{E}(\mathbf{r}, s) = -s\mathbf{B}(\mathbf{r}, s), \tag{11.1}$$

$$\nabla \times \mathbf{H}(\mathbf{r}, s) = \begin{cases} \sigma \mathbf{E}(\mathbf{r}, s) & \text{inside shield,} \quad (11.2) \\ 0 & \text{outside shield,} \quad (11.3) \end{cases}$$

with the usual boundary conditions that tangential $\mathbf{E}$ and $\mathbf{H}$ be continuous across the air–shield interfaces (see Figure 11.1). The constitutive relation between $\mathbf{H}$ and $\mathbf{B}$ is

$$\mathbf{B} = \begin{cases} \mu \mathbf{H} & \text{inside shield,} \\ \mu_0 \mathbf{H} & \text{outside shield.} \end{cases} \tag{11.4}$$

Because of (11.3) these equations can be solved expeditiously with the aid of the magnetic scalar potential Φ, where $\mathbf{H} = -\nabla\Phi$. Then we have the Laplace equation

$$\nabla^2 \Phi = 0 \tag{11.5}$$

everywhere outside the shield. Inside the shield we have to deal with (11.1) and (11.2) which, unfortunately, do not lend themselves to easy treatment. It would thus be a great simplification if we can replace the shield region by a set of scalar boundary conditions that duplicate the effects of the shield insofar as the field inside the enclosure is concerned. With such boundary conditions we simply solve the Laplace equation (11.5) inside and outside the shield without having to go into the interior of the shield region itself.

Let us now proceed to derive the equivalent boundary conditions that would reduce the three-medium vectorial problem (Figure 11.1(a)) to a two-medium scalar problem (Figure 11.1(b)). It should come as no surprise that approximations would be involved in such a drastic simplification. We will point out the approximations involved in the process of the derivation.

We now integrate (11.2) around the rectangle *abcd* inside the shield wall

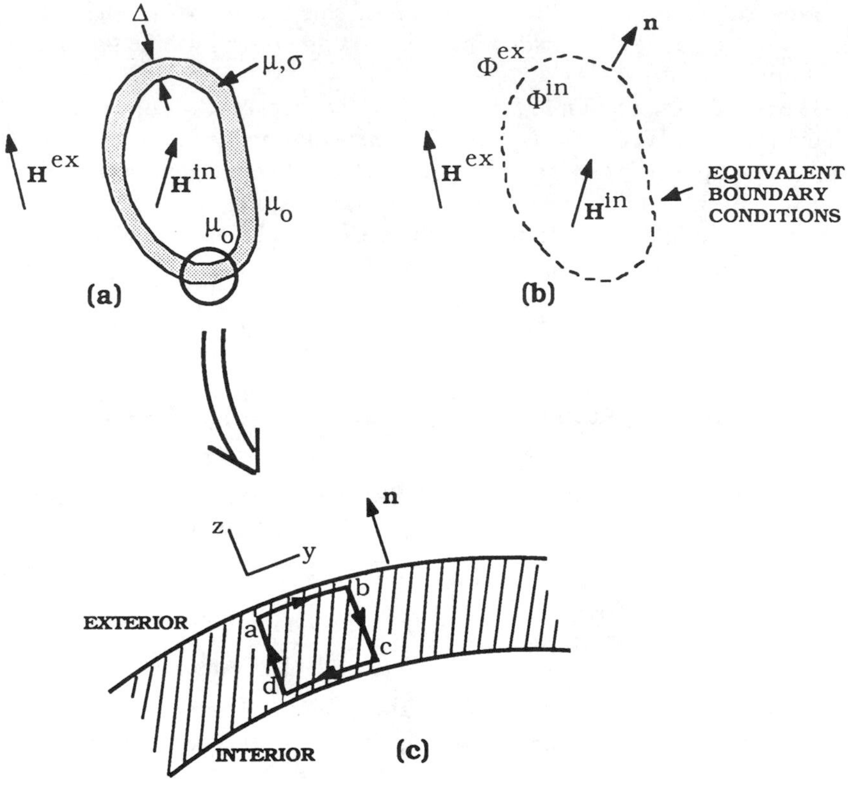

FIGURE 11.1. Geometries for deriving equivalent boundary conditions.

(Figure 11.1(c)) to obtain

$$
\begin{aligned}
H_y^{\text{ex}} - H_y^{\text{in}} &= -\sigma \int E_x \, dz, \\
H_x^{\text{ex}} - H_x^{\text{in}} &= \sigma \int E_y \, dz,
\end{aligned}
\tag{11.6}
$$

where the integrals are over the thickness Δ of the wall. In deriving (11.6) we have used the continuity of the tangential **H** across the air–shield interfaces. Also we have assumed that H_y, H_z, E_y, etc., are constant along *ab* or *cd*, and therefore the integral of H_z over *bc* cancels that over *da*.

To find E_x and E_y inside the shield we go back to (11.1) and (11.2) which give

$$(\nabla^2 - \gamma^2)E_x = 0 \tag{11.7}$$

and a similar equation for E_y. Here $\gamma^2 = s\mu\sigma$. We now make the crucial assumption that the skin depth of the shield is much smaller than the local radii of curvature. This assumption implies that the variation of the field across the thickness of the shield dominates the variations in the other directions, and hence (11.7) can be approximated by

$$\left(\frac{d^2}{dz^2} - \gamma^2\right) E_x \approx 0. \tag{11.8}$$

Substituting the solution of (11.8) and a similar one for E_y into the integrals of (11.6) and using the continuity of E_x, E_y across the air–shield interfaces we obtain

$$\begin{aligned} H_y^{\text{ex}} - H_y^{\text{in}} &= -\frac{1}{s\mu_0 p}(E_x^{\text{ex}} + E_x^{\text{in}}), \\ H_x^{\text{ex}} - H_x^{\text{in}} &= \frac{1}{s\mu_0 p}(E_y^{\text{ex}} + E_y^{\text{in}}), \end{aligned} \tag{11.9}$$

where

$$p = \frac{2}{s\mu_0 \sigma \Delta} \frac{\sqrt{st_{\text{d}}}/2}{\tanh(\sqrt{st_{\text{d}}}/2)}, \qquad t_{\text{d}} = \mu\sigma\Delta^2. \tag{11.10}$$

That is, the difference of the tangential magnetic fields is related to the sum of the tangential electric fields on both sides of the shield. Thus, (11.9) may be regarded as a generalized Ohm's law.

We now differentiate the first equation of (11.9) with respect to y and the second equation with respect to x and add the resulting equations. Then we use (11.1) to eliminate **E** in favor of **H** and introduce the magnetic scalar potential to obtain the boundary condition

$$\frac{\partial \Phi_+}{\partial n} = p\nabla_s^2 \Phi_-, \tag{11.11}$$

where $\Phi_+ = \Phi^{\text{in}} + \Phi^{\text{ex}}$, $\Phi_- = \Phi^{\text{in}} - \Phi^{\text{ex}}$, ∇_s^2 is the surface Laplacian, and $\partial/\partial n$ is the differentiation operator along the outward normal (see Figure 11.1(b)).

To find the second boundary condition we integrate (11.1) around the rectangle *abcd* inside the shield (Figure 11.1(c)) and then use the same procedure leading to (11.11) to obtain

$$\frac{\partial \Phi_-}{\partial n} = q\nabla_s^2 \Phi_+, \tag{11.12}$$

where

$$q = \frac{\mu_r \Delta}{2} \frac{\tanh(\sqrt{st_{\text{d}}}/2)}{\sqrt{st_{\text{d}}}/2}. \tag{11.13}$$

A metal enclosure is generally a good shield, i.e., an enclosure with electrically thick walls. In this case we may neglect $\nabla_s^2\Phi^{\text{in}}$ in comparison to $\nabla_s^2\Phi^{\text{ex}}$ in the first approximation. Adding (11.11) and (11.12) we obtain in the good-shield approximation

$$\frac{\partial \Phi^{\text{in}}}{\partial n} \approx \frac{q-p}{2} \nabla_s^2 \Phi^{\text{ex}}$$

$$= -\frac{1}{s\mu_0} Z_T \nabla_s^2 \Phi^{\text{ex}}, \tag{11.14}$$

where

$$Z_T = \frac{1}{\sigma \Delta} \frac{\sqrt{st_d}}{\sinh\sqrt{st_d}}. \tag{11.15}$$

The term $\nabla_s^2 \Phi^{\text{ex}}$ on the right-hand side of (11.14) can be transformed to the induced surface current $\mathbf{J}_s$ on the external surface of the enclosure by recalling the definition of Φ^{ex}. Since $\mathbf{J}_s^{\text{ex}} = \mathbf{n} \times \mathbf{H}^{\text{ex}} = -\mathbf{n} \times \nabla_s \Phi^{\text{ex}}$, we have $\mathbf{n} \times \mathbf{J}_s^{\text{ex}} = \nabla_s \Phi^{\text{ex}}$ and

$$\nabla_s^2 \Phi^{\text{ex}} = \nabla_s \cdot (\mathbf{n} \times \mathbf{J}_s^{\text{ex}}) = \mathbf{n} \cdot \nabla_s \times \mathbf{J}_s^{\text{ex}}. \tag{11.16}$$

Thus, in the good-shield approximation the normal component of the magnetic intensity on the inner surface of the shield is proportional to the surface curl of the surface current on the external surface of the shield. Specifically, combining (11.14) and (11.16) we have

$$H_n^{\text{in}} \approx \frac{1}{\mu_0 s} Z_T \nabla_s \cdot (\mathbf{n} \times \mathbf{J}_s^{\text{ex}}). \tag{11.17}$$

Knowing H_n^{in} on the inner surface of the shield we can solve the Laplace equation to find the field within the shielded volume.

Let us now consider an important case where $|st_d| \ll 1$, i.e., $\omega \ll 1/t_d$ or $\Delta \ll \delta$. This case applies to enclosures with electrically thin walls. Under this condition we have

$$p \approx \frac{2}{s\mu_0 \sigma \Delta}, \qquad \frac{\partial \Phi_-}{\partial n} \approx 0 \tag{11.18}$$

for $\mu_r \Delta \ll L$, L being a characteristic length of the enclosure. Equation (11.18) implies that the boundary conditions (11.11) and (11.12) reduce to

$$\frac{\partial \Phi^{\text{ex}}}{\partial n} = \frac{\partial \Phi^{\text{in}}}{\partial n} = \frac{1}{s\mu_0 g} \nabla_s^2 (\Phi^{\text{in}} - \Phi^{\text{ex}}), \tag{11.19}$$

where $g = \sigma \Delta$. Solving the Laplace equations for Φ^{ex} and Φ^{in}, subject to the two boundary conditions (11.19), is equivalent to solving the following vectorial equations [1]

$$\begin{aligned} \nabla \times \mathbf{E} &= -s\mu_0 \mathbf{H}, \\ \nabla \times \mathbf{H} &= g\mathbf{E}_s \delta(\mathbf{r} - \mathbf{r}'), \end{aligned} \tag{11.20}$$

where the function g defines the sheet conductance as a function of the surface coordinates (u, v), and $\mathbf{r} = \mathbf{r}'(u, v)$ defines the surface of the enclosure. In (11.20) the enclosure is treated as a mathematical surface on which currents can be induced by the time-varying magnetic field or direct current injection onto the surface by such appendages as current carrying wires. A general method for solving (11.20) can be found in [2].

11.3. Simple Canonical Shields

This section deals with magnetic field diffusion through a slab, two parallel slabs, a cylindrical shell of infinite length, and a spherical shell, as shown in Figure 11.2. These problems have been treated by various authors as three-

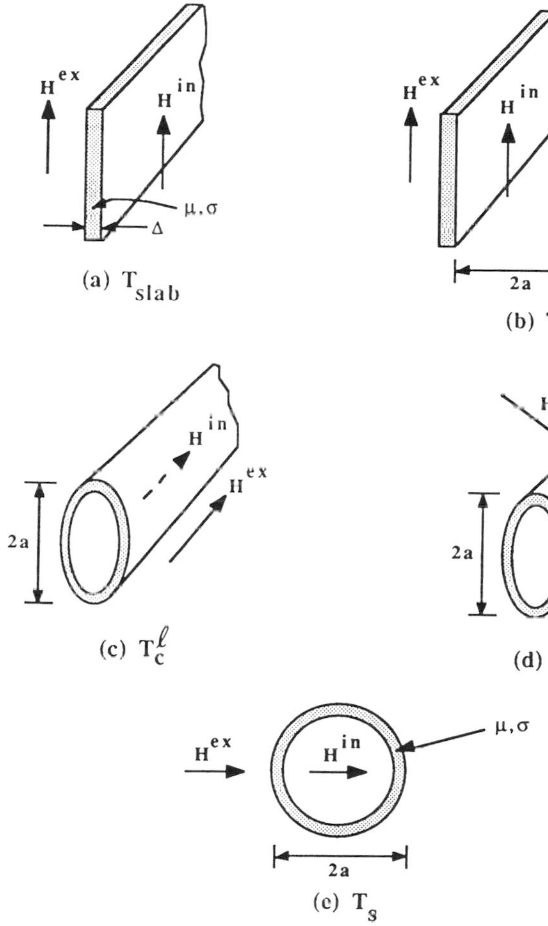

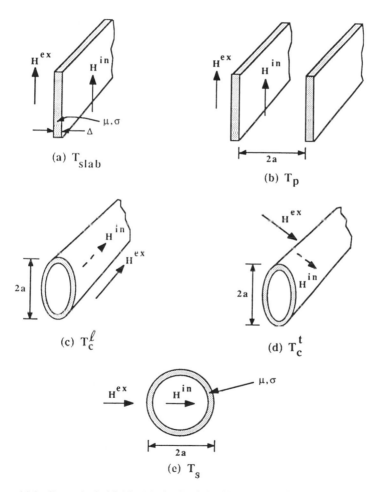

FIGURE 11.2. Canonical shields: (a) single slab, (b) two parallel plates, (c) cylindrical shell with longitudinal external field, (d) cylindrical shell with transverse external field, and (e) spherical shell.

medium vectorial boundary-value problems [3–8]. Here, we will solve these problems straightforwardly with the scalar theory introduced in the last section.

Let the transfer function T be defined as

$$T(s) = \frac{H^{\text{in}}(s)}{H^{\text{ex}}(s)} = \frac{\text{transmitted magnetic intensity}}{\text{external magnetic intensity in absence of shield}}. \qquad (11.21)$$

Taking H^{ex} to be uniform, we solve (11.5) with the boundary conditions (11.11) and (11.12) and obtain, for the geometries shown in Figure 11.2,

$$T_{\text{slab}}(z) = \frac{z}{z \cosh z + (Z_0/R) \sinh z}, \qquad (11.22)$$

$$T_{\text{p}}(z) = \frac{1}{\cosh z + Kz \sinh z}, \qquad (11.23)$$

$$T_{\text{c}}^{l}(z) = \frac{1}{\cosh z + \frac{1}{2}Kz \sinh z}, \qquad (11.24)$$

$$T_{\text{c}}^{\text{t}}(z) = \frac{1}{\cosh z + \frac{1}{2}(Kz + 1/Kz) \sinh z}, \qquad (11.25)$$

$$T_{\text{s}}(z) = \frac{1}{\cosh z + \frac{1}{3}(Kz + 2/Kz) \sinh z}, \qquad (11.26)$$

where

$$K = \frac{\mu_0 a}{\mu \Delta}, \qquad z = \sqrt{s t_{\text{d}}}, \qquad R = (\sigma\Delta)^{-1}, \qquad Z_0 = 120\,\pi \text{ ohms}. \qquad (11.27)$$

Note that the formulas (11.22)–(11.26) can be summarized in one single formula, namely,

$$T(z) = \frac{1}{\cosh z + (\alpha z + \beta/z) \sinh z}, \qquad (11.28)$$

the high-frequency approximation of which is

$$T(z) \sim \frac{2e^{-z}}{1 + \alpha z}, \qquad |z| \gg 1, \qquad (11.29)$$

whereas the low-frequency limit is

$$T(z) \sim \frac{1}{(1 + \beta) + (\alpha + \frac{1}{2} + \beta/6)z}, \qquad |z| \ll 1. \qquad (11.30)$$

Equation (11.28) is plotted in Figure 11.3 for various values of α and β. Two special values of α and β ($\alpha = 100$, $\beta = 0$ and $\alpha = 0$, $\beta = 10^4$) are shown, in which the former set of values corresponds to a nonferrous metallic enclosure of typical dimensions, whereas the latter refers to a single metallic slab with

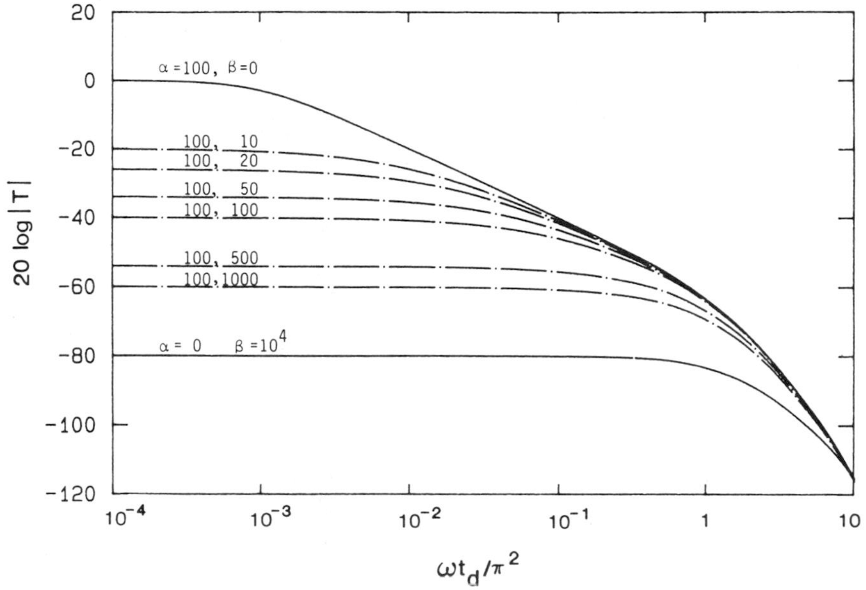

FIGURE 11.3. Transfer function versus frequency for various values of α and β.

$Z_0/R = 10^4$. It is clear from the figure that a planar approximation to an enclosure can overestimate the shielding effectiveness of the enclosure by many orders of magnitude.

The high-frequency approximation (11.29) works well for $\omega t_d > \pi^2$, while the low-frequency limit (11.30) is a good approximation for $\omega t_d < \pi^2$.

If the external field is an impulse given by

$$H^{ex} = H_0 \delta\left(t - \frac{x}{c}\right), \tag{11.31}$$

the transmitted field H^{in} can be obtained by taking an inverse Laplace transform of (11.28), namely,

$$H^{in}(t) = \frac{H_0}{2\pi j} \int_{-j\infty+c}^{j\infty+c} \frac{e^{st}\, ds}{\cosh z + (\alpha z + \beta/z) \sinh z}. \tag{11.32}$$

We will evaluate this integral for two different time regimes, namely, (a) $t > t_d/\pi^2$ and (b) $t < t_d/\pi^2$.

(a) $t > t_d/\pi^2$

In this time regime the integral (11.32) can be evaluated by first transforming s to z through $z = \pi\sqrt{st_d}$. Then the integration contour C_1 in the s-plane transforms to the hyperbola C_2 in the z-plane (Figure 11.4). After the transformation the integrand becomes an odd function of z, and therefore the lower

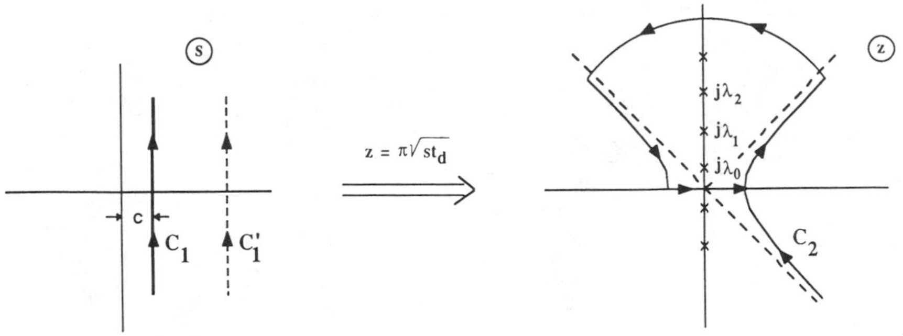

FIGURE 11.4. Integration contours for early-time and late-time evaluation of (11.32).

portion of the contour C_2 can be reflected through the origin as shown in Figure 11.4. Finally, the integral can be evaluated by the method of residues and the result of the evaluation is [9]

$$\frac{H^{\text{in}}(t)}{H_0} = \frac{2}{t_d} \sum_{n=0}^{\infty} \frac{\lambda_n e^{-\lambda_n^2 \tau}}{[1 + \alpha - 2\alpha\beta + \alpha^2 \lambda_n^2 + (\beta + \beta^2)\lambda_n^{-2}] \sin \lambda_n}, \tag{11.33}$$

where $\tau = t/t_d$ and the λ_n's are the roots of

$$\cot \lambda_n = \alpha \lambda_n - \frac{\beta}{\lambda_n}. \tag{11.34}$$

For a single slab, $\alpha = 0$ and $\beta = Z_0/R \gg 1$ for metallic slabs. In this case, the roots of (11.34) are approximately given by

$$\lambda_n \approx n\pi, \qquad \sin \lambda_n \approx \frac{(-)^{n+1} n\pi}{\beta}, \qquad n = 1, 2, 3, \ldots. \tag{11.35}$$

Substituting (11.35) in (11.33) we obtain

$$\frac{H^{\text{in}}(t)}{H_0} \approx \frac{2\pi^2}{\beta t_d} \sum_{n=1}^{\infty} (-)^{n+1} n^2 e^{-n^2\pi^2\tau} \tag{11.36}$$

$$\sim \frac{2\pi^2}{\beta t_d} (e^{-\pi^2\tau} - 4e^{-4\pi^2\tau}) \tag{11.37}$$

The expression (11.37) is usually of sufficient accuracy for $\tau \geq 1/\pi^2$ or $t \geq t_d/\pi^2$.

In the case of a nonferrous metallic enclosure we have $\alpha \gg 1$ and $\beta < 1$, and the roots of (11.34) are approximately given by

$$\lambda_0 \approx \alpha^{-1/2}, \qquad \lambda_n \approx n\pi, \qquad n = 1, 2, 3 \ldots. \tag{11.38}$$

The first two terms of (11.33) are then given by

$$\frac{H^{\text{in}}(t)}{H_0} = \frac{1}{t_f}(e^{-t/t_f} - 2e^{-\pi^2\tau}), \tag{11.39}$$

where the fall time t_f is defined as

$$t_f = \alpha t_d = \alpha\mu\sigma\Delta^2. \tag{11.40}$$

Equation (11.39) is sufficiently accurate to describe the interior (or transmitted) field for $t \geq t_d/\pi^2$.

(b) $t < t_d/\pi^2$

To evaluate (11.32) for early times we push the path of integration C_1 to the far right of the s-plane such that c becomes very large. Along the new path C_1' (Figure 11.4) we can expand the hyberbolic functions in the integrand of (11.32) in large arguments and obtain

$$H^{\text{in}}(t) \sim \frac{H_0}{2\pi j}\int_{C_1'} \frac{e^{st-z}\,ds}{\alpha z + 1 + \beta/z}. \tag{11.41}$$

An evaluation of (11.41) for a single slab ($\alpha = 0$) with the aid of a Laplace transform table gives

$$\frac{H^{\text{in}}(t)}{H_0} \sim \frac{1}{\sqrt{\pi}\beta t_d}\tau^{-5/2}e^{-1/(4\tau)}, \tag{11.42}$$

where we have invoked the usual assumption that $\beta \gg 1$ to obtain the final simple result (11.42), which is sufficiently accurate for $\tau \lesssim 1/\pi^2$ or $t \lesssim t_d/\pi^2$.

For a nonferrous enclosure with $\alpha \gg 1$ and $\beta < 1$, a similar evaluation of (11.41) yields

$$\frac{H^{\text{in}}(t)}{H_0} \sim \frac{2}{\sqrt{\pi}t_f}\tau^{-1/2}e^{-1/(4\tau)} \tag{11.43}$$

which is accurate to within 0.1% for $\tau \leq 1/\pi^2$ and $\alpha \geq 100$.

In Figures 11.5 and 11.6 we plot the transmitted pulse through a single slab and into a nonferrous enclosure, respectively. The enclosure can be either a two-parallel-plate geometry, an infinitely long cylindrical shell, or a spherical shell. There are two important features about the two figures. Figure 11.5 is valid for all values of β so long as $\beta \geq 377$ (or for the slab's resistance $R \leq 1$ ohm per square), whereas Figure 11.6 holds for nonferrous enclosures with $\alpha \geq 100$. Another feature is that the pulse width of the transmitted pulse is about t_d/π^2 for the single slab, and in the order of αt_d for the enclosure, the ratio of the latter to the former is about $\alpha\pi^2$, which accounts for the great difference in low-frequency shielding between single slabs and enclosures, as shown previously in Figure 11.3. Table 11.1 summarizes the important parameters of the transmitted pulse for the case where the incident pulse has a pulse width less than t_d/π^2.

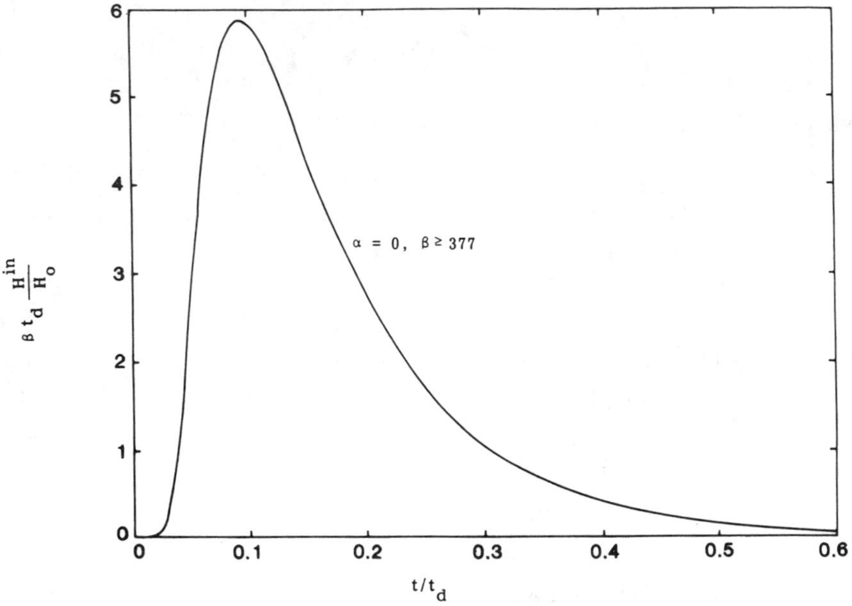

FIGURE 11.5. Universal waveform for a delta pulse through a single slab.

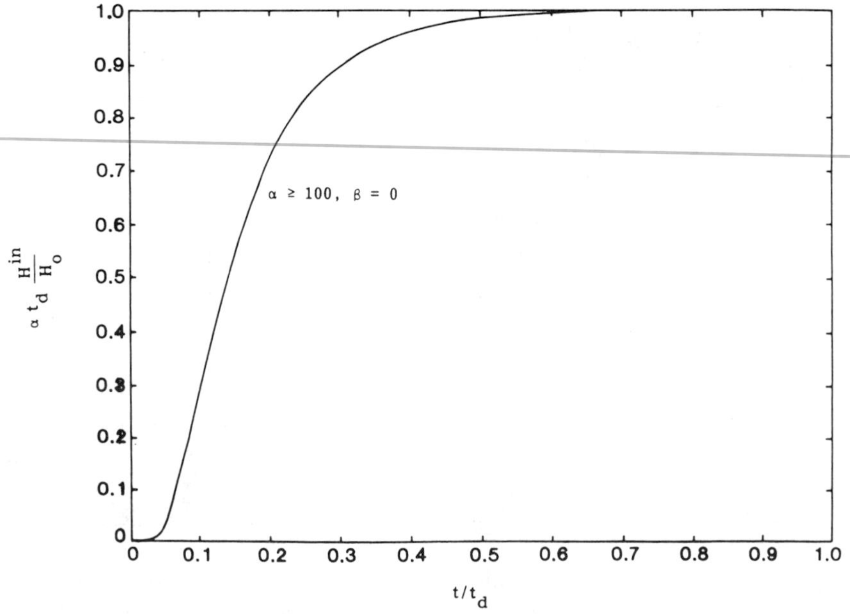

FIGURE 11.6. Universal waveform for a delta pulse into a metallic enclosure.

TABLE 11.1. Engineering parameters of the transmitted pulse ($\beta = Z_0/R$, $\alpha = V/(\mu_r \Delta S)$).

Geometry	H^{in}(peak)	$\dot{H}^{in}$(peak)	Rise time (10–90%)	Decay time (1/e)
Single slab	$\frac{6H_0}{\beta t_d}$	$\frac{120H_0}{\beta t_d^2}$	$\frac{t_d}{20}$	$\frac{t_d}{\pi^2}$
2-Parallel-plates cylinder, sphere	$\frac{H_0}{\alpha t_d}$	$\frac{6H_0}{\alpha t_d^2}$	$\frac{t_d}{4}$	αt_d

Before concluding this section it is important to point out that the transfer functions (11.23)–(11.26) for enclosures of practical applications can be summarized in one single formula, namely,

$$T = \frac{1}{\cosh z + \alpha z \sinh z}, \tag{11.44}$$

where $\alpha = V/(\mu_r \Delta S)$. For cylindrical or spherical enclosures (11.44) is accurate to within 5% for $\alpha \geq 10$, or $V/(\Delta S) \geq 10\mu_r$.

11.4. N-Layered Shields

In the preceding section the discussion was restricted to a single slab or an enclosure with only one layer of shield. We now extend the discussion to enclosures of multishields. We will start with a spherical enclosure with two-layered shields and then generalize the results to parallel plates, cylindrical enclosures, and enclosures with N-layered shields.

At the end of the last section we found that for enclosures of practical applications (namely, $V/(\Delta S) \geq 10\mu_r$), the transfer function can be accurately described by the simple formula (11.44). In this section our discussion will be restricted to this practically important case, although the general case can be solved in a straightforward manner, only making the final results more complicated.

Figure 11.7 shows a spherical enclosure with two nonferrous metallic shields immersed in a slowly varying field $H^{ex}(t)$. Solving (11.5) with the boundary conditions (11.11) and (11.12) we find, for the total transfer function, $T(s)$ to be

$$T = \frac{T_1 T_2}{1 - \gamma_{12} T_1 T_2}, \tag{11.45}$$

where T_1 and T_2 are the transfer functions of the individual shields given by (11.44) and

$$\gamma_{12} = \left(\frac{a_2}{a_1}\right)^3 \alpha_1 \alpha_2 z_1 z_2 \sinh z_1 \sinh z_2 \tag{11.46}$$

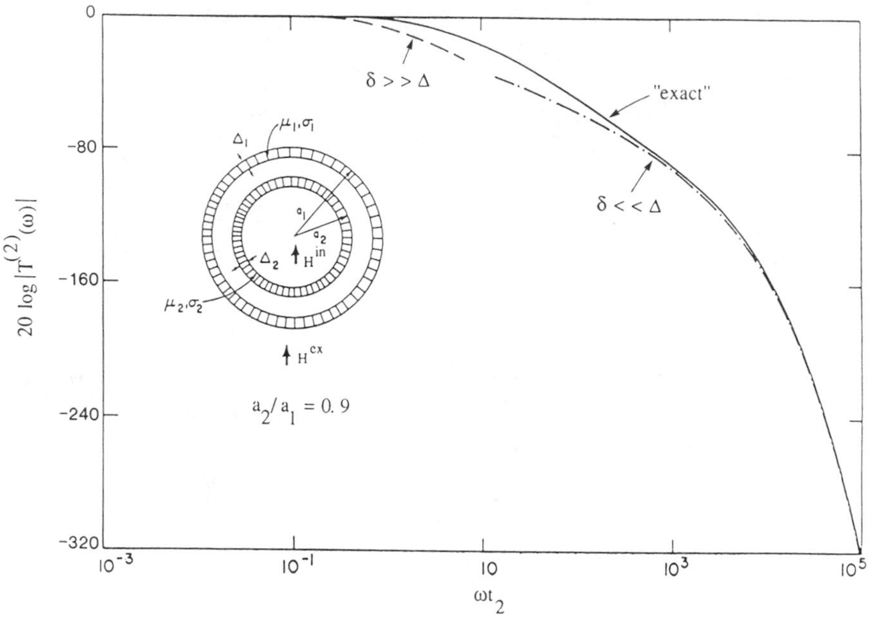

FIGURE 11.7. Transfer function of a spherical enclosure with two shields.

with $\alpha_i = a_i/(3\Delta_i)$ and $z_i^2 = s\mu_0\sigma_i\Delta_i^2$. The high-frequency limit of (11.45) is

$$T(s) \sim \frac{4e^{-(z_1+z_2)}}{(1+\alpha_1 z_1)(1+\alpha_2 z_2) - (a_2/a_1)^3\alpha_1\alpha_2 z_1 z_2} \tag{11.47}$$

for $\delta_1 \ll \Delta_1$ and $\delta_2 \ll \Delta_2$, and the low-frequency limit is

$$T(s) \sim \frac{1}{(1+st_1)(1+st_2) - (a_2/a_1)^3 s^2 t_1 t_2} \tag{11.48}$$

for $\delta_1 \gg \Delta_1$ and $\delta_2 \gg \Delta_2$, where $t_1 = \mu_0\sigma_1\Delta_1 a_1/3$, $t_2 = \mu_0\sigma_2\Delta_2 a_2/3$, and δ_1 and δ_2 are the skin depths. Equations (11.45), (11.47), and (11.48) are plotted in Figure 11.7 with $(a_2/a_1) = 0.9$ and $\Delta_1 = \Delta_2, \sigma_1 = \sigma_2$. The figure shows that (11.48) is a good approximation for $\omega t_2/\pi^2 < 1$ and (11.47) for $\omega t_2/\pi^2 > 1$. The interaction term (11.46) between shields has the effect of degrading the shielding effectiveness of the enclosure but it is small except for high frequencies ($\omega t_2 > \pi^2$) and $a_2/a_1 \geq 0.9$.

As we saw in the last section, the low-frequency approximation of the transfer function retains all the essential features of the penetrant pulse except for early times when $t < t_d/\pi^2$. We therefore expect that (11.48) will do likewise.

We now generalize (11.48) to a spherical enclosure of N-layered shields. Let (11.48) be rewritten as

$$\frac{1}{T^{(2)}} \sim 1 + (t_1 + t_2)s + \left[1 - \left(\frac{a_2}{a_1}\right)^3\right] t_1 t_2 s^2. \tag{11.49}$$

For a spherical enclosure of three nonferrous shields we use the same, although more complicated, procedure for the two-shield enclosure and find

$$\begin{aligned} \frac{1}{T^{(3)}} \sim 1 + (t_1 + t_2 + t_3)s &+ \left[1 - \left(\frac{a_2}{a_1}\right)^3\right] t_1 t_2 s^2 \\ &+ \left[1 - \left(\frac{a_3}{a_2}\right)^3\right] t_2 t_3 s^2 + \left[1 - \left(\frac{a_3}{a_1}\right)^3\right] t_1 t_3 s^2 \\ &+ \left[1 - \left(\frac{a_2}{a_1}\right)^3\right]\left[1 - \left(\frac{a_3}{a_2}\right)^3\right] t_1 t_2 t_3 s^3. \end{aligned} \tag{11.50}$$

On a close examination of (11.49) and (11.50) we can write, for an N-layered spherical enclosure with $a_1 > a_2 > a_3 \ldots$,

$$\frac{1}{T^{(N)}} \sim 1 + s \sum_{i=1}^{N} t_i + s^2 \sum_{i>j}^{N,N} \left[1 - \left(\frac{a_i}{a_j}\right)^3\right] t_i t_j + \cdots + s^N \prod_{i=1}^{N} \left[1 - \left(\frac{a_i}{a_{i-1}}\right)^3\right] t_i \tag{11.51}$$

with $a_0 = \infty$. As expected, $T^{(N)}$ has N poles lying on the negative real axis of the s-plane. If we plot $|T^{(N)}|$ on a log–log scale as a function of frequency, the asymptotes will be N straight lines joining end to end at the N roots of (11.51), each with slope -20 dB per decade except for the low-frequency line which has a zero slope.

For parallel-plate or cylindrical shields we simply replace the exponents in the ratios of radii in (11.51) by 1 or 2 accordingly.

11.5. Electric Field Penetration

The two crucial assumptions in the magnetic diffusion problem are (a) the neglect of displacement currents, and (b) the radii of curvature of the shield surface greater than the skin depth of the shield. In this section we will apply these approximations to the exact solution of a boundary-value problem and see if the corresponding results in previous sections agree with the result derived here. In addition, we will have the solution of the electric field penetration for this particular boundary-value problem and compare how it differs from the magnetic field penetration.

The particular boundary-value problem is that of a plane wave penetrating

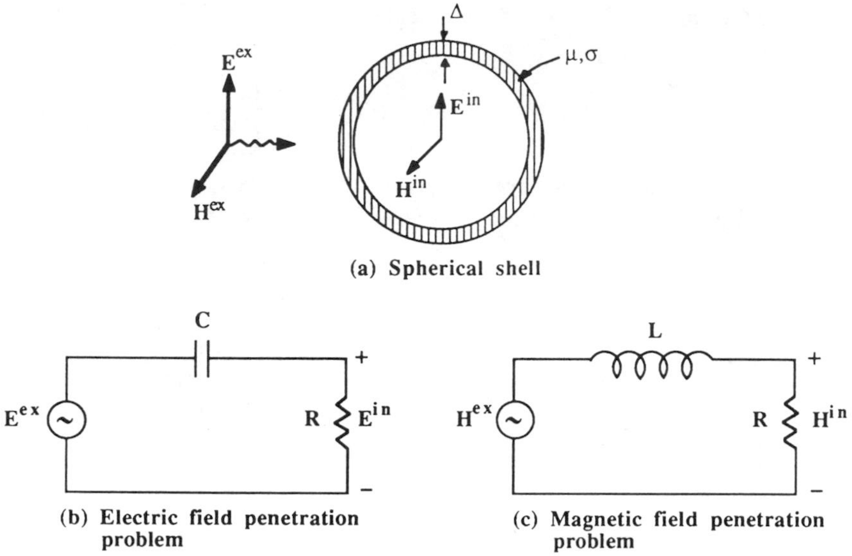

FIGURE 11.8. Equivalent circuit representations for electric and magnetic field penetration into a spherical shell.

a spherical conducting shell of outer radius a and inner radius b (Figure 11.8a)), which has been considered by Harrison and Papas [7]. At the center of the shell, Papas noticed that the field is described by a single term of a double infinite series. This important observation makes possible any further analytical treatment of the problem. Let k_0 and k be the wave numbers of free space and the shell. Assumptions (b) and (a) imply, respectively, $|ka|, |kb| \gg 1$ and $|k_0 a|, |k_0 b| \ll 1$. Making the appropriate large and small argument expansions of the spherical Bessel functions in the expressions of $\mathbf{E}^{\text{in}}$ and $\mathbf{H}^{\text{in}}$ at the center of the shell, we obtain [10]

$$T_E \equiv \frac{E^{\text{in}}}{E^{\text{ex}}} \approx \frac{1}{\cos(k\Delta) + \dfrac{1}{3}\left(\dfrac{2k}{k_0^2 a \mu_r} - \dfrac{\mu_r k_0^2 a}{k}\right)\sin(k\Delta)}, \tag{11.52}$$

$$T_H \equiv \frac{H^{\text{in}}}{H^{\text{ex}}} \approx \frac{1}{\cos(k\Delta) + \dfrac{1}{3}\left(\dfrac{2\mu_r}{ka} - \dfrac{ka}{\mu_r}\right)\sin(k\Delta)}, \tag{11.53}$$

where $\Delta = a - b$, $\mu_r = \mu/\mu_0$, $k^2 = -s\mu\sigma$, $k_0^2 = -s^2\mu_0\varepsilon_0$, and a or $b \gg \Delta$. Equation (11.53) is identically equal to (11.26), which we have set out to prove. The difference between (11.52) and (11.53) becomes more obvious for nonferrous ($\mu_r = 1$) and electrically thin shells ($|k\Delta| \ll 1$). For this case, (11.52) and

(11.53) reduce, respectively, to

$$T_E \approx \frac{sCR}{1 + sCR}, \tag{11.54}$$

$$T_H \approx \frac{1}{1 + sL/R}, \tag{11.55}$$

with $C = 3\varepsilon_0 a/2$, $L = \mu_0 a/3$, and $R = (\sigma\Delta)^{-1}$. Equations (11.54) and (11.55) can be interpreted by the simple circuits shown in Figure 11.8(b) and (c); each can be referred to as a voltage transfer to a resistance R through a capacitance C or an inductance L. The great difference in shielding against electric and magnetic fields by conducting shells at low frequencies is clearly shown in (11.54) and (11.55) and also in Figure 11.8(b) and (c).

Epilogue

Throughout this chapter we have strived for physical simplicity and mathematical elegance, the two invaluable qualities that underlie the teachings of Professor Papas. The main ideas presented here have been derived from many discussions with Professor Papas in the last two decades. I have been enjoying his continued counsel, guidance, and kindness, for which I am eternally grateful.

Acknowledgments

I have benefited a great deal in discussions of the subject for many years with two of Professor Papas' former students, Drs. F.C. Yang and R.W. Latham (now deceased).

References

[1] K.S.H. Lee, editor (1986), *EMP Interaction: Principles, Techniques, and Reference Data*, Hemisphere, New York.

[2] R.W. Latham and K.S.H. Lee (1968), Theory of inductive shielding, *Canad. J. Phys.*, **46**, 1745.

[3] L.V. King (1933), Electromagnetic shielding at radio frequencies, *Phil. Mag.* and *J. Sci.*, **15**, no. 97, 201.

[4] J.C. Jaeger (1940), Magnetic screening by hollow circular cylinders, *Phil. Mag.*, **29**, 18.

[5] H. Kaden (1959), *Wirbelstrome und Schirmung in der Nachrichtentechnik*, Springer-Verlag, Berlin.

[6] C.W. Harrison, Jr., et al. (1965), The propagation of transient electromagnetic fields into a cavity formed by two imperfectly conducting sheets, *IEEE Trans. Antennas and Propagation*, **AP-13**, no. 1, 149.

[7] C.W. Harrison, Jr. and C.H. Papas (1965), On the attenuation of transient fields by imperfectly conducting spherical shells, *IEEE Trans. Antennas and Propagation*, **AP-13**, 960.
[8] D.G. Dudley and J.P. Quintenz (1975), Transient electromagnetic penetration of a spherical shell, *J. Appl. Phys.*, **46**, no. 1, 173.
[9] K.S.H. Lee and G. Bedrosian (1979), Diffusive electromagnetic penetration into metallic enclosures, *IEEE Trans. Antennas and Propagation*, **AP-27**, 194.
[10] F.C. Yang (1979), private communication.

12
Coherence Effects in Guided Wave Optical Systems

ALAN ROLF MICKELSON*

12.1. Introduction

The optical portion of the electromagnetic spectrum is probably the spectral region in which electromagnetic sources show the greatest degree of diversity in coherence properties. Thermal optical sources (heated wire filaments) radiate totally random waves which can exhibit appreciable spectral component amplitudes throughout the optical spectrum. Highly stabilized, single-frequency lasers, on the other hand, can exhibit short-term linewidths as small as 1 Hz with the attendant coherence lengths of thousands of kilometers.

A good part of the explanation of the optical source diversity has to do with the practicality of source design. Terrestrial thermal microwave sources are ineffectual, as the room temperature blackbody curves already peak in the infrared. Further, there are few noticeable material resonances at microwave frequencies and below. Therefore, with the exception of a few maser frequencies, low-frequency electromagnetic sources tend to be "classical" in nature. That is, these low-frequency sources use electromagnetic resonators in combination with macroscopic feedback to achieve oscillation. Optical sources are the inverse case. Heating of materials such as wire filaments and gases tends to shift their blackbody peaks into the optical spectrum, and therefore thermal optical sources are commonplace. Also, atomic transitions are common in the optical spectrum, and therefore lasers can be developed at almost any optical wavelength. "Classical" optical sources are the ones lacking, due to the complexity of designing a macroscopic feedback mechanism at optical frequencies. In this sense, all optical sources will be, to some degree, dirty, that is, thermal sources in that they are broad band and laser sources in that they must exhibit some degree of spontaneous emission noise, whether it be in the form of phase or amplitude fluctuation. In this sense, all practical optical sources must be considered to be partially coherent and, further, with the degree of partial coherence varying widely from source to source.

---

* Electrical & Computer Engineering, Campus Box 425, University of Colorado, Boulder, CO 80309, USA.

Coherence takes on a special importance in guided wave systems. A usual explanation for the phenomena of mode guidance in a guiding structure is one that involves some kind of Fresnel reflection (or geometrical optical ray bending) coupled with a phase integral which describes some condition for plane waves (or rays) to interfere constructively. If the incoming radiation is incoherent, it will be hard for it to satisfy a coherence condition, at least unless the losses are sufficiently high to enhance the coherence by decreasing the radiation's entropy in that mode. A possible way to achieve this without adversely affecting total coupling to the guide is to make the guide multimoded and thereby guide "different pieces" of the radiation in different modes. However, for this technique to be successful, the coherence of the radiation must be minimal, as residual coherence between the modes will cause a time-varying interference which will tend to obscure any information content of the guided wave.

It is the purpose of this chapter to discuss in some detail the effects of source coherence on the propagation characteristics in guided wave structures, where the archetypal optical guided wave structure will be considered to be the optical fiber. This section is intended as a review, as the bulk of the material could be found in numerous texts, including [1]. In the next section, attention will be placed on spatial coherence, its definition, and its relation to temporal coherence. Section 12.3 will discuss the salient properties of light-emitting diodes and semiconductor injection lasers, the two most important sources for lightwave communication. Section 12.4 will discuss the propagation of spatial coherence, both in free space (Van Cittert–Zernicke theorem) and in various optical systems and waveguides. It will be in this section that salient features of the fiber coupling problem will be discussed. Section 12.5 will concentrate on a number of effects observed in multimode systems, including the phenomenon of modal noise and the effect of propagation and dispersion on coherence. Section 12.6 discusses coherence in single-mode systems, including the effects of polarization noise.

12.2. Spatial Coherence

To simplify monochromatic electromagnetic calculations, we generally define an auxiliary complex quantity $\hat{\mathbf{E}}$ by the relation

$$\mathbf{E}(t) = \mathrm{Re}[\hat{\mathbf{E}}e^{-i\omega t}], \tag{12.1}$$

where $\mathbf{E}(t)$ is the actual (measurable) electric field vector and Re denotes the real part. For polychromatic fields, we must generalize eq. (12.1). This is generally done by writing

$$\mathbf{E}(t) = \mathrm{Re}[\mathbf{V}(t)], \tag{12.2}$$

where the complex vector function, often referred to as the analytical signal representation of $\mathbf{E}(t)$, is defined by

$$\mathbf{E}(\omega) = \frac{1}{\pi} \int_0^\infty \mathbf{V}(t) e^{i\omega t} \, dt, \tag{12.3a}$$

$$\mathbf{E}(t) = \mathrm{Re}\left[\int_{-\infty}^{\infty} \mathbf{E}(\omega) e^{-i\omega t} \, d\omega \right], \tag{12.3b}$$

where $\mathbf{E}(\omega)$ is the (one-sided) spectrum of $\mathbf{V}(t)$ [2].

To discuss energy transport of an electromagnetic field, we generally use the Poynting vector. For monochromatic plane waves, the time-averaged Poynting vector takes on the especially simple form

$$\langle \mathbf{S}(t) \rangle = \frac{\hat{E}^* \hat{E}}{2\eta} \hat{e}_k, \tag{12.4}$$

where $\eta = \sqrt{\mu/\varepsilon}$ is the impedance of the medium, the angular brackets denote the time average, and $\hat{e}_k$ is a unit vector in the direction of the k vector, which is the direction of propagation of the plane wave (that is, the direction to which $\mathbf{E}$ and $\mathbf{H}$ are perpendicular). As eq. (12.1) had to be generalized for polychromatic waves, so does eq. (12.4). The required generalization is that

$$\langle \mathbf{S}(t) \rangle = \frac{\langle \mathbf{V}^*(t) \cdot \mathbf{V}(t) \rangle}{2\eta} \hat{e}_k, \tag{12.5}$$

where $\mathbf{V}(t)$ is now being used to denote the time variation of the electric field disturbance at whatever spatial point at which we wish to evaluate the time-averaged Poynting vector. We can use (12.5) to define the optical intensity I, together with the relation

$$\langle S(\mathbf{r}, t) \rangle = I(\mathbf{r}, t) \hat{e}_k, \tag{12.6}$$

where $\mathbf{r}$ defines the coordinate of the point at which the intensity is to be evaluated.

In what follows, there will be a need to use a quasi-monochromatic representation of the analytic signal $\mathbf{V}(t)$. For any component of $\mathbf{V}(t)$, denoted by $V_i(t)$, we will choose the representation that

$$V_i(t) = a(t) e^{i\varphi(t)} e^{-i\bar{\omega} t}, \tag{12.7}$$

where $a(t)$ and $\varphi(t)$ are real-valued random variables with φ distributed on 0 to 2π and $\bar{\omega}$ denotes the average angular frequency of the disturbance. Using a time representation such as eq. (12.7), we can represent a single polarization state (for example, y) of a z-directed quasi-monochromatic plane wave in the form

$$V_y(x, y, z, t) = a(z, y, z) e^{i\bar{k}z} e^{-i\bar{\omega} t} e^{i\varphi(x, y, z, t)}, \tag{12.8}$$

where $\bar{k}$ is defined by

$$\bar{k} = \frac{\bar{\omega} n}{c}, \tag{12.9}$$

and where a and φ vary with x and y due to initial phase front distortion of the polychromatic source and with z due to the fact that $\bar{k}$ is only an average propagation constant. Therefore, wave packets will distort (disperse) as they travel along the propagation direction. As the quasi-monochromatic approximation, however, has already been invoked in writing out $\bar{\omega}$ and $\bar{k}$, it is usually acceptable to ignore dispersion, at least for short propagation lengths, and thereby ignore the z-dependence of a and φ.

Before discussing spatial coherence, it is a good idea to discuss temporal coherence at least briefly. We could well describe temporal coherence as the quantity that is measured by a Michelson interferometer, but here we will choose not to be so operational [3]. Instead, let us say that we have a signal such as that described by eq. (12.7), we split it equally in two, we then delay one of the two by a time τ, and then we mix the two in a device which measures exactly the optical intensity I as is defined by eq. (12.6). The result is

$$\begin{aligned} I(\tau) &= \frac{1}{2\eta}\langle |V(t) + V(t-\tau)|^2\rangle \\ &= \frac{1}{2\eta}\{\langle a^2(t)\rangle + \langle a^2(t-\tau)\rangle \\ &\quad + 2\,\mathrm{Re}[e^{-i\bar{\omega}\tau}\langle a(t)a(t-\tau)e^{i(\varphi(t)-\varphi(t-\tau))}\rangle]\}. \end{aligned} \tag{12.10}$$

To better interpret the result of the gedanken experiment described above, is to define a complex degree of coherence function $\hat{\gamma}(\tau)$ by

$$\hat{\gamma}(\tau) = e^{-i\bar{\omega}\tau}\frac{\langle a(t)a(t-\tau)e^{i(\varphi(t)-\varphi(t-\tau))}\rangle}{[\langle a^2(t)\rangle\langle a^2(t-\tau)\rangle]^{1/2}} \tag{12.11}$$

such that eq. (12.10) can be rewritten as

$$I(\tau) = I_1(t) + I_2(t-\tau) + 2\sqrt{I_1(t)}\sqrt{I_2(t-\tau)}\,\mathrm{Re}[\hat{\gamma}(\tau)] \tag{12.12}$$

using the notation

$$I_1(t) = \langle a^2(t)\rangle, \tag{12.13a}$$

$$I_2(t-\tau) = \langle a^2(t-\tau)\rangle, \tag{12.13b}$$

for which the I_1 (I_2) refers to the intensity which would be measured if the delayed (nondelayed) beam were blocked. It is also possible to represent the complex degree of coherence in the form

$$\hat{\gamma}(\tau) = \gamma(\tau)e^{i\alpha(\tau)}e^{-i\bar{\omega}\tau}, \tag{12.14}$$

where $\gamma(\tau)$ is the (normalized) modulus and is known as the degree of coherence and $\alpha(\tau)$ is a phase which can be determined from eq. (12.11).

A couple of examples should bring out the meaning of the preceding equations and thereby help to illuminate the concept of temporal coherence. First, if we were to consider a totally monochromatic wave, than a and φ

would become time independent and

$$I(\tau) = 2I(0)[1 + \cos \omega\tau], \tag{12.15a}$$

$$\hat{\gamma}(\tau) = e^{-i\omega\tau}, \tag{12.15b}$$

$$\gamma(\tau) = 1. \tag{12.15c}$$

That $\gamma(\tau)$, the degree of coherence, is equal to 1 is a pleasant result, as a monochromatic wave must be completely coherent in every way for all time. A second example might be that of a phase noise dominated system. In this limit, we can take the a as independent of time to obtain the relations

$$I(\tau) = 2I(0)[1 + \mathrm{Re}[\hat{\gamma}(\tau)]], \tag{12.16a}$$

$$\hat{\gamma}(\tau) = e^{-i\bar{\omega}\tau}\langle e^{i(\varphi(t)-\varphi(t-\tau))}\rangle, \tag{12.16b}$$

$$\gamma(\tau) = |\langle e^{i(\varphi(t)-\varphi(t-\tau))}\rangle|. \tag{12.16c}$$

Certainly, for $\tau \simeq 0$, (12.16a–c) will reduce to (12.15a–c) above. However, we can easily picture that there will be a $\tau = \tau_c$ such that, at that delay, the difference $\varphi(t) - \varphi(t - \tau)$ becomes a uniformly distributed stochastic function of time. At this delay, the time average of the exponential will be zero, and therefore $\gamma(\tau_c) = 0$. This value τ_c is generally called the temporal coherence time, and the associated temporal coherence length is defined as $l_c = c\tau_c$ where c is the speed of light. This definition of temporal coherence length is not tied to phase noise but, quite in general, is defined as

$$\tau_c = \min[\tau > 0] \ni \gamma(\tau) = \varepsilon, \tag{12.17}$$

where ε is a small constant which can vary with the application.

To discuss spatial coherence, let us limit ourselves to a one-dimensional model in which we will take z to be the optic axis, y to be the direction of polarization of the electric field, and x to be the single direction in which the phase front can exhibit variation. Such a model is not really so unrealistic as it may seem, as we shall soon see. As with temporal coherence, we could define spatial coherence as that quantity which is measured by a diffractometer (the device employed in Young's experiment) [4]. Here, however, we will consider a gedanken experiment which can be approximated on a diffractometer. Consider a y-polarized, z-propagating, polychromatic plane whose analytical signal is approximately expressible as

$$V(x, z < 0, t) = a(x, t)e^{i\bar{k}z}e^{-i\bar{\omega}t}e^{i\varphi(x,t)} \tag{12.18}$$

incident on a perfectly conducting screen in the plane $z = 0$. Note that eq. (12.18) has assumed quasi-monochromaticity of the incident polychromatic disturbance. Say that the wall contains two "pin-hole" apertures at coordinates $x = \pm d/2$, $y = 0$. These "pin-holes" are small enough to pass a sufficient quantity of light to be measurable yet small enough to diffract the incident light sufficiently such that the diffracted wave "uniformly" illuminates the central portion (around $x = y = 0$) of a second plane screen at coordinate

$z = L$. In general, we take the distance L to be much greater than the distance d, and therefore the pinhole apertures can be many wavelengths and still diffract the incident waves sufficiently. Now, we need to find the disturbances due to each pinhole (which are labeled 1 for that at $x = d/2$ and 2 for that at $x = -d/2$) near the origin of the plane $z = L$. These disturbances, in analytical signal form, are approximately given by

$$V_1(x \sim 0, z = L, t) = K_1 V_1\left(x = \frac{d}{2}, z = 0, t - \tau_1\right)e^{iks_1}, \qquad (12.19a)$$

$$V_2(x \sim 0, z = L, t) = K_2 V_1\left(x = -\frac{d}{2}, z = 0, t - \tau_1\right)e^{iks_2}, \qquad (12.19b)$$

where s_1 and s_2 are the propagation distances, given by

$$s_1 \sim s_2 \sim \sqrt{\left(\frac{d}{2}\right)^2 + L^2}. \qquad (12.20)$$

For $x \sim 0$, τ_1 and τ_2 are the propagation delays, which must be given by the propagation distances divided by the light propagation velocity, and K_1 and K_2 are propagation factors, which for identical pinholes and roughly equal distances should be equal. Further, diffraction theory tells us that K_1 and K_2, as defined by eq. (12.19), should be purely imaginary.

What we wish to do now is to calculate the intensity pattern which would be observed in the plane $z = L$ if both pinholes were open and illuminated by a plane wave as described by (12.18). The result can be expressed formally as

$$I(x, z = L, t) = \frac{1}{2\eta_0}[|V_1(x, z = L, t)|^2 + |V_2(x, z = L, t)|^2 + 2\,\mathrm{Re}\{V_1^*(x, z = L, t)V_2(x, z = L, t)\}]. \qquad (12.21)$$

Substituting eq. (12.19) into eq. (12.20), we find

$$I(x, z = L, t) = \frac{1}{2\eta_0}\Bigg[\left\langle |K_1|^2 \left|V_1\left(x, \frac{d}{2}, z = 0, t - \tau_1\right)\right|^2\right\rangle + \left\langle |K_2|^2 \left|V_2\left(x = -\frac{d}{2}, z = 0, t - \tau_2\right)\right|^2\right\rangle + 2\,\mathrm{Re}\left\langle K_1 K_2^* V_1^*\left(x = \frac{d}{2}, z = 0, t - \tau_1\right) \times V_2\left(x = -\frac{d}{2}, z = 0, t - \tau_2\right)e^{ik(s_1 - s_2)}\right\rangle\Bigg]. \qquad (12.22)$$

We notice immediately that, if we were to move up the y-axis in the direction of increasing x, we would see a pattern of fringes as determined by the exponential factor in the last term of eq. (12.22), with the envelope of these fringes being a complicated function of the details of the correlation functions of the phase and amplitude of the initial wave as described by eq. (12.18). If the wave is quasi-monochromatic, however, the phase and amplitude of the wave will vary slowly enough that the difference between τ_1 and τ_2 is negligible with respect to time. We could define a single retarded time by $t' = t - \bar{\tau}$, where $\bar{\tau}$ is the average of τ_1 and τ_2. With this simplification, we can write

$$I(x, z = L, t) = \Gamma_{11}\left(\frac{d}{2}\right) + \Gamma_{22}\left(-\frac{d}{2}\right) + 2\,\mathrm{Re}[\Gamma_{12}(d, s_1 - s_2)], \quad (12.23)$$

where the Γ's are defined as

$$\Gamma_{11}\left(\frac{d}{2}\right) = \frac{1}{2\eta_0}\left\langle |K_1|^2 \left| V_1\left(x = \frac{d}{2}, z = 0, t'\right)\right|^2 \right\rangle, \quad (12.24a)$$

$$\Gamma_{22}\left(-\frac{d}{2}\right) = \frac{1}{2\eta_0}\left\langle |K_2|^2 \left| V_2\left(x = -\frac{d}{2}, z = 0, t'\right)\right|^2 \right\rangle, \quad (12.24b)$$

$$\Gamma_{12}(d, s_1 - s_2) = \frac{1}{2\eta_0} 2\,\mathrm{Re}\left[\left\langle K_1 K_2^* V_1^*\left(x = \frac{d}{2}, z = 0, t'\right) \times V_2\left(x = -\frac{d}{2}, z = 0, t'\right) e^{ik(s_1 - s_2)}\right\rangle\right], \quad (12.24c)$$

where Γ_{11} represents the intensity which would be measured in the plane $z = L$ if pinhole 2 were blocked off, Γ_{22} represents the intensity which would be measured in the plane $z = L$ if pinhole 1 were blocked off, and Γ_{12} is the interference term, which is generally called the mutual coherence function. Mathematically, the situation is quite analogous to the situation in eq. (12.10) and, although the interpretation is quite different here, here as there we can define

$$\Gamma_{12}(d, s_1 - s_2) = 2\sqrt{\Gamma_{11}\left(\frac{d}{2}\right)}\sqrt{\Gamma_{22}\left(-\frac{d}{2}\right)}\,\mathrm{Re}(\hat{\gamma}_{12}(d, s_1 - s_2)), \quad (12.25)$$

where the complex degree of coherence can be written out in the form

$$\hat{\gamma}_{12}(d, s_1 - s_2) = \gamma_{12}(d) e^{i\alpha(d)} e^{-ik(s_1 - s_2)}, \quad (12.26)$$

where the degree of coherence γ and the α can be determined from comparing eq. (12.25) with the following relation:

$$\hat{\gamma}_{12}(d, s_1 - s_2) = \frac{\langle V_1^*(x = d/2, z = 0, t') V_2(x = -d/2, z = 0, t') e^{ik(s_1 - s_2)} \rangle}{[\langle |V_1(x = d/2, z = 0, t')|^2 \rangle \langle |V_2(x = -d/2, z = 0, t')|^2 \rangle]^{1/2}}. \quad (12.27)$$

Note that K_1 and K_2 have been cancelled from eq. (12.26). The reason for the cancellation is the fact that, as previously mentioned, standard diffraction theory predicts that both K_1 and K_2 are purely imaginary, at least as defined by eq. (12.19), and therefore their magnitudes can be factored out from eq. (12.24c).

Now that we have performed some calculations and made a large number of definitions, it is time to consider the meaning of these multiple manipulations. Generally, we are interested most in quantities like the degree of coherence as defined by eq. (12.26) or eq. (12.14). In eq. (12.14), the degree of coherence is a function of only the time delay between the two waves, and its first zero determines the temporal coherence time. In eq. (12.26), due to the quasi-monochromatic approximation which was introduced just prior to eqs. (12.23) and (12.24), the degree of coherence is a function of only the distance between the two pinholes. By analogy with temporal coherence time, we can therefore define the spatial coherence length by the separation at which the function γ_{12} has its first zero. In the temporal case, the variation of a and φ in time caused these functions to decorrelate with temporal separation. Variation of a and φ across a phase front in a fixed plane is the cause of finite spatial coherence. Clearly, finite-sized polychromatic sources will have limited coherences of both types. With the realization that the physical quantity of interest depends only on d, the pinhole separation, we see that carrying along the s_1 and s_2 in eq. (12.25) is superfluous, and we might as well define new functions

$$J_{12}(d) = \Gamma_{12}(d, s_1 = s_2), \tag{12.28a}$$

$$\hat{\mu}_{12}(d) = \hat{\gamma}(d, s_1 = s_2), \tag{12.28b}$$

where J_{12} is called the mutual intensity and $\hat{\mu}_{12}$ is the complex degree of coherence, as was $\hat{\gamma}$. It is clear to see that

$$\hat{\mu}_{12} = \gamma_{12}(d)e^{i\alpha_{12}(d)}. \tag{12.29}$$

A final point that should be made here is the relation between temporal and spatial coherence. If a signal is truly monochromatic, then it is both temporally and spatially totally coherent, as the a and φ functions become time independent. The a and φ functions could still be even rapidly varying functions of space, but the main point is that the averaging brackets in eq. (12.27) disappear, and it becomes clear that there is a value of $s_1 - s_2$ such that the total phase goes to zero and the value of $\hat{\gamma}(d, s_1 - s_2 - \alpha/k) = 1$, showing that $\gamma(d) = 1$. In this sense, temporal coherence implies spatial coherence. The converse is not true. Consider the source of a plane wave to be a polychromatic point source at infinite (very large compared to a wavelength) distance. Even though the phase fronts will not be evenly spaced at $z = 0$, they will all be plane there. Delaying the wave with respect to itself will lead to a finite temporal coherence time, but spatially separated samples will always be 100% correlated, regardless of the separation.

12.3. Optical Sources

The most popular sources for guided wave systems are light emitting diodes (LEDs) and semiconductor laser diodes. These devices are small in size (hundreds of microns) yet reliable. Further, they are amenable to current modulation, which alleviates the need for external modulators, at least in systems where great temporal coherence is not necessary. This section will therefore limit attention to LEDs and multi- and single-mode semiconductor laser diodes.

An LED generally resembles a thermal source in its radiation characteristics [5]. A surface emitting LED usually has a homojunction structure—that is, it consists of a bottom plus contact bonded to a *p*-type region. From this bottom contact, current is injected toward a *p–n* junction. The *n* region will be open in the middle of the top, for this is the radiation surface. The minus contacts will surround the surface radiating area. As there is no mechanism to narrow the linewidth, the spectral bandwidth will be the total spectral width of the conduction to valence band transition. For the usual telecommunications diodes which are fabricated in GaAlAs or InGaAsP and radiate in a band around either 830 nm or 1300 nm, these spectral widths are usually on the order of 40–70 nm, which corresponds to 5% relative bandwidths. Surface emitting LEDs are going to radiate essentially as a Lambertian source, as there is no mechanism to narrow the angular distribution. In this sense, the surface emitting LED is much like a miniaturized thermal source. Its spectral bandwidth may be smaller (thermal sources may have relative bandwidths of 50%), but if we use the approximate relation that the coherence length $l_c \sim \bar{\lambda}^2/\Delta\lambda$, where λ is the average wavelength and $\Delta\lambda$ is the spectral linewidth, then we find that the coherence length of an LED is only tens of wavelengths. Further, its spatial coherence will be minimal, as there is no mechanism for one point on the surface to know about the phase at another.

Edge emitting LEDs have much narrower angular spreads and radiation patterns and, therefore, a higher degree of spatial coherence despite having essentially the same temporal coherence of their surface emitting cousins. The edge emitting LED is essentially a mirrorless semiconductor laser. The idea of this device (as well as the laser) is to fabricate a double heterostructure such that the light that is generated by electron-hole pair recombination can be guided out of the junction region. In such a device, the plus contact is attached to a *p*-region of an alloy composition such that the index is lower than that of an intrinsic region below it and roughly equal to that of an *n*-region below the intrinsic region. The different composition, higher index, intrinsic region serves as an extended junction layer which captures and confines the injected electrons and holes as well as some of the generated photons. The photons that are guided down this region are emitted out either of the edges of the device. In an edge emitting LED, the rediating surfaces are left rough so that there is no reflection of light back into the active region. As there is no feedback, there is no line narrowing mechanism, and the 40–70 nm spectral widths remain, even though a higher degree of spatial coherence is exhibited due to the guidance. In a semiconductor laser, the end faces are cleaved, and therefore there is an amount of radiation (30% in intensity) reflected from this interface back into the active area. This reflected radiation causes stimulated emission and therefore definite line narrowing. However, the guidance described above is two dimensional, and therefore there can be many transverse modes. Also, there can be many longitudinal modes. As a typical laser cavity length is 300 μm and the index of GaAs is roughly 3.5, we can show that the mode spacing, given by $c/2nl$, is on the order of a few angstroms. The widths of the gain curves in these multimode semiconductor lasers tend to be several nanometers, and therefore there can be tens of longitudinal modes, with many more transverse modes clustered about each longitudinal mode. For applications requiring high temporal coherence, these lasers are of little use, as their $l_c \sim \lambda^2/\Delta\lambda$ is still a fraction of a millimeter. (This assumes that we can consider the mode structure as a continuum, which may or may not be acceptable depending on the width of each line.) Single-mode operation is achieved by transversely guiding the photons. If we introduce a two-dimensional index structure such that the active region is also a single-moded, two-dimensional waveguide, then the laser operation can occur in a single transverse and longitudinal mode, at least for short-term fixed bias, fixed temperature operation. Such lasers, without any additional stabilization, have exhibited linewidths of less than tenths of an angstrom [6], [7]. By using stabilization techniques, such lasers have been demonstrated to exhibit linewidths of kilohertz, which is the kind of linewidth necessary for coherent optical communications. Such operation, however, requires that extraordinary stability measures be taken, and these measures, as we will discuss, can lead to a new class of problems.

12.4. Propagation of Coherence

Before discussing the propagation of coherence through fibers, it is useful to discuss first some of the salient aspects of the propagation of coherence through free space. To begin the discussion, it is useful to recall a few facts from diffraction theory. If we know the monochromatic electric field distribution in a plane z_1 (assume for simplicity that the wave is completely polarized in the y direction), then the field in a second plane z_2 can be found from the formula [8]

$$E_y(x_2, y_2, z_2) = \frac{1}{i\lambda} \int E_y(x_1, y_1, z_1) \frac{e^{ikr_{12}}}{r_{12}} \cos(\mathbf{r}_{12}, n)\, dx_1\, dy_1, \tag{12.30}$$

where r_{12} is the magnitude of a vector which points from (x_1, y_1, z_1) to (x_2, y_2, z_2) and $\cos(\mathbf{r}_{12}, \mathbf{n})$ is the cosine of the angle between the vector $\mathbf{r}_{12}$ and the unit normal which points in the z_1 direction from x_1, y_1. We often abbreviate eq. (12.30) by writing

$$E_y(x_2, y_2, z_2) = \int \Lambda_{12} \frac{e^{ikr_{12}}}{r_{12}} E_y(x_1, y_1, z_1)\, dx_1\, dy_1, \tag{12.31}$$

where Λ_{12} is often referred to as the obliquity factor. One of the most important results that can be derived from eq. (12.30) is the so-called Rayleigh criterion, which determines the diffraction limit. The calculation is carried out by considering a circular lens of radius a as being the only opening in the plane z_1 with the distance $z_2 - z_1$ being just the focal length f of that lens. Using the paraxial approximation in eq. (12.30) and noticing that the converging phase curvature induced by the lens cancels the diverging curvature in the complex exponential, we find

$$|E_y(x_2, y_2, z_2)|^2 = \left(\frac{ka^2}{2f}\right)^2 \left[2\frac{J_1(kar_2/f)}{kar_2/f}\right]^2, \tag{12.32}$$

where J_1 is the first-order Bessel function and $r_2 = \sqrt{x_2^2 + y_2^2}$. The first zero of the first-order Bessel function $J_1(x)$ lies at $x = 0.61$, giving us that the minimum achievable focal spot by a lens of radius a is given by

$$r_{\min} \simeq 0.61\frac{\lambda f}{a} \sim \frac{\lambda f}{2a} \simeq \frac{\lambda}{2NA}, \tag{12.33}$$

where NA is the numerical aperture of the lens. This last equality holds under the paraxial approximation.

Another important result which can be derived from eqs. (12.30) and (12.31) is the Van Cittert–Zernicke theorem [9]. In the second section of this chapter, the mutual intensity J_{12} was introduced. In a plane z_1, this quantity can be expressed as

$$J_{12}(x_1, y_1, \alpha_1, \beta_1, z_1) = \langle E_y^*(x_1, y_1, z_1)E_y(\alpha_1, \beta_1, z_1)\rangle, \tag{12.34}$$

where the d of eq. (12.28) is given by

$$d = \sqrt{(x_1 - \alpha_1)^2 + (y_1 - \beta_1)^2}. \tag{12.35}$$

As before, we will define the spatial coherence length as that length d for which J_{12} has its first zero. Here the problem is a bit more complicated than it was in Section 12.2, as here we have allowed J_{12} to be a function of several arguments. Therefore its functional form could vary throughout the plane z_1, whereas before its form was fixed by the diffractometer geometry. We will not let this worry us and will assume what is necessary to simplify the problem. Now, eq. (12.31) tells us how to propagate a field from one plane to another. We could just as well use this result to propagate J_{12} to z_2. Indeed, we can use eq. (12.31) in eq. (12.34) to obtain

$$J_{12}(x_2, y_2, \alpha_2, \beta_2, z_2)$$
$$= \iiiint \Lambda^*_{x_1x_2}\Lambda_{\alpha_1\alpha_2}\frac{e^{ik(r_{x_1x_2} - r_{\alpha_1\alpha_2})}}{r_{x_1x_2}r_{\alpha_1\alpha_2}} J_{12}(x_1, y_1, \alpha_1, \beta_1, z_1)\, dx_1\, dy_2\, d\alpha_1\, d\beta_1, \tag{12.36}$$

where $\Lambda^*_{x_1x_2}(\Lambda_{\alpha_1\alpha_2})$ is the obliquity factor tying $(x_1, y_1, z_1)((\alpha_1, \beta_1, z_1))$ to $(x_2, y_2, z_2)((\alpha_2, \beta_2, z_2))$ and the r's have analogous meaning. Generally, the

Van Cittert–Zernicke theorem is written for the quantity μ_{12} of eq. (12.28) but, for our purposes here, eq. (12.36) will suffice to make the necessary point. Consider that the source that is located in plane z_1 is uniform and totally incoherent such that its J_{12} is given by

$$J_{12}(x_1, y_1, \alpha_1, \beta_1, z_1) = I_0 \delta(x_1 - \alpha_1, y_1 - \beta_1), \qquad (12.37)$$

where δ is a two-dimensional delta function. Using eq. (37) in eq. (36), we find that

$$J_{12}(x_2, y_2, \alpha_2, \beta_2, z_2) = I_0 \iint \Lambda^*_{x_1 x_2} \Lambda_{x_1 \alpha_2} \frac{e^{ik(r_{x_1 x_2} - r_{x_1 \alpha_2})}}{r_{x_1 x_2} r_{x_1 \alpha_2}} dx_1\, dy_1. \qquad (12.38)$$

Further, we will assume that the plane z_2 is the focal plane of a circular lens which is the only opening located in plane z_1 and has numerical aperture NA. The point here is that, by analogy with the situation with which we obtained eq. (12.33) from eq. (12.30), we will also be able to determine that the minimum spatial coherence length achievable in plane z_2 will be given by

$$l_c(\min) \sim \frac{\lambda}{2NA}. \qquad (12.39)$$

The result in (12.39) is easily interpretable. In the plane z_1, the source was totally incoherent and therefore radiated into a very broad angle (Lambertian pattern). The rays that reach plane z_2, however, are only a select few—those that were radiated at angles less than the numerical aperture of the lens. This selection effect raises the coherence of the electromagnetic disturbance. Of course, the assumption that the source itself has zero coherence length is in itself a fiction. If each pair of points on the source were completely phase incoherent with each other, including all pairs of points spaced much more closely than a wavelength, the source could not radiate, as the process of radiation is dependent on correlations at the subwavelength level. However, the fact that the original source had a small degree of coherence is not the point. The point is that optical systems induce coherence and, further, diffraction limited spots must be spatially coherent spots. We could question how an imaging system could reduce the entropy of an electromagnetic wave. The answer here is also clear. In the above example, the lens was only able to deliver those rays from the source which had low enough initial angle to pass the system. Therefore, the transformation of coherence is taking place at a great loss of energy and inherently is a filtering process. The optical system is only passing those rays which had a predisposition to be coherent.

What does the combined diffraction limit/spatial coherence limit have to say about guided wave systems? Perhaps the easiest way to get at this question is to consider a very simple model of guided wave propagation. Say that there is a ray propagating down a layer of higher index, between two layers of lower index, such that it is totally internally reflected at each bounce. Let us say that the angle (defined with respect to the normal to the lower index region) at each bounce is θ and that the width of the guiding layer is a. Then the change in phase of the disturbance in propagating from an upper bounce through a lower bounce to the next upper bounce is going to be $2ka \sin \theta$ where $k = 2\pi/\lambda$ is the wave vector. The simplest condition we can think of for a mode to be guided is that this phase change is a multiple of 2π or

$$2ka \sin \theta = 2n\pi. \tag{12.40}$$

If we were to consider the cutoff point at which the guide became single moded, we would set $n = 1$ and $\sin \theta = NA$, as the numerical aperture will give the lowest angle at which total internal reflection can occur. Using these two analyses in eq. (12.40), we find a value for a, which gives the cutoff for single mode operation

$$a \leq \frac{\lambda}{2NA}, \tag{12.41}$$

which is exactly the same as the recurring diffraction limit/spatial coherence limit. The point is that a single-mode guide is diffraction limited. Further, a single-mode guide can only carry spatially coherent light. To couple light into a single mode requires an imaging system which can demagnify a spot to the point where it is both diffraction limited and spatially coherent. In some sense, we have come to a very basic definition of what a mode truly means. A mode must be spatially coherent with itself to be a mode. In the next two sections, we will investigate some of the ramifications of this fact.

12.5. Coherence in Multimode Fibers

In this section, coherence effects associated with propagation in multimode fibers will be discussed in light of the earlier discussions of temporal and spatial coherence. In particular, the cases of a temporally coherent source, a spatially incoherent source, and a finite bandwidth source will be considered in some detail.

As was discussed in Section 12.2 of this chapter, temporal coherence implies spatial coherence in the sense that a monochromatic source will always interfere with itself at any time delay or spatial offset. The only way to destroy the temporal coherence of a monochromatic disturbance is to broaden its spectrum. That such interference occurs with free-space propagation was well known from the early days of the laser, where one of the first phenomena noticed was that of laser speckle. Laser speckle can perhaps best be described by considering a case where a laser illuminates a rough (compared to a wavelength) surface and is observed on a card which is in line with the specular reflection from the surface. As the surface is rough, at a single point on the observation plane (card), reflections will be received from many points on the surface. If we consider each (greater than a wavelength sized) point on the surface to give rise to a "local plane wave," then the complex field, observed through a polarizer at a point x, y, z in the observation plane, will take the form

$$E_y(x, y, z) = \sum_{j=1}^{N} a_j e^{i\delta j}, \tag{12.42}$$

where N can be very large, the a_j are random, and the δ_j is random and uniformly distributed. For N large enough, we can invoke the law of large numbers to find that E_y will be exponentially distributed (two-dimensional random walk) and therefore that the intensity is a Poisson distribution. Armed with the distribution, we can then determine the average sizes of the bright and dark interference cells as a function of N, the degrees of freedom of the pattern. These interference cells are the speckles which we observe when viewing "rough" objects in coherent light.

This speckle phenomenon will also exist in multimode fibers, as is clear from writing the expression for the complex transverse field of a fiber whose optic axis coincides with the z coordinate of an x, y, z coordinate system. At a point x, y, z within the fiber core, we can write [10]

$$\begin{aligned}\mathbf{E}_t(x, y, z) = \sum_{l,m}^{N} & \mathbf{E}^1_{lm}(x, y)(A^c_{1lm} \cos l\theta + A^s_{1lm} \sin l\theta)e^{i\beta_{1lm}z} \\ & + \mathbf{E}^2_{lm}(x, y)(A^c_{2lm} \cos l\theta + A^s_{2lm} \sin l\theta)e^{i\beta_{2lm}z},\end{aligned} \tag{12.43}$$

where l and m are the azimuthal and radial mode numbers, respectively, $\mathbf{E}^{1(x,y)}_{lm}(\mathbf{E}^{2(x,y)}_{lm})$ is the mode pattern (eigenfunction) associated with the first (second) of the two independent polarization states, A^c_{1lm} and A^s_{1lm} are the complex excitation coefficients for sine and cosine parts of the first (second) of the two independent polarization states for the mode lm, and $\beta_{1lm}(\beta_{2lm})$ is the propagation constant for the first (second) of the two independent polarization states for the mode lm. Now, in general, N is quite large in a typical multimode fiber (roughly 1000 for a 6.25 μm core, 0.26 NA fiber), and the propagation constants of different modes have no set relation to one another. Therefore, the interference between the different mode patterns should not only be quite complex but should vary almost randomly with propagation distance down the fiber. This effect is compounded if we consider that fibers always have residual birefringence and therefore the degeneracy between sine and cosine modes is only an approximate one. Despite this clear prediction of fiber speckle from the basic equations, there seems to be no discussion in the literature of the effect before 1975 [11]. This could in part be due to the fact that most people who were interested in fiber at that time were interested in driving fibers with semiconductor lasers which, prior to roughly 1977, were all multimode, with linewidths of roughly 10 GHz per mode and modes extending over several nanometers. It was not until 1978 that the phenomenon of laser speckle as it relates to signal degradation was first termed modal noise [12]. By 1978, various people had attempted driving multimode systems, which were commonly employed for telephone company trunking at that point in time, with single-mode semiconductor lasers. The primary observation was that the spectral purity of the source led to drastic increases in received bit error rates. The explanation for this is basically that environmental factors, such as even tiny temperature changes (10^{-4} °C acting over 1 km can cause 2π phase changes) can cause a drastic alteration of the speckle pattern. As fiber links, by nature, will have spatially selective loss mechanisms (higher-order mode cutoff in misaligned connectors), spatially varying speckle patterns will lead to temporally varying received powers. Although it would seem that this effect could be explained pretty well in terms of conventional speckle theory as described by eq. (12.42), it turns out that the statistics of eq. (12.43) are actually much more interesting than those of eq. (12.42) [10], [13], but to discuss this in detail would be a digression from the main thrust of this chapter.

Of course, the first question to occur to us who have just realized the problems associated with coherent propagation in multimode fibers is: How do initially (at the source) incoherent waves propagate in a multimode fiber? As was discussed previously, the fundamental mode of a fiber must by nature be "self-coherent," if only because its spotsize corresponds to the minimum coherence length which can be passed by an optical system of a given numerical aperture. An incoherent excitation therefore could only poorly excite a single mode, as the mode would act as a filter for those rays with exactly the right properties. If we recall the argument that led to that conclusion, we would realize that the conclusion about "self-coherence" must extend to all the modes of a fiber. The idea is that, in its simplest definition, a mode is an entity which bounces back and forth such that its phase reinforces on successive bounces. The higher-order modes, therefore, are like the fundamental mode in that they will act as a filter of an incoherent wave and pick out only the rays that will most excite them. These rays, however, will not necessarily be ones that will be coherent with those in the fundamental. However, these rays should have some relation to those propagating in the fundamental, especially for modes that travel at very close mode angles (close propagation constants). It is the relation between these mode coherences that is of interest to us—or at least a path to finding them.

We can see rapidly that using expression (12.43) back in eq. (12.28) for the spatial coherence is a fruitless approach to the task of obtaining an analytical result, as the expression will be intractable. However, there is a considerable simplification that can be made to the formulation. For a monochromatic excitation of a multimode fiber, a spectrum of propagation constants will be excited. If the frequency of the monochromatic wave is changed, the spectrum of the propagation constants will be shifted. Considering typical values of the magnitude of this shift for 62.5 μm, 0.26 numerical aperture graded index fiber, we would find that roughly a 4 Å shift in center wavelength of a source would be sufficient to shift the new mode spectrum by one mode spacing of the old mode spectrum [14]. This means that a source with a greater than 4 Å bandwidth would excite a continuum of modes within the fiber. As LEDs have bandwidths of 40–70 nm and even multimode lasers have gain curves of 4 nm or more, all semiconductor sources with the exception of single-mode laser diodes will excite mode continua in the fiber. It is straightforward to show in this case that we can relate the near-field intensity $I(r/a)$ to a modal power distribution $p(R)$ by the relation [14]

$$I\left(\frac{r}{a}\right) = \frac{V^2}{\pi} \int_{f^{1/2}(r/a)}^{I} p(R)R\, dR, \tag{12.44}$$

where $V = \sqrt{2\Delta}\, k_0 n_1 a$ is the fiber V number, $f(r/a)$ is the fiber profile function, and R is the mode parameter, given in terms of the ray tracing variables by

$$R = \left[f\left(\frac{r}{a}\right) + \frac{\sin^2 \theta_0}{2\Delta} \right]^{1/2} \tag{12.45}$$

and in terms of the modal propagation constant β by

$$R^2 = \frac{1}{2\Delta}\left[1 - \frac{\beta^2}{n_1^2 k_0^2}\right]. \tag{12.46}$$

Here it is being assumed that the fiber's index profile is given by

$$n^2(r) = \begin{cases} n_1^2(1 - 2\Delta f(r/a)), & r < a, \\ n_1^2(1 - 2\Delta), & r > a, \end{cases} \tag{12.47}$$

where $f(r/a)$ is assumed to be zero at the origin and monotonically increasing to a value of 1 at the core-cladding interface, which is located at the coordinate $r = a$.

The $p(R)$ as defined by eq. (12.44) is a quantity resembling a specific intensity as it is a function of both position and solid angle through the mode parameter R. Indeed, if we assume that a fiber end resembles a quasi-homogeneous source with specific intensity $p(R)$, we can derive analytical expressions for the coherence function which contains the profile function as a parameter [15]. In this manner, it becomes possible to use a wavefront reversing interferometer to perform index profiling [16].

A last problem to be discussed in this section would be that of a source which is neither totally monochromatic nor totally spatially incoherent. A multimode laser which by chance had very tiny individual mode widths or a single-mode laser under current modulation could exhibit finite linewidth characteristics. The most interesting thing about such a case is perhaps the disappearance of speckle effects with propagation distance. As was discussed in Section 12.2, a source with a finite temporal coherence length will stop exhibiting interference when interfered with itself if delayed by a length equal to or greater than the temporal coherence length. At the input to the fiber, all of the modes will have a relative delay of zero, and therefore there will be speckles at the input plane of a fiber. Modes in a fiber do not propagate at the same group velocities, however, due to modal dispersion. Although fiber profiles which are close to an ideal square law profile can exhibit minimal modal dispersion, even this minimal amount is very large compared to the dispersion of, for example, a single-mode fiber. Dispersion of a typical multimode fiber can easily be on the order of 1 nsec/km, and therefore two modes can separate by as much as 30 cm in a kilometer's propagation. A 30-cm coherence length corresponds to a relative linewidth of 3×10^{-6} or, for a 1-μm center wavelength, a 0.03-Å linewidth. This corresponds to a quite coherent laser—in fact, about as small a linewidth achievable in a nonstabilized single-mode semiconductor laser. It is clear, therefore, that, over a 10-km propagation distance, most speckle in a multimode fiber will disappear if the source is even a very good semiconductor laser. However, before getting to the 10 km point, the wave could well suffer the cross-sectionally selective loss that was mentioned earlier and therefore exhibit reduced signal-to-noise ratio characteristics of modal noise. For a multimode semiconductor laser whose coherence length could well be less than 1 mm, coherence effects can well disappear in tens of meters.

12.6. Coherence in Single-Mode Fibers

If a single-mode fiber were really single-moded, then we would not expect to see any coherence effects. However, single-mode fibers (except for a few specialty ones) generally propagate at least two modes, one for each polarization state. Certainly we could consider this an advantage in coupling light in from an incoherent unpolarized source. This 3-dB improvement in coupling with respect to a single polarization fiber is little consolation with respect to the tens of decibels of excess loss over coupling from a coherent polarized source. The second polarization state, however, can prove deleterious to an information stream imposed upon a coherent carrier.

Polarization is always a consideration in any coherent propagation problem. As was discussed in Section 12.2 of this report, temporal coherence always implies spatial coherence. Any two spatial points within a temporally coherent wave train will always interfere with each other and therefore exhibit speckle. But it must also be true, by the same reasoning, that the two orthogonal polarization states must also be coherent with each other and therefore always exhibit a degree of polarization of unity. In a multimode fiber, where many modes of hybrid polarization exist, the degree of polarization is unity really only locally, as the polarization state of a given mode is not uniform across the core. If we were therefore to put a polarizer across the core and focus the transmitted light down on a photodetector, it can happen that different points on the core can add destructively and result in a zero detected polarized field. For this reason, discussion has been given to the depolarization of coherent waves in propagating in a multimode fiber. This depolarization is only due to the locality of the polarization interference effect and the observation of the polarization with equipment that samples the whole endface and is not due to spectral broadening of the disturbance. The two modes of a single-mode fiber, however, are fundamental modes and should have uniform circular polarization across the core. In an ideal fiber, these two circularly polarized modes should be degenerate, and we could just as well represent these modes as linearly polarized modes. In a real fiber, though, the degeneracy will be broken. Quite generally, the degeneracy will be broken such that the linearly polarized modes are the actual modes, as birefringence induced during the fabrication and pulling process will be linear birefringence due to induced stresses, which will have a favored direction. Were this birefringence uniform down the fiber's length, there would be two principal axes along which light launched with polarization vector along one of these axes would remain polarized. Light launched with polarization vector along any other direction would have its polarization evolve as it propagated down the fiber, as its components along the two principal axes will propagate with different phase velocities. The polarization will therefore, at any cross section, be elliptical [17], [18].

Polarization modal noise in single-mode fibers is caused by mechanisms analogous to those which cause multimode modal noise. Just as splices, connectors, fiber scattering, etc., can cause loss which is nonuniform across the core of a multimode fiber, such mechanisms can also cause polarization selective loss or, worse still, polarization coupling. Polarization selective loss will lead to polarization "fading" at the detector due to the fact that random time-varying perturbations will change the amount of energy in each polarization state, and therefore the composite polarization state will change in both direction and magnitude [19]. In multimode systems, the modal noise itself was due to interference between different modes which were temporally coherent with each other yet displaced essentially randomly in phase, leading to a time-varying spatial intensity pattern. In a single-mode system, due to the fact that different polarization states must be summed coherently yet do not interfere, the modal noise causing mechanism must be considered to be a mode interfering with itself. That this does indeed happen becomes clear if we consider that the imperfect single-mode fiber can operate as a multiple beam interferometer in the sense that light propagating in a single polarization state can be scattered into the other and propagate at that polarization state's phase velocity for some distance before being scattered back into the initial state to interfere with the initial wave. In this way, signal-to-noise ratios can be degraded from their optimal, although the effect is nowhere nearly as dramatic as the degradation in a multimode fiber with coherent excitation. It is this same multipath effect that can lead to depolarization in single-mode fibers excited by finite coherence length sources [20].

12.7. Conclusions

This chapter has reviewed some of the basic theory of partial coherence and applied, in some cases heuristically, to the phenomena of propagation in optical fibers. Considerations involving coupling phenomena as well as modal noise and mode continua have been found to follow quite simply from the presentation of the theory. These are certainly not the only coherence phenomena in guided wave optics, although they are perhaps the most straightforward to describe. The use of fiber in interometric devices (sensors, gyros, etc.) has stimulated renewed interest in linear coherence. Perhaps an even more fertile field of endeavor is the field of higher-order coherence as it can be applied to nonlinear effects. Guided wave devices can exhibit much lower power nonlinear thresholds than bulk devices due to the considerably increased energy densities in guided devices. Nonlinear coherence effects can lead to macroscopically observable quantum effects, in that coherence (correlation) causes quantum mechanical averages (the angular brackets of quantum ensembles) to take on other than classical limits. Such an effect is the one that is observable from squeezing states, which can actually lower observed quantum vacuum fluctuations. Discussion of such effects, however, are beyond the scope of this chapter, which is intended to introduce the reader to coherence effects in waveguides. These more interesting effects should perhaps be left to a chapter in a second book to be dedicated to Professor Papas.

References

[1] M. Born and E. Wolf (1975), *The Principles of Optics*, 5th ed., Pergamon Press, Oxford.

[2] M. Born and E. Wolf (1975), *The Principles of Optics*, 5th ed., Pergamon Press, Oxford. p. 494.

[3] M. Born and E. Wolf (1975), *The Principles of Optics*, 5th ed., Pergamon Press, Oxford. p. 257.

[4] M. Born and E. Wolf (1975), *The Principles of Optics*, 5th ed., Pergamon Press, Oxford. p. 260.

[5] H. Kressel and J.K. Butler (1977), *Semiconductor Lasers and Heterojunction LED's*, Academic Press, New York.

[6] M. Osinski and J. Buus (1987), Linewidth broadening factor in semiconductor lasers: An overview," *IEEE J. Quant. Elect.*, **JQE-23**, 9–28.

[7] C.H. Henry (1986), Phase noise in semiconductor lasers, *J. Light Tech.*, **JLT-4**, 298–311.

[8] M. Born and E. Wolf (1975), *The Principles of Optics*, 5th ed., Pergamon Press, Oxford. Chap. 8.

[9] M. Born and E. Wolf (1975), *The Principles of Optics*, 5th ed., Pergamon Press, Oxford. Chap. 10, Sec. 4.

[10] A.R. Mickelson and A. Weierholt (1983), Modal noise-limited signal to noise ratios in multimode optical fibers, *Appl. Opt.*, **22**, 3084–3089.

[11] B. Crosignani, B. Daino, and P. DiPorto (1975), *Appl. Phys. Lett*, **27**, 237.

[12] R.E. Epworth (1978), The phenomena of modal noise in analogue and digital fiber optic systems, *Technical Digest: Fourth European Conference on Optical Communication*, Genoa, pp. 492–501.

[13] D.R. Hjelme and A.R. Mickelson (1983), Microbending and modal noise, *Appl. Opt.*, **22**, 3874–3879.

[14] A.R. Mickelson and M. Eriksrud (1982), Mode-continuum approximation in optical fibers, *Opt. Lett.*, **7**, 572–574.

[15] S. Piazzolla and G. De Marchis (1980), Spatial coherence in optical fibers, *Opt. Comm.*, **32**, 380–382.

[16] B. Daino, S. Piazzolla, and A. Sagnotti (1979), Spatial coherence and index-profiling in optical fibres, *Optica Acta*, **26**, 923–928.

[17] R. Ulrich and A. Simon (1979), Polarization optics of twisted single mode fibers, *Appl. Opt.*, **18**, 2241–2251.

[18] A. Simon and R. Ulrich (1977), Evolution of polarization along a single-mode fiber," *Appl. Phys. Lett.*, **31**, 517–520.

[19] I.P. Kaminov (1981), Polarization in optical fibers, *IEEE J. Quant. Elect.*, **JQE-17**, 15–22.

[20] W.K. Burns, R.P. Moeller, and C.L. Chen (1983), Depolarization in a single-mode fiber, *J. Light Wave Tech.*, **JLT-1**, 44–50.

13
How Scattering Increases as an Edge Is Blunted: The Case of an Electric Field Parallel to the Edge

KENNETH M. MITZNER,* KENNETH J. KAPLIN,* and JOHN F. CASHEN*

13.1. Introduction

An excellent teacher is much more than a purveyor of facts. He instills in his students a point of view and a mode of attack, a way of approaching problems that becomes so natural to them that they may not realize until years later that they have imbibed great wisdom. Professor Papas has taught us to go back again and again to the basics, to mine them as the inexhaustible mother lode of insight. He has taught us, furthermore, to seek simple results with simple explanations, and he has helped us build up the confidence and faith we need for this most daring and radical of human endeavors, the hunt for simplicity.

In this study we present a simple but accurate closed-form solution for far-field backscatter of a plane wave from a perfectly conducting wedge with its edge blunted by a rounding arc, as illustrated in Figure 13.1, and we use basic considerations of realizability to explain the form of the solution. We restrict the direction of propagation of the incident wave to be a direction for which geometrical optics predicts backscatter from the rounded surface, that is, a direction which is normal to the rounded surface along a line parallel to the edge, the *specular line* for backscatter. We further restrict consideration to the polarization with the electric field parallel to the edge. We examine in detail the case in which the edge is rounded by a circular arc of radius a, then indicate how the results can be generalized for other rounding arcs. We also give a solution for scattering from a halfplane of finite thickness with the edge rounded, a problem which can be treated as the limit of the blunted edge problem when the wedge angle becomes 0°.

In the usual manner of modern diffraction theory [1], we characterize the far-field scattering in terms of a *diffraction coefficient*, which we designate as u. We show that numerical data for u, which we obtain by the method of moments solution of an integral equation formulation, are matched well by

---

* Northrop Corporation, B-2 Division, 8900 Washington Boulevard, Pico Rivera, CA 90660, USA.

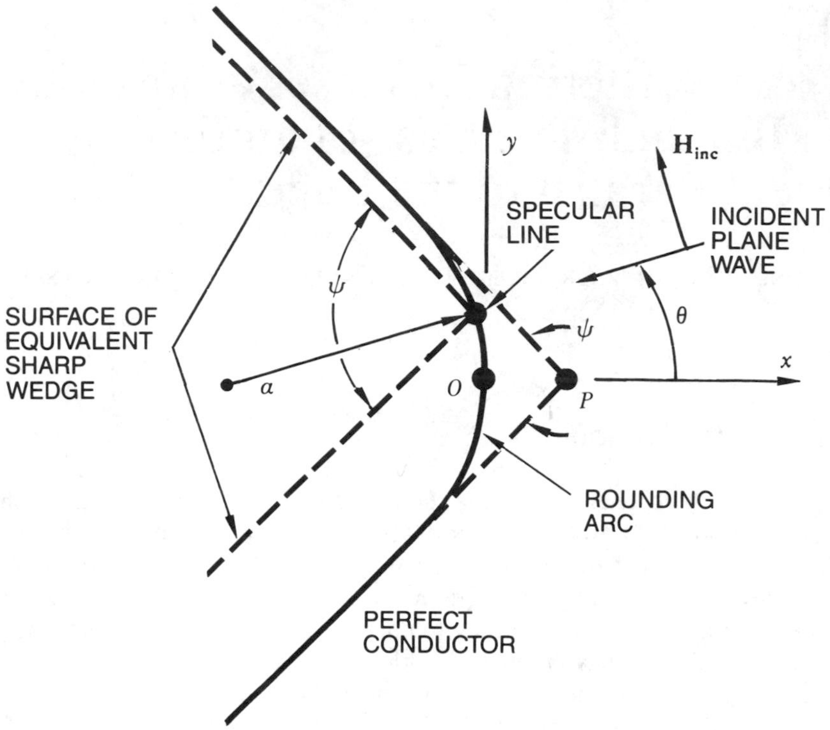

FIGURE 13.1. Blunted edge geometry: Cross section view.

the *transition function* formula

$$u = (u_{GO}^2 + u_K^2)^{1/2}$$

of eq. (13.13). Here u_{GO} is the *geometrical optics diffraction coefficient* to which u converges in the high-frequency limit. In this limit the backscatter is the same as for a circular cylinder of radius a [2] and appears to emanate from the specular line. The low-frequency limit of u is u_K, which is the *Keller diffraction coefficient* [1], [3] for the *equivalent sharp wedge* of Figure 13.1, a wedge of the same angle and orientation as the original wedge but with its edge along the specular line.

Thus the transition function simply combines the formulas for the low- and high-frequency limiting behavior in the manner of Pythagoras's theorem. This gives a gradual monotonic transition in magnitude and phase between the two limiting cases across an *intermediate frequency range* which surrounds the *crossover frequency*, the frequency at which the sharp wedge and geometrical optics solutions are equal in magnitude. The value of this crossover frequency is inversely proportional to a (which sets a scale for the geometry) and is also a function of the wedge angle ψ and the angle of incidence θ of Figure 13.1.

As shown in eq. (13.31), the transition function has a very simple frequency dependence when evaluated in a coordinate system with origin on the specular line. By using this result to derive eq. (13.41) for the transfer function of the pulse scattering problem, then examining the behavior of the transfer function in the complex-frequency plane, it is readily verified that the transition function is consistent with basic physical realizability constraints. Other closely related formulas, which can fit the numerical data about as well, fail this test.

13.2. Analysis and Results

13.2.1. *Problem Formulation*

The body in Figure 13.1 is an infinitely long perfectly conducting wedge with the edge rounded off by a circular arc of radius a. The wedge angle ψ is restricted to the range

$$0^\circ < \psi < 180^\circ \tag{13.1}$$

so that the body is convex. The origin of coordinates O is at the bisector of the rounding arc. The z-axis is parallel to the edge and points out of the page, with $\mathbf{e}_z$ the unit vector in this direction. The x-axis lies along the bisector of the wedge, and the y-axis is so oriented as to yield a right-handed orthogonal (x, y, z) coordinate system.

The incident wave is a time-harmonic plane wave propagating normal to the edge with direction of propagation given by the angle θ. The electric field

$$\mathbf{E}_{\text{inc}} = E_0 \mathbf{e}_z \exp\left\{-i\frac{2\pi}{\lambda}(x \cos\theta + y \sin\theta)\right\} \tag{13.2}$$

is parallel to the edge, and thus the magnetic field $\mathbf{H}_{\text{inc}}$ is normal to both the edge and the direction of propagation. Here λ is the wavelength and E_0 gives the amplitude of the field. The angle of incidence θ can—without loss of generality—be taken nonnegative. We shall further restrict θ to the range

$$0 \le \theta < \left(90^\circ - \frac{\psi}{2}\right) \tag{13.3}$$

so that the direction of propagation is normal to the surface along a single z-directed line which lies on the rounded portion. (The inequality in the upper limit precludes the case of normal incidence to the flat face of the wedge.) The line of normal incidence is the *specular line* for backscatter, that is, the line along which the incident rays of classical geometrical optics are reflected in the backscatter direction.

A time-dependence factor $\exp\{-i\omega t\}$, with ω the radian frequency, is assumed but, as is customary, will be omitted except in the discussion of pulse scattering. The frequency f is related to λ and ω by

$$f = \frac{\omega}{2\pi} = \frac{c}{\lambda} \tag{13.4}$$

with c the speed of light. The scattered electric field $\mathbf{E}_{\text{scat}}$ is, like the incident field, parallel to the edge and is thus completely characterized by the scalar E in the expression

$$\mathbf{E}_{\text{scat}} = E\mathbf{e}_z. \tag{13.5}$$

For all values of ψ and θ allowed by eqs. (13.1) and (13.3), the standard far-field approximation [1], [4], which can be written in the form

$$E = -uE_0\left(\frac{\lambda^{1/2}}{2\pi R^{1/2}}\right)\exp\left\{i\frac{2\pi R}{\lambda}\right\} \tag{13.6}$$

with u independent of R, is valid in the backscatter direction at a sufficiently large distance R from the z-axis. The scattered magnetic field $\mathbf{H}_{\text{scat}}$ for far-field backscatter is parallel to $\mathbf{H}_{\text{inc}}$ of Figure 13.1. The dimensionless scalar function u in eq. (13.6) is the two-dimensional *diffraction coefficient* for the scattering body of interest and contains all information about the far-field scattering that is specific to the body.

It should be noted that $u \to \infty$ as θ approaches the upper limit in eq. (13.3), that is, as the direction of incidence approaches normal to the flat surface of the wedge. This singular behavior arises because normal incidence backscatter from the semi-infinite flat wedge face is so strong that it never for any range of R settles down to the characteristic $1/R^{1/2}$ far-field dependence on distance assumed in eq. (13.6). For values of θ near the limit, the minimum value of R at which the far-field approximation is satisfactory depends on θ and increases without bound as θ approaches the limit.

For any body of infinite length and constant section there exists a diffraction coefficient u for which eq. (13.6) is valid (although, as we have just seen for the configuration of Figure 13.1, not necessarily valid at every backscatter angle when the body has a geometrical cross section of infinite extent). We shall now consider this general case briefly before returning to the specifics of the blunted edge problem.

The diffraction coefficient in eq. (13.6) has been normalized *relative to a knife edge*, that is, normalized so that

$$u = \exp\left\{i\frac{\pi}{4}\right\} \tag{13.7}$$

is the diffraction coefficient for grazing illumination ($\theta = 0$) on a horizontal halfplane ($\psi = 0$) with its edge (the so-called "knife edge") along the z-axis. The $\pi/4$ radian phase term of eq. (13.7) is the *phase anomaly* [5] associated with the knife edge problem, that is, the additional phase change beyond that corresponding to the geometrical optics path length and the effect of the boundary condition. This phase anomaly is due to the complicated behavior of the scattered field near the edge.

The radar cross section (RCS) for the two-dimensional problem of scattering from an infinitely long feature of constant section is [4]

$$\sigma_{2D} = \left(\frac{\lambda}{2\pi}\right)|u|^2, \tag{13.8}$$

which is in units of length. For practical problems it is much more convenient, however, to work with the RCS normalized relative to a knife edge

$$\sigma_{KE} = |u|^2, \tag{13.9}$$

which is dimensionless, and the RCS in dB relative to a knife edge (dBKE)

$$\sigma_{dBKE} = 10 \log_{10} \sigma_{KE} = 20 \log_{10}|u|. \tag{13.10}$$

Thus, for example, a frequently encountered three-dimensional problem is backscatter from a constant section feature of finite length L for incidence normal to the axis of the feature. When the end effects can be neglected, the RCS is simply [6]

$$\sigma = \frac{L^2}{\pi}\sigma_{KE} = \frac{L^2}{\pi}|u|^2, \tag{13.11}$$

with u the same as for infinite length. This form shows clearly how σ, which is in units of area, depends on L and how all dependence on other parameters, including wavelength, is contained in the diffraction coefficient.

Returning our attention now to the specific geometry of Figure 13.1, the only length parameter is the radius a. It then follows by dimensional analysis that u for this geometry can depend on a and λ only through the parameter

$$\bar{f} = \frac{a}{\lambda} = \frac{a}{c}f = \frac{a}{2\pi c}\omega. \tag{13.12}$$

For fixed frequency, $\bar{f}$ can be interpreted as a measure, in units of wavelength, of how much the edge has been blunted.

The emphasis here, however, shall be on an alternative interpretation in which a is considered to be fixed, so that $\bar{f}$ becomes the *normalized frequency* and $\bar{\omega} = 2\pi\bar{f}$ is the corresponding *normalized radian frequency*. The problem of interest then reduces to determining u and the corresponding σ_{KE} and σ_{dBKE} as functions of ψ, $\bar{f}$, and θ for all $\bar{f}$ and for the values of ψ and θ allowed in eqs. (13.1) and (13.3). Low, intermediate, and high frequency will in this context mean low, intermediate, and high values of $\bar{f}$.

13.2.2. *Numerical Solution*

The procedure followed in this study was empirical. Approximate values of u were obtained for various values of wedge angle ψ, normalized frequency $\bar{f}$,

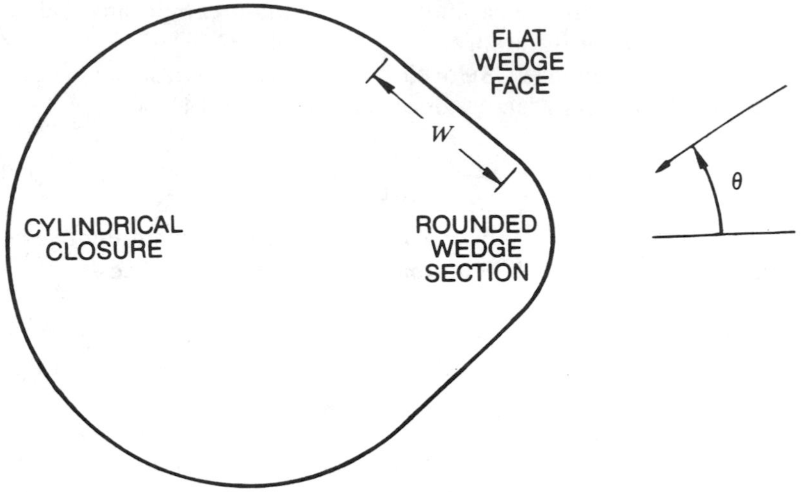

FIGURE 13.2. "Test body" for which integral equations are solved.

and angle of incidence θ by numerical solution of integral equations. The transition function was then developed by fitting a closed-form expression to the numerical data. The integral equation computations were performed using standard in-house method of moments computer codes developed by Northrop Corporation. The codes solve a matrix equation for each pair of values of ψ and $\bar{f}$, then generate results for as many values of θ as desired.

The codes are designed for application to bodies of finite extent in the (x, y)-plane, and thus it was necessary to approximate the geometry of Figure 13.1 by the "test body" of Figure 13.2, which has flat wedge faces of finite width W that are terminated by a cylindrical closure at the rear. Computations were carried out for selected values of ψ in the range 10° to 90° and, for each values of ψ, at enough values of $\bar{f}$ and θ to produce smooth curves of u versus these parameters. The face width W—and thus the overall size of the test body—was adjusted with ψ and $\bar{f}$. It was set large enough so that, for θ up to about 20°, the error in u due to the test body closure is within acceptable limits of about 15% in magnitude—which corresponds to about 1.3 dB in RCS—and a few degrees in phase. Obtaining data of the same accuracy at higher values of θ would require significant increases in solution time.

In short, reliable—though not perfect—numerical data were obtained for wedge angles 10–90°, incidence angles 0–20°, and a broad frequency range. Results for the bracketing cases, $\psi = 10°$ and 90° with $\theta = 0°$ and 20°, are shown in Figures 13.3–13.6. For each pair of ψ and θ values, the normalized RCS, σ_{dBKE}, which is related to the magnitude of u by eq. (13.10), and the phase of u are plotted versus $\bar{f}$.

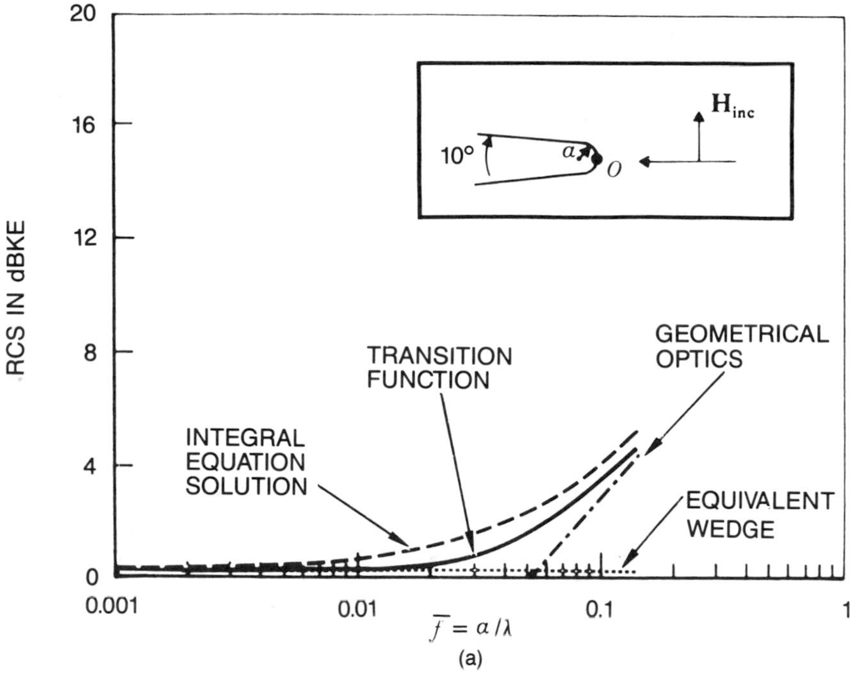

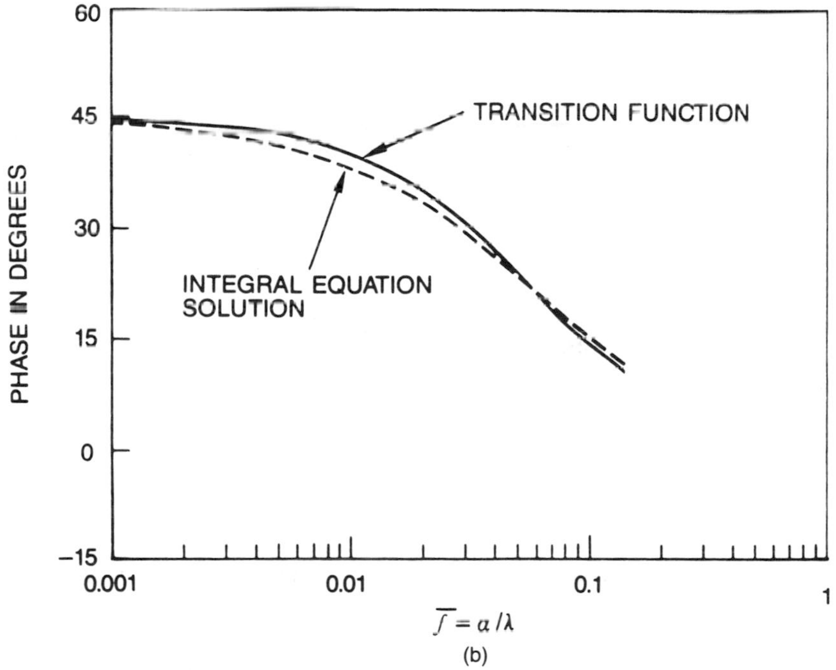

FIGURE 13.3. Backscatter from a blunted edge: 10° wedge angle; incidence along bisector $\theta = 0°$. (a) Normalized RCS σ_{dBKE}; (b) phase of the diffraction coefficient u.

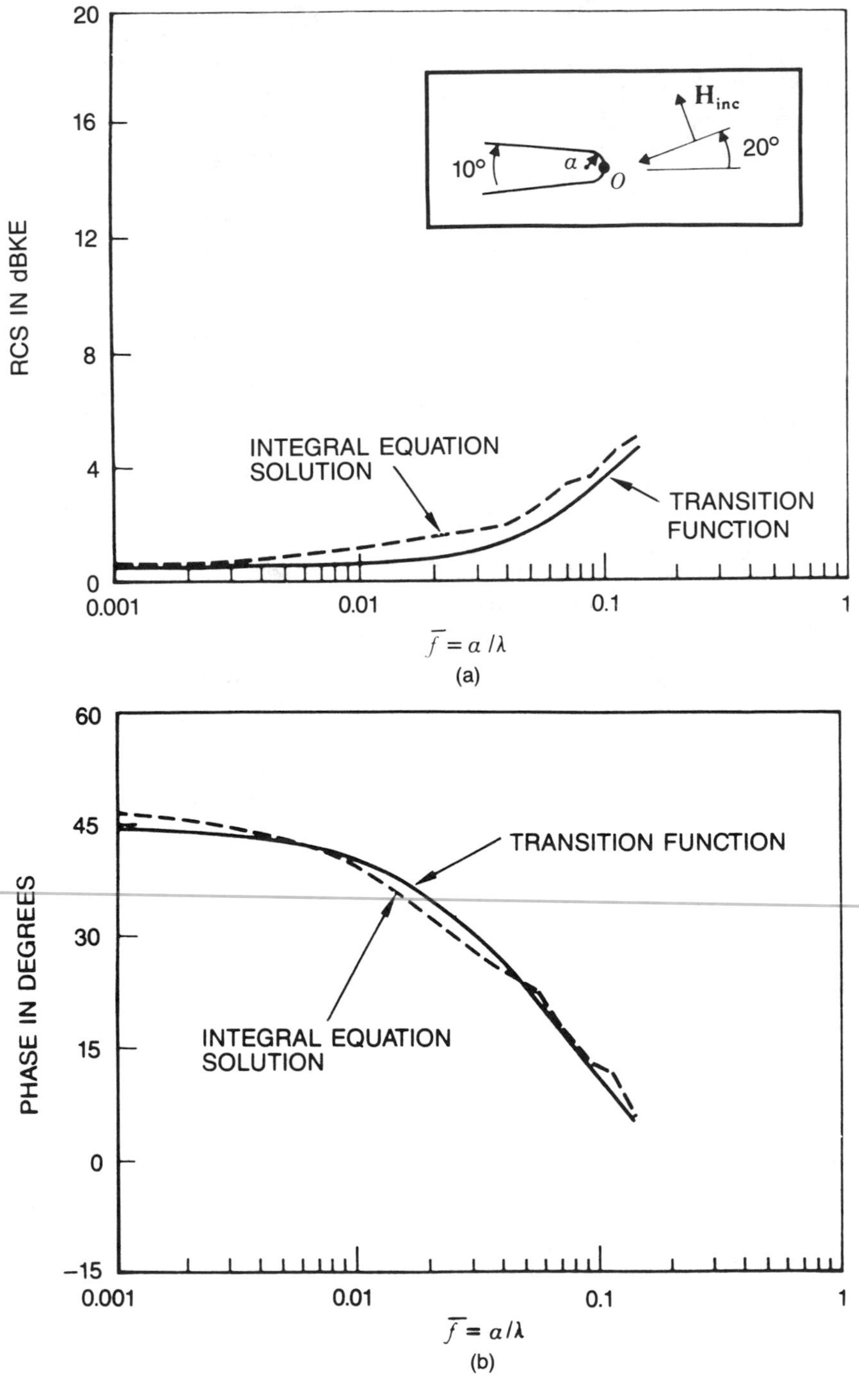

FIGURE 13.4. Backscatter from a blunted edge: 10° wedge angle; incidence 20° off bisector. (a) Normalized RCS σ_{dBKE}; (b) phase of the diffraction coefficient u.

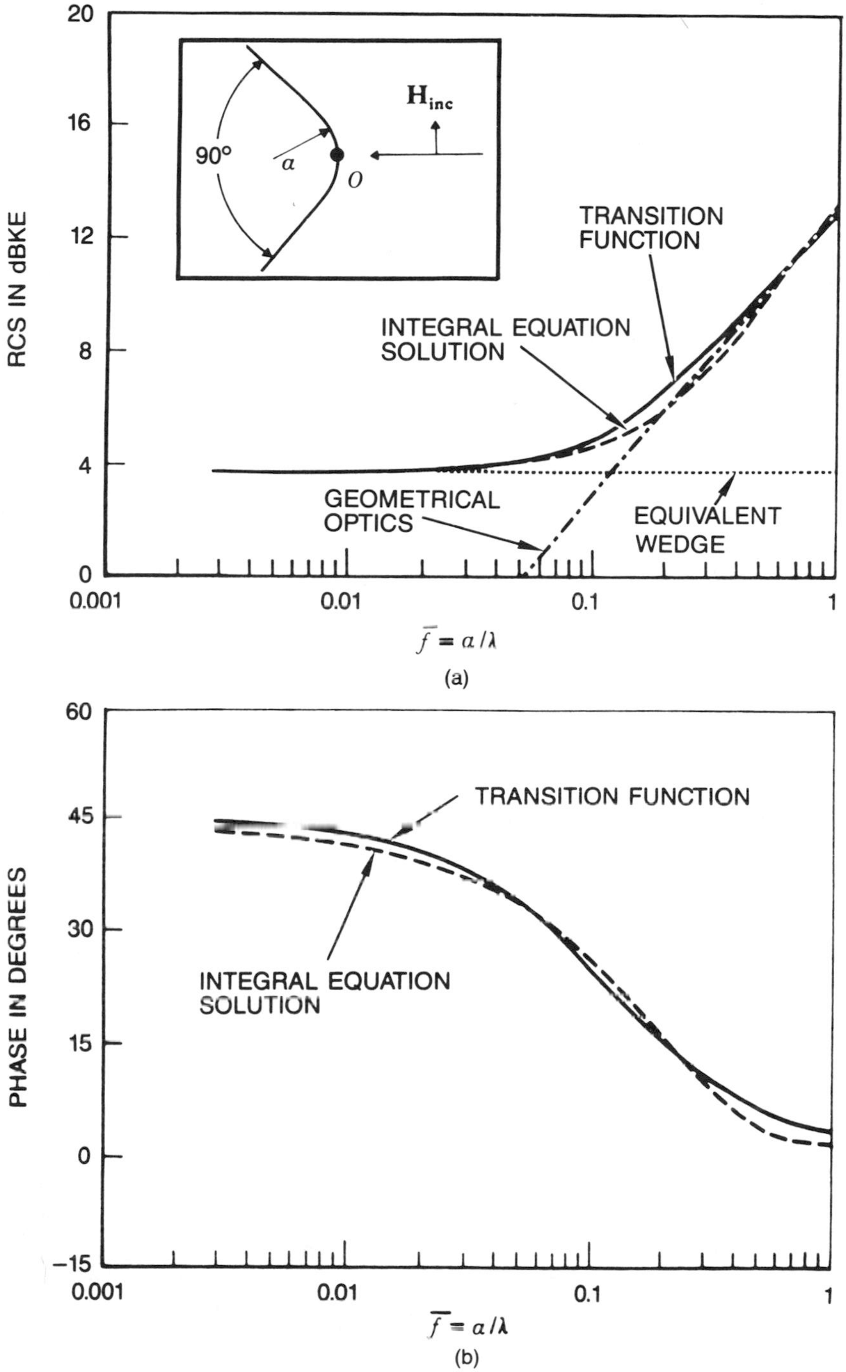

FIGURE 13.5. Backscatter from a blunted edge: 90° wedge angle; incidence along bisector $\theta = 0°$. (a) Normalized RCS σ_{dBKE}; (b) phase of the diffraction coefficient u.

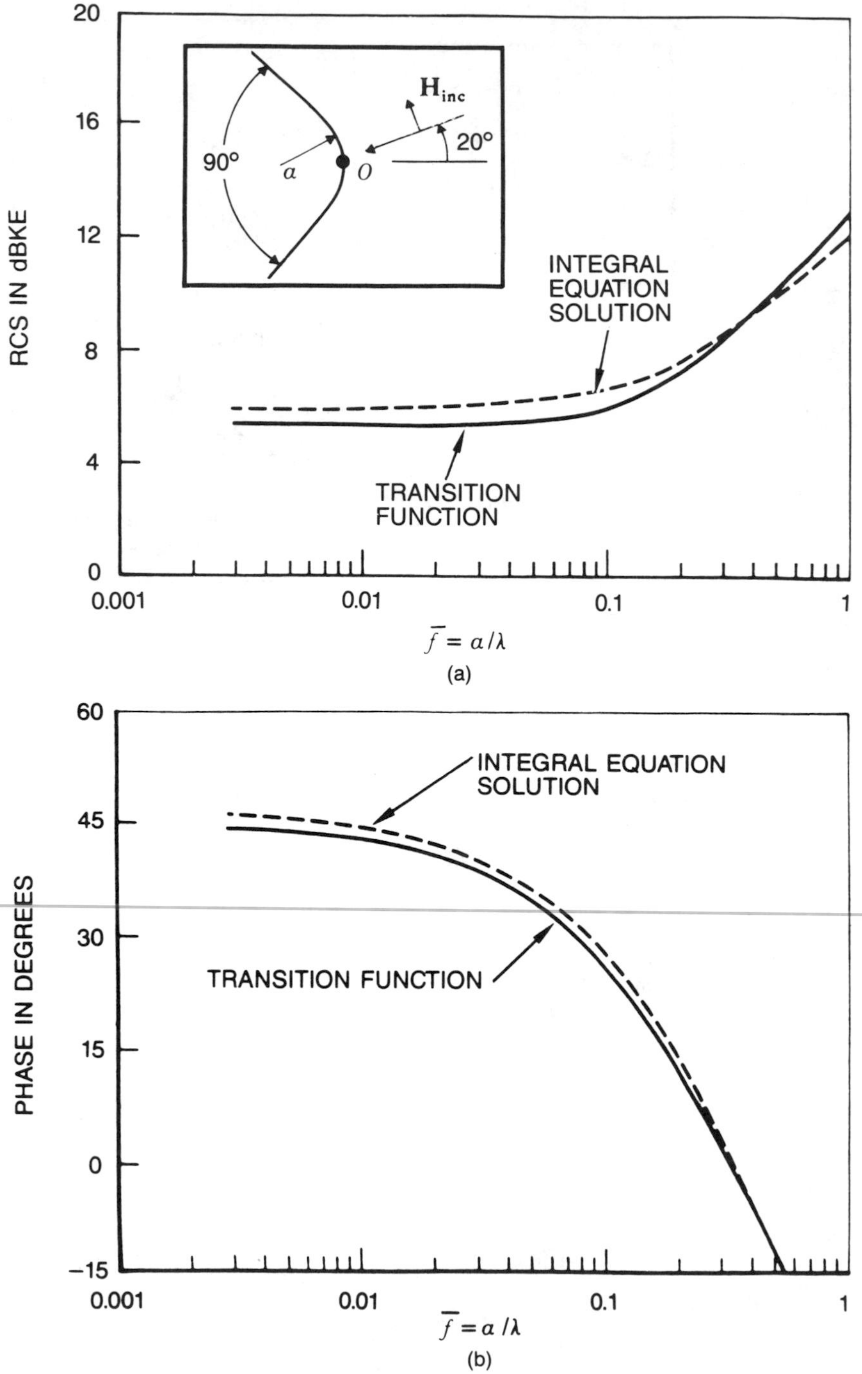

FIGURE 13.6. Backscatter from a blunted edge: 90° wedge angle; incidence 20° off bisector. (a) Normalized RCS σ_{dBKE}; (b) phase of the diffraction coefficient u.

13.2.3. *The Transition Function*

Figures 13.3–13.6 also show how well the numerical data for both RCS and phase match data calculated from the *transition function* formula

$$u = (u_{\mathrm{GO}}^2 + u_{\mathrm{K}}^2)^{1/2}. \tag{13.13}$$

Here u_{GO} is the *geometrical optics diffraction coefficient*, which characterizes the behavior of u in the high-frequency limit so that

$$u \to u_{\mathrm{GO}} \quad \text{as} \quad \bar{f} \to \infty, \tag{13.14}$$

and u_{K} is the *Keller diffraction coefficient* for the *equivalent sharp wedge* of Figure 13.1, which characterizes the behavior of u in the low-frequency limit so that

$$u \to u_{\mathrm{K}} \quad \text{as} \quad \bar{f} \to 0. \tag{13.15}$$

The small deviations between the two sets of results are consistent with the limitations we have already noted on the accuracy of the integral equation data. The systematic behavior of the deviations appears not to be of any significance but rather to be an artifact of the computational procedure, including the way the test body size is varied with $\bar{f}$.

13.2.3.1. The Geometrical Optics Term

In the geometrical optics (GO) theory, the far-field backscatter emanates from the specular line of Figure 13.1 with a magnitude that depends only on the local geometry near the specular line, that is, on the radius of curvature a. Thus the geometrical optics backscatter for the rounded wedge is the same as for the circular cylinder of radius a which has the rounding arc as part of its surface, and u_{GO} can be found from the well-known circular cylinder solution [2].

The expression for u_{GO} is

$$u_{\mathrm{GO}} = U_{\mathrm{GO}} \exp\{-i2\omega t_s\} = U_{\mathrm{GO}} \exp\{-i2\bar{\omega}\bar{t}_s\}, \tag{13.16}$$

where

$$U_{\mathrm{GO}} = \left(\frac{2\pi^2 a}{\lambda}\right)^{1/2} = (2\pi^2 \bar{f})^{1/2} \tag{13.17}$$

is the corresponding diffraction coefficient for phase normalized to an origin which lies on the specular line. In the phase term, which just compensates for the two-way time advance when the incident wave reaches the specular line before the z-axis, the one-way time advance t_s and its normalized form $\bar{t}_s$ are given by

$$t_s = \frac{a}{c}\bar{t}_s, \qquad \bar{t}_s = 2 \sin^2\frac{\theta}{2}. \tag{13.18}$$

It should be noted that there is no phase anomaly associated with the geometrical optics result. The far-field phase is completely explained by the path length and the phase reversal at the surface, where incident and scattered tangential electric fields must cancel.

The normalized RCS limit formulas corresponding to eq. (13.14) are

$$\sigma_{KE} \to 2\pi^2 \bar{f} \qquad \text{as} \quad \bar{f} \to \infty \tag{13.19}$$

and

$$\sigma_{dBKE} \to 10 \log_{10}(2\pi^2 \bar{f}) \qquad \text{as} \quad \bar{f} \to \infty. \tag{13.20}$$

Thus, in the geometrical optics limit, the normalized RCS increases with $\bar{f}$ and does not depend on any other parameters. This geometrical optics RCS is plotted in Figures 13.3(a), 13.5(a), and 13.7(a).

The limit formula for the phase of u in radians is

$$\arg u \to -2\bar{\omega}\bar{t}_s \qquad \text{as} \quad \bar{f} \to \infty \tag{13.21}$$

with the special case

$$\arg u \to 0 \qquad \text{as} \quad \bar{f} \to \infty \qquad \text{for} \quad \theta = 0. \tag{13.22}$$

13.2.3.2. The Sharp Wedge Term

The low-frequency limit of u is the Keller diffraction coefficient u_K [1], [3] for the equivalent sharp wedge of Figure 13.1, which has the same wedge angle and orientation as the original wedge but is translated so that its edge coincides with the specular line. The expression for u_K can be written in the form

$$u_K = U_K \exp\left\{i\frac{\pi}{4}\right\} \exp\{-i2\bar{\omega}\bar{t}_s\}. \tag{13.23}$$

Here the first two factors give the sharp wedge diffraction coefficient with phase normalized for an origin that lies on the edge. As in the case of u_{GO}, the phase shift of the third factor—with $\bar{t}_s$ given by eq. (13.18)—just compensates for the two-way time advance when the specular line (and thus the edge of the equivalent wedge) is displaced from the z-axis. The second factor provides the characteristic phase anomaly of $\pi/4$ radians (45°) for a sharp wedge, which has already been encountered in eq. (13.7) for the knife edge.

The first factor, U_K, is a function of ψ and θ but not of $\bar{f}$,

$$U_K = \left(\frac{1}{1 - \cos q\pi} + \frac{1}{\cos q\pi + \cos 2q\theta}\right) q \sin q\pi \tag{13.24}$$

with

$$q = \frac{\pi}{2\pi - \psi} \tag{13.25}$$

and the angles expressed here in radians. It can readily be verified that U_K is real and positive for all values of ψ and θ allowed by eqs. (13.1) and (13.3). Thus

$$|u_K| = U_K \tag{13.26}$$

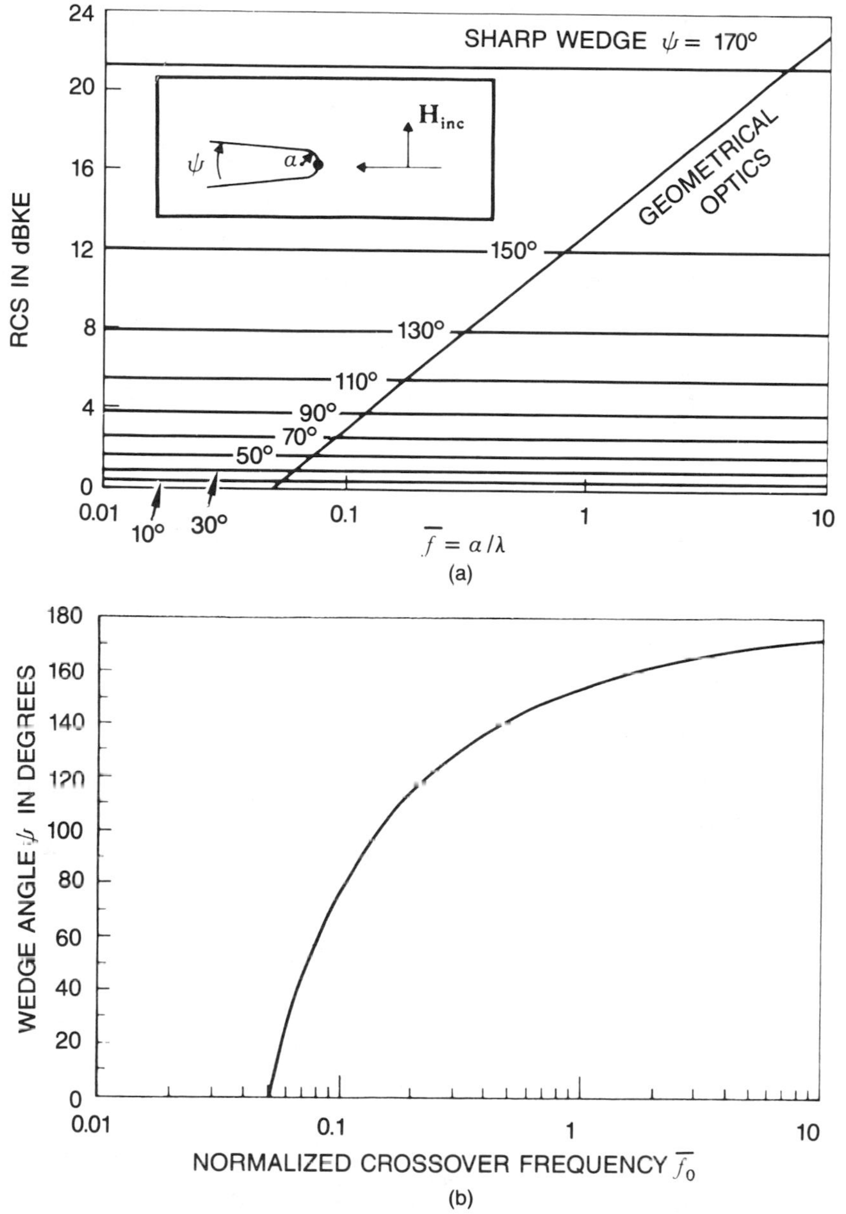

FIGURE 13.7. How the crossover frequency increases with wedge angle for incidence along the bisector. (a) The crossover phenomenon; (b) relationship between normalized crossover frequency $\bar{f}_0$ and wedge angle ψ.

and the normalized RCS limit formulas corresponding to eq. (13.15) are

$$\sigma_{\mathrm{KE}} \to U_{\mathrm{K}}^2 \qquad \text{as} \quad \bar{f} \to 0 \tag{13.27}$$

and

$$\sigma_{\mathrm{dBKE}} \to 20 \log_{10} U_{\mathrm{K}} \qquad \text{as} \quad \bar{f} \to 0. \tag{13.28}$$

The sharp wedge RCS is shown for various values of wedge angle in Figures 13.3(a), 13.5(a), and 13.7(a).

It is clear that the rounded wedge must have the same RCS as a sharp wedge at low enough frequency, but it was not obvious beforehand that the edge of the equivalent sharp wedge coincides with the specular line. That result is one of the conclusions of this study. Figures 13.3(b), 13.4(b), 13.5(b), and 13.6(b) show how accurately our transition function matches numerical phase data. Agreement is poorer with other choices of edge location, for example, at point P in Figure 13.1 where the edge would be if there were no rounding.

13.2.3.3. The Crossover Frequency

Substituting from eqs. (13.16) and (13.23) into eq. (13.13) gives

$$u = U \exp\{-i2\bar{\omega}\bar{t}_s\} \tag{13.29}$$

with

$$U = (U_{\mathrm{GO}}^2 + iU_{\mathrm{K}}^2)^{1/2}, \tag{13.30}$$

where U is the blunted edge diffraction coefficient with phase normalized for an origin which lies on the specular line. Both U_{GO} and U_{K} are real and positive, and thus the phase anomaly of U decreases monotonically from 45° at low frequency to 0° at high frequency. Equivalently, the phase anomaly at fixed frequency diminishes and eventually vanishes as the edge becomes blunter.

Some useful formulas for U are

$$U = (2\pi^2\bar{f} + iU_{\mathrm{K}}^2)^{1/2} \tag{13.31a}$$

$$= 2^{1/2}\pi(\bar{f} + i\bar{f}_0)^{1/2} \tag{13.31b}$$

$$= U_{\mathrm{K}} \exp\left\{i\frac{\pi}{4}\right\}\left(1 - i\frac{\bar{f}}{\bar{f}_0}\right)^{1/2}. \tag{13.31c}$$

Here

$$\bar{f}_0 = \frac{U_{\mathrm{K}}^2}{2\pi^2} \tag{13.32}$$

is the *normalized crossover frequency* at which the sharp wedge RCS equals the geometrical optics RCS, that is, the normalized frequency at which

$$U_{\mathrm{GO}} = U_{\mathrm{K}}. \tag{13.33}$$

It can be seen from the data of Figures 13.3–13.6, especially Figures 13.3(a) and 13.5(a) where the crossover is drawn in, that $\bar{f}_0$ lies in the middle of an *intermediate frequency range* over which the transition from wedge-like behavior to cylinder-like behavior takes place. At $\bar{f}_0$ the RCS is 1.5 dB above the value for a sharp wedge and the phase anomaly is 22.5°, halfway between the low- and high-frequency limits. Because U_K is a function of the wedge angle ψ and the incidence angle θ, so is $\bar{f}_0$. For the cases illustrated in Figures 13.3–13.6, $\bar{f}_0$ varies over roughly a 3:1 range.

The lower limit $\bar{f}_{0\,\min}$ of $\bar{f}_0$ is

$$\bar{f}_{0\,\min} = \frac{1}{2\pi^2} = 0.0507, \tag{13.34}$$

which occurs at grazing incidence to a knife edge, for which case

$$U_K = 1 \tag{13.35}$$

in conformity with eq. (13.7).

Figure 13.7 shows how $\bar{f}_0$ increases with ψ for $\theta = 0°$. This increase is without bound, that is,

$$\bar{f}_0 \to \infty \qquad \text{as} \quad \psi \to 180°. \tag{13.36}$$

As ψ increases, the angle of incidence approaches normal to both wedge faces, and this is what drives $\bar{f}_0$ higher. Indeed, $\bar{f}_0$ increases without bound whenever the angle of incidence approaches normal to a wedge face, that is,

$$\bar{f}_0 \to \infty \qquad \text{as} \quad \theta \to \left(90° - \frac{\psi}{2}\right). \tag{13.37}$$

The infinite limit for $\bar{f}_0$ arises, just as do the infinite limits we have already observed for u and for the minimum value of R at which the far-field approximation is satisfactory, because normal and near-normal incidence backscatter from the semi-infinite flat wedge face is so strong. For near-normal incidence backscatter, the effect of edge rounding is difficult to distinguish against the flat surface background scattering even at frequencies where a/λ is large and at distances of many wavelengths from the body. But for any fixed angle off normal, no matter how small, the specular scattering from the rounded edge will ultimately dominate at high enough frequency for observation at great enough distance from the body.

It should be noted that the validity of the transition function has been verified numerically only for small values of $\bar{f}_0$, indeed not for any value greater than 0.18, which corresponds to $\psi = 90°$ with $\theta = 20°$. For studying large values of $\bar{f}_0$—and for many other practical situations—it would be desirable to use a modified diffraction coefficient, obtainable by the procedures of Ufimtsev's physical theory of diffraction (PTD) [7], [8], which does not include the dominant scattering contribution from the flat wedge faces, namely the physical optics contribution.

13.2.3.4. Pulse Response

Because the transition function is defined at all frequencies, it can be used to find the pulse response. This response can be shown to be consistent with conditions for physical realizability whereas the responses corresponding to other closely related functional forms are not.

The pulse problem can be formulated by treating the time-harmonic waves of eqs. (13.2) and (13.6) as functions of a complex ω variable and integrating over a path in the ω-plane. It is thus found that, for an incident plane wave pulse

$$\mathbf{E}_{\text{inc}} = \int d\omega\, E_0(\omega)\mathbf{e}_z \exp\left\{-i\omega\left(t + \frac{x\cos\theta + y\sin\theta}{c}\right)\right\}, \tag{13.38}$$

the far-field scattering at time t and distance R from the origin is given by eq. (13.5) with

$$E = -\left(\frac{a}{2R}\right)^{1/2} \int d\omega\, Q\left(-i\frac{a}{c}\omega\right) E_0(\omega)\exp\{-i\omega T\} \tag{13.39}$$

and

$$T = t + 2t_s - \frac{R}{c}. \tag{13.40}$$

The quantity of primary interest here is the transfer function Q, which characterizes the impulse response of the body. This function is the product of U with the $\lambda^{1/2}$ term in eq. (13.6) and a normalizing factor independent of frequency. The argument of Q in eq. (13.39) can be recognized to be $-i\bar{\omega}$. If this is replaced by a normalized complex frequency variable $\bar{s}$, which is in conformity with standard Laplace transform notation, then Q takes the form

$$Q(\bar{s}) = \left(\frac{\bar{s} + 2\pi\bar{f}_0}{\bar{s}}\right)^{1/2}. \tag{13.41}$$

The branch points in $\bar{s}$-space lie on the real axis at the origin and at $-2\pi\bar{f}_0$, a configuration which assures that the pulse response is real.

The branch cut can be taken along the negative real axis between the branch points. It is then possible to construct arbitrarily large closed contours in the right half of $\bar{s}$-space which do not enclose any singularities. When the integration indicated in eq. (13.39) is performed around such contours for E_0 independent of frequency—which corresponds to an incident impulse—and $T < 0$, the integrals vanish. This is sufficient to assure that the impulse response does not travel faster than the speed of light, which in turn is sufficient to assure that Q corresponds to a causal system. (In order to proceed further and verify the conservation of energy required for a passive scattering problem, it would be necessary to consider the bistatic scattering. Rather than undertake this task, we shall simply note that there is nothing in the form of eq. (13.41) which is incompatible with the conservation of energy constraint.)

It can, in contrast, readily be confirmed that a diffraction coefficient of the form

$$u = (u_{GO}^p + u_K^p)^{1/p} \tag{13.42}$$

with p slightly greater or less than 2, which is difficult to distinguish from the transition function of eq. (13.13) on the basis of data fit, corresponds to a Q with branch points in the right halfplane. Such branch points, like right-halfplane poles, are not allowable for a causal scattering problem.

It can also be shown that a small shift in the position of the equivalent sharp wedge would move the branch point off the negative real axis, thus creating a nonphysical situation because complex branch points must occur in conjugate pairs. This is another demonstration of the key role of the phase anomaly, which establishes exactly the required phase relation between the geometrical optics and sharp wedge contributions to the transition function.

13.2.4. *Generalization to Related Problems*

The transition function can be generalized readily for application to a halfplane of finite thickness with a rounded edge and to a noncircular rounding arc.

13.2.4.1. Finite Thickness Halfplane

Figure 13.8 shows a halfplane of finite thickness $2a$ with the edge rounded off by a circular arc of radius a. This can be recognized as the limiting case $\psi = 0$ of the blunted edge configuration. The equivalent wedge with its edge at the specular point now becomes an equivalent knife edge. All the formulas given previously are still valid, with the magnitude U_K of the Keller diffraction coefficient simplifying to

$$U_K = \frac{\sec \theta + 1}{2}. \tag{13.43}$$

The crossover frequency at grazing incidence is given by eq. (13.34).

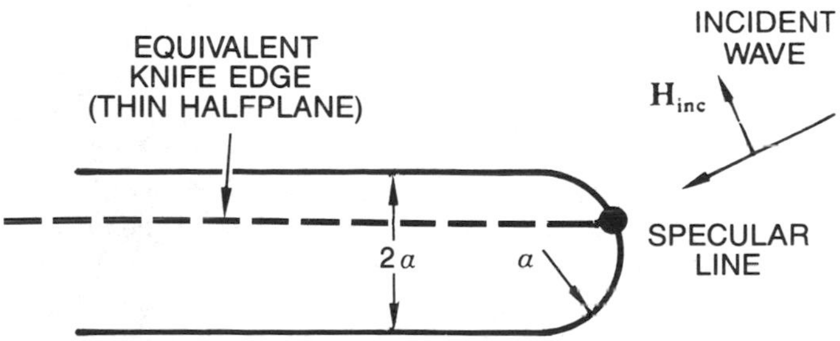

FIGURE 13.8. Thick halfplane with rounded edge.

13.2.4.2 Noncircular Rounding Arc

If, in either the wedge or the halfplane problem, the rounding is accomplished by some sufficiently smooth curve other than a circular arc, then the only modifications needed in the transition function formalism are to interpret a as the radius of curvature at the specular line—which makes a a function of θ—and to replace eq. (13.18) for the time advance by an expression appropriate to the new edge geometry.

13.3. Conclusions

The *transition function* of eq. (13.13)—further elaborated in eqs. (13.29)–(13.31) and various ancillary equations—is a *diffraction coefficient*, normalized *relative to a knife edge*, which describes the backscatter at all frequencies from a wedge with its edge blunted by a rounding arc when the direction of incidence is normal to the rounded surface along a *specular line* and the electric field **E** is parallel to the edge. Originally, this transition function was developed for the case of a circular rounding arc, but it has been generalized for application to noncircular rounding arcs and to a halfplane with a rounded edge. It has been shown that this transition function is compatible with basic physical realizability requirements, whereas formulas which appear very similar are nonphysical.

The transition function gives a smooth monotonic transition in magnitude and phase from wedge-like low-frequency behavior, characterized by the *Keller diffraction coefficient* for the *equivalent sharp wedge* of Figure 13.1, to high-frequency behavior characterized by the *geometrical optics diffraction coefficient*, which is a function of the *normalized frequency* $\bar{f} = a/\lambda$, with a the rounding arc radius of curvature (the value at the specular line in the case of noncircular rounding). The transition takes place across an *intermediate frequency range* surrounding the *crossover frequency*, the frequency at which the sharp wedge scattering is equal in magnitude to the geometrical optics curved surface scattering. The *normalized crossover frequency* $\bar{f}_0$ of eq. (13.32) and Figure 13.7, which is the value of $\bar{f}$ at crossover, can vary from as small as 0.0507 for grazing incidence on a thick halfplane to arbitrarily large.

A small value of $\bar{f}_0$ corresponds to a configuration for which the "high-frequency" geometrical optics theory remains a good approximation down to frequencies which are "low" in the sense that a/λ is small. Validity of geometrical optics down to small values of a/λ is also observed for a circular cylinder for the same polarization [2]. In both cases, geometrical optics eventually does fail as frequency is further decreased.

The transition function was determined empirically. We have verified it numerically for wedge angles of 10–90° with circular rounding arcs and incidence angles up to 20° off the bisector, and we expect it to be accurate over a much broader range of parameters. It would, of course, be desirable to find a mathematical derivation of the function. Concepts intimately related to ray optics, namely specular reflection and *phase anomaly*, have been shown here to be relevant at all frequencies, and this suggests that the modern generalization of ray theory developed by Maslov [9–11] might be applicable.

In closing, we recommend strongly to other investigators the use of normalization relative to a knife edge as for u in eq. (13.6), σ_{KE} in eq. (13.9), and σ_{dBKE} in eq. (13.10). This normalization has proven extremely helpful over the years in work with elongated scattering features. Note, for example, how the normalization results in the simple expression of eq. (13.11) for the RCS of a constant section feature of finite length. Another advantage is that σ_{KE} for a body of finite section is always finite in the zero frequency limit, which facilitates the use of quasi-statics, whereas the unnormalized σ_{2D} of eq. (13.8) can become infinite in this limit. Thus, for a thin circular cylinder of radius a and the polarization of interest here, we have [2]

$$\sigma_{KE} = \frac{\pi^2}{\ln^2(2\pi a/\lambda)}, \tag{13.44}$$

which vanishes in the limit, but

$$\sigma_{2D} = \frac{\pi\lambda}{2\ln^2(2\pi a/\lambda)}, \tag{13.45}$$

which becomes infinite.

Acknowledgments

The authors wish to thank Dr. Keith Glover, Mr. David Toth, and Miss Kathleen Nakao for their assistance with computations. The integral equation codes used in this study represent the cumulative efforts of many individuals at the Northrop Corporation over many years.

References

[1] J.B. Keller (1962), Geometrical theory of diffraction, *J. Opt. Soc. Amer.*, **52**, 116–130; reprinted in R.C. Hansen, Ed., *Geometric Theory of Diffraction*, IEEE Press, New York, 1981.

[2] J.J. Bowman, T.B.A. Senior, and P.L.E. Uslenghi, Eds. (1969), *Electromagnetic and Acoustic Scattering by Simple Shapes*, North-Holland, Amsterdam (Reprinted by Hemisphere, New York, 1987), Section 2.2.1, especially Equations (2.11), (2.31), and (2.32) and Figure 2.6.

[3] *Ibid.*, Section 6.2.1.

[4] *Ibid.*, Sections I.2.4, I.2.5.

[5] M. Born and E. Wolf (1986), *Principles of Optics*, 6th ed. (with corrections), Pergamon, Oxford, Section 8.8.4.

[6] G.T. Ruck, D.E. Barrick, W.D. Stuart, and C.K. Krichbaum (1970), *Radar Cross Section Handbook*, Plenum, New York, Volume 1, Section 4.3.1.3, especially Equation (4.3–43) as specialized to normal incidence backscatter.

[7] P.Ya. Ufimtsev (1962), *Metod Krayevykh Voiln v Fizicheskoy Teorii Difraktsii*, Sovetskoye Radio, Moscow. (Translated into English as *Method of Edge Waves in the Physical Theory of Diffraction*, United States Air Force Systems Command Foreign Technology Division Document FTD-HC-23-259-71, 1971.)

[8] E.F. Knott, J.F. Shaeffer, and M.T. Tuley (1985), *Radar Cross Section*, Artech House, Dedham, Section 5.7.

[9] V.P. Maslov (1972), *Theorie des Perturbations et Methodes Asymptotiques*, Dunod, Paris (Translation of *Teoria Vozmushchenii, Asimptoticheskie Metodi*, University of Moscow, 1965).

[10] V.P. Maslov and M.V. Fedoriuk (1981), *Semi-Classical Approximation in Quantum Mechanics*, Reidel, Dordrecht.

[11] R.W. Ziolkowski and G.A. Deschamps (1984), Asymptotic evaluation of high-frequency fields near a caustic: An introduction to Maslov's method, *Radio Science*, **19**, 1001–1025.

14
Electromagnetic Fields Determined from Two-Dimensional Infrared Thermal Patterns

JOHN D. NORGARD*

14.1. Introduction

An infrared measurement technique is described in this chapter which can be used to detect electromagnetic fields, both continuous wave and pulsed. This technique has been used to study the interaction of electromagnetic fields with conducting and lossy dielectric materials [1], [2]. This technique has been successfully applied at radio and microwave frequencies, and millimeter wavelengths [3], [4]. Of special interest has been the scattering/diffraction of electromagnetic waves from conducting objects, with complicated geometrical shapes, and the penetration of electromagnetic waves through small apertures in partially shielded enclosures.

The infrared technique involves placing a thin planar detection screen of lossy material in the region over which the electromagnetic field is to be mapped. The field is detected through the joule heating that occurs when the electromagnetic energy is absorbed by the screen material. When the local surface temperature of the screen rises to 0.1 K or higher above the ambient background temperature, the two-dimensional induced temperature distribution at the surface of the screen (which corresponds to the electromagnetic field intensities in the screen) can be detected with an infrared scanning system via emitted thermal radiation. These tests are normally performed in an anechoic chamber so that the ambient temperature can be easily controlled. The measured two-dimensional temperature profile in the screen can be directly calibrated to the electric and/or magnetic fields incident on the screen.

Continuous wave measurements by an infrared measurement technique have been demonstrated and reported over the past several years [5], [6]. While the technique requires a minimum energy deposition for sufficient heating, the electric and magnetic parameters of the detection screen can be selected, such that the thermal mass of the screen is reduced, allowing a fast

---

* Electromagnetics Laboratory, Department of Electrical and Computer Engineering, College of Engineering and Applied Science, University of Colorado, Colorado Springs, CO 80933-7150, USA.

minimally perturbing response. Infrared data acquisition to a high-speed memory has also been developed [7] to store approximately 500,000 pixels of a two-dimensional infrared image in less than three seconds. This corresponds to thirty 128 × 128 frames of data with each pixel element represented as an 8-bit word, which corresponds to the electric or magnetic field intensity at that location.

As a diagnostic tool, this technique can be used to support tests and evaluations of electronic systems in the presence of electromagnetic radiation, e.g., to determine the free-field environments around microwave sources (EMR) [8], [9], to determine the energy coupled into electronic circuits (EMI) through partially shielded enclosures [10], [11], to determine electromagnetic coupling between electronic components (EMC) [12], [13], and to verify hardening techniques. The near, far, and internal fields (HPM) [14], [15] generated by antennas and apertures can be mapped using thin lossy screens. Apertures in enclosures can be identified by placing a resistive coating on the surface of the metal in the area suspected of containing an aperture. Aperture distributions and interior modes coupled into cavities can also be determined [16], [17].

14.2. Infrared Measurement Technique

Electromagnetic fields are easily discernible with the infrared measurement technique. Relative intensities are immediately apparent on each scan of the detection screen. Absolute intensities of the electromagnetic fields can also be determined by calibrating the detection screen. A brief overview of the details of the measurement technique is given below.

14.2.1. *Detection Screen*

The infrared technique is based on Poynting's theorem for the absorption of electromagnetic energy in a lossy, complex material. In the frequency domain, the complex constitute parameters of the material (μ, ε, σ) are denoted by

$$\varepsilon(\omega) \equiv \varepsilon'(\omega) - j\varepsilon''(\omega), \tag{14.1}$$

$$\mu(\omega) \equiv \mu'(\omega) - j\mu''(\omega), \tag{14.2}$$

$$\sigma(\omega) = \sigma(\omega), \tag{14.3}$$

where $'$ ($''$) denotes the real (imaginary) part. As noted, these parameters are complex functions of the frequency of the incident electromagnetic wave.

Part of the incident energy is reflected from the surface of the material, some is transmitted through the material, and some is absorbed by the material. The absorbed energy is due to conductive losses (σ) and polarization (ε'') and magnetization (μ'') effects. The lossy material forms a detection screen for the absorbed electromagnetic energy.

For the special case of a large but thin planar screen being irradiated by an incident plane wave, the incident energy is simply related to the absorbed energy through Snell's and Fresnel's laws and the conservation of energy [18]. Specifically, the incident electric field intensity is related to the reflected, transmitted, and absorbed electric field intensities by the angles of incidence, reflection, and transmission and the reflection and transmission coefficients of the screen material. For a typical detection screen setup, the frequency-domain relationships have been developed for a lossy complex planar screen sandwiched between two other materials, usually air. These relationships have been presented elsewhere [19].

By Poynting's theorem, the absorbed power in a given volume (V) is a function of the square of the magnitudes of the electric (E) and magnetic (H) field intensities [20].

$$P_{\text{abs}} = \int_V (\sigma E^2 + \omega\varepsilon'' E^2 + \omega\mu'' H^2)\, dV, \tag{14.4}$$

where σ is the conductivity, ε'' is the imaginary permittivity, μ'' is the imaginary permeability of the detector material, and ω is the angular frequency of the incident electromagnetic radiation. It is, therefore, possible to relate surface temperature variations to E and H field intensities. As shown by the above equation, a material can absorb power via E and/or H field coupling.

The absorbed electromagnetic energy in the detection screen is converted into thermal heat energy. Part of this energy is reradiated as "blackbody" energy, which can be detected with a thermal infrared scanning system. For typical carbon-based or ferrite detection screens, the radiation can be approximated by a "gray body" radiator and also depends on the emissivity of the surface of the screen [21], [22]. The emissivity of the detection screen used in the experiments is measured. Some of the absorbed electromagnetic energy which is converted into heat energy is also convected and conducted into the surrounding materials, namely, the air around the surface of the screen and the support structure on which the screen is placed.

In general, the heat transfer problem in an electromagnetic heated homogeneous material (phase) involves solving a nonlinear second-order differential equation in both space and time, while considering radiative and convective heat losses from the phase surface, conductive heat transfer within the phase and to the surrounding surfaces in contact with the screen, and the electromagnetic power absorbed as a function of distance into the phase. For the case of the thin screens considered here, the temperature is initially considered to be constant within the phase, so that the conductive term inside the phase can be ignored, and the power absorbed is considered to be independent of the direction normal to the surface of the screen. To simplify the solution further, the conductive heat loss into the supporting structure can be made negligibly small by using a nonconducting support, e.g., a styrofoam base, which has very low thermal conductivity. Also, the time dependence will be ignored for a steady-state solution.

The problem reduces, therefore, to considering only the radiative and convective heat losses from the detector. The convective heat loss at object temperature T relative to an ambient background temperature T_{amb} is given as [22]

$$Q_c = \bar{h}(T - T_{amb})\ \mathrm{W/m^2}, \qquad (14.5)$$

where $\bar{h}$ varies between 1.4 and 1.6. The radiative loss is approximated by

$$Q_r = \varepsilon_s \sigma_0 (T^4 - T_{amb}^4)\ \mathrm{W/m^2}, \qquad (14.6)$$

where ε_s is the detector surface emissivity, σ_0 is the Stefan–Boltzman constant ($\mathrm{W/m^2 - K^4}$), and the temperatures are in degrees Kelvin. Relating eqs. (14.5) and (14.6) to the absorbed power results in the following equation at thermal equilibrium:

$$P_{abs} = Q_c + Q_r. \qquad (14.7)$$

The nonlinear eq. (14.7) can be solved for the surface temperature, T, using iterative techniques.

The theoretical infrared emission problem, therefore, reduces to the simultaneous solution of a nonlinear thermal radiation problem from a planar screen, with a corrective convection term, and an electromagnetic boundary-value problem of a plane wave incident on a lossy, complex planar screen. If the normal of the planar detection screen can be oriented in the direction of propagation of the incident plane wave, then Snell's and Fresnel's laws can be applied to the simple case of normal incidence.

14.2.2. *Calibration*

The detection screen is calibrated such that the temperature data is transformed to current densities or to electric field strengths ($J = \sigma E$). The infrared camera system provides data in the form of "ISU" or "isothermal" units; for observations of electromagnetic fields with surface temperatures which cover a wide range of values, correlation from ISU values to temperatures requires a few simple calculations in conjunction with equipment calibration. The infrared camera responds in a nearly linear fashion to small changes in object temperature (less than 5 °C in this chapter). The difference between final (steady-state, with electromagnetic field incident) and initial (no electromagnetic field incident) ISU values corresponds in an approximately linear fashion to the difference between their temperatures.

As an example, to develop a mathematical model for temperature increase as a function of electric field, the heating that occurs is treated as a resistive power loss, as in circuit theory. That is, the power dissipated by a volume carrying an electric current is $P = I^2R$ where I is the total current and R is the total resistance of the volume. For the case of electric currents in a carbon detection screen, the current density J is assumed to be uniform for small regions of the detection screen (corresponding to the area viewed by a single pixel of the infrared scanning system), and thus, can be treated as a constant rather than a vector.

Assuming that the power absorbed per unit area is dissipated primarily as heat, the temperature increase will be an approximately linear function of the power absorbed, and in turn, proportional to the square of the current density present. This implies that the increase in temperature ΔT can be modeled as

$$\Delta T = aJ^2, \tag{14.8}$$

where a is an unknown proportionality constant.

To account for imperfect agreement with such an idealized model, however, a linear correction term is included in the relation discussed above. Therefore, let

$$\Delta T = aJ^2 + bJ, \tag{14.9}$$

where b is an additional unknown proportionality constant. Using the quadratic equation,

$$J = \frac{-b + \sqrt{b^2 + 4a\Delta T}}{2a}. \tag{14.10}$$

Note here that "a" must be positive, as is ΔT; therefore, the second term under the radical is always positive (that is, the radical term is always greater than b). The positive root is also used to ensure that J (current density) is never negative.

Determination of the constants "a" and "b" is accomplished by applying an electric current through the carbon screen and measuring the resultant heating which occurs (in ISU units, which then provides a measurement of temperature). The data points obtained with this method are then used to determine the values of "a" and "b". It should be noted that the equation is constrained to intersect the point (0, 0), since zero electric current would result in zero temperature increase.

The experimental arrangement for performing this current-to-temperature correlation involves placing two parallel copper tape strips on the ends of the detection screen (to be used as electrodes). Silver paint is applied between the copper and carbon surfaces overlapping the junction to ensure that good electrical contact is made. With an identical piece of the carbon screen next to the test screen (to serve as a reference at ambient temperature), the current through the test screen is incrementally increased while the differential temperature levels between the test and reference screens are recorded by the infrared system. It is assumed that the current density is uniformly distributed throughout the carbon sheet. The final correlation to incident electric field magnitude can then be accomplished by relating J not only to the absorbed E, through the constitutive relations, but also to the incident electric field, E_0, through Snell's and Fresnel's laws. Thus, the surface differential temperature is functionally related to total sample current, current density, and power density.

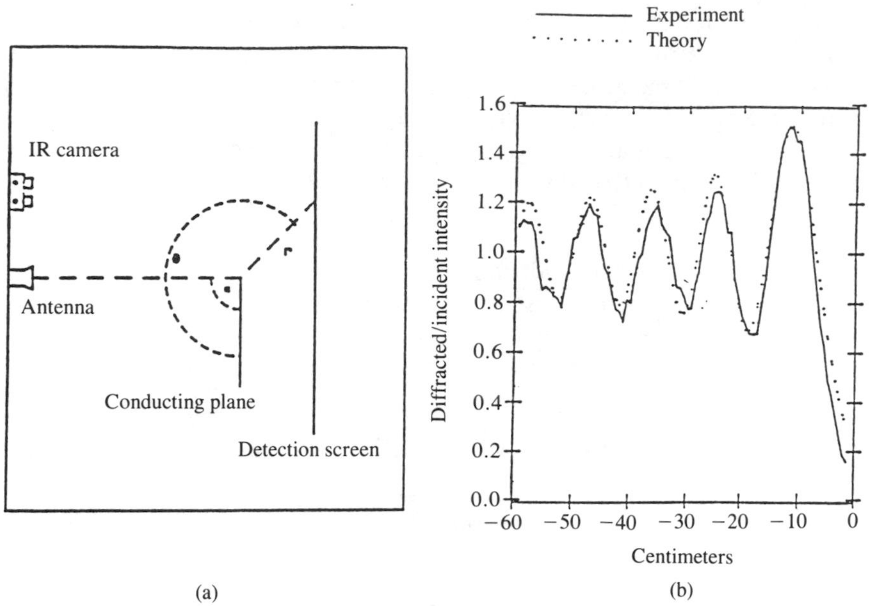

FIGURE 14.1. Fresnel diffraction from an edge. (a) Half-plane experiment; (b) theory and infrared results.

14.2.3. *Infrared Accuracy*

To demonstrate the accuracy of the infrared measurement technique, the classical problem of scattering from a conducting half-plane was studied. The experimental setup is shown in Figure 14.1(a).

The solution for the diffraction of a plane wave by a conducting half-plane was developed by Sommerfeld in 1896 [23]. This diffraction problem can be expressed exactly and simply in terms of the Fresnel integrals [24]. Sommerfeld's derivation used the *E*-polarized plane wave equation shown below. The parameters in the following equation are shown in Figure 14.1(a).

$$E_y = \exp(-jkr)\cos(\theta - \alpha), \tag{14.11}$$

where r is the distance from the edge of the conducting plane to the point on the detection screen, and θ is the angle that point makes to the plane of the conductor. For this example, $\alpha = 90°$, e.g., normal incidence on the conducting plane ($k = 2\pi/\lambda$).

In order to calculate the intensity at the screen, the Sommerfeld result is used for the *E*-field at the screen. Sommerfeld's result can be expressed in the form

$$E_y = \exp(-0.25j)\exp(jkr)[G(u) - G(v)]\sqrt{\pi}, \tag{14.12}$$

where

$$u = -\sqrt{2kr}\cos(0.5(\theta - \alpha)), \tag{14.13}$$

$$v = -\sqrt{2kr}\cos(0.5(\theta + \alpha)), \tag{14.14}$$

$$G(a) = \exp(-ja^2)F(a), \tag{14.15}$$

where "a" can represent u or v, and $F(a)$ is the complex form of the Fresnel cosine and sine integrals

$$F(a) = \int_a \cos(q^2)\,dq + j\int_a \sin(q^2)\,dq. \tag{14.16}$$

The Fresnel cosine and sine integrals, evaluated from 0 to a, can be solved by using the following series [24]

$$\int_0^x \cos(q^2)\,dq = \frac{x}{1!} - \frac{x^5}{5\cdot 2!} + \frac{x^9}{9\cdot 4!} - \cdots, \tag{14.17}$$

and

$$\int_0^x \sin(q^2)\,dq = \frac{x^3}{3\cdot 1!} - \frac{x^7}{7\cdot 3!} + \frac{x^{11}}{11\cdot 5!} - \cdots. \tag{14.18}$$

These series can only be evaluated for small values of x; however, numerical integration can be used to evaluate these integrals for large values of x.

In order to correlate this relationship to the experimental data, the intensity of the diffracted field is divided by the intensity of the free-field incident wave for the experiment. This is done so that the experimental data can be compared to the expression in (14.11) with an amplitude value of unity. The theoretical intensity relationship is then calculated by taking the square of the complex absolute value of (E_y). The theoretical values are shown in Figure 14.1(b).

The Infrared technique, in conjunction with computer processing, was also used to measure the interference pattern. The measured values are also shown in Figure 14.1(b). The detection screen used in this experiment was a carbon impregnated paper with a thickness of 80 μ, a conductivity of 8 mhos/meter, and a power transmission coefficient of 0.87. This paper has a negligibly small ε'' and μ''; and, therefore, heats due to E field coupling only through the σ term in eq. (14.4).

In Figure 14.1(b), the solid line is experimental data, and the dotted line is numerically generated theoretical data obtained using the exact Sommerfeld solution for the Fresnel diffraction problem. The error associated with the experimental data, due to thermal resistive effects (smearing) and screen nonuniformity, results in an error of approximately 6%. The Infrared measurement technique is, therefore, a relatively accurate method to measure electromagnetic field intensities.

14.2.4. *Infrared Advantages/Disadvantages*

The advantages of using an infrared measurement technique are:

(i) the technique is simple, accurate, and quick, and produces a contiguous two-dimensional distribution of the electromagnetic field;
(ii) the electromagnetic field can be due to objects with complicated geometrical shapes, since it is an experimental technique.

The disadvantages of infrared detection include limited dynamic range, response time limitations, energy dissipation through thermal conduction and convection, and the requirement for sufficient energy deposited in the observed medium. The range for the temperature distribution is a function of the material parameters, available energy, and sensitivity of the detection system. With multiple frame averaging, the discrimination in temperature between pixel elements can be as small as 0.02 K, and an observed temperature rise for a thin resistive material illuminated with an incident power density of 40 mW/cm^2 at 3 GHz has exceeded a 20 K temperature rise in less than 60 seconds for a dynamic range of 30 dB.

Standard probe measurements, having a greater dynamic range, are made in areas of interest identified initially through infrared detection. As thermal mass is increased or encapsulating layers are present at components such as resistors (circuit application), the response time to obtain a maximum steady temperature is increased. Enough information as to the locations of intense fields and current activity can, however, be obtained without achieving a steady-state condition. Caution should be exercised in the analysis of electromagnetic energy resulting in heat production, as the heat dissipation is accomplished through the thermal mechanisms of conduction and convection, as well as radiation, in the infrared detection range.

14.3. Applications

Of the several applications mentioned in the Introduction on the use of the infrared measurement technique, two of the most interesting problems are aperture penetration and coupling to interior modes of a cavity. The remainder of this chapter will concentrate on the theoretical aspects and experimental measurements associated with these related problems. Specifically, the effect of cavity and aperture geometry (resonances) on the energy coupled through a slot aperture in a hollow cylinder is investigated.

14.3.1. *Coupling Phenomenology*

Assessing the electromagnetic effects on electronic systems due to harsh radio-frequency environments requires both a detailed understanding of the interior coupling phenomena and the susceptibility or functional degradation of critical exposed components. Both of these fundamental aspects are essential in determining the survivability and vulnerability of electronic systems to intentional or unintentional radio-frequency energy throughout their operational lifetime. Determining the electromagnetic coupling and susceptibility of these complex systems is a formidable task involving extensive investigative procedures to fully characterize the electromagnetic effects on the system. The opportunity for the undesired radio-frequency energy to penetrate the operational system/equipment can be achieved either via "front"-door or "back"-door coupling. Front-door coupling involves entrance of the radio-frequency energy through antennas or sensor paths designed to deliberately pick up electromagnetic energy. These are readily understood and well-defined phenomena. Back-door coupling, on the other hand, involves the entrance or leakage of radio-frequency energy into a structure or system along paths made by discontinuities present in the conducting surface of the structure such as cracks, holes, seams, air intakes, etc. The uncertainties in identifying and defining the actual interior coupling paths are influenced by relatively minor changes in geometry (i.e., cavity shape, apertures, sizes, shapes, orientations) resulting in potentially wide differences in field distribution and wire coupling.

Predicting the interior field distribution remains a difficult problem for many analysis techniques, which tend to be restricted in defining elaborate details in system geometries and resource limited in solving complex system coupling problems. Therefore, it is essential that system coupling evaluations concentrate on radiation measurements to obtain interior field distributions and interior wire coupling data. To ensure that sufficient and adequate coupling characterization data is acquired, the overall coupling phenomena must be clearly understood. The ability to identify the coupling mechanisms can also play a vital role for system design alterations. As indicated earlier, the structural geometries are a major consideration in either mitigating or accenting the interior field distributions and wire coupling, and these must be quantified. The parameter space in which the susceptibility and coupling measurements are performed is limited. Selecting a finite set of parameters, such as frequency and aspect angle, requires an awareness of how and when aperture/cavity resonances occur. The resonance sharpness and corresponding field patterns assist in defining the appropriate resolution or interval in which to immerse the system so maximum coupling opportunities are not inadvertently missed. The maximum coupling areas can directly relate to possible system susceptibilities or functional degradations.

The infrared measurement technique can be used to map internal fields. These measurements can be used to determine the modes coupled to the interior of the cylinder and can be used to identify various coupling mechanisms. The effect of aperture resonances can be more readily identified, as well as the effects of complex geometries.

14.3.2. *Experimental Setup*

The experimental setup to measure the electromagnetic fields in an aperture and inside a empty circular cylindrical cavity is described below. These tests were performed in an anechoic chamber.

14.3.2.1. Cavity/Aperture Geometry

Consider the right circular cylinder shown in Figure 14.2. A cylinder was constructed with a thin slot aperture cut in its side parallel to the axis; one end was solid copper while the other was a copper mesh material. The wire mesh end cap acts as a solid shield at the microwave frequencies (1–18 GHz) of this experiment. While the wire mesh appears to be opaque to the energy at microwave wavelengths, it is still transparent to the energy at the infrared wavelengths, which are approximately 10,000 times shorter in length. The mesh size used was approximately 1.6 mm (1/16″) opening. A one-meter length of copper pipe, with a 10.16 cm (4″) inside diameter, was placed horizontally in an anechoic chamber on a block of styrofoam, as shown in Figure 14.2. A 7.62 cm (3″) long rectangular slot was cut into the side of the cylinder half-way between the ends of the cylinder, with the long axis of the slot parallel to the axis of the cylinder. The width of the slot was 0.64 mm (25 thousandths of an inch). This slot size was chosen to model a long thin seam or crack. A circular section of carbon paper of thickness 80 μ and conductivity 8 mhos/m was placed on a round plug of art board and placed inside the cylinder. The art board was used to stiffen the carbon paper and to keep it flat and to position the disk inside the cylinder. The carbon paper was trimmed off the outer edge of the disk so as not to touch the metal pipe and short out the induced charges or currents on the carbon paper. This disk was moved back and forth inside the cylinder to detect the circularly cylindrical electric fields inside the cavity

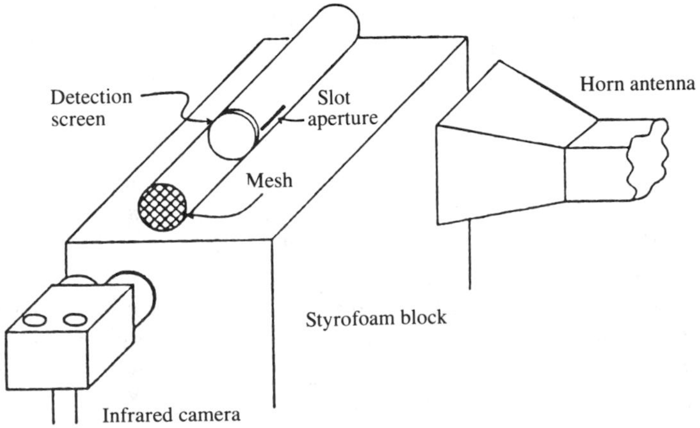

Figure 14.2. Experimental setup.

at any cross section of interest. The thickness and conductivity of the carbon paper were adjusted so the insertion of the infrared sensitive paper did not significantly change the internal field patterns by altering the resonant frequencies of the empty cavity.

14.3.2.2. Cylinder Illumination

The cylinder was initially illuminated in the microware region with the incident electric field oriented perpendicular to the thin aperture. The aperture was aligned on the bore sight of the horn pattern to produce maximum coupling. The frequency of the source was varied to produce a range of effects. Specifically, the cylinder was illuminated below and above the waveguide/cavity cutoff frequency. Also, the frequency was swept across several of the cylinder cavity resonant frequencies and the aperture resonance frequency.

14.3.2.3. Aperture Excitation

When an electromagnetic field penetrates an aperture, the coupled fields, which fringe inside the aperture, create equivalent electric and magnetic dipole moments, as depicted in Figure 14.3. These dipole moments then couple into and excite the various internal cavity modes which can exist (propagate) at the incident signal frequency. The strength of the coupling mechanism depends on the resonant frequencies of the interior and exterior of the cavity and the resonant frequencies of the aperture.

The onset of the internal cavity resonance is studied as it relates to the energy coupled through a small aperture. In Figures 14.4 and 14.5, a two-dimensional cross-sectional field thermogram is depicted at 1 cm from the center of the aperture, off-resonance and on-resonance, respectively. The presence of the dipole field due to the aperture moments is clearly evident.

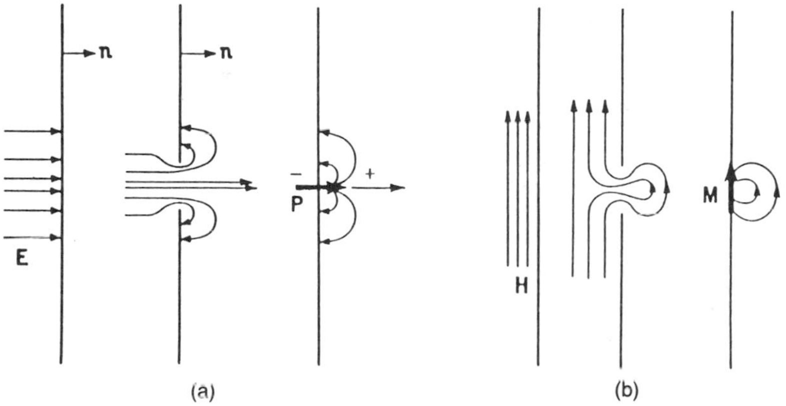

Figure 14.3. Electromagnetic field penetration through an aperture. (a) Electric dipole moment; (b) magnetic dipole moment.

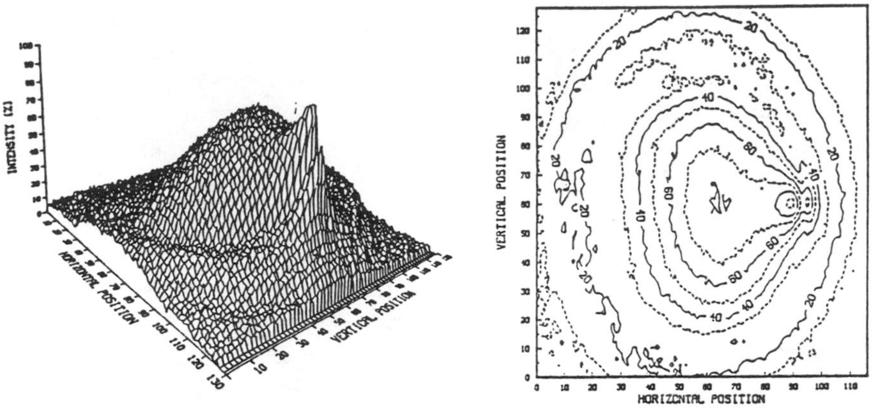

FIGURE 14.4. One centimeter off center of slot at 1.75 GHz.

The two circular areas indicate an aperture excitation of the cavity field mode in the cylinder. Several other cuts were taken (not shown) such that a three-dimensional electric field distribution could be obtained. The internal energy obtained off-resonance was much lower, and present only near the aperture, and seemingly uninteresting. The infrared results have also been compared with theoretical values for planar and cylindrical structures [25]. Even though an actual system may have a wire or other structure that could see this localized field, the frequency dependence issue is critical to a system vulnerability/susceptibility analysis.

Internal field mapping shows dramatically that the amount of energy coupled through an aperture into a given structure is strongly dependent on frequency. In particular, frequencies corresponding to the resonance of the

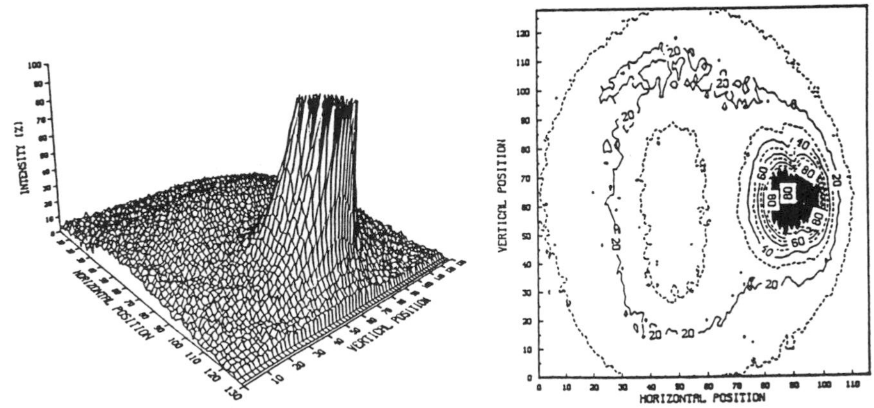

FIGURE 14.5. One centimeter off center of slot at 1.965 GHz.

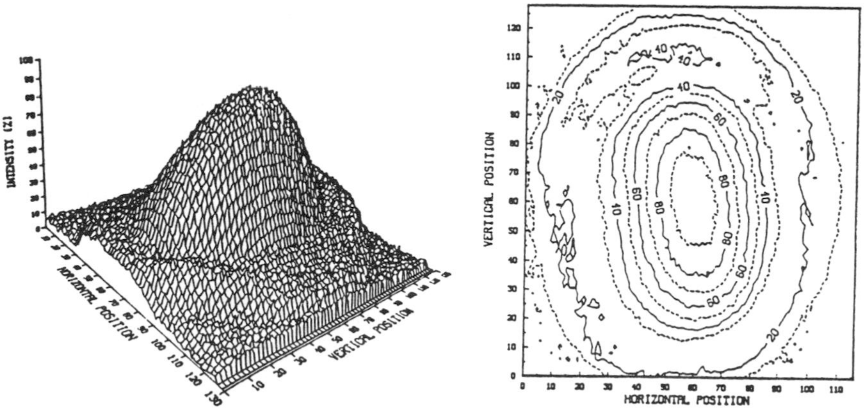

FIGURE 14.6. Four centimeters off center of slot at 1.75 GHz.

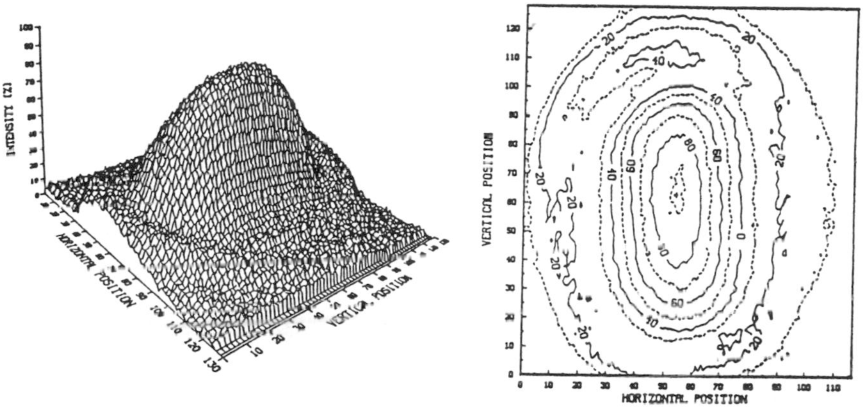

FIGURE 14.7. Four centimeters off center of slot at 1.965 GHz.

cavity and resonance of the aperture showed significant coupling into the cylinder, while those frequencies slightly off a resonance frequency show a greatly reduced internal field confined to the immediate region of the aperture. At points away from the aperture, as shown in Figures 14.6 and 14.7, only a pure circular cavity mode is excited, as expected. Which modes are excited, and with what intensity, is the subject of the next section of this chapter.

14.3.3. *Internally Excited Cavity Modes*

A modal expansion technique is now used to determine the modal coefficients of the electromagnetic fields coupled into a closed circular cylinder. The main interest is to determine which modes are excited in the cylindrical cavity, and the relative strengths of these modes. The coefficients are found by combining various orthogonality properties of the fields with experimental measurements made using an infrared technique. The measurements are made so that the magnitudes of the electric field components can be determined. The field distributions are obtained from the heating of lossy dielectric sheets in a particular cylindrical cross section of interest. The lossy material is striated to measure the magnitude of a given spatial electric field component.

14.3.3.1. Modal Extraction Technique

In order to detemine the electromagnetic fields coupled into the cylinder, Maxwell's equations must be solved in accordance with certain boundary conditions. This is most easily done with the Hertz potentials [26]. The electric and magnetic field intensity E and H can be derived from the electric and magnetic Hertz potentials Π^{e} and Π^{h} by

$$\mathscr{E} = \nabla \times \nabla \times \Pi^{e} - \zeta \nabla \times \Pi^{h}, \tag{14.19}$$

$$\mathscr{H} = \nabla \times \nabla \times \Pi^{h} - \eta \nabla \times \Pi^{e}, \tag{14.20}$$

where

$$\zeta \equiv \varepsilon \frac{\partial}{\partial t} + \sigma, \tag{14.21}$$

$$\eta \equiv \mu \frac{\partial}{\partial t}, \tag{14.22}$$

Π^{e} and Π^{h} are solutions to the wave equations

$$(\nabla^2 - \Gamma^2)\Pi^{e} = 0, \tag{14.23}$$

$$(\nabla^2 - \Gamma^2)\Pi^{h} = 0, \tag{14.24}$$

subject to the appropriate boundary conditions inside the cylinder and on the surface of the cylinder. In the above equations

$$\Gamma \equiv \mu\varepsilon \frac{\partial^2}{\partial t^2} + \mu\sigma \frac{\partial}{\partial t}. \tag{14.25}$$

For structures with axial symmetry, only the axial component of Π gives a nontrivial solution. For the cylindrical cavity, with symmetry in the $\hat{z}$ direction,

$$\mathscr{E} = -\hat{z}\nabla_t^2 \Pi_z^{e} + \frac{\partial}{\partial z}\nabla_t \Pi_z^{e} + \zeta \hat{z} \times \nabla_t \Pi_z^{h}, \tag{14.26}$$

$$\mathscr{H} = -\hat{z}\nabla_t^2 \Pi_z^{h} + \frac{\partial}{\partial z}\nabla_t \Pi_z^{h} - \eta \hat{z} \times \nabla_t \Pi_z^{e}, \tag{14.27}$$

where, in the frequency domain,

$$(\nabla^2 + k^2)\Pi_z^{e} = 0, \tag{14.28}$$

$$(\nabla^2 + k^2)\Pi_z^{h} = 0, \tag{14.29}$$

with $k \equiv \omega\sqrt{\mu\varepsilon}$.

In the cylindrical coordinate system of Figure 14.2,

$$\Pi_z^{\mathrm{e}} = \sum_m \sum_n \sum_i C_{mni}^{\mathrm{e}} J_m\left(\frac{R_{mn}\rho}{a}\right) \cos(m\varphi) \cos\left(\frac{i\pi z}{h}\right), \tag{14.30}$$

$$\Pi_z^{\mathrm{h}} = \sum_m \sum_n \sum_i C_{mni}^{\mathrm{h}} J_m\left(\frac{R'_{mn}\rho}{a}\right) \cos(m\varphi) \sin\left(\frac{i\pi z}{h}\right), \tag{14.31}$$

where R_{mn} and R'_{mn} are the nth roots of the Bessel function [24] and its derivative, respectively, of the first kind of order m, i.e., for $m = 0, 1, 2, \ldots$

$$J_m(R_{mn}) = 0 \qquad (n = 1, 2, 3, \ldots), \tag{14.32}$$

$$J'_m(R'_{mn}) = 0 \qquad (n = 1, 2, 3, \ldots). \tag{14.33}$$

The cylindrical components of the electric and magnetic fields are

$$E_z = \nabla_t^2 \Pi_z^{\mathrm{e}} = -C^{\mathrm{e}} J_m\left(\frac{R_{mn}\rho}{a}\right) \cos(m\varphi) \cos\left(\frac{i\pi z}{h}\right), \tag{14.34}$$

$$E_t = \nabla_z \nabla_t \Pi_z^{\mathrm{e}} + \zeta \hat{z} \times \nabla_t \Pi_z^{\mathrm{h}} = \hat{\rho} E_\rho + \hat{\varphi} E_\varphi, \tag{14.35}$$

where

$$\begin{aligned} E_\rho = {}& -C^{\mathrm{e}}\left(\frac{R_{mn}}{a}\right)\left(\frac{i\pi}{h}\right) J'_m\left(\frac{R_{mn}\rho}{a}\right) \cos(m\varphi) \sin\left(\frac{i\pi z}{h}\right) \\ & + j\omega\mu C^{\mathrm{h}} \frac{m}{\rho} J_m\left(\frac{R'_{mn}\rho}{a}\right) \sin(m\varphi) \sin\left(\frac{i\pi z}{h}\right), \end{aligned} \tag{14.36}$$

$$\begin{aligned} E_\varphi = {}& C^{\mathrm{e}}\left(\frac{m}{\rho}\right)\left(\frac{i\pi}{h}\right) J_m\left(\frac{R_{mn}\rho}{a}\right) \sin(m\varphi) \sin\left(\frac{i\pi z}{h}\right) \\ & + j\omega\mu C^{\mathrm{h}}\left(\frac{R'_{mn}}{a}\right) J'_m\left(\frac{R'_{mn}\rho}{a}\right) \cos(m\varphi) \sin\left(\frac{i\pi z}{h}\right). \end{aligned} \tag{14.37}$$

Similarly,

$$H_z = -\nabla_t^2 \Pi_z^{\mathrm{h}} = C^{\mathrm{h}} J_m\left(\frac{R'_{mn}\rho}{a}\right) \cos(m\varphi) \sin\left(\frac{i\pi z}{h}\right), \tag{14.38}$$

$$H_t = \nabla_z \nabla_t \Pi_z^{\mathrm{h}} - \eta \hat{z} \times \nabla_t \Pi_z^{\mathrm{e}} = \hat{\rho} H_\rho + \hat{\varphi} H_\varphi, \tag{14.39}$$

where

$$\begin{aligned} H_\rho = {}& C^{\mathrm{h}}\left(\frac{i\pi}{h}\right)\left(\frac{R'_{mn}}{a}\right) J'_m\left(\frac{R'_{mn}\rho}{a}\right) \cos(m\varphi) \cos\left(\frac{i\pi z}{h}\right) \\ & + j\omega\varepsilon C^{\mathrm{e}} \frac{m}{\rho} J_m\left(\frac{R_{mn}\rho}{a}\right) \sin(m\varphi) \cos\left(\frac{i\pi z}{h}\right), \end{aligned} \tag{14.40}$$

$$\begin{aligned} H_\varphi = {}& C^{\mathrm{h}} \frac{m}{\rho}\left(\frac{i\pi}{h}\right) J_m\left(\frac{R'_{mn}\rho}{a}\right) \sin(m\varphi) \cos\left(\frac{i\pi z}{h}\right) \\ & + j\omega\varepsilon C^{\mathrm{e}}\left(\frac{R_{mn}}{a}\right) J'_m\left(\frac{R_{mn}\rho}{a}\right) \cos(m\varphi) \cos\left(\frac{i\pi z}{h}\right). \end{aligned} \tag{14.41}$$

14.3.3.2. Orthogonality Conditions

The orthogonality properties of the eigenfunctions are direct consequences of the orthogonality properties of the field components. These properties permit the expansion of an arbitrary field in the cylinder in terms of the complete set of mode functions. This development follows the work of Borgnis and Papas [27].

If the eigenfunctions are nondegenerate, it can be shown that (let $p \equiv mni$)

$$\int_A \Pi^{\mathrm{e}}_{z,p}\Pi^{\mathrm{e}}_{z,q}\,dA = 0 \qquad (p \neq q), \tag{14.42}$$

$$\int_A \Pi^{\mathrm{h}}_{z,p}\pi^{\mathrm{h}}_{z,q}\,dA = 0 \qquad (p \neq q), \tag{14.43}$$

$$\int_A \nabla\Pi^{\mathrm{e}}_{z,p}\cdot\nabla\Pi^{\mathrm{e}}_{z,q}\,dA = 0 \qquad (p \neq q), \tag{14.44}$$

$$\int_A \nabla\Pi^{\mathrm{h}}_{z,p}\cdot\nabla\Pi^{\mathrm{h}}_{z,q}\,dA = 0 \qquad (p \neq q), \tag{14.45}$$

$$\int_A \hat{z}\cdot(\nabla\Pi^{\mathrm{e}}_{z,p}\times\nabla\Pi^{\mathrm{h}}_{z,q})\,dA = 0 \qquad (p \neq q), \tag{14.46}$$

where the integrations are extended over a cross section A of the cylinder. These orthogonality relationships yield the following orthogonality relationships for the fields, for modes p and q:

$$\int_A E^{\mathrm{e}}_{z,p}E^{\mathrm{e}*}_{z,q}\,dA = 0 \qquad (p \neq q), \tag{14.47}$$

$$\int_A H^{\mathrm{h}}_{z,p}H^{\mathrm{h}*}_{z,q}\,dA = 0 \qquad (p \neq q), \tag{14.48}$$

$$\int_A E^{\mathrm{e}}_{t,p}\cdot E^{\mathrm{e}*}_{t,q}\,dA = 0 \qquad (p \neq q), \tag{14.49}$$

$$\int_A H^{\mathrm{h}}_{t,p}\cdot H^{\mathrm{h}*}_{t,q}\,dA = 0 \qquad (p \neq q), \tag{14.50}$$

$$\int_A E^{\mathrm{h}}_{t,p}\cdot E^{\mathrm{h}*}_{t,q}\,dA = 0 \qquad (p \neq q), \tag{14.51}$$

$$\int_A H^{\mathrm{e}}_{t,p}\cdot H^{\mathrm{e}*}_{t,q}\,dA = 0 \qquad (p \neq q), \tag{14.52}$$

$$\int_A \hat{z}\cdot(E^{\mathrm{e}}_{t,p}\times H^{\mathrm{e}*}_{t,q})\,dA = 0 \qquad (p \neq q), \tag{14.53}$$

$$\int_A \hat{z}\cdot(E^{\text{h}}_{t,p} \times H^{\text{h}*}_{t,q})\,dA = 0 \qquad (p \neq q), \tag{14.54}$$

$$\int_A \hat{z}\cdot(E^{\text{e}}_{t,p} \times H^{\text{h}*}_{t,q})\,dA = 0 \qquad (p \neq q), \tag{14.55}$$

where * stands for complex conjugate.

The eigenfunctions are normalized as follows:

$$\int_A \nabla\Pi^{\text{e}}_{z,p}\cdot\nabla\Pi^{\text{e}*}_{z,q}\,dA = 1, \tag{14.56}$$

$$\int_A \nabla\Pi^{\text{h}}_{z,p}\cdot\nabla\Pi^{\text{h}*}_{z,q}\,dA = 1. \tag{14.57}$$

For $p = q$, this normalization gives

$$\int_A |\Pi^{\text{e}}_{z,p}|^2\,dA = \frac{1}{k_p^2}, \tag{14.58}$$

$$\int_A |\Pi^{\text{h}}_{z,p}|^2\,dA = \frac{1}{k_p^2}. \tag{14.59}$$

This yields

$$\int_A E^{\text{e}}_{t,p}\cdot E^{\text{e}*}_{t,p}\,dA = k_z^2, \tag{14.60}$$

$$\int_A E^{\text{h}}_{t,p}\cdot E^{\text{h}*}_{t,p}\,dA = 1, \tag{14.61}$$

$$\int_A H^{\text{e}}_{t,p}\cdot H^{\text{e}*}_{t,p}\,dA = 1, \tag{14.62}$$

$$\int_A H^{\text{h}}_{t,p}\cdot H^{\text{h}*}_{t,p}\,dA = k_z^2, \tag{14.63}$$

$$\int_A E^{\text{e}}_{z,p}\cdot E^{\text{e}*}_{z,p}\,dA = 1, \tag{14.64}$$

$$\int_A H^{\text{h}}_{z,p}\cdot H^{\text{h}*}_{z,p}\,dA = 1. \tag{14.65}$$

The unknown and arbitrary electric and magnetic fields E and H within the cavity can be expanded in terms of the complex set of mode functions.

$$\mathbf{E} = \sum_{p=1}^{\infty} [A_p(\mathbf{E}^{\text{e}}_{t,p} + \hat{z}E^{\text{e}}_z) + B_p\mathbf{E}^{\text{h}}_{t,p}], \tag{14.66}$$

$$\mathbf{H} = \sum_{p=1}^{\infty} [B_p(\mathbf{H}^{\text{h}}_{t,p} + \hat{z}\mathbf{H}^{\text{h}}_z) + A_p\mathbf{H}^{\text{e}}_{t,p}], \tag{14.67}$$

where A_p and B_p are expansion coefficients given by

$$A_p = \frac{\int_A \mathbf{E} \cdot \mathbf{E}_{t,p}^{e*}\, dA}{\int_A \mathbf{E}_{t,p}^{e} \cdot \mathbf{E}_{t,p}^{e*}\, dA} = \frac{1}{k_z^2} \int_A \mathbf{E}_t \cdot \mathbf{E}_{t,p}^{e*}\, dA, \tag{14.68}$$

$$B_p = \frac{\int_A \mathbf{E} \cdot \mathbf{E}_{t,p}^{h*}\, dA}{\int_A \mathbf{E}_{t,p}^{h} \cdot \mathbf{E}_{t,p}^{h*}\, dA} = \int_A \mathbf{E}_t \cdot \mathbf{E}_{t,p}^{h*}\, dA. \tag{14.69}$$

From a knowledge of the transverse part of the electric field vector ($\mathbf{E}_t$) over any cross section of the cylinder, the amplitudes of the electric and magnetic modes can, therefore, be determined. This tangential component of the electric field can be measured with the infrared technique.

In the experimental work, the coefficients A_p ($p = 1, 2, 3, \ldots, m$) and coefficients B_p ($q = 1, 2, 3, \ldots, n$) are normalized so that

$$\sqrt{A_1^2 + A_2^2 + \cdots + A_m^2 + B_1^2 + B_2^2 + \cdots + B_n^2} = 1. \tag{14.70}$$

14.3.3.3. Modal Coefficients

The modal expansion coefficients can be determined in conjunction with data obtained using the infrared detection technique. If the dot products inside the integrals are expanded into their separate components

$$A_p = \frac{1}{k_z^2} \int_\rho \int_\varphi (E_\rho E_{\rho,p}^{e*} + E_\varphi E_{\varphi,p}^{e*}) \rho\, d\rho\, d\varphi, \tag{14.71}$$

$$B_p = \frac{1}{\omega^2 \mu^2} \int_\rho \int_\varphi (E_\rho E_{\rho,p}^{h*} + E_\varphi E_{\varphi,p}^{h*}) \rho\, d\rho\, d\varphi. \tag{14.72}$$

Since the data obtained from the infrared camera system are in rectangular coordinates, it is convenient to convert from cylindrical components (ρ and φ) to rectangular components (x and y), giving

$$A_p = \frac{1}{k_z^2} \int_x \int_y (E_x E_{x,p}^{e*} + E_y E_{y,p}^{e*})\, dx\, dy, \tag{14.73}$$

$$B_p = \frac{1}{\omega^2 \mu^2} \int_x \int_y (E_x E_{x,p}^{h*} + E_y E_{y,p}^{h*})\, dx\, dy, \tag{14.74}$$

where

$$E_{x,p}^{e*} = E_{\rho,p}^{e*} \cos\varphi - E_{\varphi,p}^{e*} \sin\varphi, \tag{14.75}$$

$$E_{y,p}^{e*} = E_{\rho,p}^{e*} \sin\varphi + E_{\varphi,p}^{e*} \cos\varphi, \tag{14.76}$$

$$E_{x,p}^{h*} = E_{\rho,p}^{h*} \cos\varphi - E_{\varphi,p}^{h*} \sin\varphi, \tag{14.77}$$

$$E_{y,p}^{h*} = E_{\rho,p}^{h*} \sin\varphi + E_{\varphi,p}^{h*} \cos\varphi, \tag{14.78}$$

where E_t^h is purely imaginary. If the pixels given by the infrared camera system are taken to be the differential areas of integration, S_{xy}, then the above integrals can be approximated by the following summations:

$$A_p = \frac{1}{k_z^2} \sum_x \sum_y S_{xy}(E_x E_{x,p}^{e*} + E_y E_{y,p}^{e*}), \tag{14.79}$$

$$B_p = \frac{1}{\omega^2 \mu^2} \sum_x \sum_y S_{xy}(E_x E_{x,p}^{h*} + E_y E_{y,p}^{h*}), \tag{14.80}$$

where the summations are carried out so that only the cross sections of interest are included. These summations are readily implemented on the computer.

A few of the lowest-order cylindrical modes which can exist inside the cavity are shown in Figures 14.8–14.15. Gray-scale images, along with surface and contour plots, of the $H_{11}, H_{21}, H_{01}, H_{31}, E_{01}, E_{21}, E_{02}, E_{12}$ modes are shown.

14.3.3.4. Experimental Data

A circular detection screen of carbon paper strips of thickness 80 μ and conductivity 8 mhos/m was placed on a round plug of styrofoam and placed inside the cylinder. The carbon paper was cut into strips so that significant currents can be induced in only one direction, thus allowing detection of specific components of the electric field. For simplicity, the strips were placed on the styrofoam parallel to one another, as shown in Figure 14.16, so that the rectangular components of the electric field (E_x and E_y) could be detected.

The cylinder was illuminated in the microwave region of 1–4 GHz. The cavity waveguide mode was excited below and above cutoff. Also, the frequency was swept across several of the cylinder cavity resonant frequencies and the aperture resonance frequency. For the hollow cylinder tested, the cavity resonances are 1.75 GHz, 2.9 GHz, and 3.6 GHz corresponding to the H_{11}, H_{21}, and H_{01} modes. The aperture is resonant at 1.965 GHz and 3.93 GHz, corresponding to one-half and full wavelength resonances, respectively. The cylinder was illuminated with the frequencies 1.75 GHz, 1.965 GHz, and 2.9 GHz.

The detection screen was placed in several locations in the cylindrical cavity with the strips oriented first in the $\hat{x}$ direction (horizontally) and then in the $\hat{y}$ direction (vertically). The cylinder was then illuminated at various frequencies, and the temperature changes in the detection screen were recorded with the infrared camera system. The center region of the data collected on the detection screen was then mapped to a predefined circle. The thermogram results are shown in the grey-scale images and surface and contour plots of Figures 14.17–14.19. These figures contain one frame of raw data. Multiple frame averaging smooths the data; convective corrections eliminate asymmetries in the figures due to heat flow across the cross section of the screen. In addition, for each set of corresponding $\hat{x}$ and $\hat{y}$ data files, the magnitude of the transverse response was calculated by taking the square root of the sum of the squares of the corresponding data points; the grey-scale images and

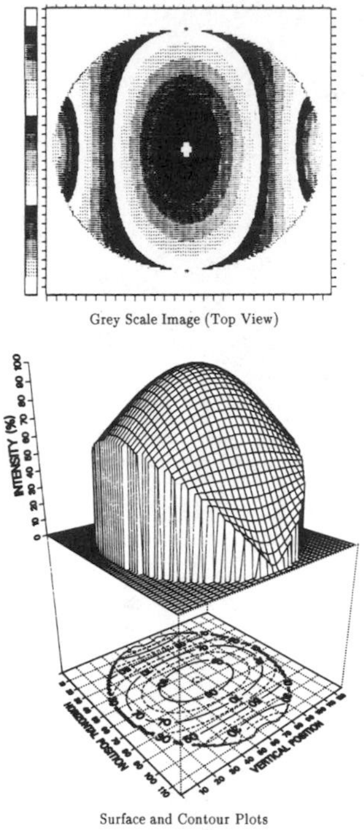

FIGURE 14.8. Theoretical prediction of the H_{11} mode.

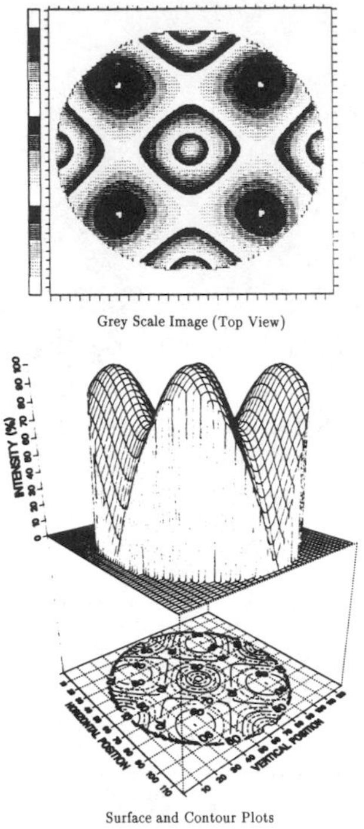

FIGURE 14.9. Theoretical prediction of the H_{21} mode.

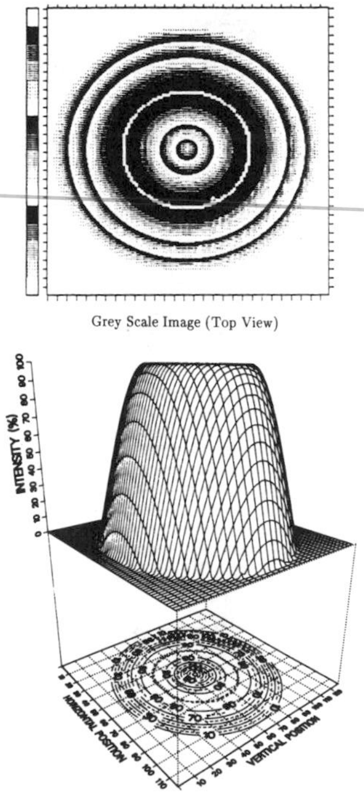

FIGURE 14.10. Theoretical prediction of the H_{01} mode.

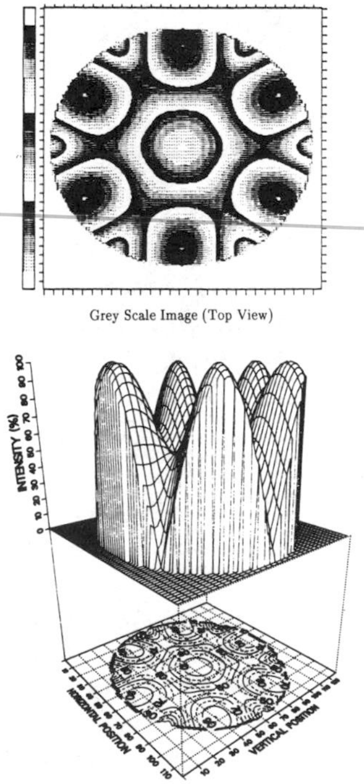

FIGURE 14.11. Theoretical prediction of the H_{31} mode.

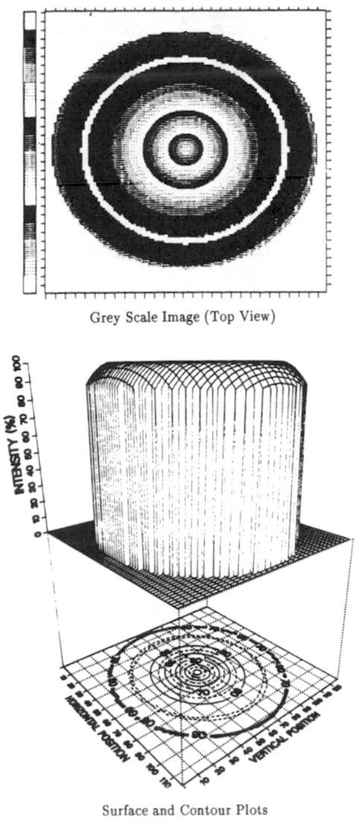

FIGURE 14.12. Theorctical prediction of the E_{01} mode.

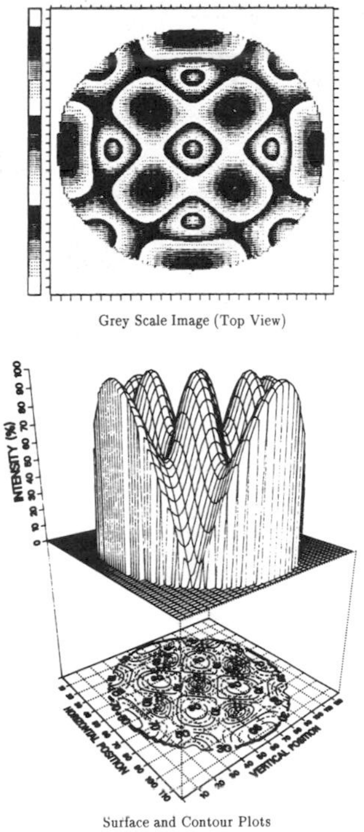

FIGURE 14.13. Theoretical prediction of the E_{21} mode.

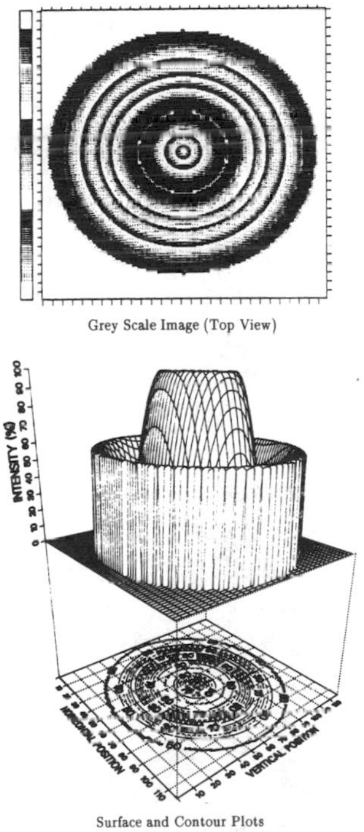

FIGURE 14.14. Theoretical prediction of the E_{02} mode.

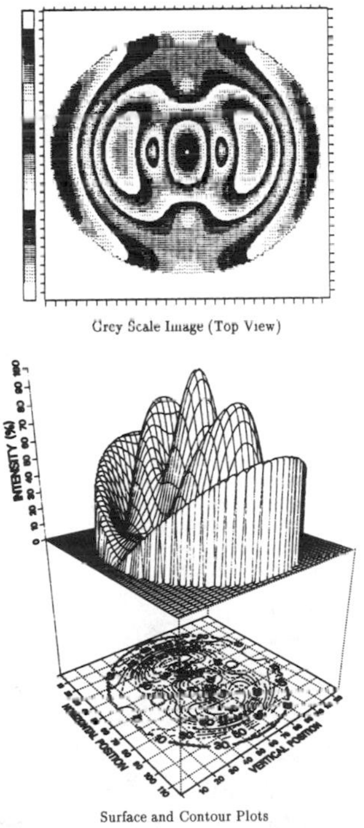

FIGURE 14.15. Theoretical prediction of the E_{12} mode.

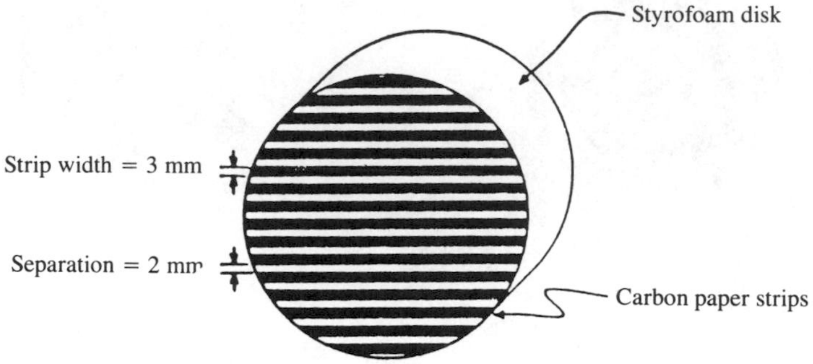

FIGURE 14.16. Detection screen of carbon paper strips.

surface and contour plots of these thermograms are also included in Figures 14.17–14.19. It is very useful to compare the grey-scale images of the measured transverse responses to the grey-scale images of the theoretically predicted mode patterns; this can provide an indication of the dominant mode present for a given experiment. The surface and contour plots serve the purpose of clarifying the intensity levels of the grey-scale images.

The infrared data associated with Figures 14.17–14.19 were used to determine the modal coefficients. The data were tested for both electric and magnetic modes whose frequencies were less than or equal to the incident frequency. For the data of Figure 14.17(a) and (b) only the dominant H_{11} mode was found, as would be expected. For the data of Figure 14.18(a) and (b), the H_{11} mode was again the only mode found to exist. For the data of Figure 14.19(a) and (b), the E_{01}, H_{11}, and H_{21} modes were found to be present with normalized coefficients $C_{01}^{e} = 0.0435$, $C_{11}^{h} \cong 0$, and $C_{21}^{h} = 0.999$, respectively. This indicates that the H_{21} mode is the main higher-order mode coupled into the cylinder from an incident wave of 2.9 GHz, as expected.

14.4. Summary

In this chapter, an infrared measurement technique was described that can be used to map electromagnetic fields and their interaction with conducting objects. This technique is nondestructive and nonperturbing and can be used to observe the effects of electromagnetic coupling of energy to the exterior or interior of complicated geometrical structures. The applications, advantages, and disadvantages of this new infrared technology were discussed.

As an example, the internal field distribution from a plane wave illumination of a hollow cylinder having a narrow, slot aperture was presented using an infrared detection technique. The technique used a thin lossy carbon material placed in the cross section of the cylinder. The induced temperature distribution was observed by an infrared scanning system through a wire mesh end

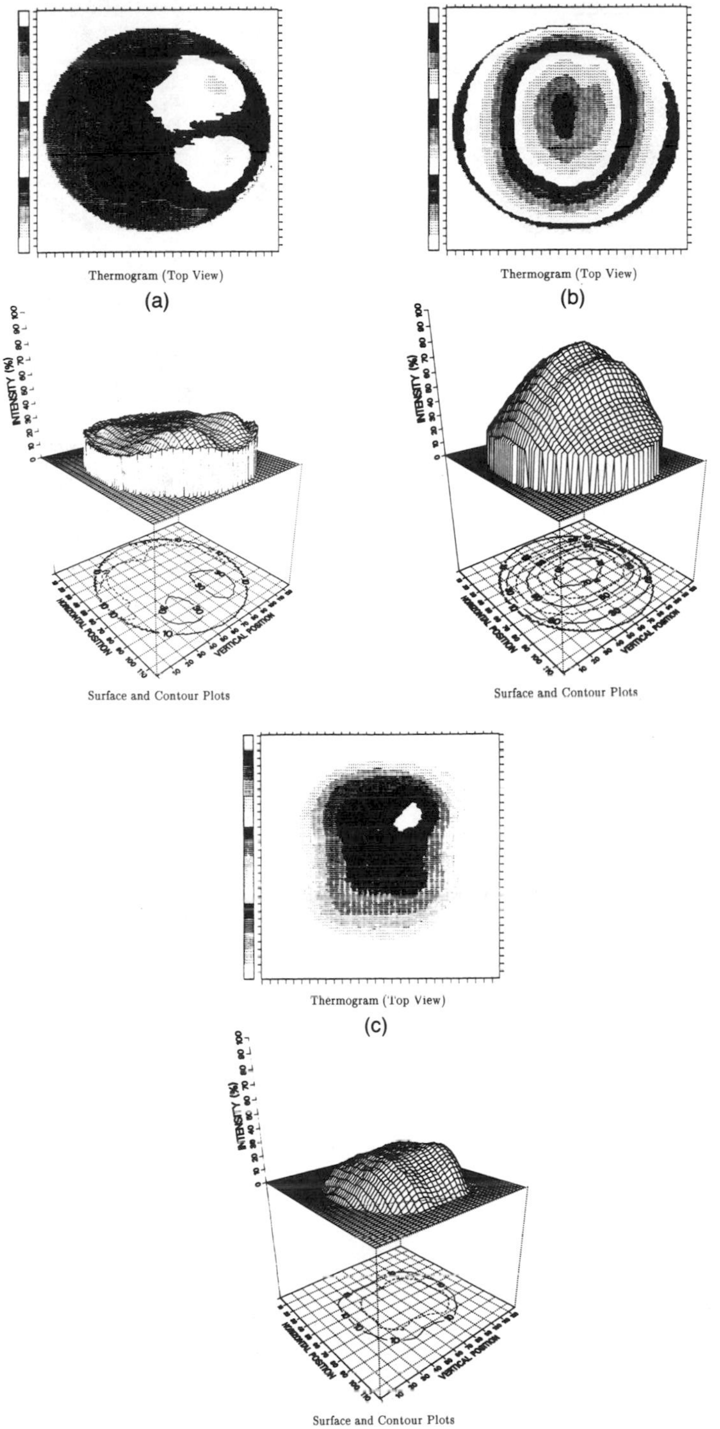

FIGURE 14.17. Heating pattern for the detection screen positioned at the edge of the aperture at 1.75 GHz. (a) $\hat{x}$ oriented; (b) $\hat{y}$ oriented; (c) transverse.

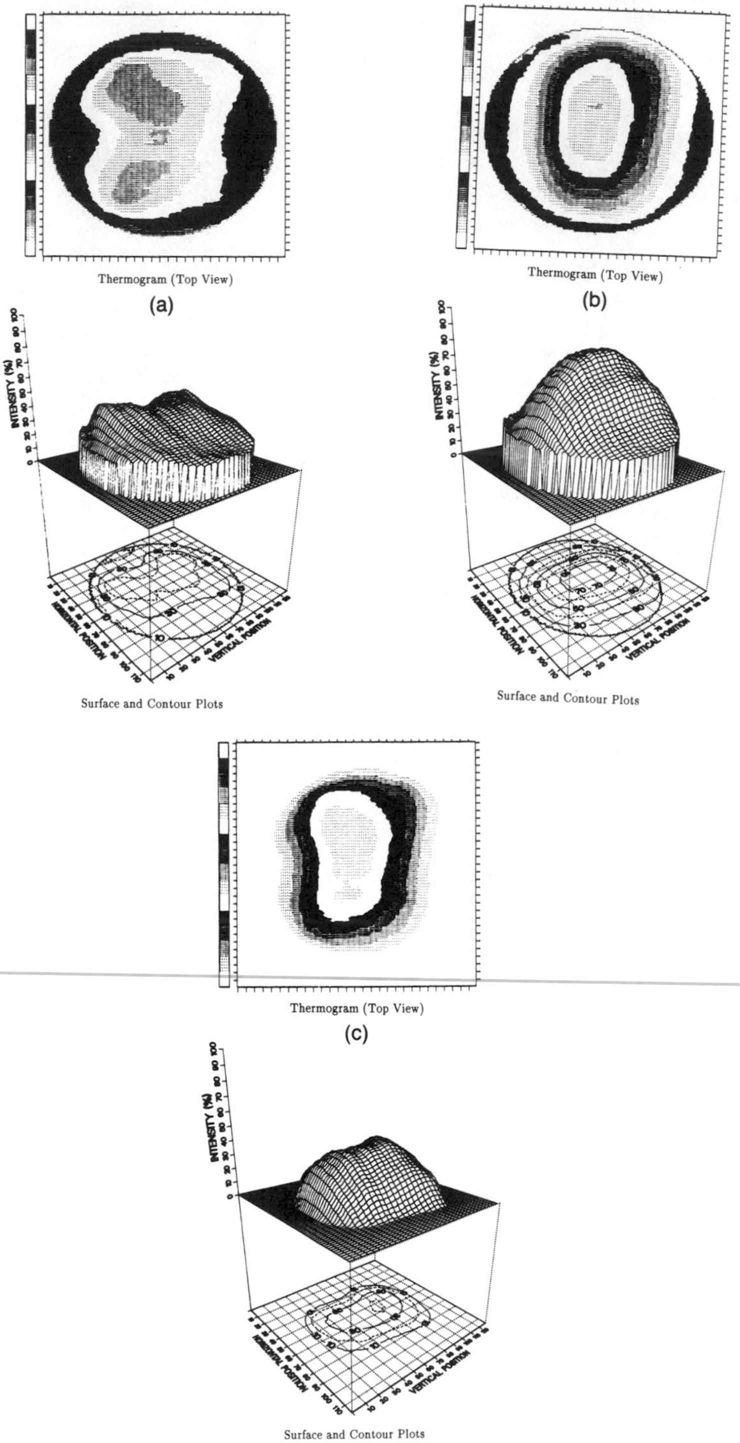

FIGURE 14.18. Heating pattern for the detection screen positioned at the edge of the aperture at 1.96 GHz. (a) $\hat{x}$ oriented; (b) $\hat{y}$ oriented; (c) transverse.

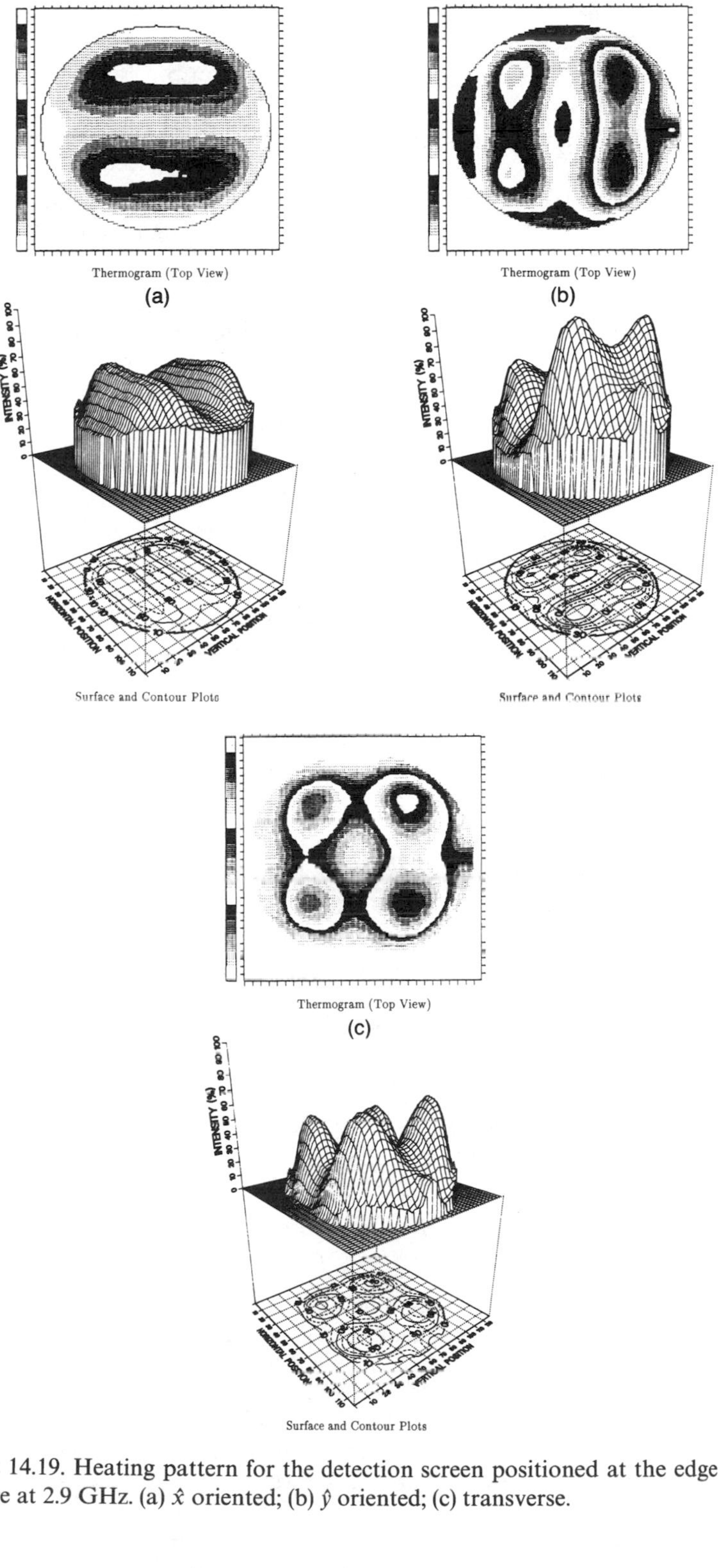

FIGURE 14.19. Heating pattern for the detection screen positioned at the edge of the aperture at 2.9 GHz. (a) $\hat{x}$ oriented; (b) $\hat{y}$ oriented; (c) transverse.

cap. This end cap preserved the electromagnetic continuity of the problem while allowing the infrared information from the interior of the cylinder to be transmitted through the wire mesh. The internal electric fields were mapped, the cavity and aperture resonances were identified, and typical results were presented. The incident electromagnetic wave frequency was scanned to locate the resonance frequencies of the cavity and aperture and to determine the sharpness of the peak intensities. The calibration of temperature to electric field was also described.

The implications of this work point to a need for understanding the coupling phenomenology for a particular structure in terms of various system resonance frequencies. Future work will involve cylinders like those described above with the addition of internal structures, such as boxes and wires. Where feasible, the positions for the detection screen will be identical with that of the hollow cylinder. The aperture resonance frequency is expected to be the same but the internal field will be altered. The cavity resonances should be shifted in frequency, strength, and sharpness of on-set. Particular emphasis will be placed on decoupling the cavity excitation from the propagating mode.

This work describes one step in achieving confidence in characterizing the susceptibility response of systems/equipment involving the introduction of wire/bundle assemblies to ascertain the actual field/wire coupling at terminations. This data can then be compared to actual system susceptibility data. Other factors in the susceptibility investigations involve achieving high field levels on test articles. In many situations, the distance between the test article and radio-frequency source is reduced to achieve higher field intensity levels. One obvious concern centers on having far field (plane wave) simulation fidelity. The effects on coupling and susceptibility and subsequent resonance due to the localized illumination must be assessed. The variations in coupling response between localized and full target illumination can be readily addressed by infrared field mapping techniques.

Acknowledgment

The technical support from my colleague, Professor Sega, at the University of Colorado is gratefully acknowledged.

References

[1] G.D. Wetlaufer, R.M. Sega, and J.D. Norgard (1985), Optimizing thin magnetic material for the thermographic detection of microwave induced surface currents, *Proceedings of the 1985 APS/URSI Symposium*, Vancouver, British Columbia, Canada.

[2] R.M. Sega and J.D. Norgard (1987), Expansion of an IR detection technique using conductive mesh in microwave shielding applications, *Proceedings of the 1987 SPIE Symposium*, San Diego, CA, August 1987.

[3] R.M. Sega, C.A. Benkelman, and J.D. Norgard (1985), Measurement of antenna patterns at 94 GHz using infrared detection, *SPIE Transactions*, Washington, DC, June 1985.

[4] C. Benkelman, J.D. Norgard, and R.M. Sega (1986), Infrared detection of millimeter wave antenna patterns, *Proceedings of the Millimeter Wave/Microwave Measurements & Standards Meeting*, Huntsville, AL, November 1986.

[5] J.D. Norgard and R.M. Sega (1987), Microwave fields determined from thermal patterns, *Proceedings of the 1987 SPIE Symposium*, Orlando, FL, May 1987.

[6] R.M. Sega and J.D. Norgard (1985), An infrared measurement technique for the assessment of electromagnetic coupling, *Proceedings of the Nuclear and Space Radiation Effects Conference*, Monterey, CA, July 1985 and *IEEE Trans. Nuclear Science*, **NS-32**, No. 6, 4330–4332.

[7] D. Fredal, P. Bussey, R.M. Sega, and J.D. Norgard (1987), Hardware and software advancement for infrared detection of microwave fields, *Proceedings of the 1987 SPIE Symposium*, Orlando, FL, May 1987.

[8] D.W. Metzger, R.M. Sega, and J.D. Norgard (1988), Numerical calculation and experimental verification of near fields from horns, *Proceedings of the URSI National Radio Science Meeting*, Boulder, CO, January 1988.

[9] D.W. Metzger, R.M. Sega, J.D. Norgard, and P. Bussey (1986), Experimental and theoretical techniques for determining coupling through apertures in cylinders, *1986 Nuclear Electromagnetic Meeting*, Albuquerque, NM, May 1986.

[10] J.D. Norgard and R.M. Sega (1986), Infrared measurement of scattering and electromagnetic penetrations through apertures, *Nuclear and Space Radiation Effects Conference*, Providence, RI, July 1986 and *IEEE Trans. Nuclear Science*, **NS-33**, No. 6, 1658–1663.

[11] J.D. Norgard and R.M. Sega (1987), Measured internal coupled electromagnetic fields related to cavity and aperture resonance, *Proceedings of the 1987 NSRE Conference*, Snowmass, CO, July 1987 and *IEEE Trans. Nuclear Science*, **NS-34**, No. 6, 1502–1507.

[12] J.D. Norgard and R.M. Sega (1987), Three-dimensional determination of cavity resonance and internal coupling, *Proceedings of the 1987 URSI Winter Meeting*, Boulder CO, January 1987.

[13] R.M. Sega and J.D. Norgard (1986), Infrared detection of microwave scattering from cylindrical structures, *Proceedings of the 1986 URSI Winter Meeting*, Boulder, CO, January 1986.

[14] R.M. Sega and J.D. Norgard (1986), Infrared diagnostic techniques for high-power microwave measurements, *Proceedings of the High Power Microwave Technology Meeting*, Albuquerque, NM (AFWL), December 1986.

[15] R.M. Sega, D. Fredal, and J.D. Norgard (1987), Initial feasibility test of an infrared diagnostic for high power microwave application, *Proceedings of the 1987 SPIE Symposium*, Orlando, FL, May 1987.

[16] P.E. Bussey, J.D. Norgard, and R.M. Sega (1988), Three-dimensional theoretical and experimental analysis of internal cylindrical fields coupled through a slot aperture, *Proceedings of the URSI National Radio Science Meeting*, Boulder, CO, January 1988.

[17] D.C. Fromme, R.M. Sega, and J.D. Norgard (1988), Experimental determination of scattering from *E*-pol and *H*-pol slit cylinders, *Proceedings of the APS/URSI Symposium*, Syracuse, NY, June 1988.

[18] S. Ramo, R.R. Whinnery, and T. Van Duzer (1984), *Fields and Waves in Communication Electronics*, 2nd ed., Wiley, New York, pp. 137–143.
[19] G.D. Wetlaufer (1985), Optimization of thin-screen material used in infrared detection of microwave induced surface currents at 2–3 GHz, M.S. thesis, University of Colorado, 1985.
[20] R.F. Harrington (1977), *Time-Harmonic Electromagnetic Fields*, McGraw-Hill, New York.
[21] R.M. Sega (1982), Infrared detection of microwave induced surface currents on flat plates, RADC-TR-82-308, 1982.
[22] R. Seigel and J.R. Howell (1972), *Thermal Radiation Heat Transfer*, 2nd ed., McGraw-Hill, New York, pp. 384–445, 679–711.
[23] M. Born and E. Wolf (1965), *Principles of Optics*, Pergamon Press, Oxford, pp. 560–570.
[24] M. Abramowitz and A. Stegun (1972), *Handbook of Mathematical Functions*, Dover, New York; pp. 297–329.
[25] N. Marcuvitz (1951), *Waveguide Handbook*, McGraw-Hill, New York.
[26] C.H. Papas (1965), *Theory of Electromagnetic Wave Propagation*, McGraw-Hill, New York.
[27] F.E. Borgnis and C.E. Papas (1958), Electromagnetic waveguides and resonators, *Encyclopedia of Physics*, Springer-Verlag, Heidelberg.

15
Dielectric Waveguide Theory

C. YEH*

15.1. Introduction

This chapter on dielectric waveguide theory is dedicated to Professor C.H. Papas who was my thesis advisor and mentor. He inspired me to be a daring and original researcher. Professor Papas has been truly a scholar and a gentleman, a rare breed indeed. I am honored to have known him.

As transmission line, dielectric waveguide has always been a laboratory curiosity in contrast with metal-based waveguide until the advent of low-loss optical fibers in the late 1960s. Consequently, the dielectric waveguide theory lagged behind that of the metal-based waveguide theory. But, since the early 1970s, because of the necessity of understanding the guiding characteristics of waves along low-loss fibers and integrated optical circuits, as well as the realization that as signal frequency increases, loss in metal also increases, several important new techniques have been developed to treat the dielectric waveguide problem. The purpose of this chapter is to assess these modern analytical/numerical techniques which have been used successfully in obtaining the propagation characteristics and field distributions of guided modes in dielectric waveguides. Some historical background information on the analysis of dielectric waveguides will be given first. Then the "pros" and "cons", as well as the limitations of various available analytical/numerical techniques, will be discussed. Finally, presentation will be given on several selected promising modern techniques together with illustrations.

15.2. Historical Background

The concept of guiding electromagnetic waves in a dielectric guide is not new. Hondros and Debye [1], in 1910, showed analytically that a circularly sym-

* Electrical Engineering Department, University of California, Los Angeles, CA 90024, USA.

metric transverse-magnetic (TM) mode can be guided by a dielectric cylinder with dielectric constant ε_1, situated in free space with dielectric constant ε_0 ($\varepsilon_1 > \varepsilon_0$). The existence of this wave was demonstrated experimentally by Zahn, Rüter, and Schriever in 1915 [2]. The complete treatment of all guided modes that can be supported by a dielectric cylinder in free-space was carried out by Carson, Mead, and Schelkunoff in 1936 [3]. They were the first to show that all noncircularly symmetric modes in a circular fiber are hybrid modes (i.e., longitudinal electric and magnetic fields must both be present for asymmetric modes), that only one mode, the lowest-order hybrid mode HE_{11}, has zero cutoff frequency, and that all other modes have finite cutoff frequencies below which they cease to exist. Numerical results for the propagation constants of several lower-order modes in circular fiber and the experimental verification were carried out by Elsasser [4] and Chandler [5] in 1949. Snitzer [6], in 1961, resolved the circular dielectric waveguide problem and applied the result to the case of light propagation along a homogeneous core circular fiber. Recognizing the fact that the index difference between the inner core region and the outer cladding region of an optical is quite small (or the order of a few percent), Snyder [7] and Gloge [8] presented simplified approximate expressions for the propagation parameters in a circular fiber in 1969 and in 1971, respectively. The first complete analysis of noncircular elliptical fiber was given by Yeh in 1962 [9]. He showed that all modes must be of the hybrid type in a noncircular fiber and that there exist two dominant modes which possess zero cutoff frequencies. It was not until 1969, that Goel [10] presented his circular-harmonics computer analysis of a rectangular homogeneous dielectric waveguide, and Marcatilli [11] presented his approximate analysis of the rectangular homogeneous dielectric structures. In 1970, Knox and Toulios [12] presented the effective index method to treat integrated circuit-type dielectric structures. Because of its simplicity, this method has been used extensively by various authors [13–16]. Most recently, Chiang [17] provided an improved version of this effective index method.

Significant progress has also been made in solving problems dealing with radially inhomogeneous circular guides. Surveys of this subject are included in papers by Yeh and Lindgren [18], and Dil and Blok [19]. As optical fiber technology and millimeter wave technology matured, in the late 1970s, demand for more sophisticated and versatile analytical/numerical techniques to solve problems dealing with arbitrarily shaped, inhomogeneous dielectric waveguides or microstrip lines, and their associated components also grows. The recent successful development of the scalar wave–fast Fourier transform (FFT) technique [20–22], the finite element technique [23], the extended boundary condition technique [24], and the transmission line matrix (TLM) technique [25] will certainly expand our ability in predicting and understanding the wave behavior in these complex guiding structures.

It should be noted that this brief historical background is by no means a complete or exhaustive survey; it is meant to be used as an introduction to the subject of dielectric waveguide theory.

15.3. Discussion of Selected Analytic/ Numerical Techniques

The governing equation for the guided-wave field of a fiber structure is the vector-wave equation

$$\nabla \times \nabla \times E - \omega^2 \mu \varepsilon(r) E = 0, \tag{15.1}$$

where ω and E are, respectively, the frequency and the electric-field vector of the guided wave, while μ and $\varepsilon(r)$ are, respectively, the permeability and the permittivity of the guiding medium. The propagation characteristics of the guided wave are obtained by requiring that the guided-wave field must satisfy the proper boundary conditions at the interface of two different media (i.e., tangential electric- and magnetic-field vectors must be continuous across the boundary) and the radiation condition for the field that extends to infinity [26].

When the guided wave propagates along a perfect straight-line path, we may assume that every component of the electromagnetic wave may be represented in the form

$$f(u, v) e^{-i\beta z} e^{i\omega t} \tag{15.2}$$

in which z is chosen as the propagation direction and u, v are generalized orthogonal coordinates in a transverse plane. β is the propagation constant and ω is the frequency of the wave. Under this assumption, the transverse-field components in homogeneous isotropic medium (ε, μ) are

$$E_u = \frac{-i}{\gamma^2}\left(\frac{\beta}{h_1}\frac{\partial E_z}{\partial u} + \frac{\omega\mu}{h_2}\frac{\partial H_z}{\partial v}\right), \tag{15.3}$$

$$E_v = \frac{-i}{\gamma^2}\left(\frac{\beta}{h_2}\frac{\partial E_z}{\partial v} - \frac{\omega\mu}{h_1}\frac{\partial H_z}{\partial u}\right), \tag{15.4}$$

$$H_u = \frac{-i}{\gamma^2}\left(\frac{\beta}{h_1}\frac{\partial H_z}{\partial u} - \frac{\omega\varepsilon}{h_2}\frac{\partial E_z}{\partial v}\right), \tag{15.5}$$

$$H_v = \frac{-i}{\gamma^2}\left(\frac{\beta}{h_2}\frac{\partial H_z}{\partial v} + \frac{\omega\varepsilon}{h_1}\frac{\partial E_z}{\partial u}\right), \tag{15.6}$$

with

$$\gamma^2 = k^2 - \beta^2, \tag{15.7}$$

and

$$k^2 = \omega^2 \mu \varepsilon, \tag{15.8}$$

where ε is the permittivity of the medium and μ is the permeability of the medium, and the longitudinal-field components satisfy the following equation:

$$\left[\frac{1}{h_1 h_2}\left(\frac{\partial}{\partial u}\frac{h_2}{h_1}\frac{\partial}{\partial u} + \frac{\partial}{\partial v}\frac{h_1}{h_2}\frac{\partial}{\partial v}\right) + (k^2 - \beta^2)\right]\begin{Bmatrix} E_z \\ H_z \end{Bmatrix} = 0, \tag{15.9}$$

where h_1 and h_2 are the metric coefficients for the orthogonal curvilinear coordinates. Only discrete values of β will satisfy the boundary conditions. These allowed β values are called eigenvalues, and corresponding to these eigenvalues are the eigenfunctions. Each eigenvalue β corresponds to the propagation constant of a certain guided mode. It is pointed out here that TM guided modes refer to waves having $H_z = 0$, TE guided modes refer to waves having $E_z = 0$, and HE or EH guided modes (hybrid modes) refer to waves having all field components not equal to zero.

Two general types of dielectric structures are of practical interest: the multimode dielectric structures and the single-mode dielectric structures. The multimode structures are capable of supporting many guided modes (say > 10 modes) while the single-mode structures may support only a few lower-order guided modes (say < 10 modes). In the following, we shall divide our discussion of available analytic/numerical techniques into these two general categories.

15.3.1. *The Single-Mode Case*

15.3.1.1. Separation of Variables Method

For a homogeneous medium, or some special inhomogeneous medium, the vector-wave equation is separable in three coordinate systems: the rectangular coordinates, the cylindrical coordinates, and the spherical coordinates. The rectangular coordinates are particularly suited for slab-type guiding structures (see Figure 15.1) and the cylindrical coordinates are suited for the step-index circular cylindrical dielectric guide or radially graded index circular cylindrical dielectric guide or step-index elliptical cylindrical dielectric guide (see Figure 15.1). These geometrics are of great practical importance.

Basically, the separation of variables method starts with the appropriate eigensolutions of the wave equations in the core and the cladding regions of the guiding structure. Then, by matching the boundary conditions at the interface, which fits the contour of one of the coordinate surfaces, we may obtain a dispersion relation from which the propagation constant β behavior may be found. Typically, the circular cylindrical fiber structure can support a family of circularly symmetric transverse-electric TE_{0m} or transverse-magnetic TM_{0m} modes (whose fields are independent of the azimuthal coordinate) and a family of hybrid HE_{nm} or EH_{nm} modes. The subscripts n and m denote, respectively, the number of cyclic variation with the azimuthal coordinate and the mth root of the dispersion relation which is obtained by satisfying the appropriate boundary conditions. The symbol HE refers to the mode with the ratio $(\mu_0\omega/\beta)(H_z/E_z) = -1$ far from the cutoff frequency, while the symbol EH refers to modes with the ratio $(\mu_0\omega/\beta)(H_z/E_z) = +1$ far from the cutoff frequency.

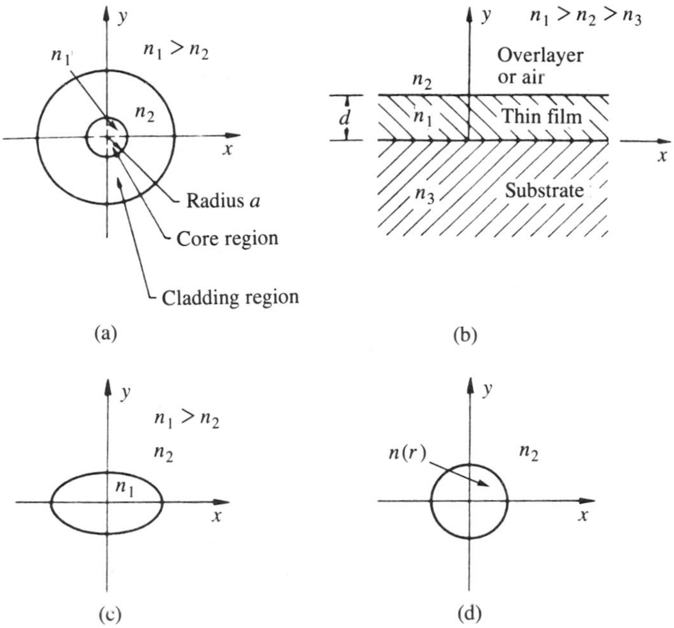

FIGURE 15.1. Cross-sectional views of the dielectric guiding structures that could be treated by the separation of variable technique. (a) Step-index circular fiber; (b) slab guide; (c) step-index elliptical fiber; (d) graded-index circular fiber.

The pros and cons of this method are summarized as follows:

Pros:

conceptually simple to use, analytic expressions are available;
results are exact and applicable to small as well as large index differences for the core and cladding regions

Cons:

may be applied to only a small family of geometries such as the step-index circular cylindrical dielectric guides, the radially graded index circular cylindrical dielectric guides, the step-index elliptical cylindrical dielectric guides, the slab dielectric guides;
dispersion relations are transcendental functions, hence difficult to use.

15.3.1.2. CIRCULAR HARMONICS EXPANSION METHOD

This method was first used by Goell [10] in his treatment of the step-index rectangular fibers. His computer analysis is based on an expansion of the guided electromagnetic field in terms of a series of circular harmonics. For example, the expressions for the longitudinal electric and magnetic fields in the core and cladding regions are, respectively,

$$E_z^{(\text{core})} = \sum_{n=-\infty}^{\infty} A_n J_n(\gamma_1 r) e^{in\theta} e^{-i\beta z + i\omega t}, \tag{15.10}$$

$$H_z^{(\text{core})} = \sum_{n=-\infty}^{\infty} B_n J_n(\gamma_1 r) e^{in\theta} e^{-i\beta z + i\omega t}, \tag{15.11}$$

and

$$E_z^{(\text{cladding})} = \sum_{n=-\infty}^{\infty} C_n K_n(\gamma_2 r) e^{in\theta} e^{-i\beta z + i\omega t}, \tag{15.12}$$

$$H_z^{(\text{cladding})} = \sum_{n=-\infty}^{\infty} D_n K_n(\gamma_2 r) e^{in\theta} e^{-i\beta z + i\omega t}, \tag{15.13}$$

where

$$\gamma_1 = (k_1^2 - \beta^2)^{1/2}, \qquad \gamma_2 = (\beta^2 - k_2^2)^{1/2};$$
$$k_1^2 = \omega^2 \mu \varepsilon_1, \qquad k_2^2 = \omega^2 \mu \varepsilon_2;$$

ε_1 and ε_2, are, respectively, the core dielectric constant and the cladding dielectric constant; A_n, B_n, C_n, and D_n are arbitrary constants; and J_n and K_n are the nth order Bessel functions and modified Bessel functions, respectively. All transverse fields can be found from (15.3)–(15.6). Point matching of the tangential electric and magnetic fields on the boundary of the core and cladding regions yields a set of linear algebraic equations containing the unknown coefficients A_n, B_n, C_n, and D_n. Setting the determinant of this set of linear equations to zero gives a dispersion relation from which the propagation constants of various modes may be computed. The size of the determinant depends on the number of points that were used for matching.

Pros:

may be used to treat noncircular guides with arbitrary core/cladding index difference.

Cons:

convergence of the results is not guaranteed for shapes that deviate significantly from the circular guide shape, i.e., good for square guide but not good for rectangular guides;

only applicable to step-index dielectric guides, i.e., core and cladding regions must contain uniform dielectrics;

this is a purely numerical approach, demand on computer (time and money) may be excessive if many matching points were used.

15.3.1.3. Marcatilli's Approximate Approach

Recognizing the fact that most of the guided energy is confined within the dielectric core region for a wide range of parameters, and very little energy is guided in the corner regions of a rectangular dielectric guide, Marcatilli [11] formulated an approximate solution to the problem of waveguiding by rec-

FIGURE 15.2. Geometry of the rectangular dielectric guide immersed in different dielectrics.

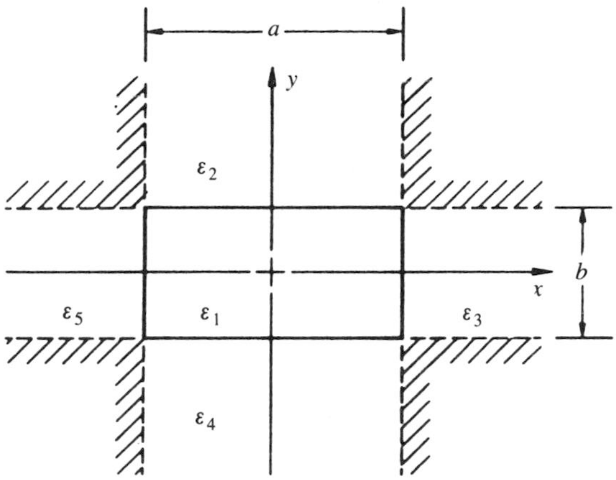

tangular dielectric structures by ignoring the matching of fields along the edges of the shaded area in Figure 15.2. By matching the tangential electric and magnetic fields only along the four sides of region 1, and assuming that the field components in region 1 vary sinusoidally in the x and y directions, those in 2 and 4 vary sinusoidally along x and exponentially along y, and those in regions 3 and 5 vary sinusoidally along y and exponentially along x, we may obtain a dispersion relation from which the propagation constants of various modes may be calculated.

Pros:

- resultant mathematical expressions are easy to manipulate since only simple sinusodial and exponential functions have been used;
- since analytic expressions of the fields were obtained, they provide easy insight and physical interpretation of the results;
- may be used to treat any rectangular shaped homogeneous core dielectric guides;
- results are quite accurate as long as the guided energy is mostly confined within the core region.

Cons:

- only applicable to step-index rectangular structures;
- results are unreliable near the cutoff region of the mode, where the guided fields are not necessarily confined within the core region;
- this is an heuristic approach, regions of validity of the results are unknown.

15.3.1.4. FINITE ELEMENT APPROACH

This is the single most powerful technique [23] developed in recent years to treat the single-mode problems dealing with arbitrarily shaped dielectric waveguides with inhomogeneous index variations in the cross-sectional plane. The governing longitudinal fields of the guided wave are first expressed as a functional as follows:

$$
\begin{aligned}
I &= \sum_{p=1} I_p \\
&= \sum_{p=1} \iint \Biggl\{ \tau_p |\nabla H_z^{(p)}|^2 + \gamma^2 \tau_p \frac{\varepsilon_p}{\varepsilon_0} \left| \frac{1}{\gamma} \left(\frac{\varepsilon_0}{\mu_0} \right)^{1/2} \nabla E_z^{(p)} \right|^2 \\
&\quad + 2\tau_p \gamma^2 \hat{e}_z \cdot \left[\frac{1}{\gamma} \left(\frac{\varepsilon_0}{\mu_0} \right)^{1/2} \nabla E_z^{(p)} \cdot \nabla H_z^{(p)} \right] \\
&\quad - \Gamma^2 \left[(H_z^{(p)})^2 + \gamma^2 \frac{\varepsilon_p}{\varepsilon_0} \left[\frac{1}{\gamma} \left(\frac{\varepsilon_0}{\mu_0} \right)^{1/2} E_z^{(p)} \right]^2 \right] \Biggr\} \, dx \, dy, \qquad (15.14)
\end{aligned}
$$

where

$$
\gamma = \frac{\beta c}{\omega}, \qquad \tau_p = \frac{\gamma^2 - 1}{\gamma^2 - \varepsilon_p/\varepsilon_0}, \qquad \Gamma^2 = \left(\frac{\omega}{c} \right)^2 (\gamma^2 - 1),
$$

the symbol p represents the pth region when we divide the guiding structure into many appropriate regions. Minimizing the above surface integral over the whole region is equivalent to satisfying the wave equation (15.1) and the boundary conditions for E_z and H_z. In the finite element approximation, the primary dependent variables are replaced by a system of discretized variables over the domain of consideration. Therefore, the initial step is a discretization of the original domain into many subregions. For the present analysis, there are a number of regions in the composite cross section of the waveguide for which the permittivity is distinct. Each of these regions is discretized into a number of smaller triangular subregions interconnected at a finite number of points called nodes. Appropriate relationships can then be developed to represent the waveguide characteristics in all triangular subregions. These relationships are assembled into a system of algebraic equations governing the entire cross section. Taking the variation of these equations with respect to the nodal variable leads to an algebraic eigenvalue problem from which the propagation constant for a certain mode may be determined.

Pros:

- may be used to treat any arbitrarily shaped, inhomogeneous dielectric guides;
- numerical results can be generated very efficiently;
- results are based on the exact Maxwell equations.

Cons:

- no analytic expression is available.

15.3.1.5. Extended Boundary Condition Method

For arbitrarily shaped step-index fibers, integral representations for the longitudinal electric and magnetic fields can be derived which satisfy the appropriate wave equations and all the necessary boundary conditions. By expanding the longitudinal fields in terms of a complete set of circular harmonics and by making use of the analytic continuation technique, we may reduce the integral representations to a set of linear algebraic equation [24]. Setting its determinant to zero and finding its roots yields the propagation constants.

Pros:

may be used to treat any arbitrarily shaped step-index guides;
numerical results can be generated very efficiently;
results are based on the exact Maxwell equations.

Cons:

this technique cannot be used to treat the inhomogeneous index structure.

15.3.1.6. The TLM Approach

None of the above techniques is capable of treating waveguides whose dimensions or composition can vary along the direction of propagation of the guided waves. According to the transmission line matrix (TLM) approach [25], half of Maxwell's equations are fully accounted for by three shunt nodes oriented in the x–y, y–z, and x–z planes, while the remaining half of the Maxwell's equations are satisfied by three series nodes oriented in the same planes. The six nodes are interconnected to form a three-dimensional node in space. In each coordinate plane there is one shunt node and one series node. The nodes are named to correspond to the field quantity they represent. Thus, the common voltage at shunt node E_x corresponds to the x component of the electric field. The common current at series node H_x corresponds to the x component of the magnetic field, and so on. To represent a three-dimensional propagation space a number of these three-dimensional nodes are connected to form a three-dimensional mesh network. The full Maxwell's equations are thus satisfied at each three-dimensional node. The continuity of tangential electric and magnetic fields across a dielectric/dielectric boundary is automatically satisfied in the TLM model when the three-dimensional nodes are joined up by elementary sections of ideal transmission lines. For example, for a dielectric/dielectric boundary in the x–z plane, since the common voltages at the shunt nodes correspond to the electric field and the common currents at the series nodes correspond to the magnetic field, the following equations (valid for a transmission-line element joining the nodes on either side of the boundary) are applicable:

$$E_{z1} = E_{z2} + \frac{\partial E_{z2}}{\partial y} \Delta l,$$

$$E_{x1} = E_{x2} + \frac{\partial E_{x2}}{\partial y} \Delta l,$$

$$H_{x1} = H_{x2} + \frac{\partial H_{x2}}{\partial y} \Delta l,$$

$$H_{z2} = H_{z2} + \frac{\partial H_{z2}}{\partial y} \Delta l.$$

Since the voltage and current in the transmission lines are smooth functions of position along the line, the continuity of the tangential fields across a boundary placed in between the nodes is assured. In the TLM method, the numerical procedure involves determination of the impulse response of the network. Dispersion analysis by the TLM technique involves resonating a section of the transmission line by placing shorting planes along the axis of propagation (the z-axis in this case), such that the images of the line in the shorting planes appear to be continuations of the structure. Each separation of the shorting planes then equals half of the guides wavelength for the fundamental mode at the frequency given by the resonant frequency of the cavity. If the distance between the shorting planes is $2L$, the phase constant is given by $\beta = \pi/2L$.

Pros:

capable of analyzing arbitrarily shaped, inhomogeneous, anisotropic three-dimensional dielectric structures such as couplers, horns, tapers, or discontinuities;

results are based on the exact Maxwell's equations.

Cons:

demand on computational time may be large.

15.3.1.7. Scalar Wave–FFT Beam Propagation Approach

All of the above techniques were based on the exact vector-wave equations. Hence they were valid for large as well as small-index differences between core and cladding regions. We recognize, however, that for most practical optical waveguides the index difference between the core and cladding regions are only of the order of a few percent [20–22]. It has been shown recently that if certain limiting conditions (such as small-index differences and gentle-index profiles) are satisfied [22], then the scalar-wave approximation will yield valid results for single-mode structures. These limiting conditions are usually satisfied by many practical fiber or integrated optics structures.

Starting with the scalar-wave equation and making use of the paraxial approximation, we may develop a propagation code based on the FFT technique to trace the evolution of the transverse field as it propagates down a guide structure. This technique not only provides information on the field distribution, it also yields information on the propagation constant of the dominant mode.

Pros:

may be used to treat arbitrarily shaped, inhomogeneous guide structures such as guide couplers, horns, or tapers;

efficient computational time.

Cons:

index variation must be gentle and index differences must be small.

Knowing the pros and cons of the above techniques in their treatment of single-mode structures, we may choose the most efficient method in dealing with the problem at hand. In Section 15.4, several examples of contemporary interest will be given.

15.3.2. *The Multimode Case*

In principle, all of the single-mode techniques discussed earlier may be used to deal with the multimode structures. Contributions from each mode may be summed according to the principle of superposition to yield the correct result for the multimode case. However, this approach may be extremely cumbersome and sometimes unworkable when hundreds or even thousands of propagating modes are supportable by the multimode structure. Hence, the treatment of multimode structures based on the modal concept, perhaps should be altered. In the following, we shall discuss two approaches which are not based on this modal concept.

15.3.2.1. Geometrical Optics Approach

Because of its conceptual simplicity, the ray-tracing geometrical optics technique has always been of importance whenever optics problems are encountered. By tracing the ray paths according to the theory of geometrical optics, we are able to calculate the evolution of an input optical signal as it propagates down the waveguide. It can be seen, however, that this technique could become extremely laborious and that, since diffraction phenomenon was not taken into account, the result may also be rather inaccurate.

Pros:
- calculation involves elementary concepts and procedures.

Cons:
- calculations may be tedious;
- results may be grossly inaccurate due to oversimplification of the propagation phenomenon.

15.3.2.2. Scalar Wave–FFT Beam Propagation Approach

This technique was discussed earlier. It is seen that this scalar wave–FFT approach [20–21] does not depend on the decomposition of fields into individual eigenmodes. It deals completely with total field quantities and not with individual modes. In other words, the evolution of the total field is calculated according to the scalar wave equation as it propagates down the guiding structure. Since the wave nature of the fields is taken into consideration, diffraction phenomenon is automatically included. Hence, this approach provides an accurate description of the guided wave in a multimode structure.

Pros:

- decomposition of fields into modes is not necessary for multimode structures;
- provides accurate wave description of the propagation field;
- may be used to treat complex arbitrarily shaped, inhomogeneous multimode structures such as multimode dielectric couplers, guides with general index profiles, horns, tapers, and branches, etc.

Cons:

- index variation must be gentle and index differences must be small;
- depolarization phenomenon is ignored.

15.4. Selected Examples

According to the previous discussions, the most promising new techniques in dealing with general shaped optical waveguide structures having inhomogeneous index profiles are the finite element method, the scalar wave–FFT method, and the TLM technique. In the following, we shall provide several illustrative examples of these techniques. In order to demonstrate the versatility of these techniques, we have purposely chosen examples concerning rather unusual guiding structures.

15.4.1. *Examples Based on the Finite Element Method*

It has been mentioned earlier that the finite element method is a very powerful technique. It can be used to solve single-mode problems dealing with guiding structures whose cores may be of arbitrary cross-sectional shape and whose material media may be inhomogeneous in more than one transverse direction. Since this technique is based on the exact Maxwell's equations, its results are valid for large or small index variations.

15.4.1.1. The Optical Stripline Guide

The optical stripline is a planar waveguide with a strip of slightly lower index on a higher index thin film. An advantage of this guide is a relaxation of the stringent requirement for smoothness of waveguide side walls because a small portion of the field strength impinges on the side walls of the strip.

The cross-sectional geometry of the optical stripline is given in Figure 15.3. It also shows the refractive index distribution where n_1, n_2, n_3, and n_0 are the refractive index of thin film, substrate, strip, and air, respectively. These refractive indexes satisfy one of the following relations for the stripline:

$$n_1 > n_2 \geq n_3 > n_0,$$

$$n_1 > n_3 \geq n_2 > n_0.$$

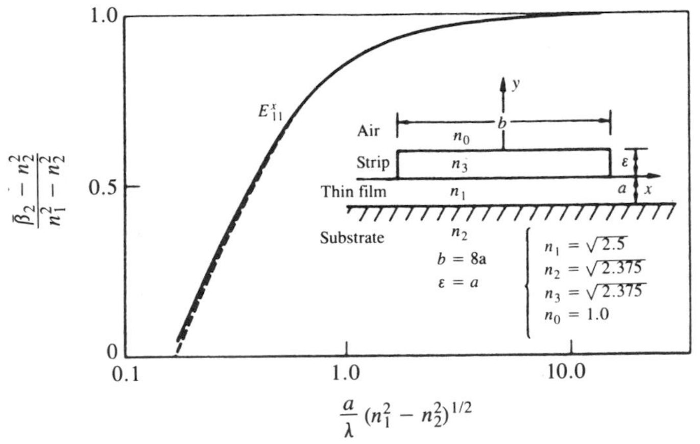

FIGURE 15.3. Dispersion curves for the dominant E_{11}^y mode in an optical stripline waveguide. The solid line represents results found according to the finite element method, while the dashed line represents results found according to the vector variational method.

In the present example, the numerical data of the refractive indexes is chosen to be $n_1 = \sqrt{2.5}$, $n_2 = n_3 = \sqrt{2.375}$, and $n_0 = 1.0$. This refractive index distribution corresponds to the case where the thin film and the substrate are glass and the ambient is air.

The fundamental modes of this stripline guide are the E_{11}^y mode and E_{11}^x mode. The principal transverse-field components of the E_{11}^y mode are E_y and H_x and those of the E_{11}^x mode are E_x and H_y. The present analysis involves solving only the E_{11}^x mode.

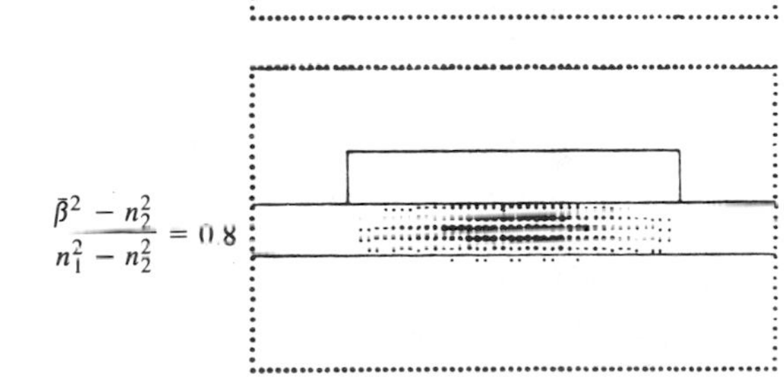

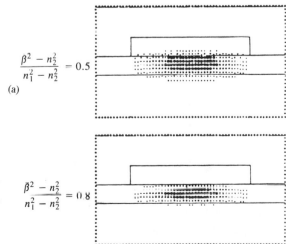

FIGURE 15.4. Grey-scale plots for the power intensity distributions of the dominant E_{11}^y mode in an optical stripline waveguide at two different frequencies.

The finite element model employs 900 elements and 928 nodes in one-half of the cross section because of the symmetry on the y-axis. The dispersion curve for the E^x_{11} mode is computed with the boundary condition, $E_z = 0$ on the y-axis. The curve is plotted in Figure 15.3.

Figure 15.4 shows the power intensity distribution on the grey scale taken at the normalized propagation constant = 0.5 and 0.8. It is seen that the intensity distribution is well confined in the x direction by the dielectric strip because the width b of the strip is larger compared with the thin film thickness a and the wavelength λ. It is also confined in the y direction of the thin film. Therefore, the optical stripline minimizes scattering loss and undesired mode conversion, caused by imperfections of the etched side walls of the optical waveguide.

15.4.1.2. Triangular Fiber Guide

Shown in Figure 15.5 are the propagation characteristics and the power distribution of the dominant mode in a triangular optical fiber guide. These

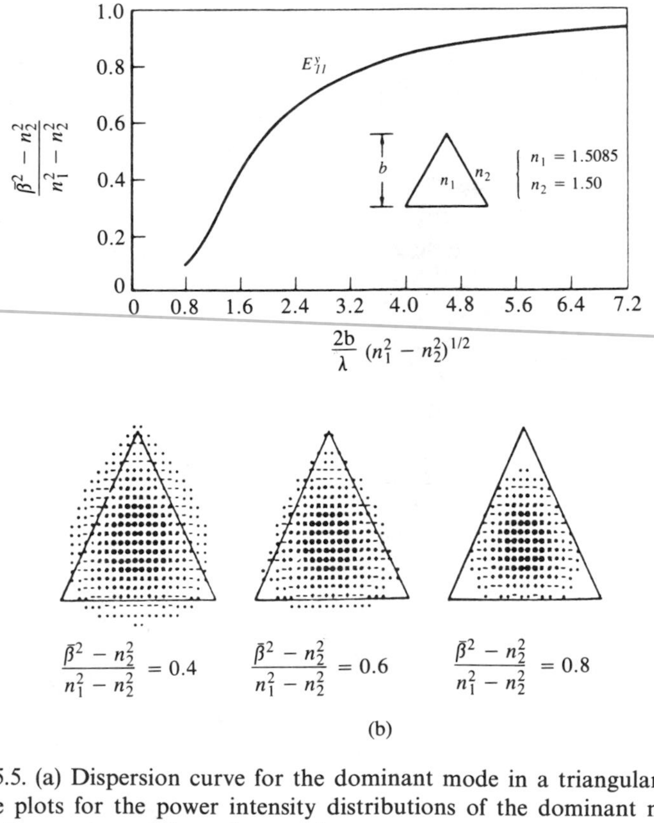

Figure 15.5. (a) Dispersion curve for the dominant mode in a triangular fiber. (b) Grey-scale plots for the power intensity distributions of the dominant mode in a triangular fiber at three different frequencies.

results were obtained by the finite element method employing 689 elements and 659 nodes.

15.4.1.3. Single Material Fiber Guide

Typically, optical fibers are constructed with a central glass core surrounded by a glass cladding with a slightly lower refractive index. The single material fiber is created by a structural form that uses only a single low-loss material. The guided energy is concentrated primarily in the central enlargement. The fields decay exponentially outward from the central enlargement. The guided-wave fields at the outside cylinder fused-channel guide were obtained and displayed in Figure 15.6.

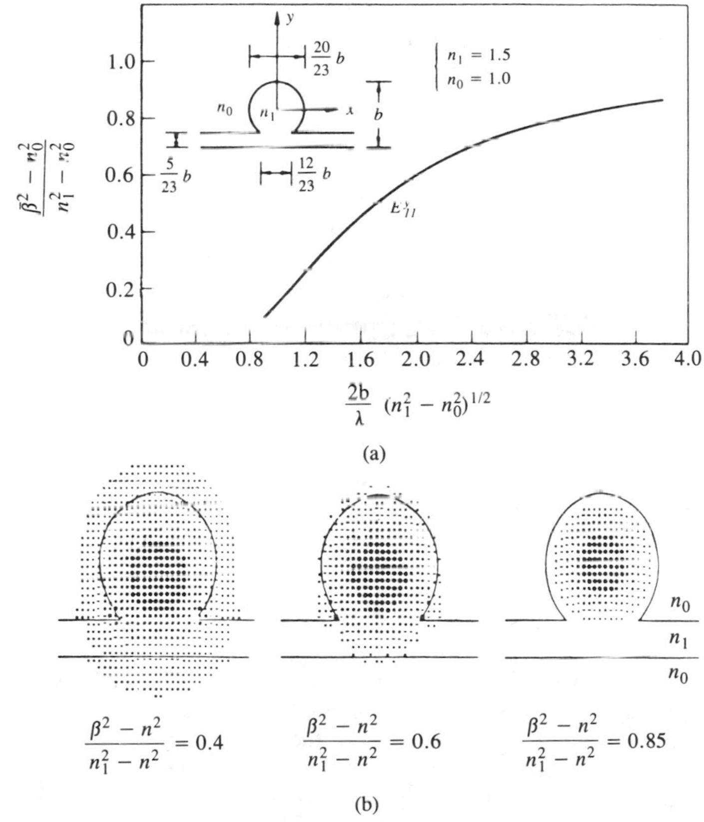

Figure 15.6. (a) Dispersion curve for the dominant mode in a single material fiber. (b) Grey-scale plots for the power intensity distributions of the dominant mode in a single material fiber at three different frequencies.

15.4.2. *Examples Based on the Scalar Wave–FFT Method*

Multimode optical structures are of great practical importance. As noted earlier, the only efficient method which incorporates the wave nature of the guided energy in multimode structure is the scalar wave–FFT method. The only significant restriction to keep in mind is that index variations of the structure under consideration must be gentle.

15.4.2.1. Truncated Gaussian-Beam Propagation in Multimode Inhomogeneous Fiber Guide

It is recognized that analytic solutions for Gaussian-beam propagation problems exist only for certain specific radial-index profiles. Even for these cases, the solutions are often involved and cumbersome to use. By simply specifying the transverse-index variation of the guiding structure, we can readily calculate the propagation characteristics of an incident beam in this medium using the scalar wave–FFT technique. Shown in Figure 15.7 is the evolution of intensity patterns of an incident truncated Gaussian beam as it propagates down a multimode fiber structure with parabolic-index profile.

15.4.2.2. Multimode Inhomogeneous Fiber Couplers

To further demonstrate the versatility and the power of this scalar wave–FFT technique, the exceedingly complex coupling problem dealing with two neighboring multimode graded-index fibers was treated. Shown in Figure 15.8 are the evolution of the power density distribution for the guided beam as it travels down the coupled structure and the percent of total power in one-half of the structure as a function of the axial distance.

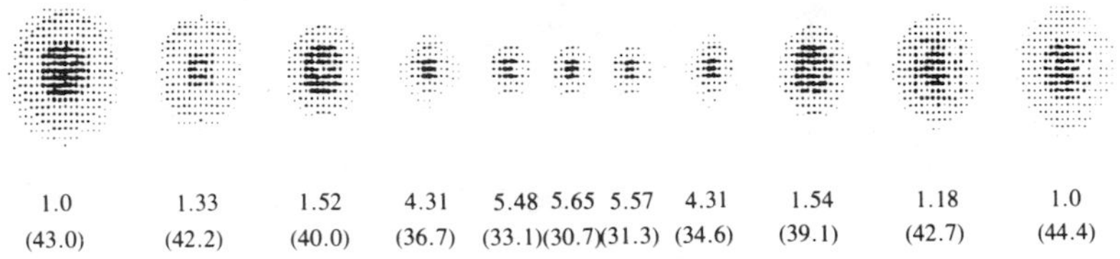

FIGURE 15.7. Selected grey-scale intensity patterns along the fiber axis for a truncated Gaussian beam propagating in a radially inhomogeneous fiber with parabolic-index profile. The values in the brackets represent the beam waists in micrometers of the beam while the other unbracketed values represent the highest intensity values for these patterns. Note that the truncated Gaussian profile of the beam is not preserved along the propagation path. At some points along the path, the intensity at the center of the beam takes a dip. Owing to the presence of the diffraction effects, the beam is not completely symmetrical about a certain "focal point" where the beam achieves the smallest beam waist.

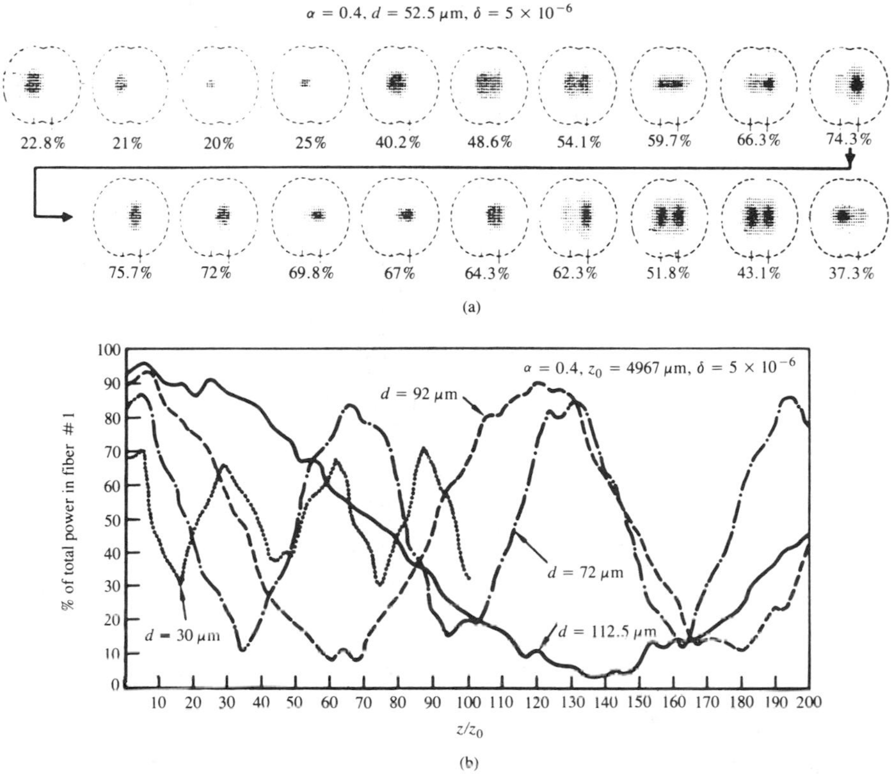

FIGURE 15.8. (a) A cross-sectional view of the power density distribution for the guided wave. Note the power exchange phenomenon as the wave propagates along the coupled structure. The percent value indicates the percent of total power contained in the right-hand half of the structure (i.e., fiber 2). The distance between each frame is 9934 μm. (b) Percent of total power in the left-hand half of the structure (i.e., fiber 1) as a function of the axial distance along the coupling structure. This is the multimode coupling case.

15.4.3. *Example Based on the TLM Method. Microstrip on Substrate with Periodically Stratified Dielectrics*

The versatility of the TLM method can be seen in this example. Shown in Figure 15.9 is a microstripline placed on a substrate made up of alternate layers of isotropic dielectric material of relative permittivities ε_1 and ε_2. The length of the periodic cell is d. Note that the two end layers in the cavity are of width $d/4$ so that the images of the structure in the end shorting planes appear to be continuations of the structure.

According to Floquet's theorem, an infinite set of spatial harmonics exists for guided waves along a periodic structure. These spatial harmonics must be present simultaneously in order that the total field may satisfy the boundary

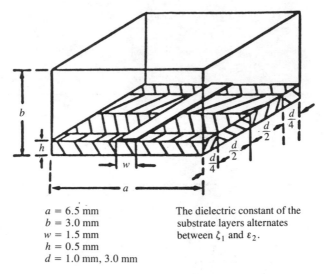

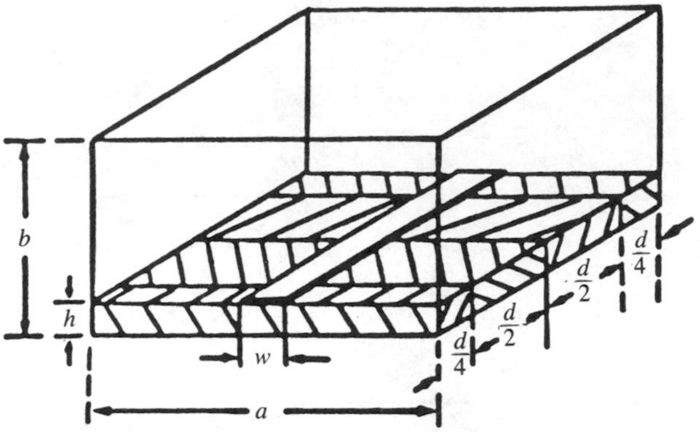

$a = 6.5$ mm
$b = 3.0$ mm
$w = 1.5$ mm
$h = 0.5$ mm
$d = 1.0$ mm, 3.0 mm

The dielectric constant of the substrate layers alternates between ζ_1 and ε_2.

FIGURE 15.9. Geometry of microstrip on substrate with periodically stratified index of refraction.

conditions. The eigenvalue equation for β for a periodic structure will always yield solutions $\beta_n = \beta + 2n\pi/d$, in addition to the fundamental solution. These other possible solutions are clearly the propagation constants of the spatial harmonics. A complete ω–β diagram thus exhibits $k_0 d$ as a periodic function of βd, that is, the βd curve is continued periodically outside the range

$$-\pi \leq \beta d \geq \pi.$$

The principal values of βd are plotted in Figure 15.10 for various values of ε_1 and ε_2. The length of the unit cell in curves 1, 2, and 3 is $d = 1.0$ mm. The cutoff frequency for the low-frequency passband is given by the value of $k_0 d$ when $\beta d = \pm\pi$. Note that Figure 15.10 shows only the first passband of the periodic structure. Examination of the diagram shows that the cutoff frequency may be reduced by increasing ε_1 or ε_2. Although the value of $k_0 d$ at cutoff was somewhat increased when the magnitude of d was tripled, the overall effect of increasing d was to reduce the cutoff frequency since $k_0 d$ did not triple at cutoff. In general, we may deduce the following conclusions:

(1) To increase the upper cutoff frequency, we should lower the dielectric constant of the material. The phase velocity of the wave is lowered by reducing the length of the periodic cell. Hence, by controlling the width of the cell, it is possible to adjust the phase velocity of the wave.
(2) Microstrip on a substrate with a periodically stratified index of refraction exhibits the slow-wave and filtering properties common to all periodic waveguiding structures.

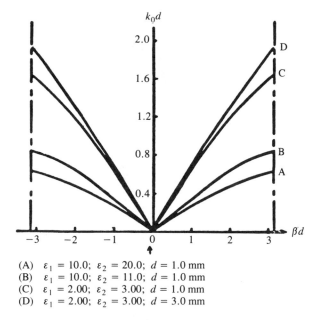

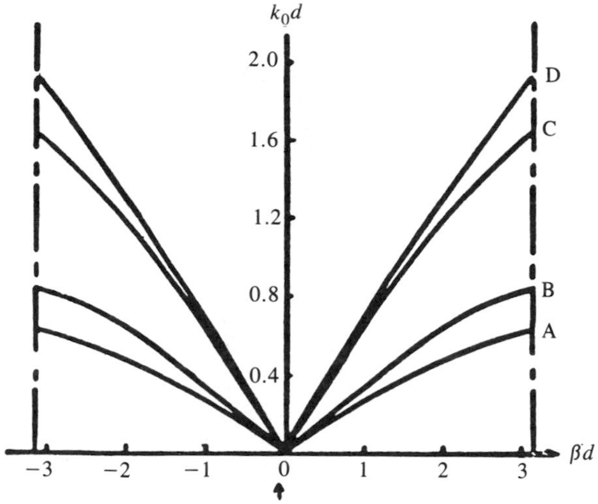

(A) $\varepsilon_1 = 10.0$; $\varepsilon_2 = 20.0$; $d = 1.0$ mm
(B) $\varepsilon_1 = 10.0$; $\varepsilon_2 = 11.0$; $d = 1.0$ mm
(C) $\varepsilon_1 = 2.00$; $\varepsilon_2 = 3.00$; $d = 1.0$ mm
(D) $\varepsilon_1 = 2.00$; $\varepsilon_2 = 3.00$; $d = 3.0$ mm

FIGURE 15.10. ω–β diagram for microstrip on periodically stratified index of refraction.

15.5. Conclusions

Dielectric structures are becoming increasingly important as modern waveguides. These waveguides take on many different shapes which usually cannot be analyzed analytically. It is unavoidable to conclude from this chapter that significant advances in the theoretical treatment of dielectric waveguide problems in the last decade have been in the numerical/computational areas. This is not to minimize the traditional thought that analytic expressions for simplified/idealized structures provide a great deal of insight into many physical phenomena of fundamental importance. However, from the engineering standpoint, it is not sufficient just to understand the physical phenomenon of wave interaction taking place in an idealized structure but rather, we must be able to generate solid quantitative data for practical complex guiding structures. Furthermore, the trend toward data bus multiplexing [27] (the interconnection of a number of spatially distributed terminals via fiber optic waveguides cables) as well as integration of optical circuits for data processing (integrated optics), and the rapid development of mm wave technology point to the development of dielectric waveguide components such as couplers, branches, tapers, and integrated optical or mm wave circuits. The only viable means of analyzing these complex structures is to make use of the highly versatile techniques using computers.

References

[1] D. Hondros and P. Debye (1910), *Ann. Physik*, **32**, 465.
[2] H. Zahn (1916), *Ann. Physik*, **49**, 907.
[3] J.K. Carson, S.P. Mead, and S.A. Schelkunoff (1936), *Bell System Tech. J.*, **15**, 310.
[4] W. Elsasser (1949), *J. Appl. Phys.*, **20**, 1193.
[5] C. Chandler (1949), *J. Appl. Phys.*, **20**, 1188.
[6] E. Snitzer (1961), *J. Opt. Soc. Amer.*, **51**, 491.
[7] A.W. Snyder (1969), *IEEE Trans. Microwave Theory Tech.*, **MTT-17**, 1138.
[8] D. Gloge (1971), *Appl. Opt.*, **10**, 2442.
[9] C. Yeh (1962), *J. Appl. Phys.*, **33**, 3235.
[10] J.E. Goell (1969), *Bell System Tech. J.*, **48**, 2133.
[11] E.A.J. Marcatilli (1969), *Bell System Tech. J.*, **48**, 2071.
[12] R.M. Knox and P.P. Toulios 1970, *Proc. Symp. Submilimeter Waves*, Brooklyn, NY, Polytechnic Press, New York, p. 497.
[13] W.V. McLevige, T. Itoh, and R. Mittra (1975), *IEEE Trans. Microwave Theory Tech.*, **MIT-23**, 788.
[14] T. Itoh (1976), *IEEE Trans. Microwave Theory Tech.*, **MIT-24**, 821.
[15] G.B. Hocker and W.K. Burns (1977), *Appl. Opt.*, **16**, 113.
[16] H. Furuta, H. Noda, and A. Ihaya (1974), *Appl. Opt.*, **13**, 322.
[17] K.S. Chiang (1986), *Appl. Opt.*, **25**, 348.
[18] C. Yeh and G. Lindgren (1977), *Appl. Opt.*, **16**, 483.
[19] J.G. Dil and H. Blok (1973), *Opto.-Elec.*, **5**, 415.
[20] C. Yeh, L. Casperson, and B. Szejn (1978), *J. Opt. Soc. Amer.*, **68**, 989.
[21] C. Yeh, W.P. Brown, and R. Szejn (1979), *Appl. Ipt.*, **18**, 489.
[22] C. Yeh and F. Manshadi (1985), *J. Lightwave Tech.*, **LT-3**, 199.
[23] C. Yeh, K. Ha, S.B. Dong, and W.P. Brown (1979), *Appl. Opt.*, **18**, 1596.
[24] L. Eyges, P. Gianino, and P. Wintersteiner (1979), *J. Opt. Soc. Amer.*, **69**, 1226.
[25] G.E. Mariki and C. Yeh (1985), *IEEE Trans. Microwave Theory Tech.*, **MIT-33**, 789.
[26] C. Yeh (1975), Advances in communication through light fibers, in *Advances in Communication Systems*, vol. 4, *Theory and Applications*, Academic Press, New York.
[27] M. Gerla, P. Rodriques, and C. Yeh (1985), *J. Lightwave Tech.*, **LT-3**, 586.

Author Index

Subject Index